The Burgee

Premier Marina Guidebook

By David Kutz

Fifth Edition

Pierside Publishing
23911 Newell Ln. N.E.
Kingston, WA 98346
Tel: (360) 297-2935
Fax: (360) 297-3505
E-Mail: DavidKutz@aol.com
www.northstarsportswear.com/theburgeebook

Dedication:
This Edition is dedicated to Abbey Gale Rose (1993-2006), our beloved Labrador boat dog. For 13 years she cruised with us in Northwest waters without a whimper and always a cheerful wag of her tail. She personally judged all the pet areas in this book. We will have another boat dog, but there will never be another Abbey Rose.

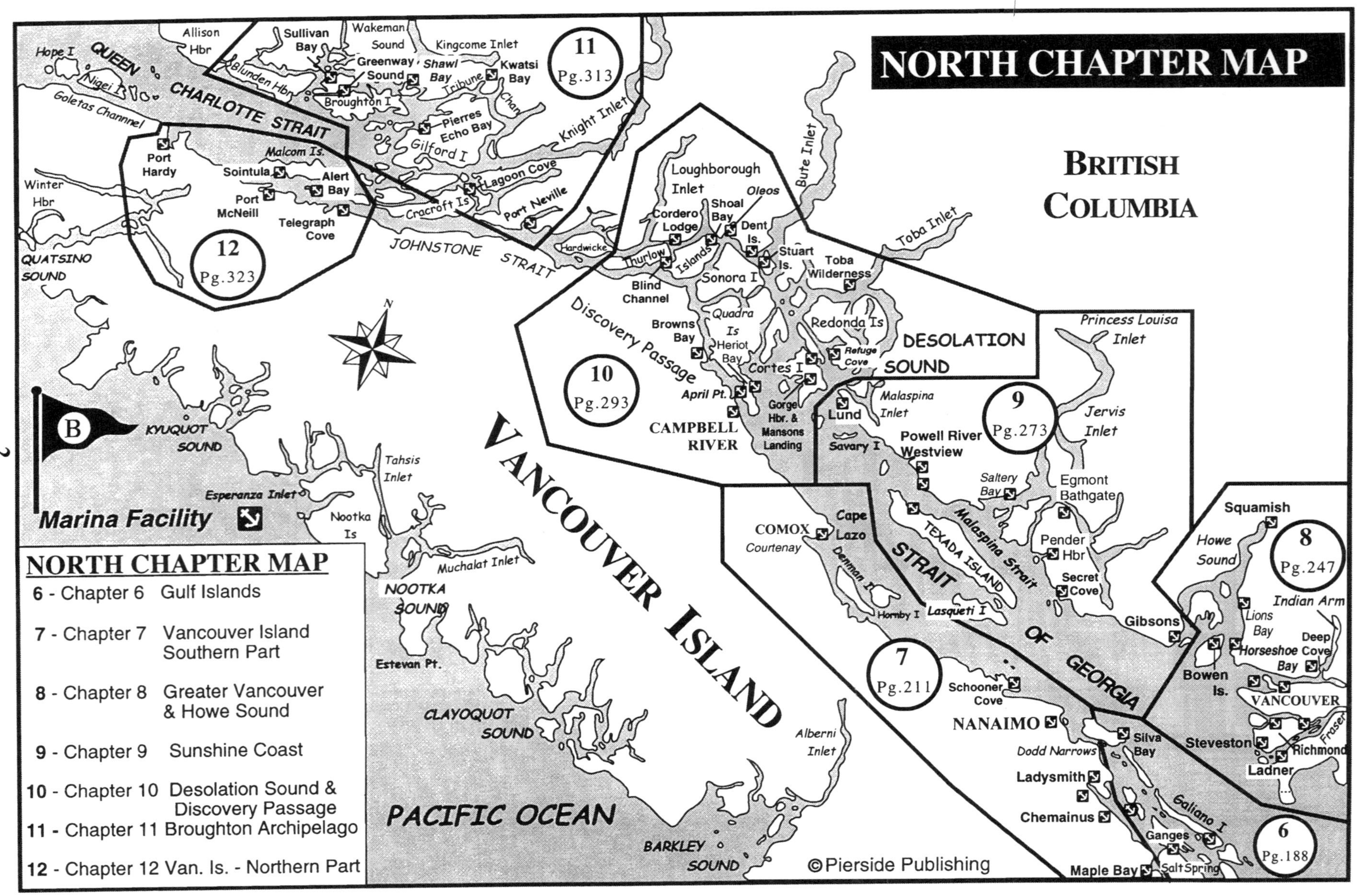

NORTH CHAPTER MAP
BRITISH COLUMBIA
VANCOUVER ISLAND
PACIFIC OCEAN
Allison Hbr
Hope I
QUEEN CHARLOTTE STRAIT
Nigei I
Goletas Channel
Sullivan Bay
Blunden Hbr
Wakeman Sound
Greenway Sound
Broughton I
Kingcome Inlet
Shawl Bay
Kwatsi Bay
Tribune Chan
11 Pg.313
Pierres Echo Bay
Gilford I
Knight Inlet
Lagoon Cove
Port Neville
Cracroft Is
Malcom Is.
Port Hardy
Sointula
Alert Bay
Port McNeill
Telegraph Cove
Winter Hbr
QUATSINO SOUND
12 Pg.323
JOHNSTONE STRAIT
Hardwicke
Loughborough Inlet
Cordero Lodge
Shoal Bay
Oleos
Dent Is.
Thurlow
Islands
Blind Channel
Sonora I
Stuart Is.
Bute Inlet
Toba Inlet
Toba Wilderness
Discovery Passage
Browns Bay
Quadra Is
Heriot Bay
Redonda Is
Refuge Cove
DESOLATION SOUND
Cortes I
10 Pg.293
April Pt.
CAMPBELL RIVER
Gorge Hbr. & Mansons Landing
Lund
Malaspina Inlet
Savary I
Powell River
Westview
9 Pg.273
Princess Louisa Inlet
Jervis Inlet
Saltery Bay
Egmont
Bathgate
Pender Hbr
Secret Cove
Malaspina Strait
TEXADA ISLAND
COMOX
Courtenay
Cape Lazo
Denman I
Hornby I
Lasqueti I
STRAIT OF GEORGIA
Gibsons
7 Pg.211
Schooner Cove
NANAIMO
Dodd Narrows
Ladysmith
Chemainus
Silva Bay
Galiano I
Ganges
Salt Spring
Maple Bay
6 Pg.188
Squamish
Howe Sound
8 Pg.247
Indian Arm
Lions Bay
Horseshoe Bay
Deep Cove
Bowen Is.
VANCOUVER
Fraser
Steveston
Richmond
Ladner
Tahsis Inlet
Nootka Is
Esperanza Inlet
KYUQUOT SOUND
Muchalat Inlet
NOOTKA SOUND
Estevan Pt.
CLAYOQUOT SOUND
Alberni Inlet
BARKLEY SOUND
B
Marina Facility
©Pierside Publishing
NORTH CHAPTER MAP
6 - Chapter 6 Gulf Islands
7 - Chapter 7 Vancouver Island Southern Part
8 - Chapter 8 Greater Vancouver & Howe Sound
9 - Chapter 9 Sunshine Coast
10 - Chapter 10 Desolation Sound & Discovery Passage
11 - Chapter 11 Broughton Archipelago
12 - Chapter 12 Van. Is. - Northern Part

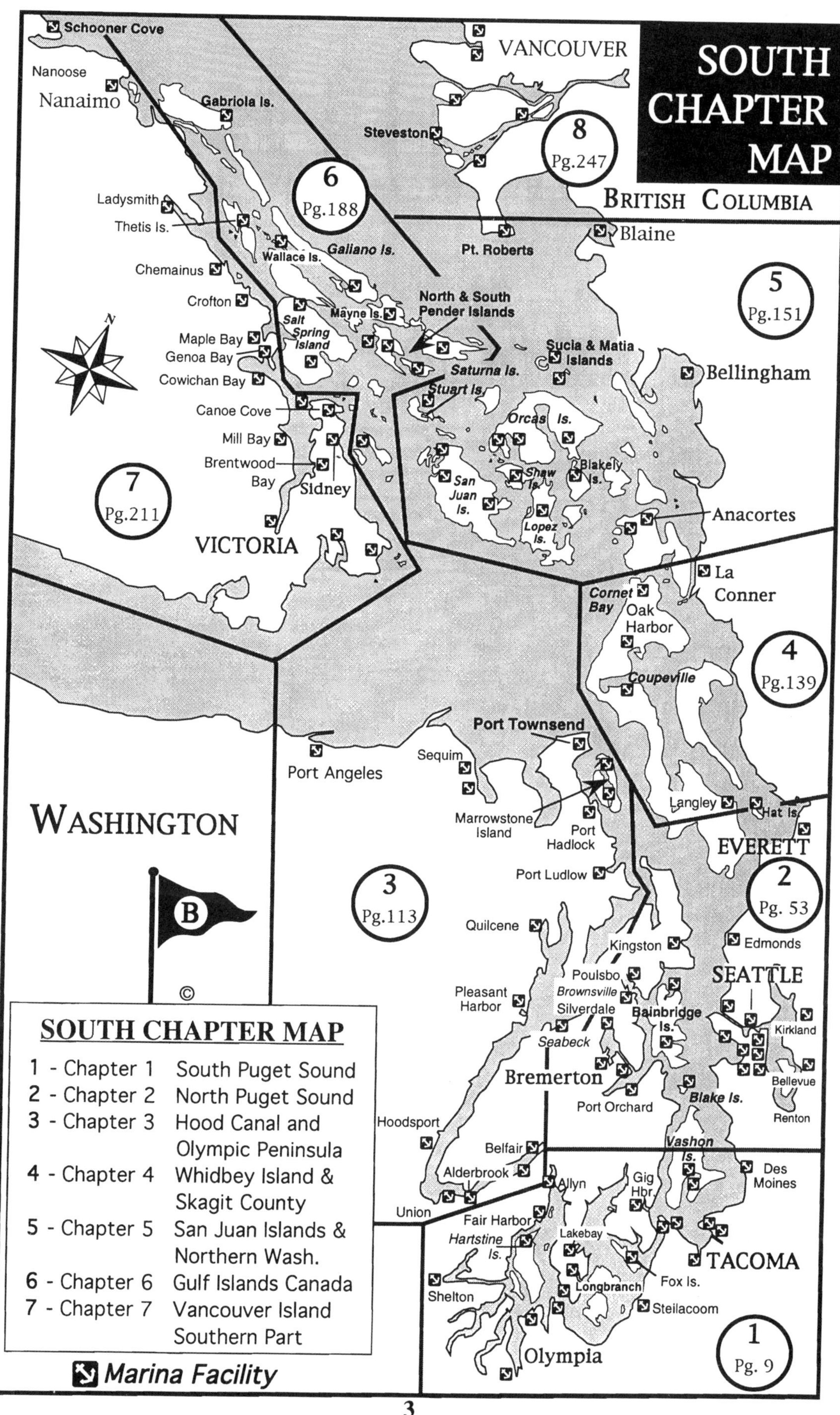
SOUTH CHAPTER MAP
BRITISH COLUMBIA
WASHINGTON
Schooner Cove
Nanoose
Nanaimo
Gabriola Is.
VANCOUVER
Steveston
8 Pg.247
6 Pg.188
Ladysmith
Thetis Is.
Wallace Is.
Galiano Is.
Pt. Roberts
Blaine
Chemainus
Crofton
Salt Spring Island
Mayne Is.
North & South Pender Islands
5 Pg.151
Maple Bay
Genoa Bay
Cowichan Bay
Saturna Is.
Sucia & Matia Islands
Bellingham
Stuart Is.
Canoe Cove
Mill Bay
Orcas Is.
Brentwood Bay
Sidney
San Juan Is.
Shaw Is.
Blakely Is.
7 Pg.211
Lopez Is.
Anacortes
VICTORIA
La Conner
Cornet Bay
Oak Harbor
4 Pg.139
Coupeville
Port Townsend
Sequim
Port Angeles
Marrowstone Island
Port Hadlock
Langley
Hat Is.
EVERETT
Port Ludlow
3 Pg.113
2 Pg. 53
Quilcene
Kingston
Edmonds
Poulsbo
Brownsville
SEATTLE
Pleasant Harbor
Silverdale
Bainbridge Is.
Kirkland
Seabeck
Bremerton
Bellevue
Port Orchard
Blake Is.
Renton
Hoodsport
Vashon Is.
Belfair
Alderbrook
Des Moines
Allyn
Gig Hbr.
Union
Fair Harbor
Hartstine Is.
Lakebay
TACOMA
Fox Is.
Shelton
Longbranch
Steilacoom
1 Pg. 9
Olympia
B
©
SOUTH CHAPTER MAP
1 - Chapter 1 South Puget Sound
2 - Chapter 2 North Puget Sound
3 - Chapter 3 Hood Canal and Olympic Peninsula
4 - Chapter 4 Whidbey Island & Skagit County
5 - Chapter 5 San Juan Islands & Northern Wash.
6 - Chapter 6 Gulf Islands Canada
7 - Chapter 7 Vancouver Island Southern Part
Marina Facility

The Burgee

Premier Marina Guidebook

By David Kutz

Published by:

Pierside Publishing
23911 Newell Ln. N.E.
Kingston, WA 98346 USA
Telephone: (360) 297-2935
Fax: (360) 297-3505
E-Mail: DavidKutz@aol.com

First Edition First Printing 1994
Second Edition First Printing 1996
Third Edition First Printing 2000
Fourth Edition First Printing 2003
Fifth Edition First Printing 2007

Printed in the United States of America

Cover and illustrations by David Kutz

Editing assistance by Janice Kutz

Library of Congress Cataloging of Publication Data
Kutz, David J.
The Burgee: Premier Marina Guidebook / by David Kutz.— 5th Edition
Includes index.
Library of Congress Control Number 2006095093
ISBN 0-9641934-8-5: **US $26.95 Softcover**

Table of Contents

Individual Marina Facilities Listed in Index in Back of Book

A WORD FROM THE AUTHOR

For those new to the world of boating, a ***Burgee*** is a small pennant representing the logo of a Yacht Club or Boating Club. It is generally triangular in shape, sometimes with a swallow tail. A Burgee is flown from the bow staff of power vessels or from the mast of a sailing vessel. Since one of the unique features of this book is yacht club reciprocal facilities, the title *"**Burgee**" was* chosen.

The five Burgee Book editions (this the Fifth Edition) were created over the course of many years of extensive research and computer work. The chartlets and data are designed with Macintosh formats. Thank goodness it is allowable to have fun while one works. We still have many days of enjoyment compiling and updating this book's data on our 35' single engine trawler, *Shine*.

To find this book useful, it is not necessary to belong to a specific yacht or boating club. There are hundreds of public boating facilities included in the Burgee Book. We are sure will find the information contained in this book helpful to promote your enjoyment of the great sport of recreational boating. Thank you and good cruising! *-David Kutz*

ABOUT THE AUTHOR

David Kutz and his wife Janice have been avid recreational boaters in Washington and Canadian waters for many years. In addition to the Burgee Book, Kutz spends countless hours of volunteer work to improve our enjoyment of the water, and protect and promote the interests of recreational boaters. He is an active member and Past Commodore of the Kingston Cove Yacht Club and currently serving as an officer of the Recreational Boating Association of Washington (RBAW). He also serves on the Washington State Boating Safety Council, the Executive Committee of the National Boating Federation (NBF) and is a member of the International Order of the Blue Gavel (IOBG).

In 1990, Kutz saw a niche to fill in the boating book market, started a small publishing company and completed the First Edition of the Burgee Book in 1994. The Second Edition followed in 1996, the Third in 2000, the Fourth in 2004, and Fifth Edition in 2007. In addition to Pierside Publishing, Kutz and his wife are the former owners of Northstar Sportswear Corporation located in downtown Kingston, Washington.

HOW TO USE THIS GUIDEBOOK

"The Burgee Book" includes marinas and marine parks that provide overnight moorage available to the general boating public. To have a listing, the facility must have actual docking facilities.

•Map and Locations: ***Marinas are listed by town or city generally from south to north in each regional chapter as depicted on the Chapter Maps on Pages 2 & 3.***

Daily Rate: Overnight moorage rates change frequently and only approximate costs are posted in the Burgee Book. Marina moorage rates are categorized in the following price ranges:

Economical – Under 75¢ per foot
Moderate – 75¢ to $1.25 per foot
Premium – Over $1.25 per foot

•Yacht Club Reciprocals: The Burgee Book also features Yacht Club Reciprocal Moorage information volunteered by many clubs in Washington State and British Columbia. ***If you utilize yacht club reciprocal moorage, you must hold a current membership with the reciprocating club.*** Your yacht club should have a list of clubs you are reciprocal with.

Accuracy: The information listed in this book is intended to be as accurate as possible, but please remember things change frequently, especially rates and available services.

•Chartlets: The chartlets depicting the marina facilities are **not to scale** and only intended to be a general guide to give you an idea where the guest facilities are located.

•Lat./Lons: The latitudes and longitudes are not exact and only intended to be used as a general guide.

•Compasses: The compasses shown in the chartlets are in the general direction of **Magnetic North.** They are not exact and only intended to be used for general direction.

DISCLAIMER

"The Burgee Book" is designed to provide only general information for the marine facilities listed. It is sold with the understanding that the publisher and author are not engaged in rendering legal navigational or informational services. The chartlets (marina maps) contained in this book are not to scale and not intended for use in navigation. Selected laws require all ships to have on board, maintain, and use appropriate navigational charts. Our chartlets (marina maps) of the marina facilities do not meet that requirement. Every effort has been made to make this guidebook as accurate as possible but the book contains current information only up to the printing date. There may be mistakes both typographical and in content. Therefore, this text should be used only as a general guide and not as the ultimate source of acquiring details about marina facilities. The Author, Pierside Publishing, or Bookseller shall have neither liability nor responsibility to any person or entity with respect to any loss or damage caused, or alleged to be caused, directly or indirectly by the information contained in this book.

BOATING SERVICE ORGANIZATIONS

In addition to the many yacht clubs shown in this book, attention should be given to the importance of two boating organizations that look after the interests of not only Northwest boaters but all recreational boaters who enjoy this great sport across the country. These organizations are made up of hard working volunteers who are boaters just like you.

The Burgee Book encourages Northwest boaters to join these organizations with an individual membership. Personal involvement is totally optionally. Your membership dues will get you a special periodic newsletter to keep you informed of pertinent issues affecting boating and your support will help support our boating lobbyists. Membership fees are nominal.

RECREATIONAL BOATING ASSOCIATION OF WASHINGTON
WWW.RBAW.ORG

For the past 50 years, the Recreational Boating Association of Washington (RBAW) has been a forum to bring together a common voice, the ideas, goals and objectives of Northwest Boaters. As a unified group, RBAW conveys positions of sound reasoning to the Washington State Legislature. This group implements these communications more effectively than could be done by an individual. RBAW's goal is to represent the interests of Northwest boaters effectively. RBAW has a professional lobbyist who makes sure the collective voice is heard by Washington State lawmakers, especially on issues concerning taxation. RBAW also works to enhance Washington State Marine Parks, boating safety and educational programs, environmental issues, and works on Federal issues that affect recreational boating. RBAW has about 50 boating and yacht club memberships along with individual members to constitute a membership of over 10,000 boaters. Membership is economical and encouraged for everyone who enjoys boating in the Northwest. For membership information, please write RBAW, P.O. Box 23601, Federal Way, WA 98093 or E-Mail: info@rbaw.org

NATIONAL BOATING FEDERATION WWW.N-B-F.ORG

The National Boating Federation (NBF) consists of 20 Boating Associations (including RBAW) from across the country. Founded in 1966, NBF is the largest nationwide volunteer alliance of recreational boaters and works exclusively on Federal issues that concern boating. Effectively the NBF works on subjects to enhance and protect boating like boating safety, fuel taxes, Coast Guard, NOAA, and the FCC on VHF radio matters. The NBF is counseled by a professional lobbyist who represents boating interests from several organizations. They speak for 2 million boaters nationwide. For membership information please E-Mail: info@n-b-f.org

CHAPTER 1

SOUTH PUGET SOUND

Chapter Map - Page 3

Olympia

NAME OF MARINA: ***PERCIVAL LANDING PARK*** RADIO: None
TELEPHONE: **360-753-8380** MGR: City of Olympia
E-MAIL: Go to Website FAX: 360-753-8334 (Fax)
ADDRESS: 222 Columbia St. Olympia, WA 98501
SHORT DESCRIPTION & LOCATION: **www.ci.olympia.wa.us/par/parks/moorage.asp**
47°02.8'-122°54.3' The 3 moorage floats offer one of the most interesting boating stops in the Northwest. Located at the head of Budd Inlet adjacent to downtown Olympia, this well kept facility is convenient to many shops, restaurants, & sights.

GUEST BOAT CAPACITY: ....Appx. 45-60 boats
DOCKSIDE DEPTH AT ZERO TIDE: ....8 ft.
SEASON: ....All year
RESERVATION POLICY: ....None
AMT W/ELECTRICITY: ....None
FUEL DOCK: ....None
MARINE REPAIRS: ....Close by
TOILETS: ....Yes
HOT SHOWERS: ....Yes
RESTAURANT: ....Close by
PICNIC AREA: ....Yes
BASIC STORE: ....Close by
BROADBAND/WI-FI: ....None
DAILY RATE:.............Economical (Under 75¢/foot)

GUEST DOCK: ....1200 ft. total
GUEST SLIPS: Guest docks only
WATER: ....Yes
AMPS: ....30 A
PUMP OUT STATION ....Yes
HAUL OUT: ....At Swantown
BOAT RAMP: ....None
LAUNDRY: ....None
BAR: ....Close by
POOL: ....None
GOLF: ....None
PET FRIENDLY: ....Fair
OTHER: State Capitol Building & related tourist sights, boating & fishing supplies closeby. Porta pump-out avail. in bldg by Landing.

NAME OF YACHT CLUB: ***OLYMPIA YACHT CLUB***
CLUB ADDRESS: 201 North Simmons St, Olympia, WA 98501
CLUB TELEPHONE: 360-357-6767 PERSON IN CHARGE: Caretaker
LOCATION & SPECIAL NOTES: **www.olympiayachtclub.org**
Located adjacent to Percival Landing directly in front of Yacht Club Building. Sign in and acquire key card from Caretakers (deposit required). Occasional vacated member slips available if needed. Check w/Caretakers. Walking distance to City of Olympia businesses, services and attractions.

RECIPROCAL BOAT CAPACITY:...........3-5 boats
DOCKSIDE DEPTH AT ZERO TIDE: ....6 ft.
RECIPROCAL SEASON:........All year
RESERVATION POLICY: ....None
TOILETS: ....Yes
HOT SHOWERS: ....Yes
RESTAURANT: ....Close by
DAILY RATE: $3 per day/power. 48 hours free. 50¢ per foot after free 48 hours.

RECIPROCAL DOCK: ....230 ft.
RECIPROCAL SLIPS: ....None
WATER: ....Yes
AMT W/ELECTRICITY: ....All
AMPS: ....20-30 A
BAR: ....Close by
OTHER: No Rafting
5 DAY LIMIT/YEAR

NOTE: THIS IS PRIVATE MOORAGE ONLY AVAILABLE TO MEMBERS OF RECIPROCAL YACHT CLUBS! YOUR CLUB MUST HAVE RECIPROCAL PRIVILEGES AND YOU MUST FLY YOUR BURGEE!

Olympia

CAUTION! This chartlet not intended for use in navigation.

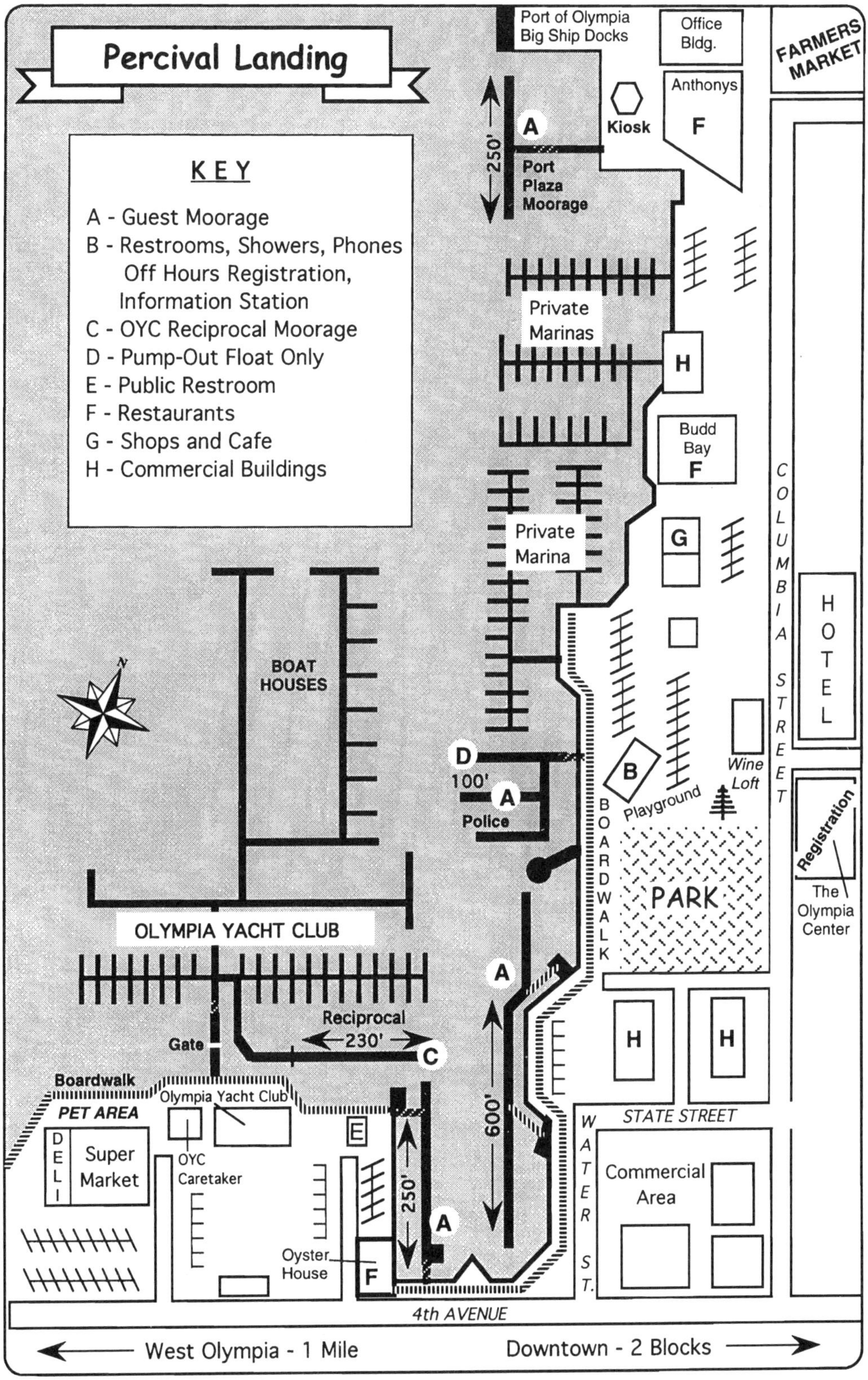

Olympia

NAME OF MARINA: ***SWANTOWN MARINA*** (Formerly East Bay) **RADIO:** VHF 65A
TELEPHONE: **360-528-8049** MGR: Cheryl Maynard
E-MAIL: marina@portolympia.com FAX: 360-528-8094 (Fax)
ADDRESS: 1022 Marine Drive NE Olympia, WA 98501
SHORT DESCRIPTION & LOCATION: **www.portolympia.com**

47°03.50' - 122°53.80' Modern & spacious well maintained marina located on east side of Budd Inlet approx. 1 mile north of downtown Olympia. Within walking distance or short cab ride to capitol city attractions, shops, & restaurants.

GUEST BOAT CAPACITY: ....Appx. 50-100 boats
DOCKSIDE DEPTH AT ZERO TIDE: ......9-12 ft.
SEASON: ....All year
RESERVATION POLICY: ....Groups only
AMT W/ELECTRICITY: ....All
FUEL DOCK: ....Close by
MARINE REPAIRS: ....On premises
TOILETS: ....Yes
HOT SHOWERS: ....Yes
RESTAURANT: ....Close by
PICNIC AREA: ....Yes
BASIC STORE: ....None
BROADBAND/WI-FI: ....Yes
DAILY RATE: ....Economical (Under 75¢/foot)

GUEST DOCK: 660 ft. plus slips
GUEST SLIPS: ....Appx. 50 plus
WATER: ....Yes
AMPS: ....30-50 A
PUMP OUT STATION ....Yes
HAUL OUT: ....Travel-Lift
BOAT RAMP: ....Yes
LAUNDRY: ....Yes
BAR: ....Close by
POOL: ....None
GOLF: ....Close by
PET FRIENDLY: ....Excellent
OTHER: Boatyard, 77 ton travel lift, security, bicycle trails. State Capitol, Museum, Farmers Market & related downtown attractions.

CAUTION! This chartlet not intended for use in navigation.

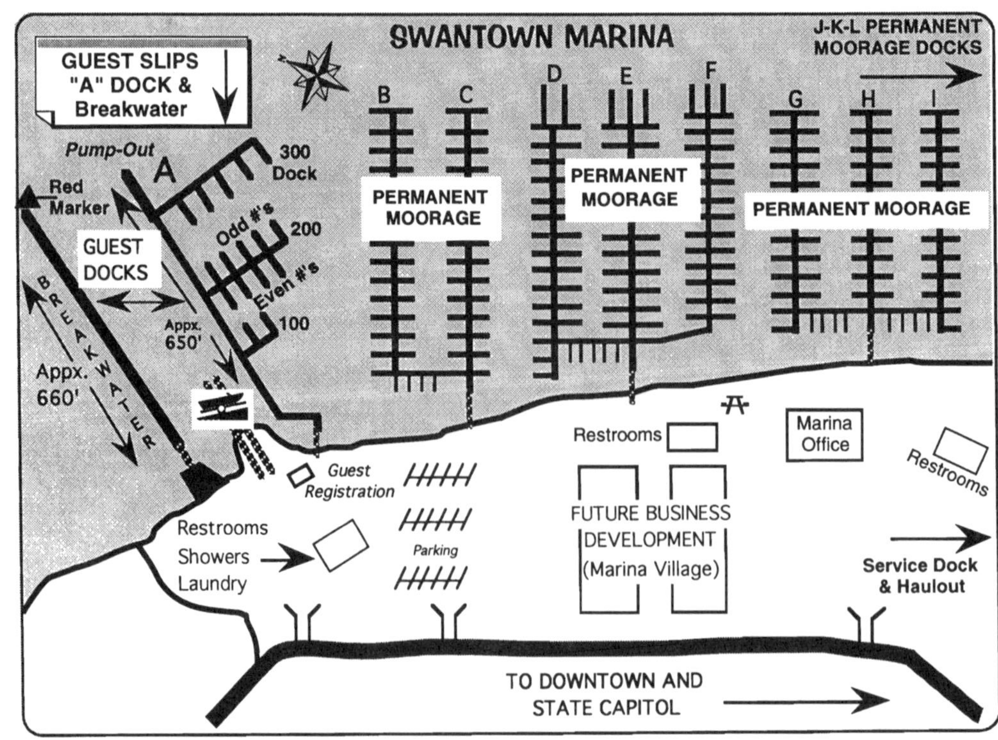

Olympia

NAME OF MARINA: ***BOSTON HARBOR MARINA*** **RADIO:** VHF Ch.16
TELEPHONE: **360-357-5670** MGR: Pam McHugh, Co-Owner
E-MAIL: bhm@bostonharbormarina.com FAX: 360-352-2816 (Fax)
ADDRESS: 312 73rd Ave. N.E. Olympia, WA 98506
SHORT DESCRIPTION & LOCATION: **www.bostonharbormarina.com**

47°08.60' - 122°54.25' Boat length to 50 ft. - Small quiet & friendly marina in neighborhood location in crescent shaped cove at Dofflemeyer Point at entrance to Budd Inlet. About 8 miles north of downtown Olympia. Fresh clams & oysters, salmon in season.

GUEST BOAT CAPACITY: ....Appx. 15 boats
DOCKSIDE DEPTH AT ZERO TIDE: ....9 ft.
SEASON: ....All year
RESERVATION POLICY: ....Accepts
AMT W/ELECTRICITY: ....All
FUEL DOCK: ....Gas, Dsl, & CNG
MARINE REPAIRS: ....Close by
TOILETS: ....Yes
HOT SHOWERS: ....None
RESTAURANT: ....Cafe/Deli
PICNIC AREA: ....Yes
BASIC STORE: ....Yes
BROADBAND/WI-FI: ....Yes
DAILY RATE: ....Economical (Under 75¢/foot)

GUEST DOCK: 200 ft. plus slips
GUEST SLIPS: ....6 small slips
WATER: ....Yes
AMPS: ....20 A
PUMP OUT STATION ....None
HAUL OUT: ....None
BOAT RAMP: ....Yes
LAUNDRY: ....None
BAR: ....None
POOL: ....None
GOLF: ....Close by
PET FRIENDLY: ....Fair
OTHER: Pay phones, dinghy area, fishing, beach combing, bicycling, gift shop, local seafood in season. Sunday Breakfast served.

CAUTION! This chartlet not intended for use in navigation.

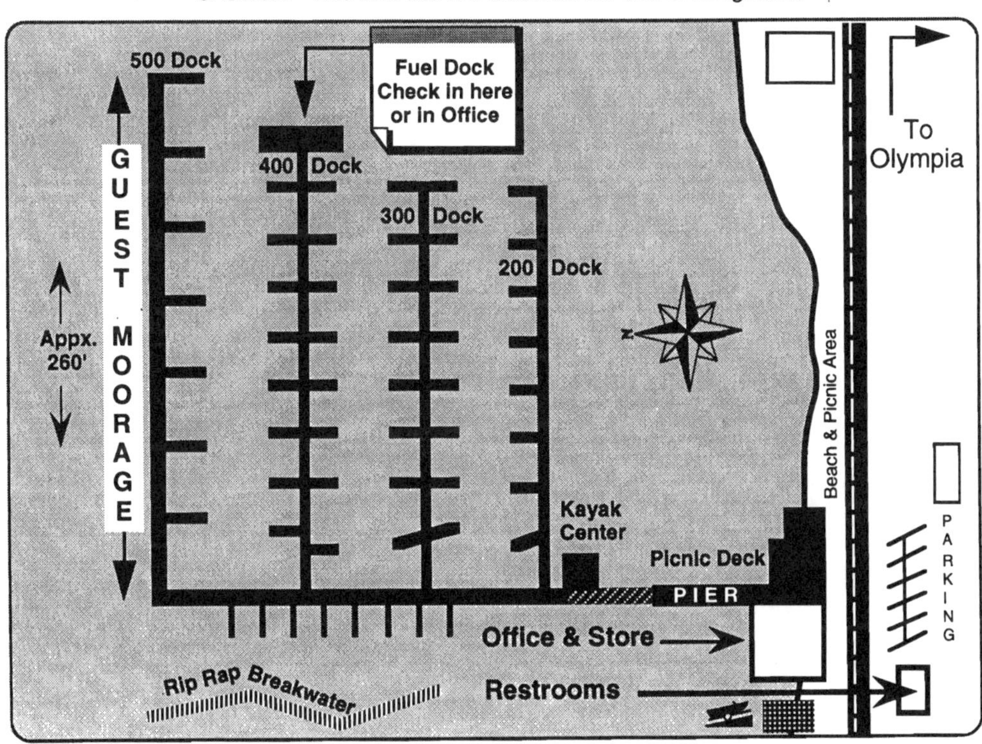

Olympia

NAME OF MARINA: ***ZITTEL'S MARINA INC.*** RADIO: None
TELEPHONE: **360-459-1950** MGR: Mike Zittel
E-MAIL: kzittel@comcast.net FAX: 360-459-8984 (Fax)
ADDRESS: 9144 Gallea St. N.E. Olympia WA 98516
SHORT DESCRIPTION & LOCATION: **www.zittelsmarina.com**

47°09.98' - 122°48.50' Located appx. 1 mile SSE of Johnson Point at foot of Case Inlet & top of Nisqually Reach. Local use marina catering to sport fishermen. Guest Dock is exposed to northerly winds. Rural location about 10 driving miles from Olympia..

GUEST BOAT CAPACITY: ....Appx. 10-15 boats
DOCKSIDE DEPTH AT ZERO TIDE: ......8-10 ft.
SEASON: ....All year
RESERVATION POLICY: ....Accepts
AMT W/ELECTRICITY: Limited-Inside slips only
FUEL DOCK: ....Gas & Diesel
MARINE REPAIRS: Boat Yard & Bottom Painting
TOILETS: ....Yes
HOT SHOWERS: ....None
RESTAURANT: ....None
PICNIC AREA: ....None
BASIC STORE: ....Yes
BROADBAND/WI-FI: ....None
DAILY RATE: ....Economical (Under 75¢/foot)
NOTE: OPEN SLIPS WITH POWER RENTED WHEN AVAILABLE

GUEST DOCK: ....Appx. 150 ft.
GUEST SLIPS: ....None
WATER: ....Yes
AMPS: ....20-30 A
PUMP OUT STATION ....Yes
HAUL OUT: ....25 Ton
BOAT RAMP: ....Yes
LAUNDRY: ....None
BAR: ....None
POOL: ....None
GOLF: ....None
PET FRIENDLY: ....Good
OTHER: No power on Guest Dock. Snack Bar, fishing supplies, storage, boat rentals, & marine supplies.

CAUTION! This chartlet not intended for use in navigation.

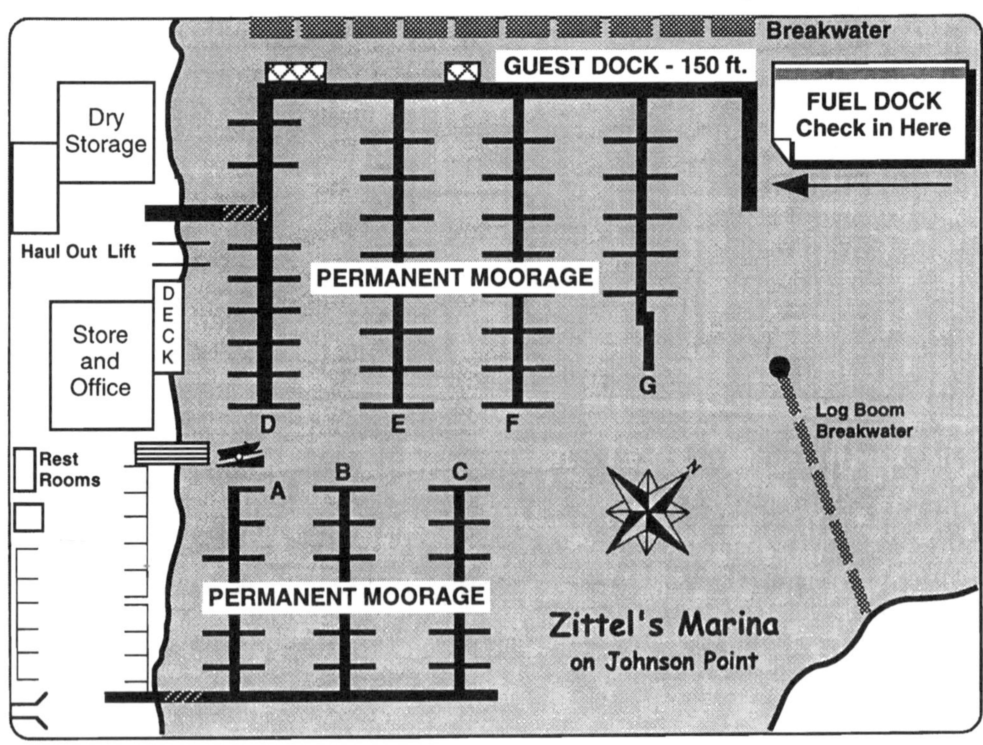

Shelton

NAME OF MARINA: *PORT OF SHELTON -Shelton Yacht Club* **RADIO:** None
TELEPHONE: **360-426-6435** **MGR:** Mike Byrne
E-MAIL: webmaster@sheltonyachtclub.com **FAX:** None
ADDRESS: P.O. Box 2270 Shelton, WA 98584
SHORT DESCRIPTION & LOCATION: **http://www.sheltonyachtclub.com**

47°12.85' - 123°05.10' The Shelton Marina is located at the head of Hammersly Inlet 6 blocks from downtown Shelton with shopping, restaurants, and marine services. The Marina owned by Port & open to public, but managed & maintained by Shelton Yacht Club.

GUEST BOAT CAPACITY: ....Appx. 4 boats
DOCKSIDE DEPTH AT ZERO TIDE: ....13 ft.
SEASON: ....All year
RESERVATION POLICY: ....None
AMT W/ELECTRICITY: ....All
FUEL DOCK: ....None - closest fuel 14 miles
MARINE REPAIRS: ....Close by
TOILETS: ....Porta-Potty
HOT SHOWERS: ....None
RESTAURANT: ....1 Mile
PICNIC AREA: ....None
BASIC STORE: ....1 Mile
BROADBAND/WI-FI: ....None
DAILY RATE: Public Moorage: Economical **Shelton Y.C.** offers reciprocal moorage free for first 48 hrs if your club is reciprocal. Power additional.

GUEST DOCK: ....Appx. 90 ft.
GUEST SLIPS: ....Dock only
WATER: ....Yes
AMPS: ....30 A
PUMP OUT STATION ....Yes
HAUL OUT: ....None
BOAT RAMP: ....None
LAUNDRY: ....Close by
BAR: ....1 Mile
POOL: ....None
GOLF: ....Close by
PET FRIENDLY: ....Good
OTHER: Club rental available with reservation,360-426-9476. Visit Forest Festival in early June & Shelton Oysterfest in early Oct.

CAUTION! This chartlet not intended for use in navigation.

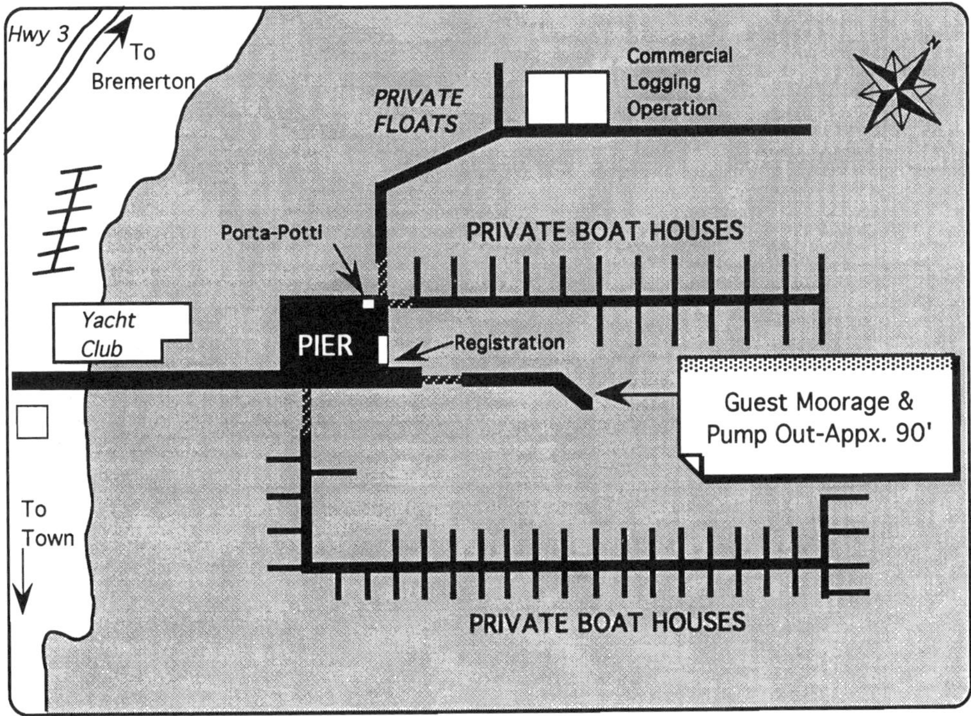

Harstine Island

NAME OF MARINA: ***JARRELL'S COVE MARINA*** **RADIO:** None
TELEPHONE: 360-426-8823 **MGR:** Lorna & Gary Hink, Owners
E-MAIL: Toll Free Tel: 800-362-8823 **FAX:** 360-432-8494 (Fax)
ADDRESS: E. 220 Wilson Rd. Shelton, WA 98584

SHORT DESCRIPTION & LOCATION:

47°17.00' - 122°53.40' Located on N.W. corner of Harstine Island in picturesque Jarrell Cove. One of South Sound's finest marina resorts with many amenities. Guest moorage consists of open slips of permanent marina tenants. Pleae get advance slip assignment.

GUEST BOAT CAPACITY: ....Appx. 5-7 boats
DOCKSIDE DEPTH AT ZERO TIDE: ....5-6 ft.
SEASON: ....Winter-On Call
RESERVATION POLICY: First come, first serve
AMT W/ELECTRICITY: ....All
FUEL DOCK: ....Gas, Dsl, & LP
MARINE REPAIRS: ....None
TOILETS: ....Yes
HOT SHOWERS: ....Yes
RESTAURANT: ....None
PICNIC AREA: ....Yes
BASIC STORE: ....Yes
BROADBAND/WI-FI: ....None
DAILY RATE:..........Economical (Under 75¢/foot)

GUEST DOCK: ....Varies
GUEST SLIPS: ....Varies
WATER: ....Yes
AMPS: ....30 A
PUMP OUT STATION ....Yes
HAUL OUT: ....None
BOAT RAMP: ....None
LAUNDRY: ....Yes
BAR: ....None
POOL: ....None
GOLF: ....Appx. 10-15 miles
PET FRIENDLY: ....Good
OTHER: Shore side RV sites, picnic & game areas, clams, fishing, dinghy to State Park.

CAUTION! This chartlet not intended for use in navigation.

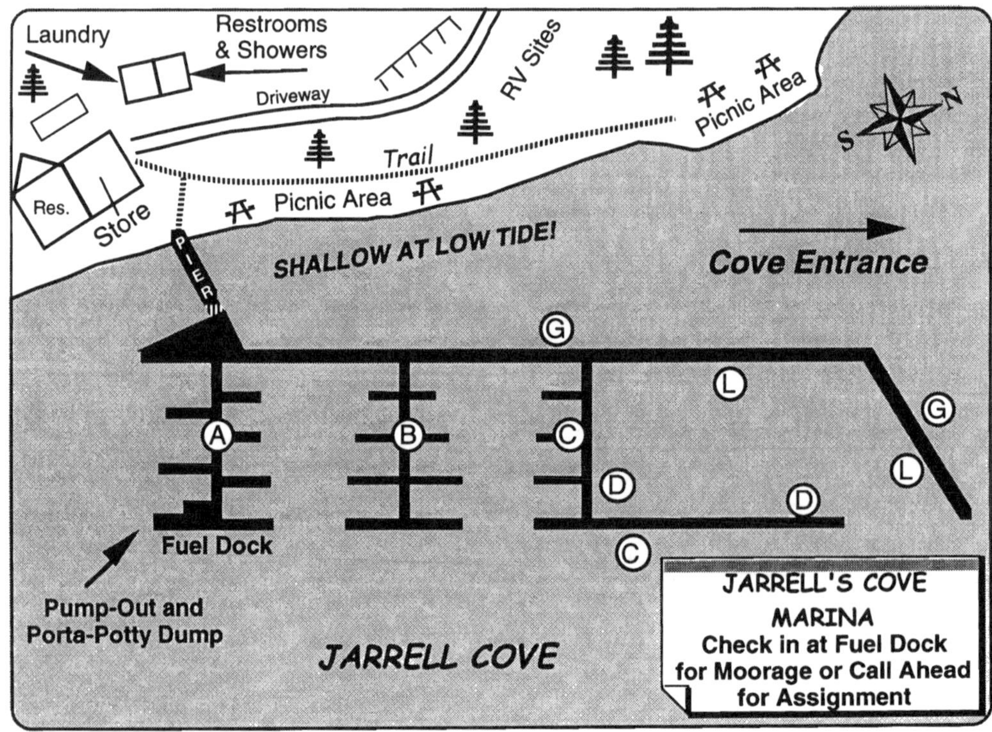

Harstine Island

NAME OF PARK: *JARRELL COVE STATE PARK*
ADDRESS: E. 391 Wingert Rd. Shelton, WA 98584
TELEPHONE: 360-426-9226 **MGR:** Mischa Cowles
SHORT DESCRIPTION & LOCATION: **www.parks.wa.gov/moorage/parks**

47°17.00'-122°53.10' Located on E side of Jarrell Cove on NW corner of Harstine Is. This is a nice wooded marine park with 42 acres & 3056 feet of shoreline. Park is busy in summer & has two good floats, mooring buoys & campsites to serve boaters.

GUEST BOAT CAPACITY: ....Appx. 45 boats
DOCKSIDE DEPTH AT ZERO TIDE:.........4-8 ft.
SEASON: ....All year
AMT W/ELECTRICITY: ....None
TOILETS: ....Yes
HOT SHOWERS: ....Yes
PICNIC AREA: ....Yes
PLAY AREA: ....Yes
BASIC STORE: ....Close by
DAILY RATE: Dock Charge..........50¢/foot
Mooring Buoys..............$10.00/night
Minimum Charge: $10.00

GUEST DOCK: 2 Floats = 680 ft.
MOORING BUOYS: 14
WATER: ....Ashore
PAY PHONES: ....None
BOAT RAMP: ....None
PICNIC SHELTER: ....Yes
BBQ: ....Yes
PUMP OUT STATION: ....Yes
PET FRIENDLY: ....Excellent
OTHER: Campsites, trails, picnic tables, fishing, hiking, bird watching, canoeing, kayaking, and clamming in season.

CAUTION! This chartlet not intended for use in navigation.

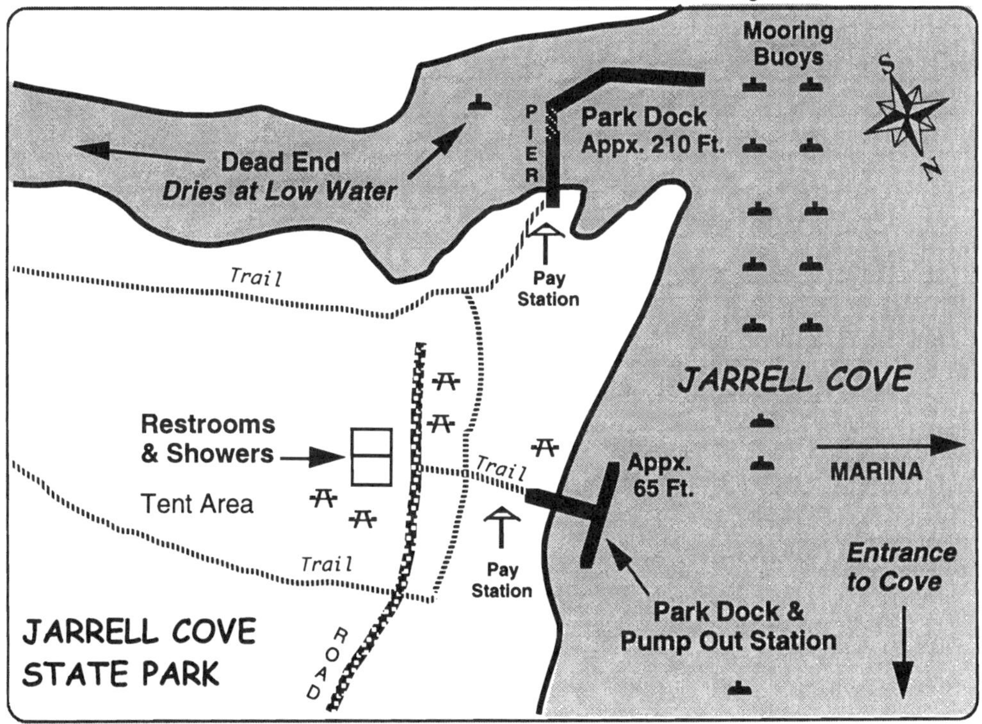

Fair Harbor

NAME OF MARINA: ***FAIR HARBOR MARINA*** RADIO: None
TELEPHONE: **360-426-4028** MGR: Vern & Susan Nelson
E-MAIL: info@fairharbormarina.us FAX: 360-277-0827 (Fax)
ADDRESS: P.O. Box 160 Grapeview, WA 98546
SHORT DESCRIPTION & LOCATION: **www.fairharbormarina.us**

47°20.30' - 122°49.80' Located on the N.W. side of Case Inlet just N. of Stretch Island behind Reach Island. This charming full service marina offers ample moorage in a sheltered Fair Harbor. The area has golfing and sightseeing attractions for boaters.

GUEST BOAT CAPACITY: ....Appx. 40-45 boats
DOCKSIDE DEPTH AT ZERO TIDE: ....25 ft.
SEASON: ....All year
RESERVATION POLICY: ....Accepts
AMT W/ELECTRICITY: ....All
FUEL DOCK: ....Gas only
MARINE REPAIRS: ....Limited
TOILETS: ....Yes
HOT SHOWERS: ....Yes
RESTAURANT: ....None
PICNIC AREA: ....Yes
BASIC STORE: ....Yes
BROADBAND/WI-FI: ....Yes
DAILY RATE: ....Economical (Under 75¢/foot)

GUEST DOCK: ....350 ft. TTL
GUEST SLIPS: ....Ends of docks
WATER: ....Yes
AMPS: ....30 A
PUMP OUT STATION ....None
HAUL OUT: ....None
BOAT RAMP: ....Yes
LAUNDRY: ....None
BAR: ....None
POOL: ....None
GOLF: ....Close by with shuttle
PET FRIENDLY: ....Excellent
OTHER: Country store & espresso bar on dock, gifts, bait & marine hardware, Landscaped rose gardens.

CAUTION! This chartlet not intended for use in navigation.

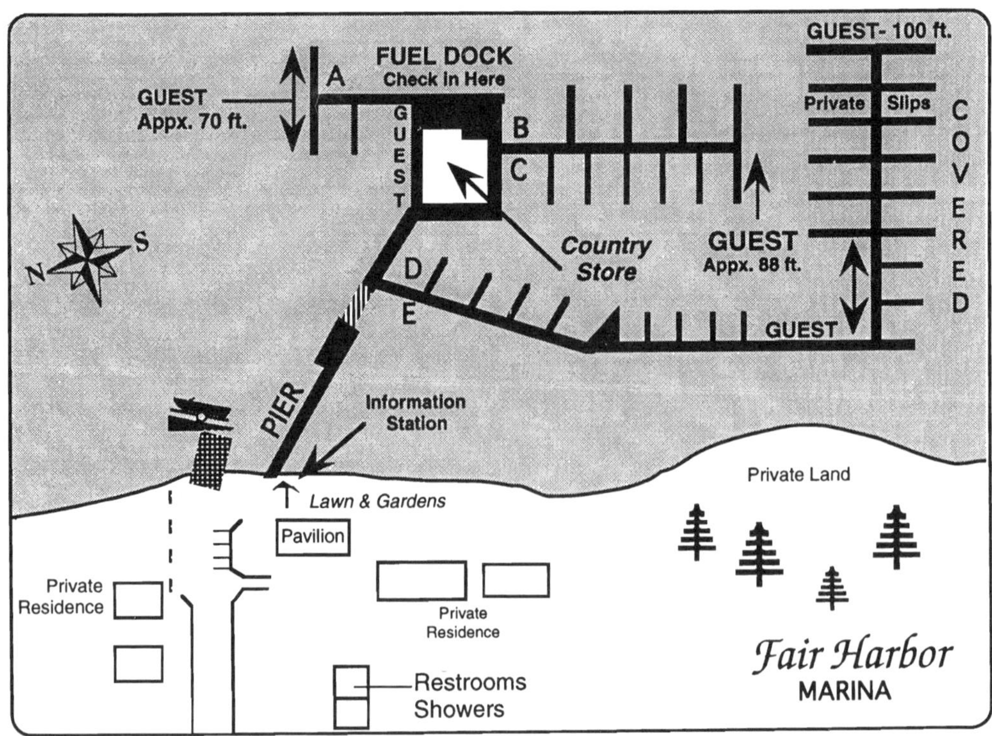

Allyn

NAME OF MARINA: *PORT OF ALLYN* RADIO: None
TELEPHONE: 360-275-2430 MGR: Bonnie Knight
E-MAIL: PortofAllyn@aol.com FAX: 360-275-2455 (Fax)
ADDRESS: P.O. Box 1, 18560 S.R. 3 Allyn, WA 98524
SHORT DESCRIPTION & LOCATION: www.portofallyn.com

47°23.90' - 122°49.60' Small guest/service float on waterfront in town of Allyn. **Dock is open to southerly winds. Moorage is available year round. Low water hazards exist for large boats at minus tides.**

GUEST BOAT CAPACITY: Appx. 10 boats
DOCKSIDE DEPTH AT ZERO TIDE: 4 ft.
SEASON: All year
RESERVATION POLICY: None
AMT W/ELECTRICITY: None
FUEL DOCK: None
MARINE REPAIRS: None
TOILETS: Yes
HOT SHOWERS: None
RESTAURANT: 1 Block away
PICNIC AREA: Yes
BASIC STORE: 1 Block away
BROADBAND/WI-FI: None
DAILY RATE: Economical (Under 75¢/foot)

GUEST DOCK: Appx. 200 ft.
GUEST SLIPS: 10 large spaces
WATER: At head of pier
AMPS: None
PUMP OUT STATION: Yes
HAUL OUT: None
BOAT RAMP: Yes-2 ea.
LAUNDRY: None
BAR: Close by
POOL: None
GOLF: Close by
PET FRIENDLY: Excellent
OTHER: Nice waterfront park with playground. Groceries, shopping, liquor store & restaurants within one block of park.

CAUTION! This chartlet not intended for use in navigation.

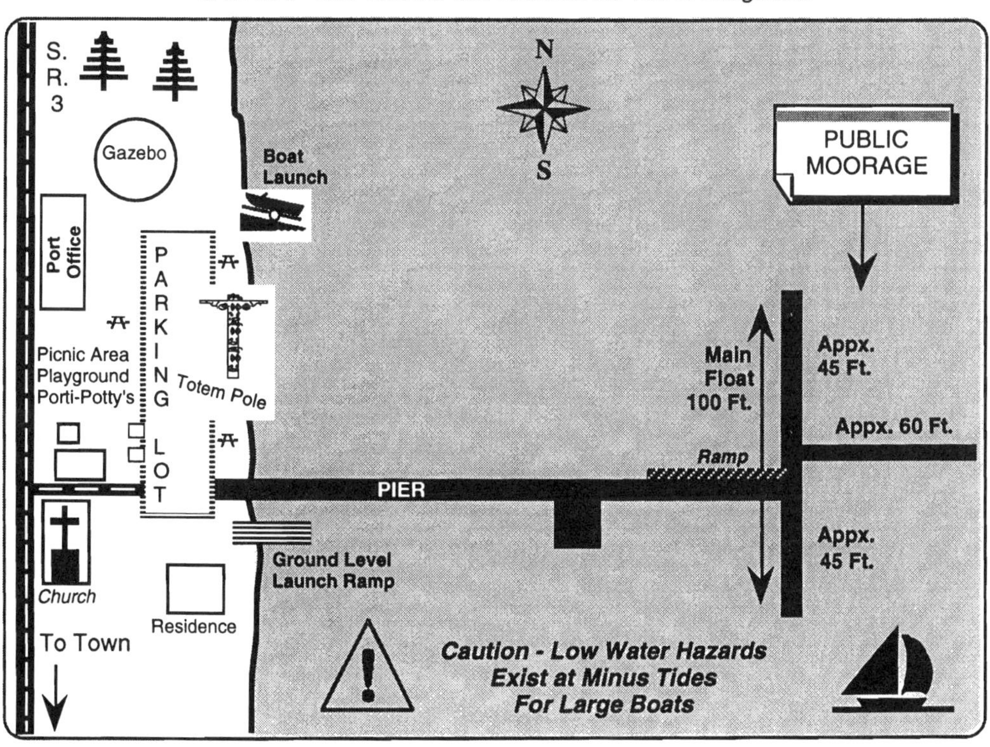

Key Peninsula

NAME OF PARK: *JOEMMA BEACH STATE PARK*
ADDRESS: P.O. Box 898 Lakebay, WA 98349
TELEPHONE: 253-884-1944 MGR: Tom Pew
SHORT DESCRIPTION & LOCATION: **www.parks.wa.gov/moorage/parks**

47°13.45' - 122°48.66' Located on S.E. end of Case Inlet just N.W. of Whiteman Cove. The Park floats offer good depth & ample moorage. A long pier with an observation deck provides access to floats from the 122 acre upland park. Open May-Sept.

GUEST BOAT CAPACITY: ....Appx. 10-20 boats
DOCKSIDE DEPTH AT ZERO TIDE: Appx. 10 ft.
SEASON: ....Closed winter
AMT W/ELECTRICITY: ....None
TOILETS: ....Yes - Vault toilets
HOT SHOWERS: ....None
PICNIC AREA: ....Yes
PLAY AREA: ....None
BASIC STORE: ....None
DAILY RATE: Dock Charge....................50¢/foot
Mooring Buoys...............$10.00/night
$10 Minimum

GUEST DOCK: Appx. 320 ft. TTL
MOORING BUOYS: ....5 Buoys
WATER: ....Yes
PAY PHONES: ....No
BOAT RAMP: ....Yes
PICNIC SHELTER: Must reserve
BBQ: ....Yes
PUMP OUT STATION: ......None
PET FRIENDLY: ....Good
OTHER: Fishing, marine trail, bicycling, shell fishing, camping. Docks exposed to southeast winds.

CAUTION! This chartlet not intended for use in navigation.

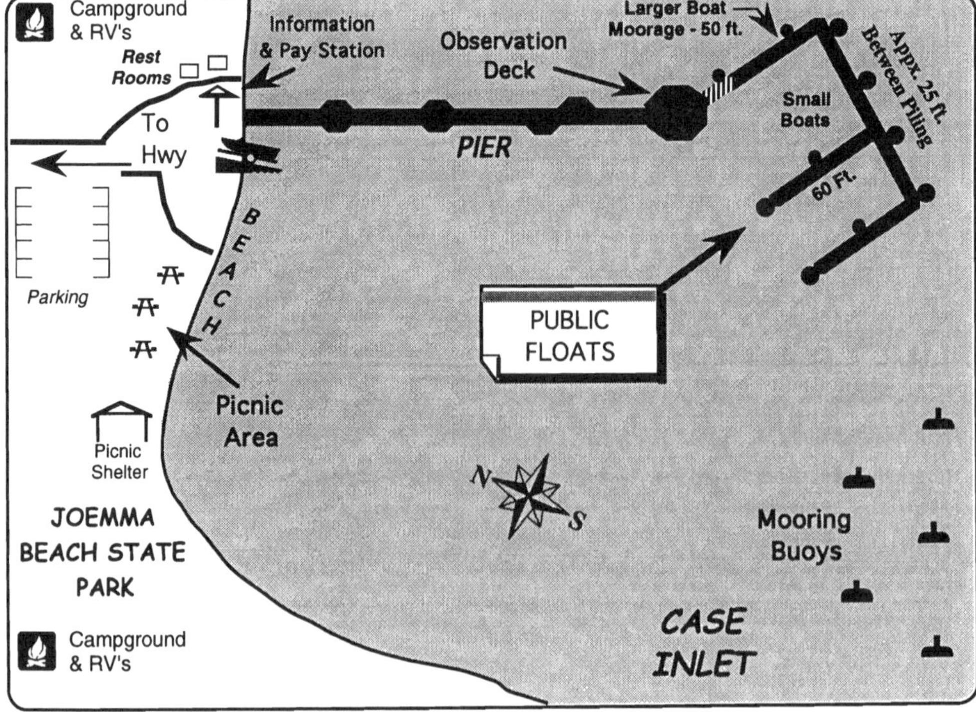

Anderson Island

NAME OF YACHT CLUB: *ORO BAY YACHT CLUB*

ADDRESS: 13312 Agate Beach Road Anderson Island, WA 98303

TELEPHONE: 253-884-6755 PERSON IN CHARGE: Duty Hrbormaster

SHORT DESCRIPTION & LOCATION: **WEBSITE??**

47°08.10' - 122°42.15' Located in SE quadrant of Anderson Is. Entrance to inner Oro Bay is between red & green buoys. Marina is located off the port beam, beyond the Tacoma & Bremerton Y.C. outstations. Appx 55 ft. of moorage along NW corner is available for reciprocal clubs. Upon arrival at dock, follow instructions on phone. NO RAFTING.

RECIPROCAL BOAT CAPACITY: ....2-4 Boats

DOCKSIDE DEPTH AT ZERO TIDE: ....7 ft.

SEASON: ....All year

RESERVATION POLICY: First come first served

TOILETS: ....Porta-Potty

HOT SHOWERS: ....None

RESTAURANT: ....At Golf Course

BROADBAND/WI-FI: ....None

DAILY RATE: Moorage................................No Charge
Electricity Charge..............$2.00/day
Length of Stay:.....2 days maximum

RECIPROCAL DOCK: ....55 ft.

RECIPROCAL SLIPS: Dock only

WATER: ....Limited

AMT W/ELECTRICITY: ....All

AMPS: ....20 A

BAR: ....None

PET FRIENDLY: ....Good

OTHER: Very quiet & peaceful. Gen. Store & golf 4 mi. by car, rock crabs in bay.

***NOTE:* THIS IS PRIVATE MOORAGE AND ONLY AVAILABLE TO MEMBERS OF RECIPROCAL YACHT CLUBS. YOUR CLUB *MUST* HAVE RECIPROCAL PRIVILEGES AND YOU MUST FLY YOUR BURGEE.**

CAUTION! This chartlet not intended for use in navigation.

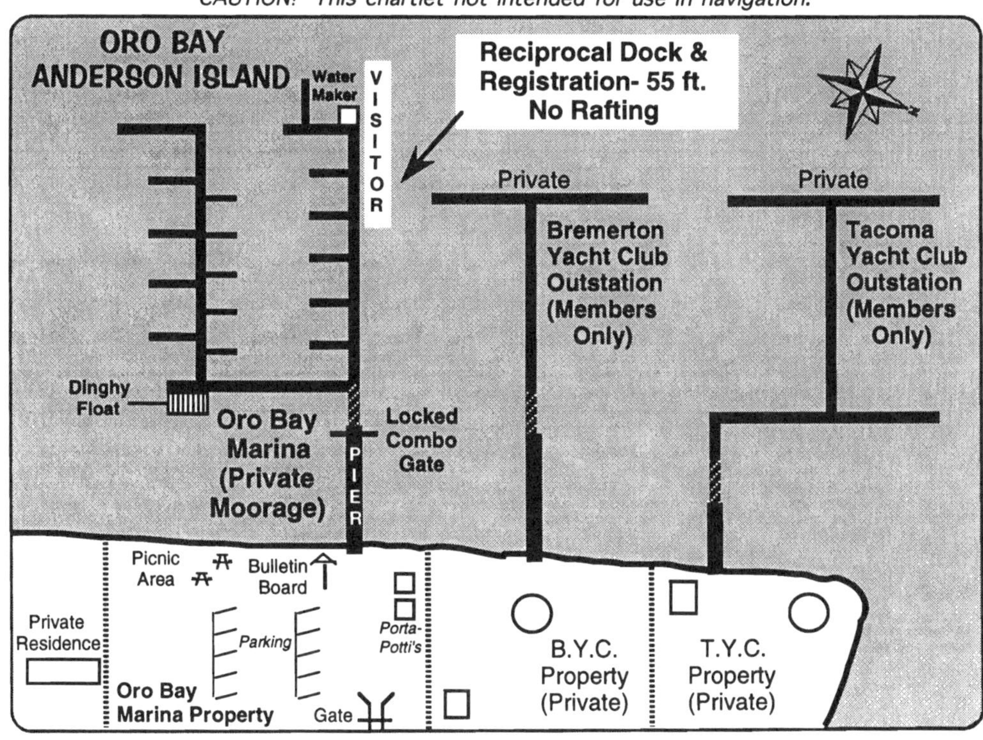

Longbranch

NAME OF YACHT CLUB: *CARLING YACHT CLUB*

ADDRESS: P.O. Box 272 Tacoma, WA 98401

TELEPHONE: 253-884-3013 PERSON IN CHARGE: Past Commodore

SHORT DESCRIPTION & LOCATION: **www.carlingyachtclub.com**

Carling offers reciprocal moorage at Longbranch Marina. Reciprocal visitors offered 80' per month on a first come basis. Visiting reciprocal club members are required to pay regular moorage charges to Dockmaster & mail moorage receipt with a copy of current club membership card to Carling within 30 days. Contact P.C. (253-884-3013) for availability.

RECIPROCAL BOAT CAPACITY: 80' per mo.
DOCKSIDE DEPTH AT ZERO TIDE: ....8 ft.
SEASON: ....All year
RESERVATION POLICY: First come, first serve
TOILETS: ....Porta-Potty
HOT SHOWERS: ....None
RESTAURANT: ....4 Miles
BROADBAND/WI-FI: ....None
DAILY RATE: 1 visit per boat per calendar year. **Power charge unreimbursed.**

RECIPROCAL DOCK: As available
RECIPROCAL SLIPS: ....Varies
WATER: ....Yes
AMT W/ELECTRICITY: ....All
AMPS: ....30 A
BAR: ....4 Miles
PET FRIENDLY: ....Good
OTHER: Services same as **Longbranch Marina** listing on Page 23.

***NOTE:* THIS IS A PRIVATE RECIPROCAL AGREEMENT TOTALLY INDEPENDENT OF LONGBRANCH MARINA OPERATIONS. YOUR CLUB *MUST* HOLD RECIPROCAL PRIVILEGES TO REQUEST REIMBURSEMENT.**

SEE DATA & CHARTLET ON PAGE: 23

NOTES

Longbranch

NAME OF MARINA: ***LONGBRANCH MARINA*** **RADIO:** VHF Ch.16
TELEPHONE: 253-884-5137 MGR: Mark Jones, Dockmaster
E-MAIL: lic@longbranchimprovementclub.org FAX: 253-884-6256 (Fax)
ADDRESS: P.O. Box 111 Longbranch, WA 98349
SHORT DESCRIPTION & LOCATION: **www.longbranchimprovementclub.org**

47°12.60' - 122°45.00' Well maintained friendly & low key marina in lovely Filucy Bay located on east side of Key Peninsula. Sheltered cove provides very comfortable moorage. This is a popular location for boating club cruises. 8 Road miles to Key Center services.

GUEST BOAT CAPACITY: ....Appx. 50 boats
DOCKSIDE DEPTH AT ZERO TIDE: ....8 ft.
SEASON: ....All year
RESERVATION POLICY: ....None
AMT W/ELECTRICITY: ....All
FUEL DOCK: ....None
MARINE REPAIRS: ....None
TOILETS: ....Porta-Potty
HOT SHOWERS: ....None
RESTAURANT: ....None
PICNIC AREA: ....Function Shelter on Main Dock
BASIC STORE: ....None
BROADBAND/WI-FI: ....None
DAILY RATE: ....Economical (Under 75¢/foot)

NOTE: Cell phone recepption is marginal inside Filucy Bay.

GUEST DOCK: ....760 ft. total
GUEST SLIPS: ....3 Large docks
WATER: ....Yes
AMPS: ....30 A
PUMP OUT STATION ....None
HAUL OUT: ....None
BOAT RAMP: ....None
LAUNDRY: ....None
BAR: ....None
POOL: ....None
GOLF: ....None
PET FRIENDLY: ....Good
OTHER: Fishing, good dinghy area, bicycling, and rock crabbing. No retail services. Ice on dock. Taxi service to Key Center.

CAUTION! This chartlet not intended for use in navigation.

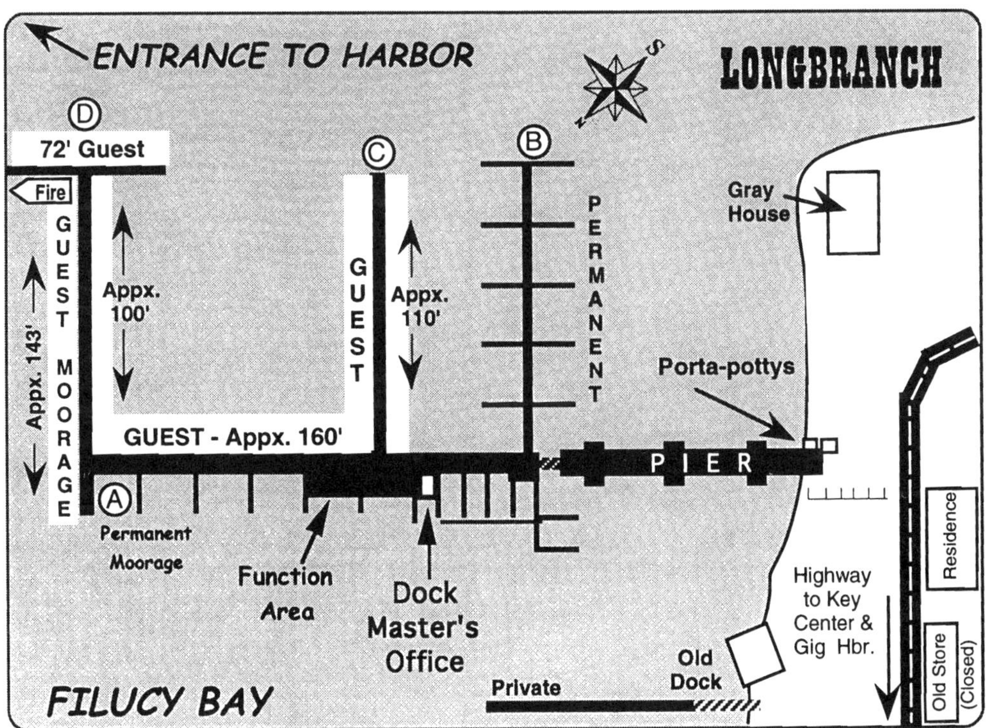

Lakebay

NAME OF PARK: *PENROSE POINT STATE PARK*
ADDRESS: 321 158th Ave. KPS Lakebay, WA 98349
TELEPHONE: 253-884-2514 **MGR:** Dave Roe/Current Park Manager
SHORT DESCRIPTION & LOCATION: **www.parks.wa.gov/moorage/parks**
47°15.50' - 122°45.10' Located on the E. shore of Mayo Cove on W. side of Carr Inlet 3 miles N. of Long Branch. The park has a large shoreline in a beautiful setting and is an excellent destination in South Sound. The floats go dry at extreme low tides.

GUEST BOAT CAPACITY: ....Appx 9-12 boats
DOCKSIDE DEPTH AT ZERO TIDE:..........3-5 ft.
SEASON: ....All year
AMT W/ELECTRICITY: ....None
TOILETS: ....Yes
HOT SHOWERS: ....Closed in winter
PICNIC AREA: ....Yes
PLAY AREA: ....None
BASIC STORE: ....Close by
DAILY RATE: Dock Charge..........50¢/foot
Mooring Buoys..........$10.00/night

GUEST DOCK: ....280 Ft. total
MOORING BUOYS: ....8 Buoys
WATER: ....Yes
PAY PHONES: ....Yes
BOAT RAMP: ....Close by
PICNIC SHELTER: ....Yes
BBQ: ....Yes
PUMP OUT STATION: ....Yes
PET FRIENDLY: ....Good
OTHER: Nature & hiking trails with interpretative signs, 2 miles of clamming, beachcombing & campsites

CAUTION! This chartlet not intended for use in navigation.

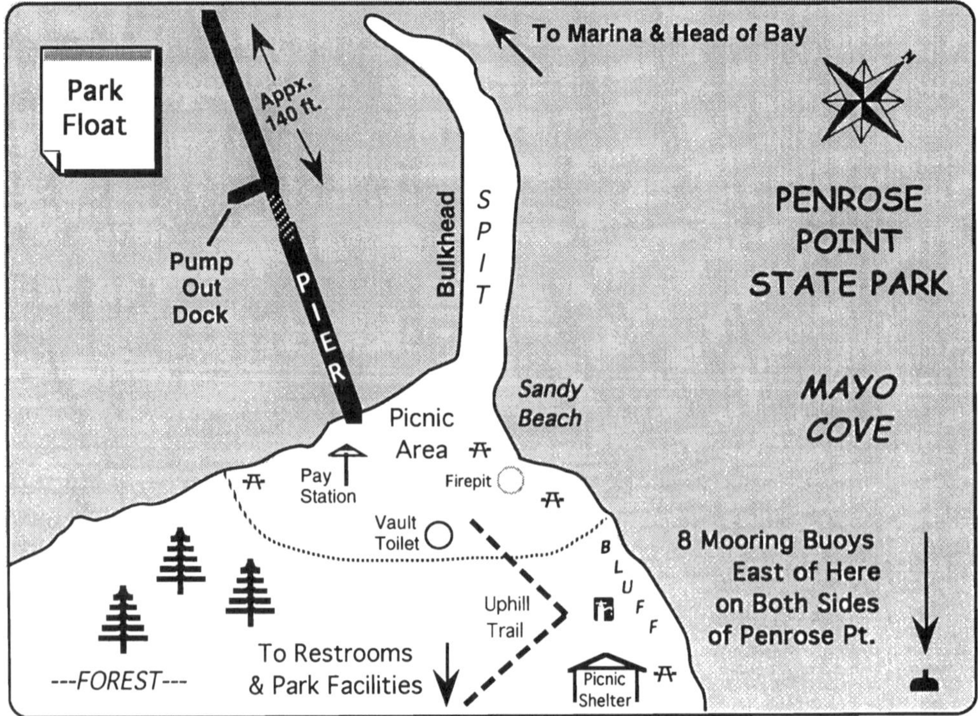

Lakebay

NAME OF MARINA: ***LAKEBAY MARINA*** RADIO: None
TELEPHONE: 253-884-3350 MGR: Dewey Hostetler
E-MAIL: None FAX: None
ADDRESS: 15 Lorenz Rd. K.P.S. Lakebay, WA 98349

SHORT DESCRIPTION & LOCATION:

47°15.40' - 122°45.20' Older rustic marina located at the head of Mayo Cove on the SW shore of Carr Inlet in a village with a store & several small private piers. The channel to the marina is difficult to navigate at low tide & caution & local knowledge are advised.

GUEST BOAT CAPACITY: ....Appx. 6-10 boats
DOCKSIDE DEPTH AT ZERO TIDE: 6 ft. 10 in.
SEASON: ....All year
RESERVATION POLICY: ....Accepts
AMT W/ELECTRICITY: ....Limited
FUEL DOCK: ....Gas only/Diesel planned
MARINE REPAIRS: ....None
TOILETS: ....Yes
HOT SHOWERS: ....None
RESTAURANT: ....Close by-1.5 miles
PICNIC AREA: ....Yes
BASIC STORE: ....Yes
BROADBAND/WI-FI: ....None
DAILY RATE: ....Economical (Under 75¢/foot)

GUEST DOCK: ....Varies
GUEST SLIPS: ....Varies
WATER: ....Yes
AMPS: ....15 A
PUMP OUT STATION ....None
HAUL OUT: ....None
BOAT RAMP: ....Close by
LAUNDRY: ....Close by
BAR: ....Close by
POOL: ....None
GOLF: ....Close by
PET FRIENDLY: ....Good
OTHER: Caters to fishermen, bait & tackle sold in store, close to Park, great dinghy area, taxi service to stores.

CAUTION! This chartlet not intended for use in navigation.

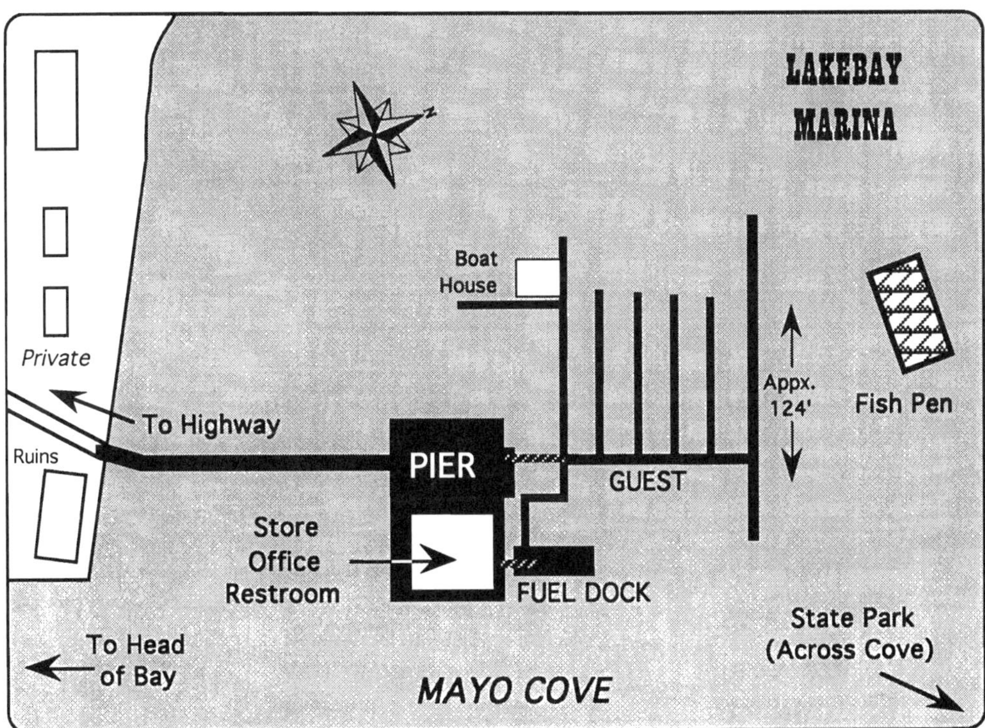

Fox Island

NAME OF YACHT CLUB: *FOX ISLAND YACHT CLUB*

ADDRESS: P.O. Box 1, 1061 12th Ave. Fox Island, WA 98333

TELEPHONE: 253-549-4194 PERSON IN CHARGE: Harbor Master

SHORT DESCRIPTION & LOCATION: **www.fiyc.com**

47°14.50' - 122°36.00' (Entrance) Located in pristine Cedrona Cove on NE side of Fox Is. about 3 mi. SW of Narrows Bridge. Modern clubhouse & grounds offer a pleasant atmosphere. Caution is necessary upon entering Cove & a large scale chart advised. In addition to recip. dock, FIYC has 3 mooring buoys inside harbor & 1 buoy outside harbor.

RECIPROCAL BOAT CAPACITY: .Appx 2-4 Boats

DOCKSIDE DEPTH AT ZERO TIDE: ....3 ft.

SEASON: ....All year

RESERVATION POLICY: ....None

TOILETS: ....In clubhouse when open

HOT SHOWERS: ....In clubhouse when open

RESTAURANT: ....None

BROADBAND/WI-FI: ....None

DAILY RATE: ....$3.00 Electricity charge

Moorage fee:............None

Length of Stay:.........2 days max

RECIPROCAL DOCK: ....60 ft.

RECIPROCAL SLIPS: Dock only

WATER: ....Yes

AMT W/ELECTRICITY: ....All

AMPS: ....30 A

BAR: ....None

PET FRIENDLY: ....Excellent

OTHER: 4 Reciprocal mooring buoys, playground, BBQ, pickle ball, horsehoes & boat launch ramp.

***NOTE:* THIS IS PRIVATE MOORAGE AND ONLY AVAILABLE TO MEMBERS OF RECIPROCAL YACHT CLUBS. YOUR CLUB *MUST* HAVE RECIPROCAL PRIVILEGES AND YOU MUST FLY YOUR BURGEE.**

CAUTION! This chartlet not intended for use in navigation.

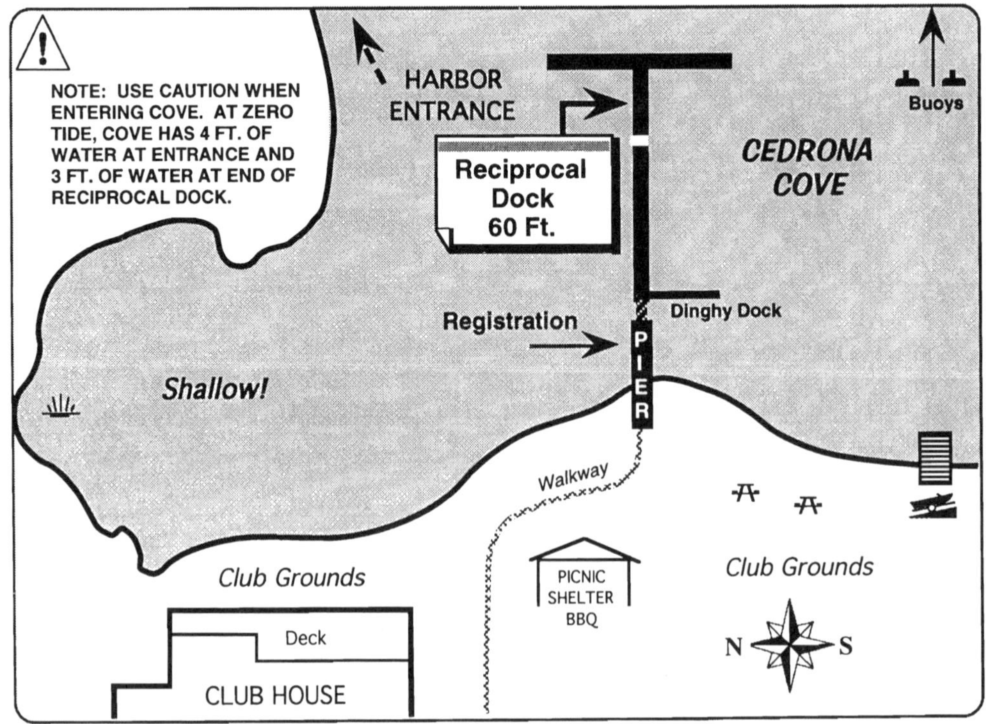

Tacoma

NAME OF YACHT CLUB: *DAY ISLAND YACHT CLUB*
ADDRESS: 2120 91st Ave W. University Place, WA 98466
TELEPHONE: 253-565-3777 PERSON IN CHARGE: Caretakers
SHORT DESCRIPTION & LOCATION: **www.dayislandyc.org**
47°14.70' - 122°33.60' DIYC lies in a lagoon w/several private marinas about 1 mi. S. of the Tacoma Narrows Bridge on the Tacoma side of the Narrows. Entry to lagoon is only possible at plus 4 ft. tide level. Watch depth marker at entrance to basin. See route map sent to your club & keep 3 green day markers to your port side. Register with caretakers.

RECIPROCAL BOAT CAPACITY: Apx 4-6 boats
DOCKSIDE DEPTH AT ZERO TIDE: ....5 ft.
SEASON: ....All year
RESERVATION POLICY: ....None
TOILETS: ....Yes
HOT SHOWERS: ....Yes
RESTAURANT: ....Close by
BROADBAND/WI-FI: ....None
DAILY RATE: ....No Charge
48 Hour Limit - Please register with caretakers upon arrival.

RECIPROCAL DOCK: Appx 70 ft.
RECIPROCAL SLIPS: Dock only
WATER: ....Yes
AMT W/ELECTRICITY: ....All
AMPS: ....30 A
BAR: ....Close by
PET FRIENDLY: ....Good
OTHER: Fuel & marine supplies nearby. Restaurant, pub, grocery & laundry 1 mile.

***NOTE:* THIS IS PRIVATE MOORAGE AND ONLY AVAILABLE TO MEMBERS OF RECIPROCAL YACHT CLUBS. YOUR CLUB *MUST* HAVE RECIPROCAL PRIVILEGES AND YOU MUST FLY YOUR BURGEE.**

CAUTION! This chartlet not intended for use in navigation.

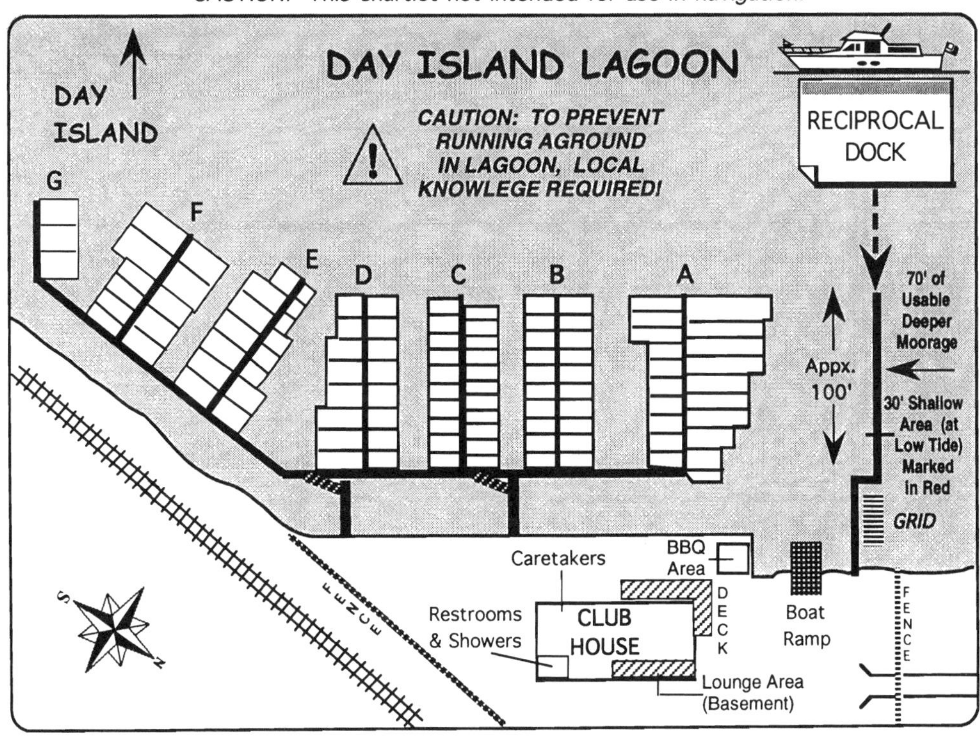

Tacoma

NAME OF MARINA: ***BREAKWATER MARINA*** **RADIO:** None
TELEPHONE: **253-752-6663** MGR: Michael Marchetti
E-MAIL: michael@breakwatermarina.com FAX: 253-752-8291 (Fax)
ADDRESS: 5603 Waterfront Drive Tacoma, WA 98407
SHORT DESCRIPTION & LOCATION: **www.breakwatermarina.com**

47°18.30' - 122°30.69' Located on E. side of ferry terminal at Pt. Defiance in the Tacoma Yacht Club Boat Basin about 4 mi. NW of downtown Tacoma. Although marina has no designated guest moorage, they generally have a few vacant slips for overnighters.

GUEST BOAT CAPACITY: ....Varies
DOCKSIDE DEPTH AT ZERO TIDE: ....18 ft.
SEASON: ....All year
RESERVATION POLICY:..........Accepts
AMT W/ELECTRICITY: ....All
FUEL DOCK: ....Gas, Dsl, & LP
MARINE REPAIRS: ....On premises
TOILETS: ....Yes
HOT SHOWERS: ....Yes
RESTAURANT: ....Close by
PICNIC AREA: ....Close by
BASIC STORE: ....Yes
BROADBAND/WI-FI: ....BroadbandXpress
DAILY RATE:..........Moderate (75¢-$1.25/foot)

GUEST DOCK: ....None
GUEST SLIPS: ....Varies
WATER: ....Yes
AMPS: ....20-30 A
PUMP OUT STATION ....Yes
HAUL OUT: ....None
BOAT RAMP: ....Close by
LAUNDRY: ....Yes
BAR: ....Close by
POOL: ....None
GOLF: ....Close by
PET FRIENDLY: ....Good
OTHER: Tackle & bait, Pt. Defiance park, & zoo nearby, bus service to city.

NAME OF YACHT CLUB: *TACOMA YACHT CLUB*
CLUB ADDRESS: 5401 N. Waterfront Dr., Tacoma, WA 98407
CLUB TELEPHONE: 253-752-3555 PERSON IN CHARGE: Tim Hummel
LOCATION & SPECIAL NOTES: **www.tacomayachtclub.com**

47°18.50' - 122°30.80' The TYC boat basin is situated at the NW entrance to Commencement Bay at Pt. Defiance about 4 mi NW of downtown Tacoma. Guest moorage is located just beyond the breakwater to your port as you enter the marina. Please complete registration envelope on dock station or in club house.

RECIPROCAL BOAT CAPACITY: Appx 4 boats
DOCKSIDE DEPTH AT ZERO TIDE: ....18 ft.
RECIPROCAL SEASON:..........All year
RESERVATION POLICY: ....None
TOILETS: ....Yes
HOT SHOWERS: ....Yes
RESTAURANT: ....Yes
DAILY RATE: Electrical service - $2.00 per night ....Moorage - N/C for first 48 hrs.

RECIPROCAL DOCK: ....130 ft.
RECIPROCAL SLIPS: Dock only
WATER: ....Yes
AMT W/ELECTRICITY: ....All
AMPS: ....30 A
BAR: ....Yes
OTHER: Nice restaurant, full menu. Visa/Master only.

***NOTE:* THIS IS PRIVATE MOORAGE *ONLY* AVAILABLE TO MEMBERS OF RECIPROCAL YACHT CLUBS! YOUR CLUB *MUST* HAVE RECIPROCAL PRIVILEGES AND YOU MUST FLY YOUR BURGEE!**

Tacoma

CAUTION! This chartlet not intended for use in navigation.

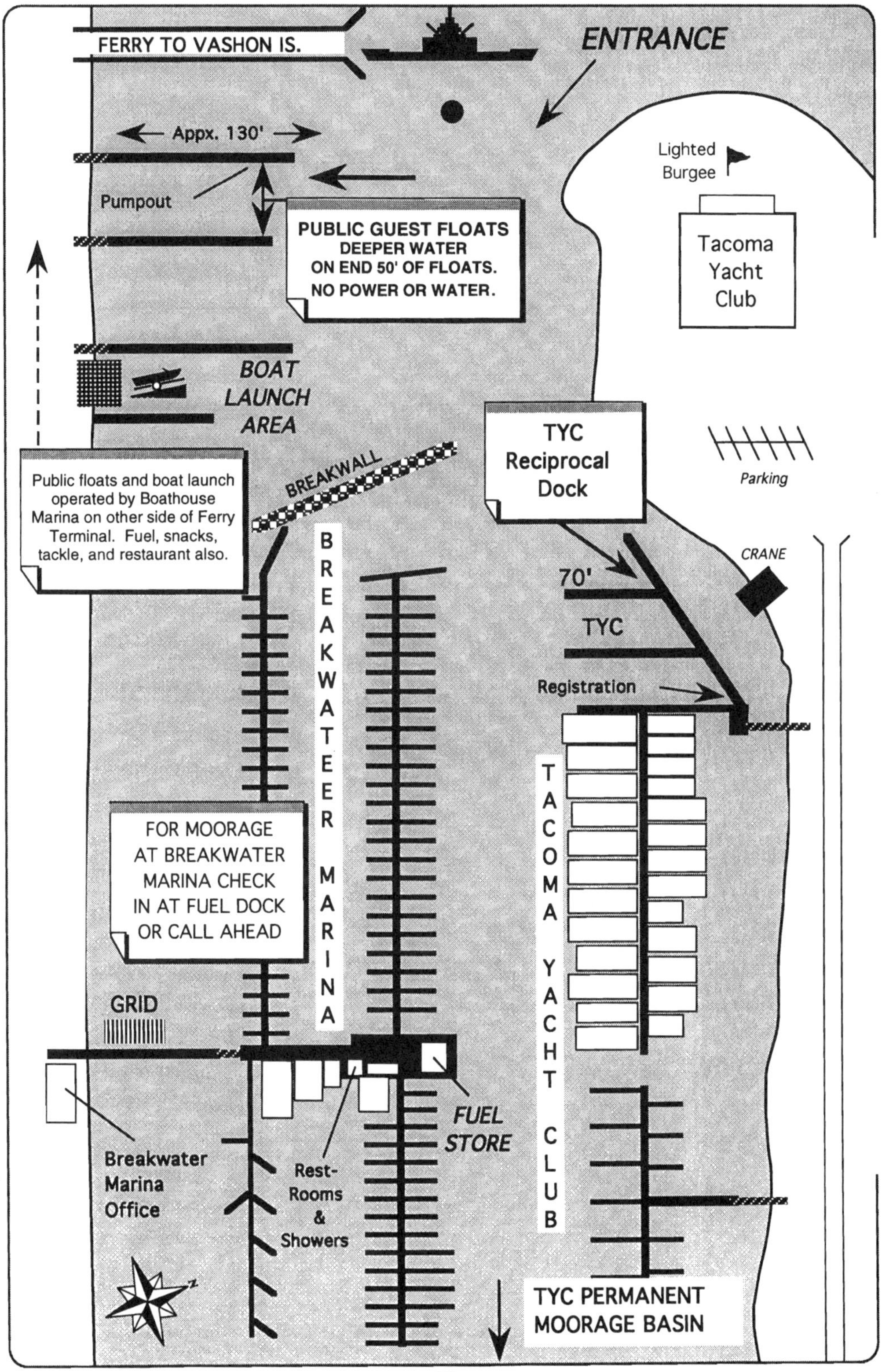

FERRY TO VASHON IS.

ENTRANCE

Appx. 130'

Lighted Burgee

Pumpout

PUBLIC GUEST FLOATS
DEEPER WATER
ON END 50' OF FLOATS.
NO POWER OR WATER.

Tacoma Yacht Club

BOAT LAUNCH AREA

TYC Reciprocal Dock

Parking

Public floats and boat launch operated by Boathouse Marina on other side of Ferry Terminal. Fuel, snacks, tackle, and restaurant also.

BREAKWALL

CRANE

70'

TYC

BREAKWATEER MARINA

Registration

TACOMA YACHT CLUB

FOR MOORAGE AT BREAKWATER MARINA CHECK IN AT FUEL DOCK OR CALL AHEAD

GRID

FUEL STORE

Breakwater Marina Office

Rest-Rooms & Showers

TYC PERMANENT MOORAGE BASIN

Tacoma

NAME OF YACHT CLUB: *FIRCREST YACHT CLUB*

ADDRESS: P.O. Box 1972 Tacoma, WA 98401

TELEPHONE: 253-752-6663 Marina PERSON IN CHARGE: Vice Commodore

SHORT DESCRIPTION & LOCATION: **www.fircrestyachtclub.org**

Fircrest Y.C. reciprocal moorage is located at **Breakwater Marina** inside the Yacht Club basin at Pt. Defiance. Advance reservations are recommended (253) 752-6663. Please check-in at fuel dock for slip assignment. Within walking distance are Anthony's Home Port Restaurant, Tacoma Yacht Club, Pt. Defiance Zoo & Aquarium and antique shops.

RECIPROCAL BOAT CAPACITY....................Varies

DOCKSIDE DEPTH AT ZERO TIDE: Appx. 18'

SEASON: ....All year

RESERVATION POLICY: ....Accepts

TOILETS: ....At Marina

HOT SHOWERS: ....At Marina

RESTAURANT: ....Close by

BROADBAND/WI-FI: ....BroadbandXpress

DAILY RATE: ....Utility Fee $3.00 per night
Two night maximum stay, max 4 visits per month.

RECIPROCAL DOCK: .....Slips only

RECIPROCAL SLIPS: ....Varies

WATER: ....Yes

AMT W/ELECTRICITY: ....All

AMPS: ....15-30 A

BAR: ....Close by

PET FRIENDLY: ....Good

OTHER: Services same as marina listing for Breakwater Marina on Page 28-29.

***NOTE:* THIS IS PRIVATE MOORAGE AND ONLY AVAILABLE TO MEMBERS OF RECIPROCAL YACHT CLUBS. YOUR CLUB MUST HAVE RECIPROCAL PRIVILEGES AND YOU MUST FLY YOUR BURGEE.**

SEE DATA & CHARTLET ON PAGE: 28-29

NOTES

Tacoma

NAME OF PARK: *OLD TOWN DOCK - Metro Parks Tacoma*
ADDRESS: 4702 S. 19th Tacoma, WA 98405
TELEPHONE: 253-591-5325 MGR: Boathouse Marina
SHORT DESCRIPTION & LOCATION: **www.metroparkstacoma.org**

47°15.60' - 122°27.80' Located on SW shore of Commencement Bay about 2 miles NW of downtown Tacoma in upscale Old Town Historic District. Aging pier & floats. Guest moorage floats situated behind the breakwater as well as one large float on N. side of pier.

GUEST BOAT CAPACITY: ....Appx. 12 boats
DOCKSIDE DEPTH AT ZERO TIDE: Appx 20 ft
SEASON: ....Closed winter
AMT W/ELECTRICITY: ....None
TOILETS: ....Yes/Not Recommended
HOT SHOWERS: ....None
PICNIC AREA: ....Close by
PLAY AREA: ....Close by
BASIC STORE: ....Close by
DAILY RATE: ....Under 19 ft. - $10/night
............19 ft, to 29 ft. - $15/night
............Over 30 ft. - $20/night
72 Hour Limit

GUEST DOCK: 2 lg. floats, 8 slips
MOORING BUOYS: ....2 Buoys
WATER: ....None
PAY PHONES: ....Yes
BOAT RAMP: ....None
PICNIC SHELTER: ....None
BBQ: ....Close by
PUMP OUT STATION: ......None
PET FRIENDLY: ....Fair
OTHER: Fresh fish market, bus service, close to several restaurants on Ruston Way. Dock renovations planned (hopefully soon).

CAUTION! This chartlet not intended for use in navigation.

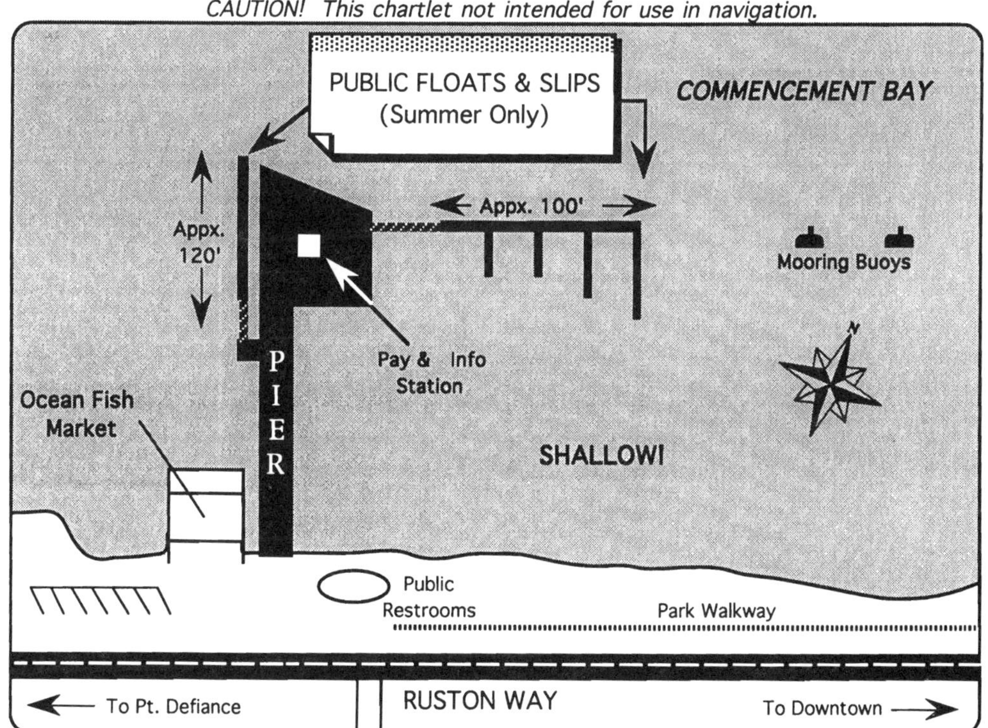

Tacoma

NAME OF MARINA: ***FOSS WATERWAY MARINA*** **RADIO:** None
TELEPHONE: **253-272-4404** MGR: Tracy McKendry
E-MAIL: None FAX: 253-272-0367 (Fax)
ADDRESS: 821 Dock St. Tacoma, WA 98402
SHORT DESCRIPTION & LOCATION: **www.fosswaterwaymarina.com**
47°15.35' - 122°26.03' (Formerly Totem Marina) Located in downtown Tacoma on the W. side of Thea Foss Waterway just before the 11th Street Bridge. Walking distance to attractions and public services. Major upland development and upgrades planned.

GUEST BOAT CAPACITY: ....Appx. 20 boats
DOCKSIDE DEPTH AT ZERO TIDE: ....26 ft.
SEASON: ....All year
RESERVATION POLICY:..........None
AMT W/ELECTRICITY: ....All
FUEL DOCK: ....Close by
MARINE REPAIRS: ....Close by
TOILETS: ....Yes
HOT SHOWERS: ....Yes
RESTAURANT: ....Close by
PICNIC AREA: ....Yes
BASIC STORE: ....Yes
BROADBAND/WI-FI: ....None
DAILY RATE:..........Economical (Under 75¢/foot)

GUEST DOCK: (2)......400 ft. TTL
GUEST SLIPS: ....Varies
WATER: ....Yes
AMPS: ....30-50 A
PUMP OUT STATION ....Yes
HAUL OUT: ....Yes
BOAT RAMP: ....Sm. boat sling
LAUNDRY: ....Yes
BAR: ....Close by
POOL: ....None
GOLF: ....Close by
PET FRIENDLY: ....Fair
OTHER: Convenience store, bait, tackle, marine supplies. Parks, museum & shopping close by.

NAME OF YACHT CLUB: ***TOTEM YACHT CLUB***
CLUB ADDRESS: 5045 N. Highland, Tacoma, WA 98407
CLUB TELEPHONE: 253-759-9062 PERSON IN CHARGE: Yeoman
LOCATION & SPECIAL NOTES: **www.totemyachtclub.org**
Totem Yacht Club provides 100 ft. of reciprocal moorage at the Foss Waterway Marina. The moorage is located within the general marina guest dock area on Piers D & E. Please register in the marina office upon arrival between the hours of 0900-1700. A $10.00 Refundable key deposit is required.

RECIPROCAL BOAT CAPACITY:...........Appx. 2-3
DOCKSIDE DEPTH AT ZERO TIDE: ....26 ft.
RECIPROCAL SEASON:..........All year
RESERVATION POLICY: ....None
TOILETS: ....At Marina
HOT SHOWERS:..........At Marina
RESTAURANT: ....Close by
DAILY RATE: ....$4.50 daily charge for power
Maximum stay is 2 days/year

RECIPROCAL DOCK: ....100 ft.
RECIPROCAL SLIPS: ...Dock only
WATER: ....Yes
AMT W/ELECTRICITY: ....All
AMPS: ....20-30 A
BAR: ....Close by
OTHER: ....Same as marina listing

***NOTE:* THIS IS PRIVATE MOORAGE *ONLY* AVAILABLE TO MEMBERS OF RECIPROCAL YACHT CLUBS! YOUR CLUB *MUST* HAVE RECIPROCAL PRIVILEGES AND YOU MUST FLY YOUR BURGEE!**

Tacoma

CAUTION! This chartlet not intended for use in navigation.

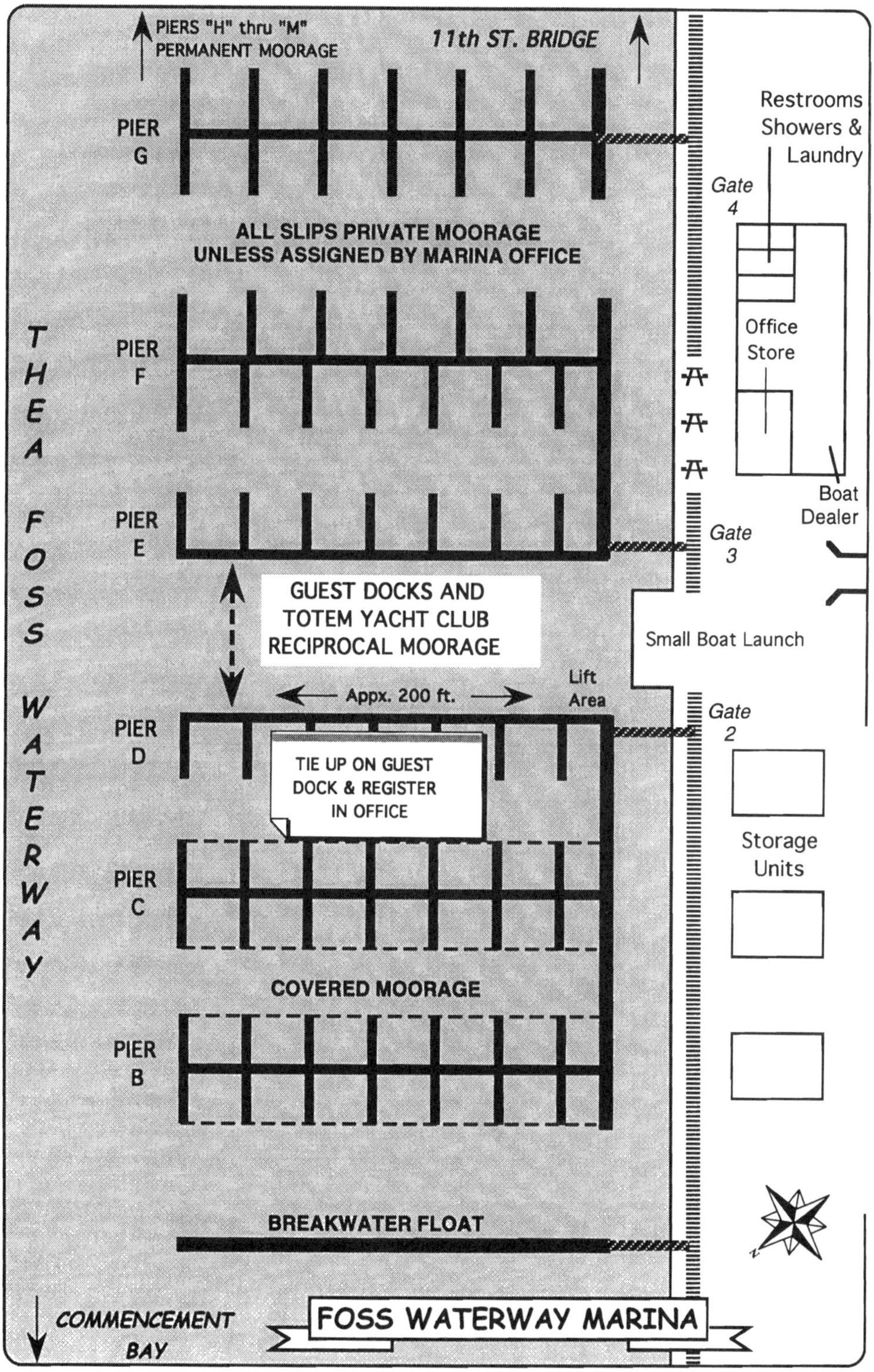

PIERS "H" thru "M"
PERMANENT MOORAGE

11th ST. BRIDGE

PIER G

Restrooms
Showers &
Laundry

Gate 4

ALL SLIPS PRIVATE MOORAGE
UNLESS ASSIGNED BY MARINA OFFICE

THEA FOSS WATERWAY

PIER F

Office
Store

PIER E

Boat
Dealer

Gate 3

GUEST DOCKS AND
TOTEM YACHT CLUB
RECIPROCAL MOORAGE

Small Boat Launch

Lift
Area

Appx. 200 ft.

Gate 2

PIER D

TIE UP ON GUEST
DOCK & REGISTER
IN OFFICE

Storage
Units

PIER C

COVERED MOORAGE

PIER B

BREAKWATER FLOAT

COMMENCEMENT BAY

FOSS WATERWAY MARINA

Tacoma

NAME OF MARINA: ***DOCK STREET MARINA*** **RADIO:** VHF 78A
TELEPHONE: **253-272-4352** **MGR:** Doug Hicks, Dockmaster
E-MAIL: info@dockstreetmarina.com **FAX:** 253-572-2768
ADDRESS: 1817 Dock Street Tacoma, WA 99402
SHORT DESCRIPTION & LOCATION: **www.dockstreetmarina.com**

48°14.66' - 122°25.94' Tacoma's new state-of-the-art downtown marina is located 1 mile down Thea Foss Waterway, directly in front of Museum of Glass. Walk or Light Rail to world class museums, restaurants, Theatre District, & Antique Row. Very boater friendly.

GUEST BOAT CAPACITY: ....Varies but ample
DOCKSIDE DEPTH AT ZERO TIDE: ......15 ft. +
SEASON: ....All year
RESERVATION POLICY: ....Recommended
AMT W/ELECTRICITY: ....All
FUEL DOCK: ....Close by
MARINE REPAIRS: ....Close by
TOILETS: ....Yes
HOT SHOWERS: ....Yes
RESTAURANT: ....Close by
PICNIC AREA: ....Yes
BASIC STORE: ....Close by
BROADBAND/WI-FI: ....BroadbandXpress
DAILY RATE: ....Moderate to Premium ($1.00 - $1.50+/foot depending on boat size) **Weekday group discounts available.**

GUEST DOCK: ....127 Ft. end ties
GUEST SLIPS: ....Appx. 30-40
WATER: ....Yes
AMPS: ....30-50-100 A
PUMP OUT STATION ....Yes
HAUL OUT: ....Close by
BOAT RAMP: ....Close by
LAUNDRY: ....Yes
BAR: ....Close by
POOL: ....None
GOLF: ....Close by
PET FRIENDLY: ....Excellent
OTHER: Cruising Clubs welcome! Galleries, shops, deli, both Museum of Glass & History close by. Slip-side pumpouts by marina staff.

NOTES

Tacoma

CAUTION! This chartlet not intended for use in navigation.

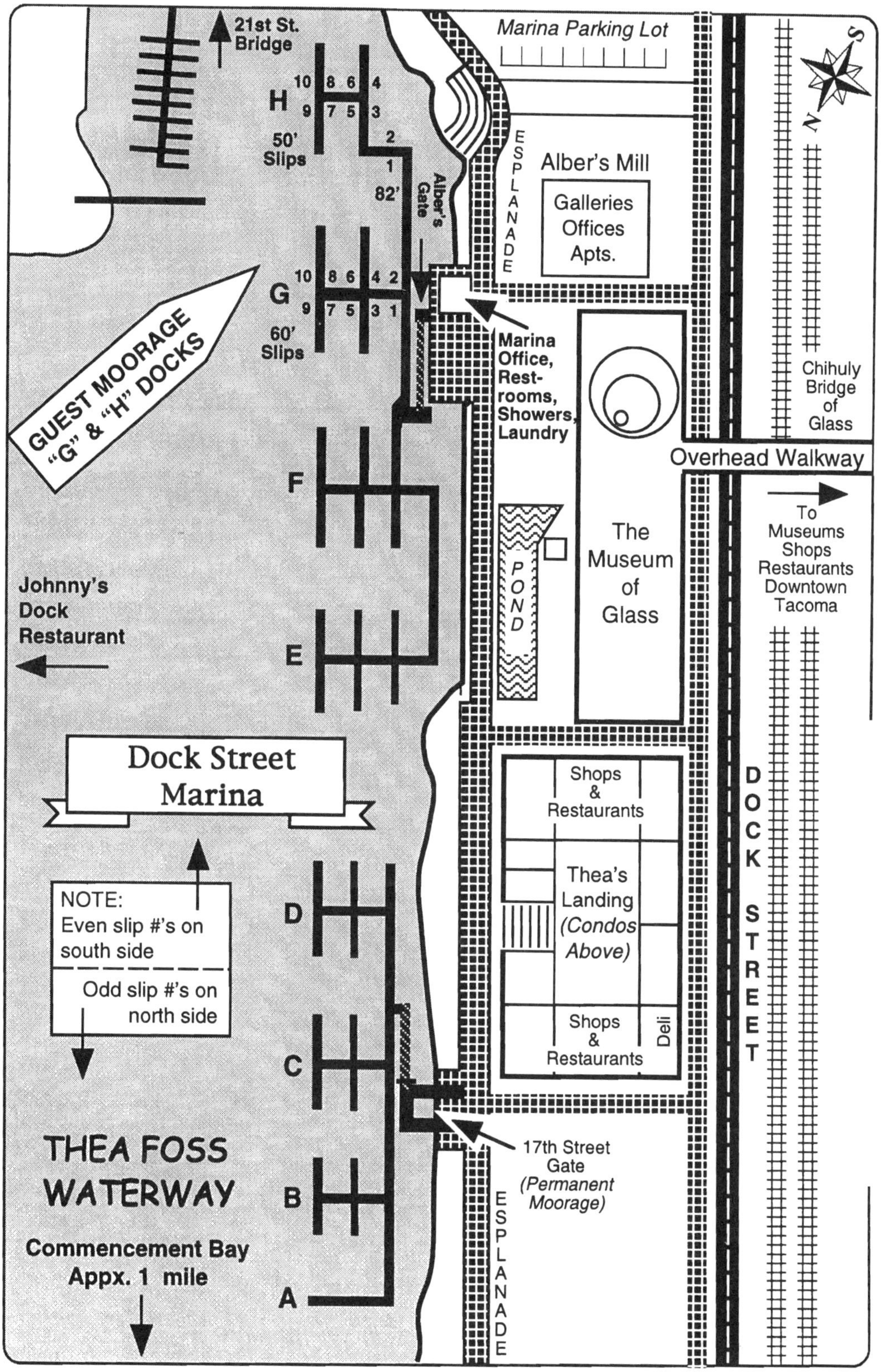

21st St. Bridge

Marina Parking Lot

H
10 8 6 4
9 7 5 3
2
1
50' Slips
82'

Alber's Gate

ESPLANADE

Alber's Mill

Galleries
Offices
Apts.

G
10 8 6 4 2
9 7 5 3 1
60' Slips

GUEST MOORAGE
"G" & "H" DOCKS

Marina Office, Rest-rooms, Showers, Laundry

Chihuly Bridge of Glass

Overhead Walkway

F

POND

The Museum of Glass

To Museums Shops Restaurants Downtown Tacoma

Johnny's Dock Restaurant

E

Dock Street Marina

Shops & Restaurants

Thea's Landing *(Condos Above)*

Shops & Restaurants

Deli

DOCK STREET

NOTE:
Even slip #'s on south side
Odd slip #'s on north side

D

C

17th Street Gate *(Permanent Moorage)*

THEA FOSS WATERWAY

B

ESPLANADE

Commencement Bay
Appx. 1 mile

A

Tacoma

NAME OF MARINA: ***JOHNNY'S DOCK RESTAURANT*** **RADIO:** None
TELEPHONE: **253-627-3186** **MGR:** Restaurant Marina Manager
E-MAIL: winston3609@yahoo.com **FAX:** 253-627-7457 (Fax)
ADDRESS: 1900 East D Street Tacoma, WA 98421
SHORT DESCRIPTION & LOCATION: **www.johnnysdock.com**

48°14.80' - 122°25.80' Located on east side of downtown Tacoma's Thea Foss Waterway about 1/2 mile south of 11th Street Bridge & just before new 21st St. Bridge. Basic tie up dock free of charge to restaurant patrons. Restaurant personnel issues gate key.

GUEST BOAT CAPACITY: ....Appx. 2-4 boats
DOCKSIDE DEPTH AT ZERO TIDE: ....12 ft.
SEASON: ....All year
RESERVATION POLICY: ....None
AMT W/ELECTRICITY: ....All
FUEL DOCK: ....Close by
MARINE REPAIRS: ....Close by
TOILETS: ....At restaurant
HOT SHOWERS: ....None
RESTAURANT: ....Yes
PICNIC AREA: ....None
BASIC STORE: ....Close by
BROADBAND/WI-FI: ....BroadbandXpress
DAILY RATE:.......No charge to restaurant patrons

Check Johnny's website for special meal coupons.

GUEST DOCK: ....Appx. 115 ft.
GUEST SLIPS: ....Yes
WATER: ....No
AMPS: ....None
PUMP OUT STATION ....None
HAUL OUT: ....None
BOAT RAMP: ....None
LAUNDRY: ....None
BAR: ....Yes
POOL: ....None
GOLF: ....Close by
PET FRIENDLY: ....Fair
OTHER: Sunday brunch, lunch and dinner daily, private rooms for groups. Chihuly Glass Museum within walking or short cab ride.

CAUTION! This chartlet not intended for use in navigation.

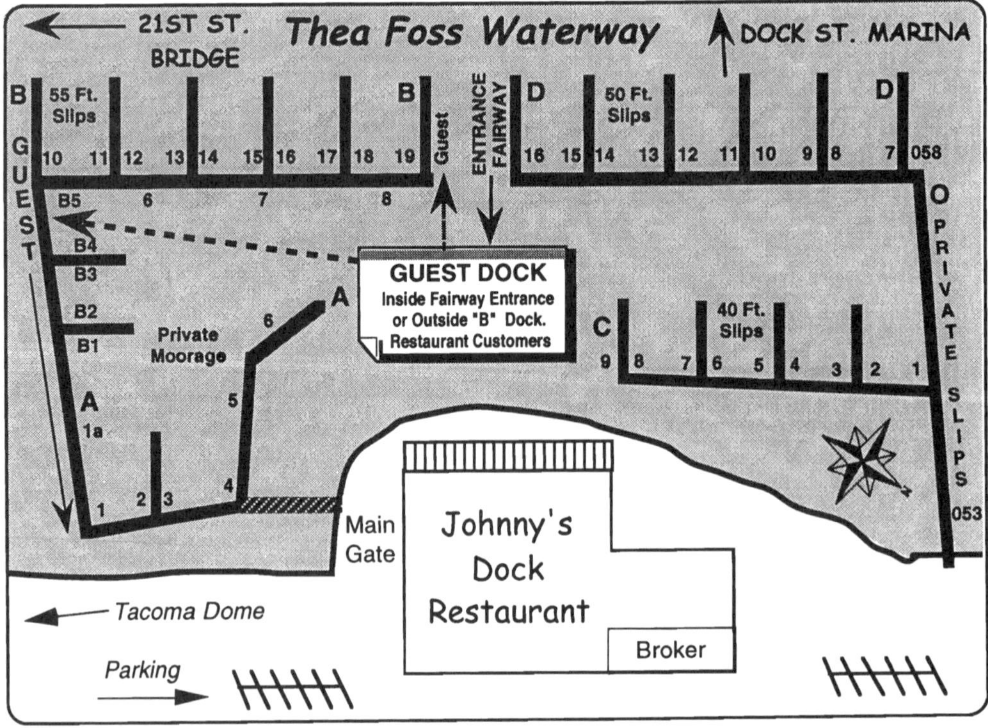

Tacoma

NAME OF MARINA: ***CHINOOK LANDING MARINA*** RADIO: VHF Ch. 79
TELEPHONE: **253-627-7676** MGR: Dennis LaPointe
E-MAIL: clm@puyallupinternational.com FAX: 253-779-0576 (Fax)
ADDRESS: 3702 Marine View Drive Tacoma, WA 98422
SHORT DESCRIPTION & LOCATION: **http://www.nwboat.com/chinook/**

47°16.90' - 122°24.20' This modern marina is located just inside the entrance of Hylebos Waterway about 2 miles SE of Browns Pt. When approaching you will see the Sign done with Native American Coastal Design- Chinook. The large guest dock is on "A" dock.

GUEST BOAT CAPACITY: ....Appx. 25-30 boats
DOCKSIDE DEPTH AT ZERO TIDE: ....8 ft.
SEASON: ....All year
RESERVATION POLICY: ....Accepts
AMT W/ELECTRICITY: ....Limited
FUEL DOCK: ....None
MARINE REPAIRS: ....Close by
TOILETS: ....Yes
HOT SHOWERS: ....Yes
RESTAURANT: ....Close by
PICNIC AREA: ....None
BASIC STORE: ....Yes
BROADBAND/WI-FI: ....None
DAILY RATE: ....Moderate (75¢-$1.25/foot)

GUEST DOCK: ....430 ft. plus slips
GUEST SLIPS: ....Varies
WATER: ....Yes
AMPS: ....30-50 A
PUMP OUT STATION ....Yes
HAUL OUT: ....Close by
BOAT RAMP: ....None
LAUNDRY: ....Yes
BAR: ....Close by
POOL: ....None
GOLF: ....Close by
PET FRIENDLY: ....Good
OTHER: Chandlery, gifts, books, espresso, small provisions, 24 hour security, close to many marine services.

CAUTION! This chartlet not intended for use in navigation.

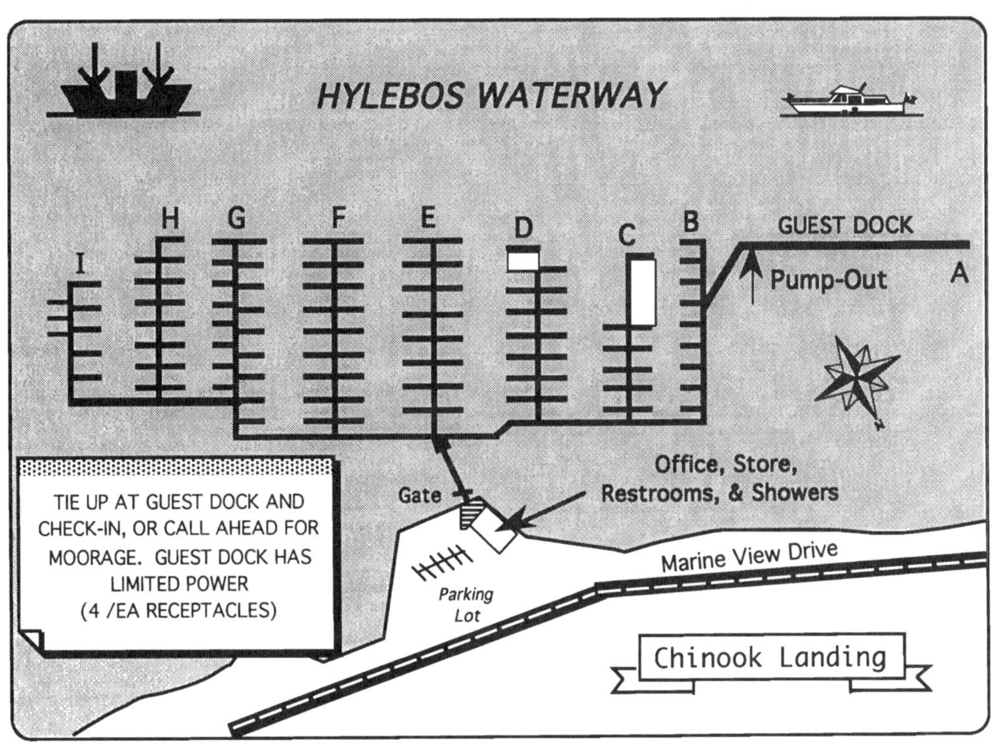

Tacoma

NAME OF YACHT CLUB: *VIKING YACHT CLUB*

ADDRESS: P.O. Box 4461 Federal Way, WA 98063

TELEPHONE: 253-863-3226 PERSON IN CHARGE: Vice Commodore

SHORT DESCRIPTION & LOCATION:

V.Y.C. offers up to 2 nights reciprocal moorage located at **Chinook Landing Marina.** Visiting reciprocal yacht club members are requested to present current membership card, fly burgee, pay the moorage charges to the harbormaster & mail the moorage receipt to Viking for reimbursement. **Marina phone: 253-627-7676 - VHF: Channel 79.**

RECIPROCAL BOAT CAPACITY .... Varies

DOCKSIDE DEPTH AT ZERO TIDE: .... 8 ft.

SEASON: .... All year

RESERVATION POLICY: .... None

TOILETS: .... At Chinook Landing Marina

HOT SHOWERS: .... At Chinook Landing Marina

RESTAURANT: .... None

BROADBAND/WI-FI: .... None

DAILY RATE: No Charge for two nights moorage

RECIPROCAL DOCK: .... At marina

RECIPROCAL SLIPS: ... At marina

WATER: .... Yes

AMT W/ELECTRICITY: .... Partial

AMPS: .... 20-30 A

BAR: .... None

PET FRIENDLY: .... Good

OTHER: Services same as Chinook Landing Marina listing on Page 37.

***NOTE:* THIS IS PRIVATE MOORAGE AND ONLY AVAILABLE TO MEMBERS OF RECIPROCAL YACHT CLUBS. YOUR CLUB *MUST* HAVE RECIPROCAL PRIVILEGES AND YOU MUST FLY YOUR BURGEE.**

SEE DATA & CHARTLET ON PAGE: 37

NOTES

Tacoma

NAME OF YACHT CLUB: *CORINTHIAN YACHT CLUB OF TACOMA*

ADDRESS: 5624 Marine View Dr. Tacoma, WA 98422

TELEPHONE: 253-383-5321 Marina Office PERSON IN CHARGE: Rear Commodore

SHORT DESCRIPTION & LOCATION: **www.cyct.com**

47°17.68' - 122°25.17' CYCT moorage is located at Tyee Marina on NE side of Commencement Bay about 1 mi. SE of Browns Pt. Upon entering Tyee go to End Tie on the 600 Dock. Slips 652 & 653 are marked with CYCT sign. Walk to CYCT container at top of Tyee 2. Instructions posted on how to register & obtain key card in display case.

RECIPROCAL BOAT CAPACITY: ....1-2 boats
DOCKSIDE DEPTH AT ZERO TIDE: 8 ft. min.
SEASON: ....All year
RESERVATION POLICY: First come, first serve
TOILETS: ....Yes
HOT SHOWERS: ....Yes
RESTAURANT: ....One mile
BROADBAND/WI-FI: ....None
DAILY RATE: ....No Charge for 48 hours

RECIPROCAL DOCK: ....54 Ft.
RECIPROCAL SLIPS: 1-2 Slips
WATER: ....Yes
AMT W/ELECTRICITY: ....All
AMPS: ....20 A
BAR: ....Close by
PET FRIENDLY: ....Good
OTHER: Puget Sound Sailing Institute next door.

***NOTE:* THIS IS PRIVATE MOORAGE AND ONLY AVAILABLE TO MEMBERS OF RECIPROCAL YACHT CLUBS. YOUR CLUB *MUST* HAVE RECIPROCAL PRIVILEGES AND YOU MUST FLY YOUR BURGEE.**

CAUTION! This chartlet not intended for use in navigation.

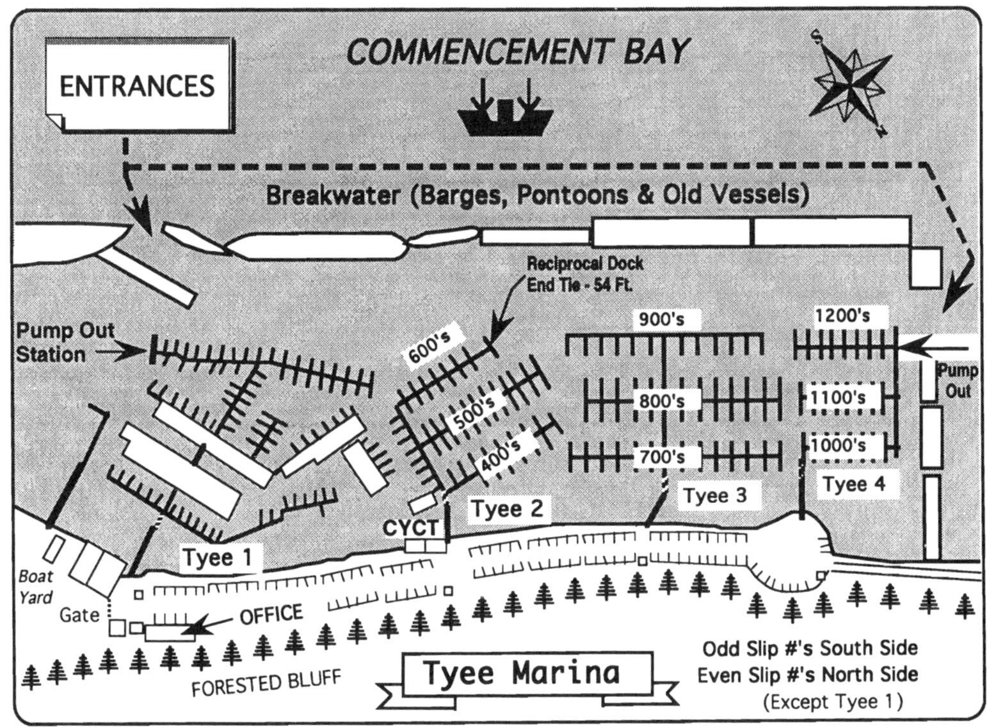

Gig Harbor

NAME OF MARINA: ***TIDES TAVERN*** **RADIO:** None
TELEPHONE: **253-858-3982** MGR: Kathy Davis
E-MAIL: tides@tidestavern.com FAX: 253-858-3914 (Fax)
ADDRESS: 2925 Harborview Drive Gig Harbor, WA 98335
SHORT DESCRIPTION & LOCATION: **www.tidestavern.com**

47°19.68' - 122°34.50' The fun Tide's guest float is just inside the entrance of Gig Harbor and is one of the first few docks on the south side upon entering. Overnight moorage is complimentary to restaurant patrons and rafting is mandatory upon request.

GUEST BOAT CAPACITY: ....Appx. 10-20 boats
DOCKSIDE DEPTH AT ZERO TIDE: ....10-12 ft.
SEASON: ....All year
RESERVATION POLICY: ....None
AMT W/ELECTRICITY: ....None
FUEL DOCK: ....Close by
MARINE REPAIRS: ....Close by
TOILETS: ....None
HOT SHOWERS: ....None
RESTAURANT: ....Yes
PICNIC AREA: ....Close by
BASIC STORE: ....Close by
BROADBAND/WI-FI: ....Planned
DAILY RATE: ....No charge to restaurant & tavern customers. **24 hours or less.**

GUEST DOCK: ....120 ft.
GUEST SLIPS: ....None
WATER: ....None
AMPS: ....None
PUMP OUT STATION ....None
HAUL OUT: ....Close by
BOAT RAMP: ....Close by
LAUNDRY: ....None
BAR: ....Beer & wine only
POOL: ....None
GOLF: ....Close by
PET FRIENDLY: ....Fair
OTHER: Central town location is close to many shops and attractions. **Inside dock goes partially dry at minus tides.**

CAUTION! This chartlet not intended for use in navigation.

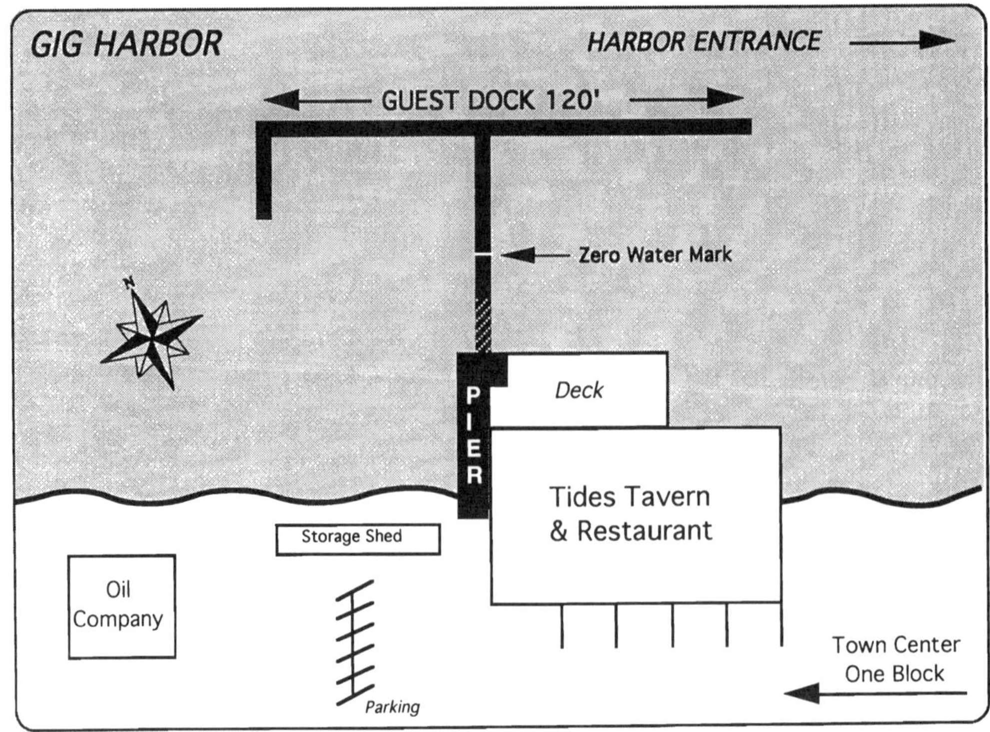

Gig Harbor

NAME OF MARINA: ***GIG HARBOR PUBLIC DOCK*** **RADIO:** None
TELEPHONE: **253-851-6170** MGR: Dave Brereton, Director
E-MAIL: Go to Website FAX: 253-851-7597 (Fax)
ADDRESS: 3510 Grandview Gig Harbor, WA 98335
SHORT DESCRIPTION & LOCATION: **www.cityofgigharbor.net**

47°19.75' - 122°34.70' Located in heart of historic Gig Harbor at base of giant USA flag in Jerisich Park. Nice park setting & close to many shops & restaurants. New docks with dinghy dock on W. side near the shore. Caution-Shallow close to shore at low tide.

GUEST BOAT CAPACITY: ....Appx. 20 boats
DOCKSIDE DEPTH AT ZERO TIDE: ......0-18 ft.
SEASON: ....All year
RESERVATION POLICY: ....None
AMT W/ELECTRICITY: ....None
FUEL DOCK: ....Close by
MARINE REPAIRS: ....Close by
TOILETS: ....Yes
HOT SHOWERS: ....None
RESTAURANT: ....Close by
PICNIC AREA: ....Yes
BASIC STORE: ....Close by
BROADBAND/WI-FI: ....None
DAILY RATE: ....No Charge
24 Hour Maximum Stay

GUEST DOCK: ....Appx. 450 ft.
GUEST SLIPS: ....None
WATER: ....None
AMPS: ....None
PUMP OUT STATION ....Yes
HAUL OUT: ....Close by
BOAT RAMP: ....Close by
LAUNDRY: ....Close by
BAR: ....Close by
POOL: ....None
GOLF: ....Close by
PET FRIENDLY: ....Fair
OTHER: Rafting prohibited, no barbecues allowed on dock, nice picnic area on pier at head of dock.

CAUTION! This chartlet not intended for use in navigation.

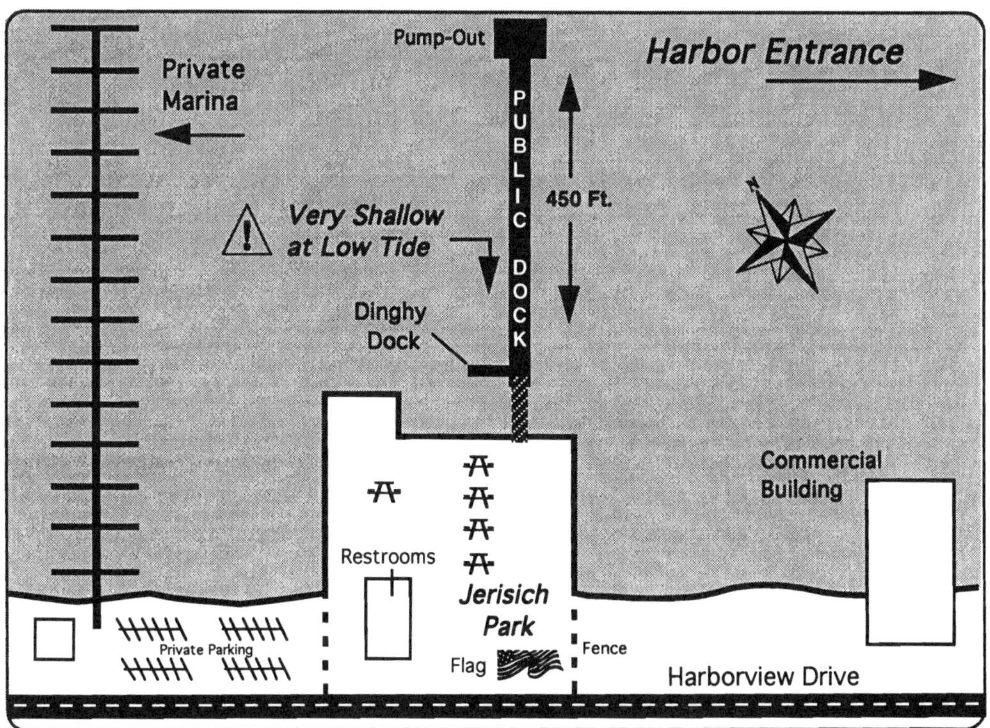

Gig Harbor

NAME OF MARINA: ***ARABELLA'S LANDING MARINA*** RADIO: None
TELEPHONE: 253-851-1793 MGR: John Moist
E-MAIL: Go to web site below FAX: 253-851-1793 (Fax)
ADDRESS: 3323 Harborview Dr. Gig Harbor, WA 98332
SHORT DESCRIPTION & LOCATION: **www.arabellaslanding.com**

47°19.95' - 122°34.90' Located near the center of historic Gig Harbor just beyond Bayview Dock on your port side. This modern & upscale marina is close to several shops & restaurants and offers many amenities for boaters. Can accommodate boats up to 160 ft.

GUEST BOAT CAPACITY: ....Appx. 10-30 boats
DOCKSIDE DEPTH AT ZERO TIDE: ....8 ft. +
SEASON: ....All year
RESERVATION POLICY: ....Preferred
AMT W/ELECTRICITY: ....All
FUEL DOCK: ....None
MARINE REPAIRS: ....Close by
TOILETS: ....Yes
HOT SHOWERS: ....Yes
RESTAURANT: ....Close by
PICNIC AREA: ....Yes
BASIC STORE: ....Close by
BROADBAND/WI-FI: ....BroadbandXpress
DAILY RATE:........Moderate *(75¢-$1.25/foot)*
Min. Charge $22.50. Outside 60' slips are a flat rate of $45/night.

GUEST DOCK: ....230 ft. + slips
GUEST SLIPS: 9
WATER: ....Yes
AMPS: ....30-50 A
PUMP OUT STATION ....Yes
HAUL OUT: ....Close by
BOAT RAMP: ....Close by
LAUNDRY: ....Yes
BAR: ....Close by
POOL: ....None
GOLF: ....Close by
PET FRIENDLY: ....Good
OTHER: Meeting & function room rental for groups. Beautifully landscaped grounds and wheel chair access to docks.

CAUTION! This chartlet not intended for use in navigation.

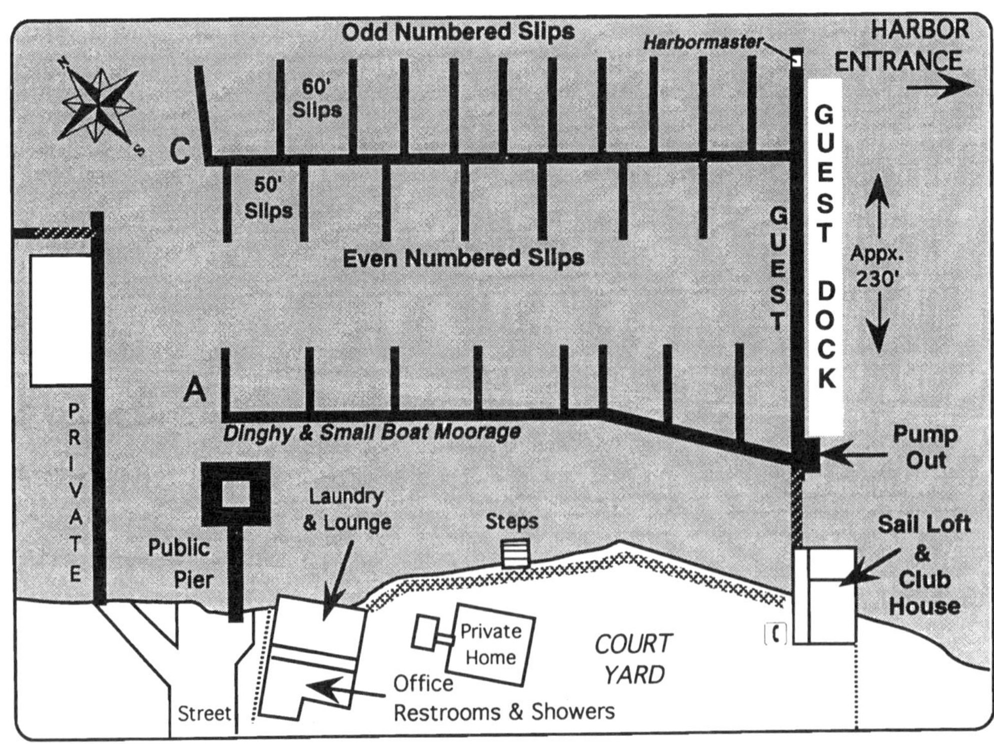

Gig Harbor

NAME OF MARINA: *ANTHONYS HOMEPORT* RADIO: None
TELEPHONE: 253-853-6353 MGR: Restaurant Manager
E-MAIL: gigharbor@anthonys.com FAX: 253-853-6955 (Fax)
ADDRESS: 8827 N. Harborview Dr., Gig Harbor, WA 98335
SHORT DESCRIPTION & LOCATION: **www.anthonys.com**

47°20.30' - 122°35.22' Typically upscale Anthonys restaurant and lounge with complimentary moorage for restaurant patrons; located at head of Gig Harbor next to Peninsula Yacht Basin. **Caution: Guest docks go partially dry at low water.**

GUEST BOAT CAPACITY: ....Appx. 6 boats
DOCKSIDE DEPTH AT ZERO TIDE: ....0-4 ft.
SEASON: ....All year
RESERVATION POLICY: ....None
AMT W/ELECTRICITY: ....None
FUEL DOCK: ....Close by
MARINE REPAIRS: ....Close by
TOILETS: ....None
HOT SHOWERS: ....Close by
RESTAURANT: ....Yes
PICNIC AREA: ....Yes
BASIC STORE: ....Close by
BROADBAND/WI-FI: ....BroadbandXpress
DAILY RATE: ....No charge for overnight moorage if dining at restaurant.

GUEST DOCK: ....Appx. 100 ft.
GUEST SLIPS: 2 lg. guest docks
WATER: ....None
AMPS: ....None
PUMP OUT STATION ....None
HAUL OUT: ....None
BOAT RAMP: ....Close by
LAUNDRY: ....Close by
BAR: ....Yes
POOL: ....None
GOLF: ....Close by
PET FRIENDLY: ....Fair
OTHER: Appx. 3/4 mile from heart of Gig Harbor. Walking distance to shops & marine services.

CAUTION! This chartlet not intended for use in navigation.

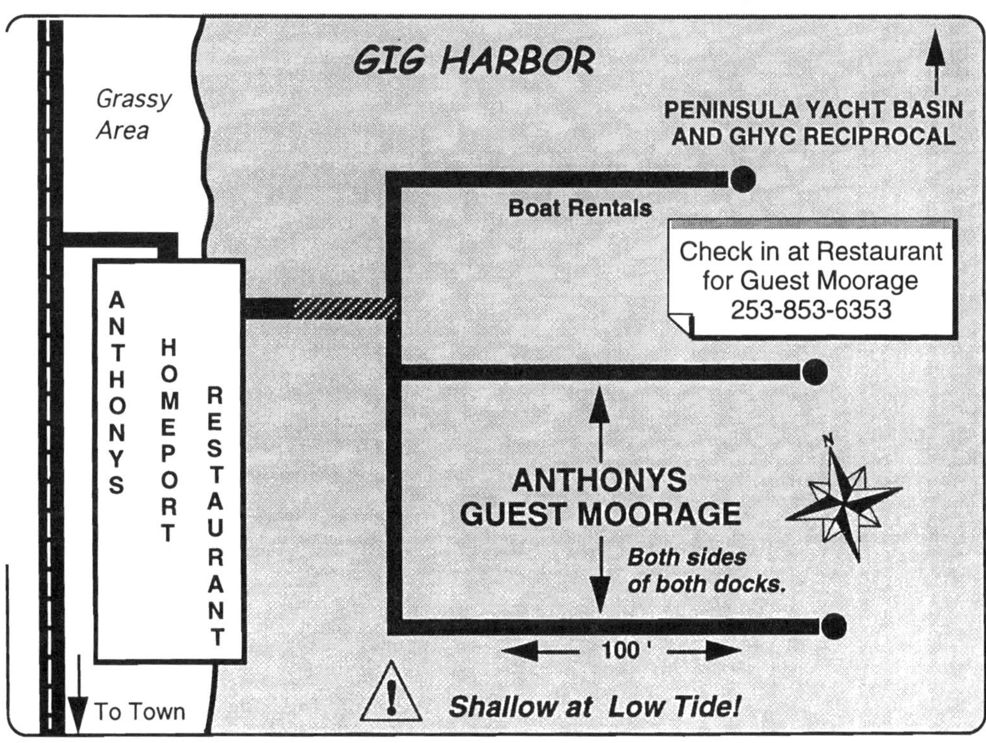

Gig Harbor

NAME OF MARINA: ***PENINSULA YACHT BASIN*** RADIO: VHF Ch 68
TELEPHONE: **253-858-2250** MGR: Steven Harbaugh
E-MAIL: peninsulayachtbasin@msn.com FAX: None
ADDRESS: 8913 N. Harborview Dr. Gig Harbor, WA 98335
SHORT DESCRIPTION & LOCATION:
47°20.32' - 122°35.22' Marina is located at the head of beautiful and protected Gig Harbor within walking distance to the heart of the historic boating town. Although no designated guest dock, a list of open slips is posted at marina office. Max length 80 ft.

GUEST BOAT CAPACITY: ....Up to 10 boats
DOCKSIDE DEPTH AT ZERO TIDE: ....9 ft. +
SEASON: ....All year
RESERVATION POLICY: ....None
AMT W/ELECTRICITY: ....All
FUEL DOCK: ....Close by
MARINE REPAIRS: ....Close by
TOILETS: ....Yes
HOT SHOWERS: ....Yes
RESTAURANT: ....Close by
PICNIC AREA: ....Close by
BASIC STORE: ....Close by
BROADBAND/WI-FI: ....BroadbandXpress
DAILY RATE: ....Moderate
(75¢-$1.25/foot)

GUEST DOCK: ....None
GUEST SLIPS: ....Varies
WATER: ....Yes
AMPS: ....30 A
PUMP OUT STATION ....None
HAUL OUT: ....Close by
BOAT RAMP: ....Close by
LAUNDRY: ....Close by
BAR: ....Close by
POOL: ....None
GOLF: ....Close by
PET FRIENDLY: ....Fair
OTHER: Groups & individuals welcome. Security gate, many nice shops and restaurants close by.

NAME OF YACHT CLUB: *GIG HARBOR YACHT CLUB*
CLUB ADDRESS: P.O. Box 22, Gig Harbor, WA 98335
CLUB TELEPHONE: 253-851-1807 PERSON IN CHARGE: Commodore
LOCATION & SPECIAL NOTES: **www.gigharboryc.com**
GHYC has an 80 ft. guest dock for reciprocal use at Peninsula Yacht Basin located on the shore side of the "100" dock, basically from the fire box to the dock end. Rafting is permitted. Security gate instructions found in annual letter sent to your Y.C. by GHYC.

RECIPROCAL BOAT CAPACITY: ....2-4 Boats
DOCKSIDE DEPTH AT ZERO TIDE: ....9 ft.
RECIPROCAL SEASON: ....All year
RESERVATION POLICY: ....Groups only
TOILETS: ....At Marina
HOT SHOWERS: ....At Marina
RESTAURANT: ....Next door
DAILY RATE: ....Up to 48 hours free moorage.
.........Nominal charge for electricity.

RECIPROCAL DOCK: ....80 ft.
RECIPROCAL SLIPS: See above
WATER: ....Yes
AMT W/ELECTRICITY: 2
AMPS: ....20-30 A
BAR: ....Next door
OTHER: ....Same as marina listing

NOTE: THIS IS PRIVATE MOORAGE ONLY AVAILABLE TO MEMBERS OF RECIPROCAL YACHT CLUBS! YOUR CLUB MUST HAVE RECIPROCAL PRIVILEGES AND YOU MUST FLY YOUR BURGEE!

Gig Harbor

CAUTION! This chartlet not intended for use in navigation.

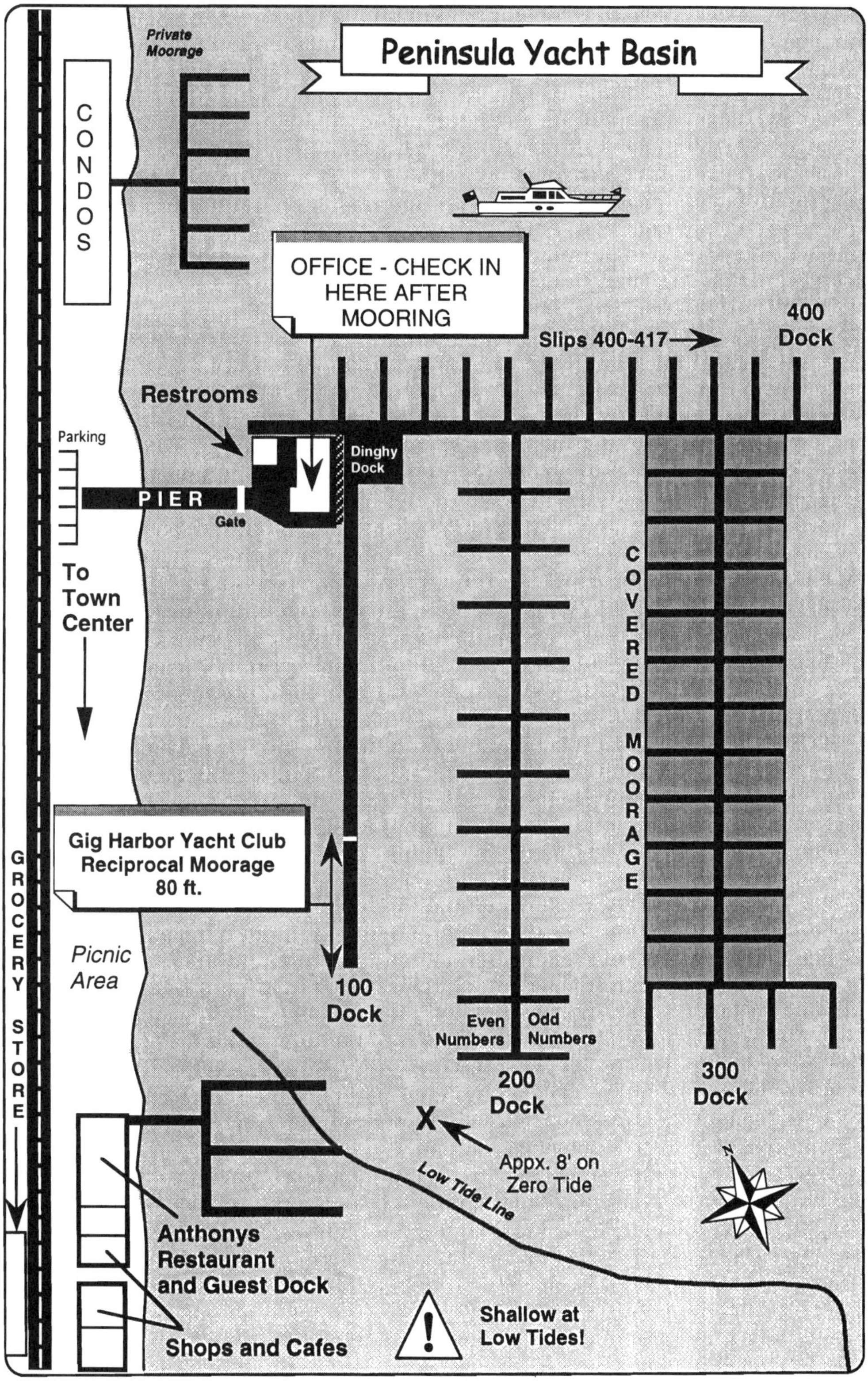

Vashon Island

NAME OF PARK: *DOCKTON PARK (King County)*
ADDRESS: P.O. Box 11 Vashon, WA 98070
TELEPHONE: 206-463-2947 **MGR:** King County Parks
SHORT DESCRIPTION & LOCATION: **www.metrokc.gov/parks**

47°22.33' - 122°27.23' Actually located on Maury Is. in a bight on the E. side of Quartermaster Harbor about 2.5 miles N. of the Harbor entrance. The modern and well kept floats and facilities make this an excellent destination for a family or group cruise.

GUEST BOAT CAPACITY: ....Appx. 60 boats
DOCKSIDE DEPTH AT ZERO TIDE: Appx 20 ft.
SEASON: ....All year
AMT W/ELECTRICITY: ....None
TOILETS: ....Yes
HOT SHOWERS: ....Yes
PICNIC AREA: ....Yes
PLAY AREA: ....Yes
BASIC STORE: ....None
DAILY RATE: ....$18.00 - Under 26 ft.
............$22.00 - 26 ft. & over
3 Night Limit w/3 nights between visits.

GUEST DOCK: ....Appx. 300 ft.
MOORING BUOYS: ....None
WATER: ....None
PAY PHONES: ....No
BOAT RAMP: ....Yes
PICNIC SHELTER: ....None
BBQ: ....Yes
PUMP OUT STATION: ....Yes
PET FRIENDLY: ....Excellent
OTHER: re pits, unguarded beach, kayaking, hiking trails. No commercial services within walking distance.

CAUTION! This chartlet not intended for use in navigation.

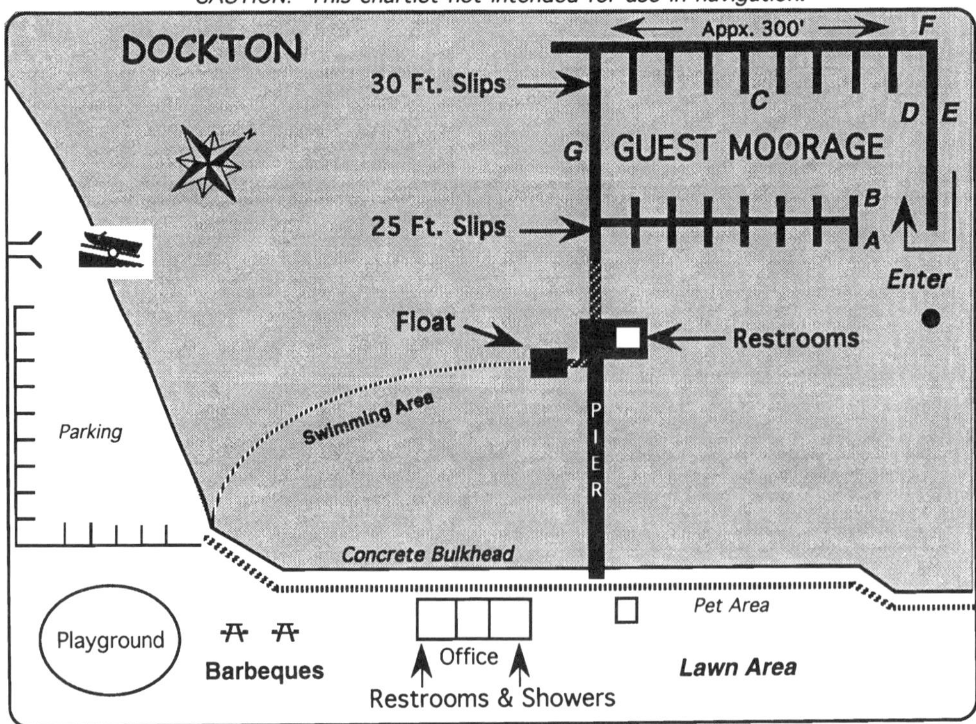

Vashon Island

NAME OF YACHT CLUB: *QUARTERMASTER YACHT CLUB*

ADDRESS: P.O. Box 2927 Vashon, WA 98070

TELEPHONE: 206-463-3309 PERSON IN CHARGE: Caretaker

SHORT DESCRIPTION & LOCATION: **www.qyc1.org**

47°23.60' - 122°27.80' Docks located on the S.W. side of inner Quartermaster Harbor. Reciprocal moorage on east side of the float nearest to shore & approached from the port (south) side of the facility when entering the area. Use caution at low tide. Register at gray registration box, foot of ramp. Questions: See caretaker aboard SARA in Slip B-06.

RECIPROCAL BOAT CAPACITY: ....Appx. 6-8

DOCKSIDE DEPTH AT ZERO TIDE: ....2-5 ft.

SEASON: ....All year

RESERVATION POLICY: First come first serve

TOILETS: ....Yes

HOT SHOWERS: ....None

RESTAURANT: ....Close by

BROADBAND/WI-FI: ....None

DAILY RATE: $4/day for all boats for 48 hours. Add'l days- 25¢ per ft. Write R.C. for group visits. Max 4 days/mo.

RECIPROCAL DOCK: ....125 ft.

RECIPROCAL SLIPS: ....Varies

WATER: ....Yes

AMT W/ELECTRICITY: ....All

AMPS: ....30 A

BAR: ....Close by

PET FRIENDLY: ....Good

OTHER: Rafting permitted, walking distance to store & restaurant in Burton. NO GARBAGE DROP

***NOTE:* THIS IS PRIVATE MOORAGE AND ONLY AVAILABLE TO MEMBERS OF RECIPROCAL YACHT CLUBS. YOUR CLUB *MUST* HAVE RECIPROCAL PRIVILEGES AND YOU MUST FLY YOUR BURGEE.**

CAUTION! This chartlet not intended for use in navigation.

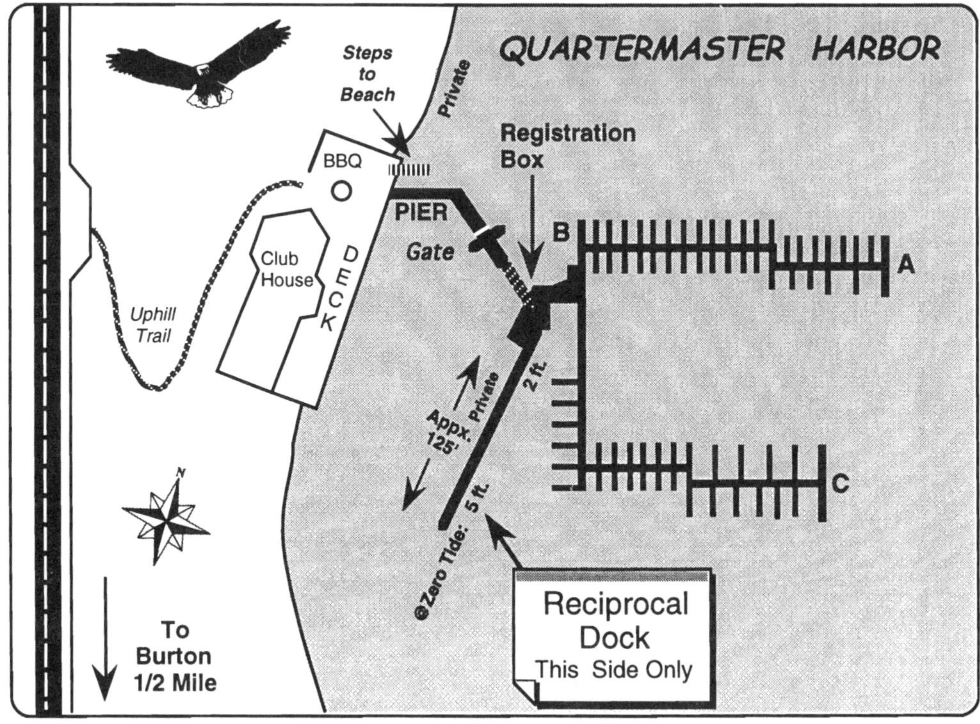

Des Moines

NAME OF MARINA: ***DES MOINES MARINA*** RADIO: VHF Ch.16
TELEPHONE: **206-824-5700** MGR: Joe Dusenbury
E-MAIL: See below web site FAX: 206-878-5940
ADDRESS: 22307 Dock St. S. Des Moines, WA 98198
SHORT DESCRIPTION & LOCATION: **www.desmoineswa.gov**
47°24.10' - 122°19.80' Located on the E. shore of East Passage about 4 miles SE of Three Tree Pt. between Seattle & Tacoma. The modern & large marina offers 1500 ft. of guest moorage behind a 2200 ft. rock breakwater close to shopping ctr. & public services.

GUEST BOAT CAPACITY: ....Appx. 65 boats
DOCKSIDE DEPTH AT ZERO TIDE: ....12 ft.
SEASON: ....All year
RESERVATION POLICY: ....Call
AMT W/ELECTRICITY: ....All
FUEL DOCK: ....Gas, Dsl, & LP
MARINE REPAIRS: ....On premises
TOILETS: ....Yes
HOT SHOWERS: ....Yes
RESTAURANT: ....Yes
PICNIC AREA: ....Yes
BASIC STORE: ....Close by
BROADBAND/WI-FI: ....Planned
DAILY RATE: ....**Economical**
(Under 75¢/foot)

GUEST DOCK: ....1500 ft. TTL
GUEST SLIPS: ....40 + Docks
WATER: ....Yes
AMPS: ....20-30 A
PUMP OUT STATION ....Yes
HAUL OUT: ....Travel-Lift
BOAT RAMP: ....Sling hoist
LAUNDRY: ....Close by
BAR: ....Yes
POOL: ....Close by
GOLF: ....Close by
PET FRIENDLY: ....Fair
OTHER: Fishing pier, antique shops, museum, Cable T.V. on guest dock.

B

NAME OF YACHT CLUB: ***DES MOINES YACHT CLUB***
CLUB ADDRESS: P.O. Box 13004, Des Moines, WA 98198
CLUB TELEPHONE: 206-878-7220 PERSON IN CHARGE: Club Steward
LOCATION & SPECIAL NOTES: **www.desmoinesyachtclub.com**
DMYC extends privileges of club facilities to members of reciprocal boating clubs. All visitors must register with club steward upon arrival for gate key. Dock space is available on the S. side of "A" dock. The club dock may be used temporarily while registering and obtaining keys. Maximum of 3 visiting boats allowed at one time, **up to 14 ft. beam.**

RECIPROCAL BOAT CAPACITY: ....3 boats
DOCKSIDE DEPTH AT ZERO TIDE: ....10 ft.
RECIPROCAL SEASON: ....All year
RESERVATION POLICY: ....None
TOILETS: ....Yes
HOT SHOWERS: ....Yes
RESTAURANT: ....Close by
DAILY RATE: ....Moorage free for 48 Hours
..........Electrical $2.00/night

RECIPROCAL DOCK: Appx. 100'
RECIPROCAL SLIPS: Dock only
WATER: ....Yes
AMT W/ELECTRICITY: ....All
AMPS: ....20 A
BAR: ....Close by
OTHER: ....Same as marina listing

***NOTE:* THIS IS PRIVATE MOORAGE *ONLY* AVAILABLE TO MEMBERS OF RECIPROCAL YACHT CLUBS! YOUR CLUB *MUST* HAVE RECIPROCAL PRIVILEGES AND YOU MUST FLY YOUR BURGEE!**

Des Moines

CAUTION! This chartlet not intended for use in navigation.

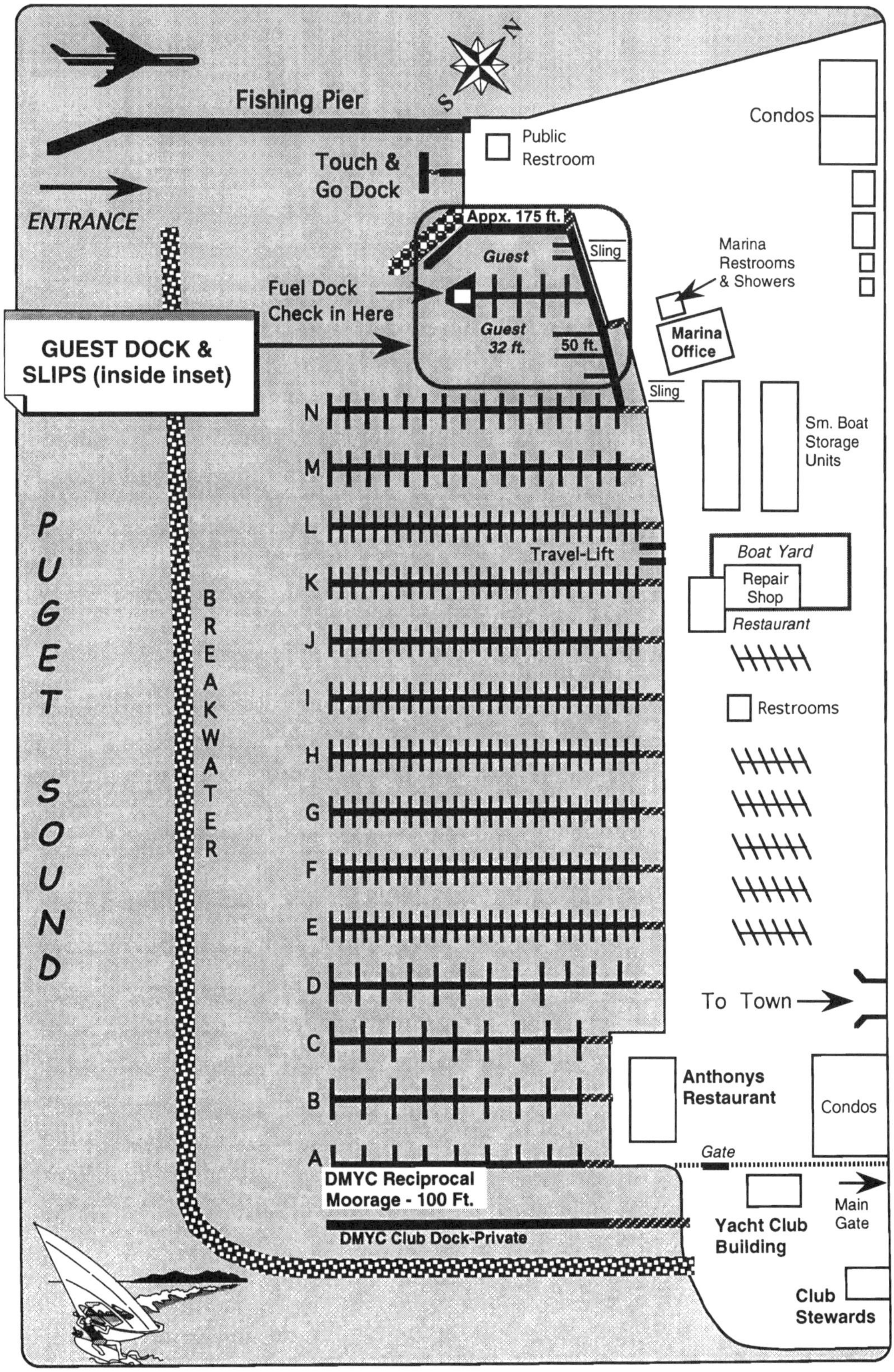

Fishing Pier

Touch & Go Dock

ENTRANCE

Public Restroom

Condos

Appx. 175 ft.

Guest

Sling

Fuel Dock Check in Here

Marina Restrooms & Showers

Marina Office

Guest 32 ft.

50 ft.

GUEST DOCK & SLIPS (inside inset)

Sling

N

M

L

K

J

I

H

G

F

E

D

C

B

A

Sm. Boat Storage Units

Travel-Lift

Boat Yard

Repair Shop

Restaurant

Restrooms

PUGET SOUND

BREAKWATER

To Town

Anthonys Restaurant

Condos

Gate

DMYC Reciprocal Moorage - 100 Ft.

DMYC Club Dock-Private

Yacht Club Building

Main Gate

Club Stewards

Des Moines

NAME OF YACHT CLUB: *SOUTH SOUND ELKS YACHT CLUB*

ADDRESS: 2402 Auburn Way No. Auburn, WA 98002

TELEPHONE: 206-833-1808 PERSON IN CHARGE: Commodore

SHORT DESCRIPTION & LOCATION: **WEBSITE?:**

The South Sound Elks Yacht club offers reciprocal moorage located at the **Des Moines Marina** Guest Dock when available. The reciprocal member must pay the moorage charges to the Marina, then send receipt to SSEYC. for reimbursement within 30 days with your name, date of visit, boat name, and copy of your Club membership card.

RECIPROCAL BOAT CAPACITY....................Varies
DOCKSIDE DEPTH AT ZERO TIDE: ....12 ft.
SEASON: ....All year
RESERVATION POLICY: ....None
TOILETS: ....At Des Moines Marina
HOT SHOWERS: ....At Des Moines Marina
RESTAURANT: ....Close by
BROADBAND/WI-FI: ....None
DAILY RATE: Moorage free, power $3/night. LIMIT: 1 day in 30 day period per boat. Max 2 boats from same club.

RECIPROCAL DOCK: ....At marina
RECIPROCAL SLIPS: ...At marina
WATER: ....Yes
AMT W/ELECTRICITY: ....All
AMPS: ....20-30 A
BAR: ....Close by
PET FRIENDLY: ....Good
OTHER: Services same as Des Moines Marina listing on Page 48 & 49

***NOTE:* THIS IS PRIVATE MOORAGE AND ONLY AVAILABLE TO MEMBERS OF RECIPROCAL YACHT CLUBS. YOUR CLUB MUST HAVE RECIPROCAL PRIVILEGES AND YOU MUST FLY YOUR BURGEE.**

SEE DATA & CHARTLET ON PAGE: 48 & 49

NOTES

Des Moines

NAME OF YACHT CLUB: *THREE TREE POINT YACHT CLUB*

ADDRESS: P.O. Box 98700 Des Moines, WA 98198

TELEPHONE: 206-824-3674 PERSON IN CHARGE: Recip. Chair

SHORT DESCRIPTION & LOCATION: **www.ttpyc.org**

Three Tree Point Y.C. offers members of reciprocating clubs 2 nights free moorage at the Des Moines Marina general guest dock, space available, limited to 2 boats per night. One moorage per month is allowed. Upon arrival check in at the fuel dock. Show your current Y.C. membership card to Marina Office Manager to be assigned moorage.

RECIPROCAL BOAT CAPACITY ....2 boats

DOCKSIDE DEPTH AT ZERO TIDE: ....10 ft.

SEASON: ....All year

RESERVATION POLICY: ....None

TOILETS: ....At Des Moines Marina

HOT SHOWERS: ....At Des Moines Marina

RESTAURANT: ....Numerous close by

BROADBAND/WI-FI: ....None

DAILY RATE: ....Electricity charge - $3.00/night
Moorage - Free for 2 nights max.
Limit: 1 visit per month per boat.

RECIPROCAL DOCK: ....At marina

RECIPROCAL SLIPS: ...At marina

WATER: ....Yes

AMT W/ELECTRICITY: ....All

AMPS: ....20-30 A

BAR: ....Close by

PET FRIENDLY: ....Fair

OTHER: Services same as Des Moines Marina listing on Pages 48 & 49.

***NOTE:* THIS IS PRIVATE MOORAGE AND ONLY AVAILABLE TO MEMBERS OF RECIPROCAL YACHT CLUBS. YOUR CLUB *MUST* HAVE RECIPROCAL PRIVILEGES AND YOU MUST FLY YOUR BURGEE.**

SEE DATA & CHARTLET ON PAGE: 48 & 49

NOTES

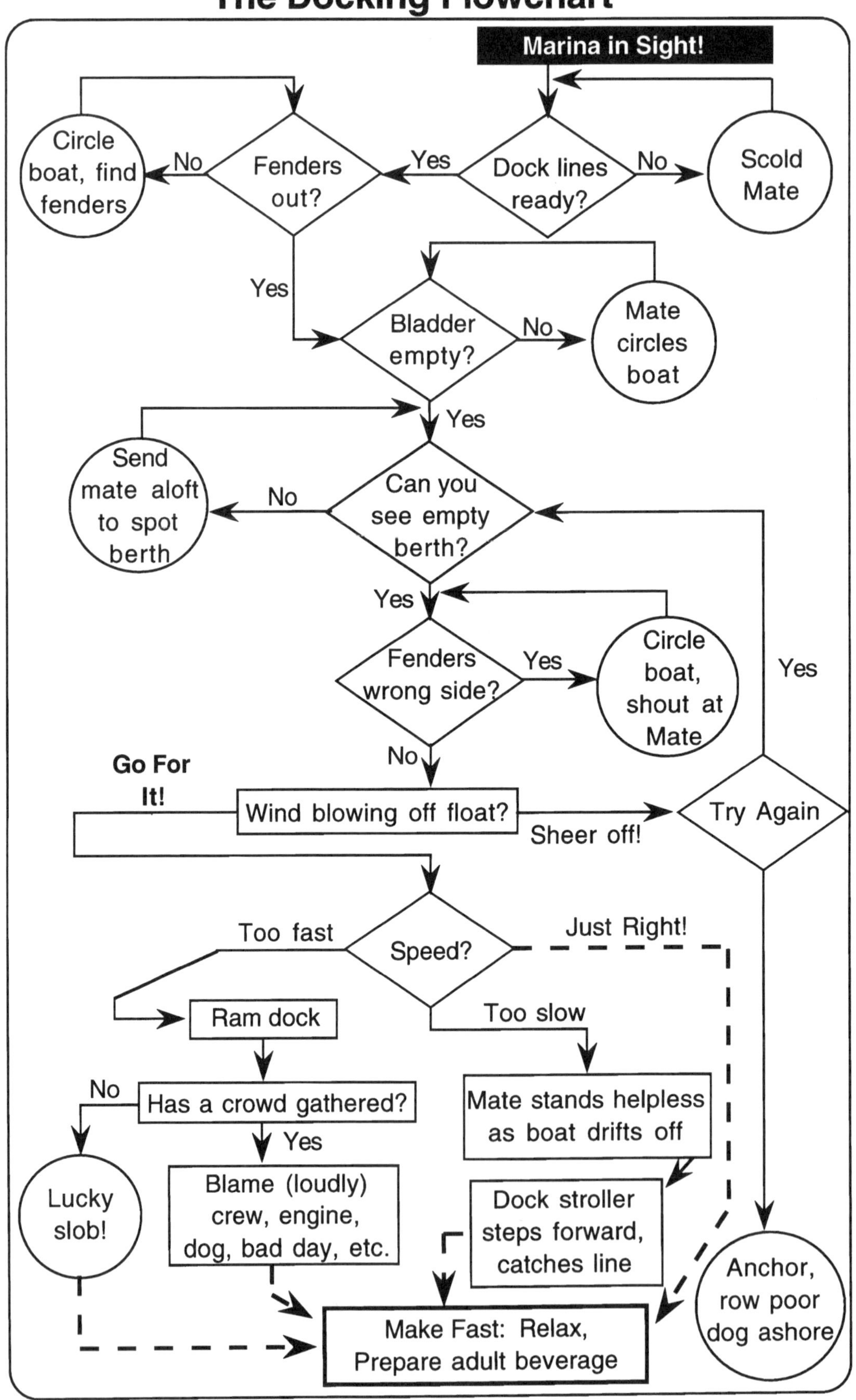
The Docking Flowchart
Marina in Sight!
Dock lines ready?
No
Scold Mate
Yes
Fenders out?
No
Circle boat, find fenders
Yes
Bladder empty?
No
Mate circles boat
Yes
Can you see empty berth?
No
Send mate aloft to spot berth
Yes
Fenders wrong side?
Yes
Circle boat, shout at Mate
No
Wind blowing off float?
Go For It!
Sheer off!
Try Again
Yes
Speed?
Too fast
Ram dock
Has a crowd gathered?
No
Lucky slob!
Yes
Blame (loudly) crew, engine, dog, bad day, etc.
Too slow
Mate stands helpless as boat drifts off
Dock stroller steps forward, catches line
Just Right!
Make Fast: Relax, Prepare adult beverage
Anchor, row poor dog ashore

CHAPTER 2

NORTH PUGET SOUND

Chapter Map - Page 3

Blake Island

NAME OF PARK: *BLAKE ISLAND STATE MARINE PARK*
ADDRESS: P.O. Box 277 Manchester, WA 98353
TELEPHONE: 360-731-8330 MGR: Paul Ruppert
SHORT DESCRIPTION & LOCATION: **www.parks.wa.gov/moorage/parks**

47°32.75' - 122°28.50' Located 2 mi. S. of Bainbridge Is. & 5 miles W. of Seattle, this 476 acre Island park offers serene beauty & wildlife in the heart of Puget Sound. Densely wooded & only accessible by boat. 200 ft. Linear moorage system on NW corner of island.

GUEST BOAT CAPACITY: ....Appx. 40 boats
DOCKSIDE DEPTH AT ZERO TIDE:..........20 ft.
SEASON: ....All year
AMT W/ELECTRICITY: ....Power on Docks
TOILETS: ....Yes
HOT SHOWERS: ....Yes
PICNIC AREA: ....Yes
PLAY AREA: ....Yes
BASIC STORE: Snack bar/gift shop-summer only
DAILY RATE: Dock Charge..........................50¢/foot
Mooring Buoys..............$10.00/night
Extra charge for power if desired.

GUEST DOCK: ....1500 Ft. Total
MOORING BUOYS: ....21 Buoys
WATER: ....Yes
PAY PHONES: ....None
BOAT RAMP: ....None
PICNIC SHELTER: ....Yes
BBQ: ....Yes
PUMP OUT STATION: ....Yes
PET FRIENDLY: ....Excellent
OTHER: Beach combing, camp-sites, hiking trails, Tillicum Village salmon dinner & auth. Indian stage show in season-check schedule.

NOTES

CAUTION RE RACCOONS & SEA OTTERS.
Important: Do not feed any wildlife!
Do not leave any food or beverages overnight on the docks or anywhere on the outside decks of your boat, including your fly bridge. This includes garbage and ice chests. This is a good rule for all Marine Parks, especially Blake Island.

Blake Island

CAUTION! These chartlets not intended for use in navigation.

MARINA CHARTLET

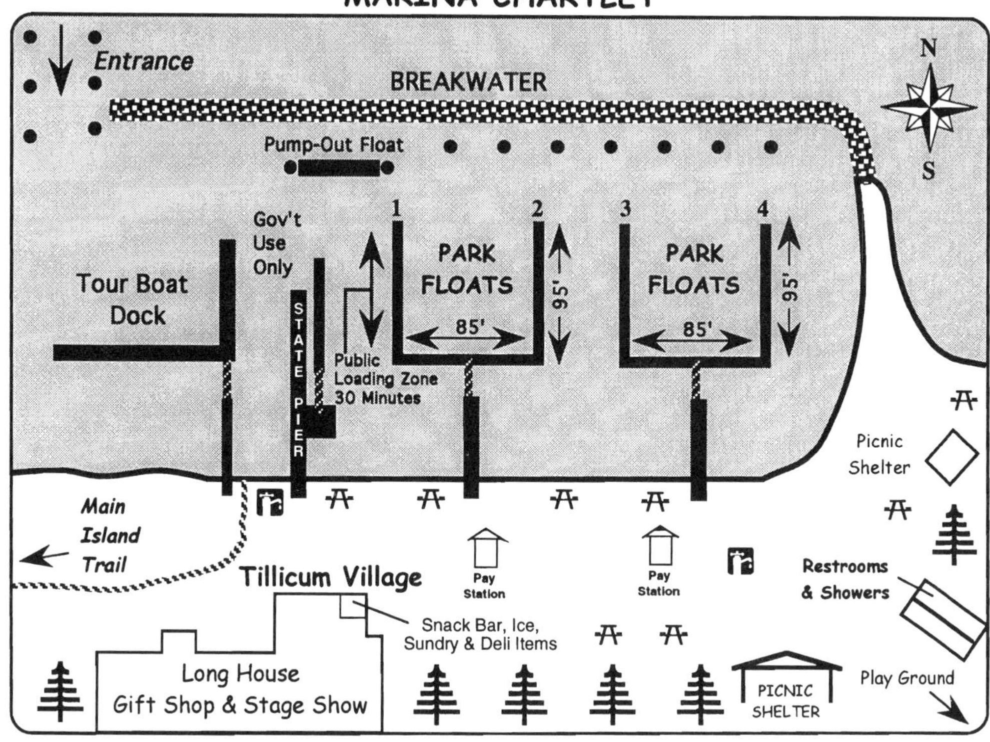

BUOY & TRAIL CHARTLET

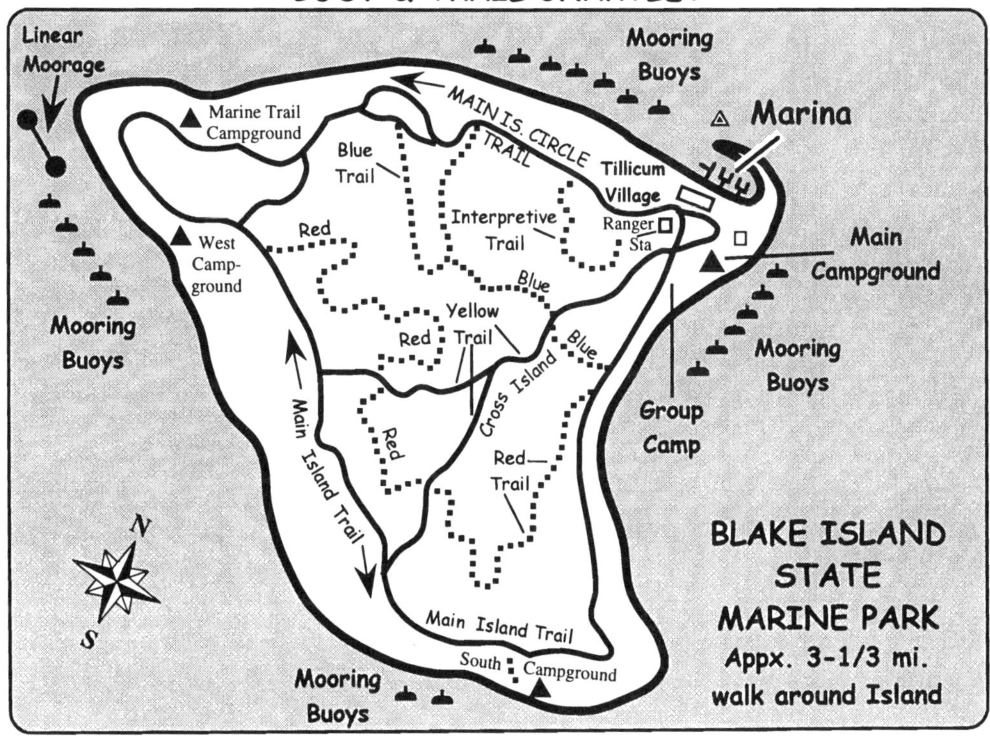

Port Orchard

NAME OF MARINA: ***PORT ORCHARD MARINA*** **RADIO:** VHF 66A
TELEPHONE: **360-876-5535** MGR: Steve Slaton, Marina Manager
E-MAIL: guest@portofbremerton.org FAX: 360-895-0291 (Fax)
ADDRESS: 8850 State Hwy 3 Port Orchard WA 98366
SHORT DESCRIPTION & LOCATION: **www.portofbremerton.org**

47°32.35' - 122°38.20' Located on south shore of Sinclair Inlet directly across from City of Bremerton and adjacent to small Bremerton foot ferry. The marina is large and modern and close to many shops & restaurants in historic downtown Port Orchard.

GUEST BOAT CAPACITY: ....Appx. 180 boats
DOCKSIDE DEPTH AT ZERO TIDE: ......12 ft. +
SEASON: ....All year
RESERVATION POLICY: ....Accepts
AMT W/ELECTRICITY: ....All
FUEL DOCK: ....Gas & Diesel
MARINE REPAIRS: ....Close by
TOILETS: ....Yes
HOT SHOWERS: ....Yes
RESTAURANT: ....Close by
PICNIC AREA: ....Close by
BASIC STORE: ....Close by
BROADBAND/WI-FI: ....Yes
DAILY RATE: ....Economical (Under 75¢/foot)

OTHER: Internet Terminal Available

GUEST DOCK: 3000 ft. plus slips
GUEST SLIPS: 46
WATER: ....Yes
AMPS: ....30-50 A
PUMP OUT STATION ....Yes
HAUL OUT: ....Close by
BOAT RAMP: ....Close by
LAUNDRY: ....Yes
BAR: ....Close by
POOL: ....None
GOLF: ....Close by
PET FRIENDLY: ....Good
OTHER: Activity float for functions, portable BBQ, Sat. Farmers Market, museum, art gallery, library, foot ferry to Bremerton.

NAME OF YACHT CLUB: *SINCLAIR INLET YACHT CLUB*
CLUB ADDRESS: P.O. Box 1197, Port Orchard, WA 98366
CLUB TELEPHONE: PERSON IN CHARGE: Commodore
LOCATION & SPECIAL NOTES: **www.sinclairyc.com**

SIYC provides some reciprocal benefit in the Port Orchard Marina guest dock area. Tie up & check in with harbormaster. SIYC. will partially pay your moorage fee for 24 hours. One reciprocal use allowed in a 60 day period and no more than 2 boats per night, up to 10 boats total for month. Electricity additional. You must have your current Y.C. card.

RECIPROCAL BOAT CAPACITY: 2 per night
DOCKSIDE DEPTH AT ZERO TIDE: ......12 ft. +
RECIPROCAL SEASON: ....All year
RESERVATION POLICY: ....None
TOILETS: ....At Marina
HOT SHOWERS: ....At Marina
RESTAURANT: ....Close by
DAILY RATE: ....Elecricity $4.00/night plus partial reimbursement if available.

RECIPROCAL DOCK: See above
RECIPROCAL SLIPS: See above
WATER: ....Yes
AMT W/ELECTRICITY: ....All
AMPS: ....30-50 A
BAR: ....Close by
OTHER: ....Same as marina listing

***NOTE:* THIS IS PRIVATE MOORAGE *ONLY* AVAILABLE TO MEMBERS OF RECIPROCAL YACHT CLUBS! YOUR CLUB *MUST* HAVE RECIPROCAL PRIVILEGES AND YOU MUST FLY YOUR BURGEE!**

Port Orchard

CAUTION! This chartlet not intended for use in navigation.

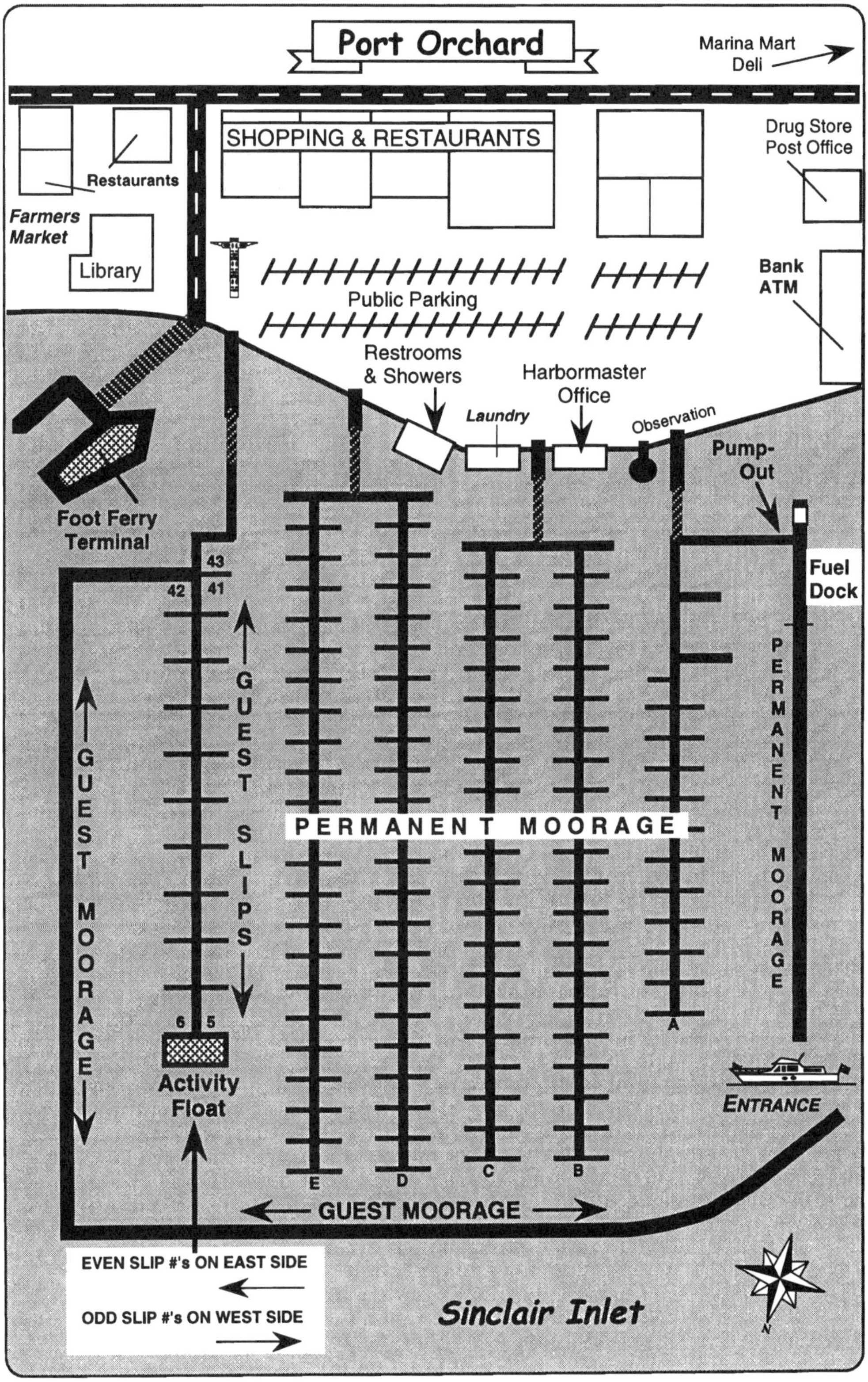

Port Orchard
Marina Mart Deli
SHOPPING & RESTAURANTS
Drug Store Post Office
Restaurants
Farmers Market
Library
Bank ATM
Public Parking
Restrooms & Showers
Harbormaster Office
Laundry
Observation
Pump-Out
Foot Ferry Terminal
Fuel Dock
43
42
41
GUEST SLIPS
GUEST MOORAGE
PERMANENT MOORAGE
PERMANENT MOORAGE
6
5
Activity Float
A
E
D
C
B
ENTRANCE
GUEST MOORAGE
EVEN SLIP #'s ON EAST SIDE
ODD SLIP #'s ON WEST SIDE
Sinclair Inlet
N

Port Orchard

NAME OF YACHT CLUB: *WEST SOUND CORINTHIAN YACHT CLUB*

ADDRESS: P.O. Box 1111 Port Orchard, WA 98366

TELEPHONE:

PERSON IN CHARGE: Rear Commodore

SHORT DESCRIPTION & LOCATION: **www.wscyc.net**

Moorage is located on the guest docks at the **Port Orchard Marina.** Usage is limited to 24 hours & not more than 3 boats from same yacht club or 6 boats total in one night. WSCYC covers up to $15 per boat per visit. (Pays for appx. 35' boat). Check in with harbormaster & ask for WSCYC reciprocal. Pay port for any excess length plus electricity.

RECIPROCAL BOAT CAPACITY Appx 6 boats

DOCKSIDE DEPTH AT ZERO TIDE: ......12 ft. +

SEASON: ....All year

RESERVATION POLICY: ....None

TOILETS: ....At Port Orchard Marina

HOT SHOWERS: ....At Port Orchard Marina

RESTAURANT: ....Close by

BROADBAND/WI-FI: ....Planned at Marina

DAILY RATE: ....Elecricity $4.00/night Plus moorage over $15.

RECIPROCAL DOCK: ....Varies

RECIPROCAL SLIPS: 6 per night

WATER: ....Yes

AMT W/ELECTRICITY: ....All

AMPS: 30-50A

BAR: ....Close by

PET FRIENDLY: ....Good

OTHER: Services same as Port Orchard Marina services listed on Pages 56 & 57.

NOTE: ***THIS IS PRIVATE MOORAGE AND ONLY AVAILABLE TO MEMBERS OF RECIPROCAL YACHT CLUBS. YOUR CLUB MUST HAVE RECIPROCAL PRIVILEGES AND YOU MUST FLY YOUR BURGEE.***

SEE DATA & CHARTLET ON PAGE: 56 & 57

NOTES

Port Orchard

NAME OF YACHT CLUB: *PORT ORCHARD YACHT CLUB*

ADDRESS: P.O. Box 3 Port Orchard, WA 98366

TELEPHONE: 360-876-9010/874-9366 PERSON IN CHARGE: Commodore

SHORT DESCRIPTION & LOCATION: **www.poyc.org**

47°32.25' - 122°38.45' Located about 1/4 mi. W. of downtown Port Orchard on the S. side of Sinclair Inlet. The reciprocal dock is on the outside of the breakwater float marked in yellow and indicated by signs. See posted instructions on docks to register & obtain gate key.

RECIPROCAL BOAT CAPACITY: ....Appx. 10-15

DOCKSIDE DEPTH AT ZERO TIDE: ....25 ft.

SEASON: ....All year

RESERVATION POLICY: ....None

TOILETS: ....Yes

HOT SHOWERS: ....Yes

RESTAURANT: ....Close by

BROADBAND/WI-FI: ....None

DAILY RATE: $3.00/day (for power) 1st 3 days. Add'l days @$5/day with approval. Maximum days allowed - 14 per yr.

RECIPROCAL DOCK: ....388 ft.

RECIPROCAL SLIPS: Dock only

WATER: ....Yes

AMT W/ELECTRICITY: ....All

AMPS: ....30 A

BAR: ....Close by

PET FRIENDLY: ....Good

OTHER: Walking distance to down town Port Orchard shops, restaurants, museum.

***NOTE:* THIS IS PRIVATE MOORAGE AND ONLY AVAILABLE TO MEMBERS OF RECIPROCAL YACHT CLUBS. YOUR CLUB *MUST* HAVE RECIPROCAL PRIVILEGES AND YOU MUST FLY YOUR BURGEE.**

CAUTION! This chartlet not intended for use in navigation.

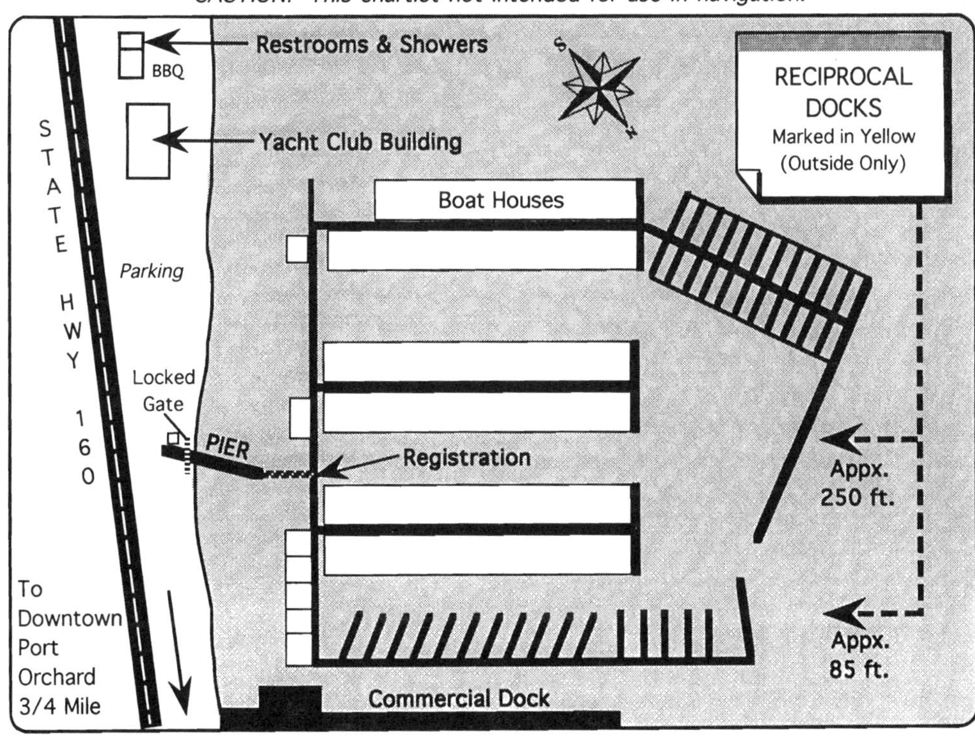

Bremerton

NAME OF MARINA: ***THE BREMERTON MARINA*** RADIO: VHF 66A
TELEPHONE: **360-373-1035** MGR: Steve Slaton
E-MAIL: steves@portofbremerton.org FAX: 360-895-0291 (Fax)
ADDRESS: 8850 SW State Hwy 3 Port Orchard, WA 98366
SHORT DESCRIPTION & LOCATION: **www.portofbremerton.org**

47°33.90' - 122°37.20' Located just N of downtown Bremerton ferry terminal next to USS Turner Joy floating museum. New upscale Harborside facility offers fine restaurants & attractions for boaters. Call ahead for slip assignment during reconstruction of Marina.

GUEST BOAT CAPACITY: ....Appx. 123 boats
DOCKSIDE DEPTH AT ZERO TIDE: 10-30 ft.
SEASON: ....All year
RESERVATION POLICY: ....Accepts
AMT W/ELECTRICITY: ....All
FUEL DOCK: ....Close by
MARINE REPAIRS: ....Close by
TOILETS: ....Yes
HOT SHOWERS: ....Yes
RESTAURANT: ....Yes
PICNIC AREA: ....Yes
BASIC STORE: ....Close by
BROADBAND/WI-FI: ....Yes
DAILY RATE: ....To be determined

GUEST DOCK: ....1170 ft. total
GUEST SLIPS: Appx. 84 Slips
WATER: ....Yes
AMPS: ....30-50 A
PUMP OUT STATION ....Yes
HAUL OUT: ....None
BOAT RAMP: ....Close by
LAUNDRY: ....Yes
BAR: ....Yes
POOL: ....None
GOLF: ....Close by
PET FRIENDLY: ....Fair
OTHER: Shopping, galleries, theaters & museums. Activity float, BBQ, board-walks, & parks. Ferries to Seattle & Port Orchard.

NAME OF YACHT CLUB: ***BREMERTON BOATING CLUB***
CLUB ADDRESS: P.O. Box 2003, Bremerton, WA 98310
CLUB TELEPHONE: None PERSON IN CHARGE: Secretary
LOCATION & SPECIAL NOTES:
BBC offers reciprocal moorage in the Bremerton Marina. Three slips are provided in the general marina for recip. members. No more than 2 boats from same club at one time are permitted and there is a monthly limit of 10 reciprocal boats. 24 Hour reciprocal limit.

RECIPROCAL BOAT CAPACITY:...............3 boats
DOCKSIDE DEPTH AT ZERO TIDE: Per marina
RECIPROCAL SEASON:....................All year
RESERVATION POLICY: ....None
TOILETS: ....At Marina
HOT SHOWERS: ....At Marina
RESTAURANT: ....Close by
DAILY RATE: .....$4.00 per day electricity charge

RECIPROCAL DOCK: ....Slips
RECIPROCAL SLIPS: .....3 Or less
WATER: ....Yes
AMT W/ELECTRICITY: ....All
AMPS: ....20-50 A
BAR: ....Close by
OTHER: ....Same as marina listing

***NOTE:** THIS IS PRIVATE MOORAGE ONLY AVAILABLE TO MEMBERS OF RECIPROCAL YACHT CLUBS! YOUR CLUB MUST HAVE RECIPROCAL PRIVILEGES AND YOU MUST FLY YOUR BURGEE!*

Bremerton

CAUTION! This chartlet not intended for use in navigation.

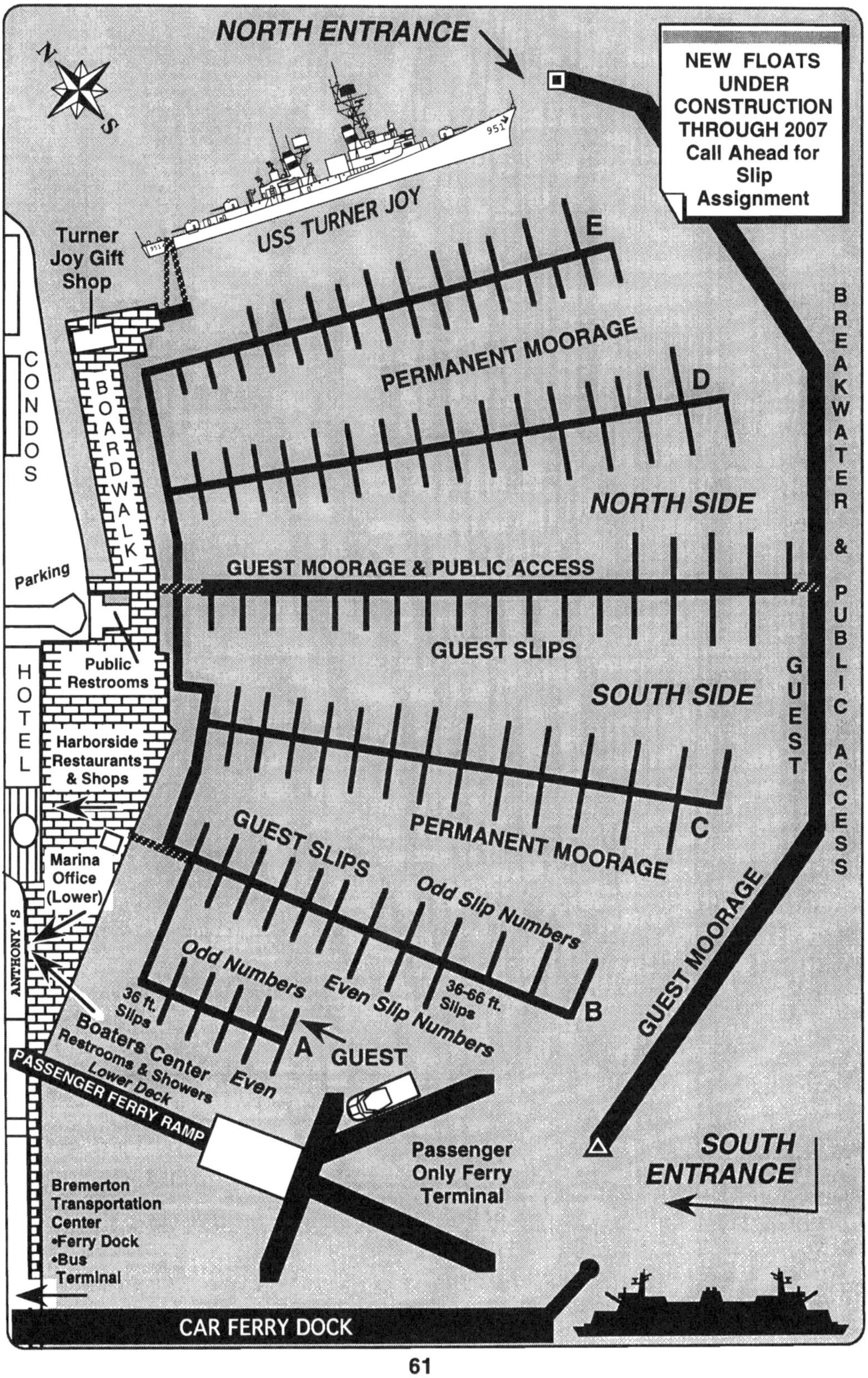

NORTH ENTRANCE

NEW FLOATS UNDER CONSTRUCTION THROUGH 2007 Call Ahead for Slip Assignment

N

S

USS TURNER JOY

Turner Joy Gift Shop

E

PERMANENT MOORAGE

D

CONDOS

BOARDWALK

BREAKWATER & PUBLIC ACCESS

NORTH SIDE

Parking

GUEST MOORAGE & PUBLIC ACCESS

GUEST SLIPS

Public Restrooms

GUEST

HOTEL

SOUTH SIDE

Harborside Restaurants & Shops

C

PERMANENT MOORAGE

GUEST SLIPS

Marina Office (Lower)

Odd Slip Numbers

ANTHONY'S

Odd Numbers

Even Slip Numbers

36-66 ft. Slips

B

GUEST MOORAGE

36 ft. Slips

A

GUEST

Boaters Center Restrooms & Showers Lower Deck

Even

PASSENGER FERRY RAMP

Passenger Only Ferry Terminal

SOUTH ENTRANCE

Bremerton Transportation Center
- Ferry Dock
- Bus Terminal

CAR FERRY DOCK

Bremerton

NAME OF YACHT CLUB: *ROCHE HARBOR YACHT CLUB*

ADDRESS: P.O. Box 94426 Seattle, WA 98124

TELEPHONE: None PERSON IN CHARGE: Rear Commodore

SHORT DESCRIPTION & LOCATION: **www.rhyc.org**

RHYC provides 80 feet of moorage space & offer 48 hours of complimentary moorage to Reciprocal Club members at the **Bremerton Marina.** Upon arrival check in at the Harbormasters's office and request Roche Harbor Y.C. reciprocal. Harbormaster may be contacted on VHF 66A or by calling 360-373-1035.

RECIPROCAL BOAT CAPACITY............1-3 boats
DOCKSIDE DEPTH AT ZERO TIDE: ....20 ft.
SEASON: ....All year
RESERVATION POLICY: ....None
TOILETS: ....At Marina
HOT SHOWERS: ....At Marina
RESTAURANT: ....Close by
BROADBAND/WI-FI: ....Yes
DAILY RATE: Electricity charge - $4.00 per day

RECIPROCAL DOCK: ....80 ft.
RECIPROCAL SLIPS: ....80 ft.
WATER: ....Yes
AMT W/ELECTRICITY: ....All
AMPS: ....20-30 A
BAR: ....Close by
PET FRIENDLY: ....Fair
OTHER: Services same as Bremerton Marina listing on Page 60 & 61.

NOTE: ***THIS IS PRIVATE MOORAGE AND ONLY AVAILABLE TO MEMBERS OF RECIPROCAL YACHT CLUBS. YOUR CLUB MUST HAVE RECIPROCAL PRIVILEGES AND YOU MUST FLY YOUR BURGEE.***

SEE DATA & CHARTLET ON PAGE: 60 & 61

NOTES

Bremerton

NAME OF YACHT CLUB: *BREMERTON YACHT CLUB* RADIO: VHF Ch.16/71
ADDRESS: 2700 Yacht Haven Way Bremerton, WA 98312
TELEPHONE: 360-479-2662 PERSON IN CHARGE: Caretakers

SHORT DESCRIPTION & LOCATION: **www.bremertonyachtclub.org**
47°35.30' - 122°39.80' BYC moorage is located in Phinney Bay on the S. side of Port Washington Narrows. Register w/caretakers at clubhouse on Wed. thru Sun. (Mon. & Tue. contact any Bridge Officer-Info at gate). Your current reciprocal Yacht Club membership card is required upon check-in.

RECIPROCAL BOAT CAPACITY: Appx. 4-5 boats
DOCKSIDE DEPTH AT ZERO TIDE: ....7 ft.
SEASON: ....All year
RESERVATION POLICY: ....None
TOILETS: ....Yes
HOT SHOWERS: ....Yes
RESTAURANT: ....None
BROADBAND/WI-FI: ....None
DAILY RATE: 1st 72 hrs complimentary moorage
Add'l days - $5 each day,
Electricity - $1.00 every day/cord.

RECIPROCAL DOCK: Appx. 165'
RECIPROCAL SLIPS: ....Varies
WATER: ....Yes
AMT W/ELECTRICITY: ....All
AMPS: ....1 @30 A/ 4 @20 A
BAR: ....None
PET FRIENDLY: ....Excellent
OTHER: Ice. Security gate requires $5.00 key deposit.

***NOTE:* THIS IS PRIVATE MOORAGE AND ONLY AVAILABLE TO MEMBERS OF RECIPROCAL YACHT CLUBS. YOUR CLUB *MUST* HAVE RECIPROCAL PRIVILEGES AND YOU MUST FLY YOUR BURGEE.**

CAUTION! This chartlet not intended for use in navigation.

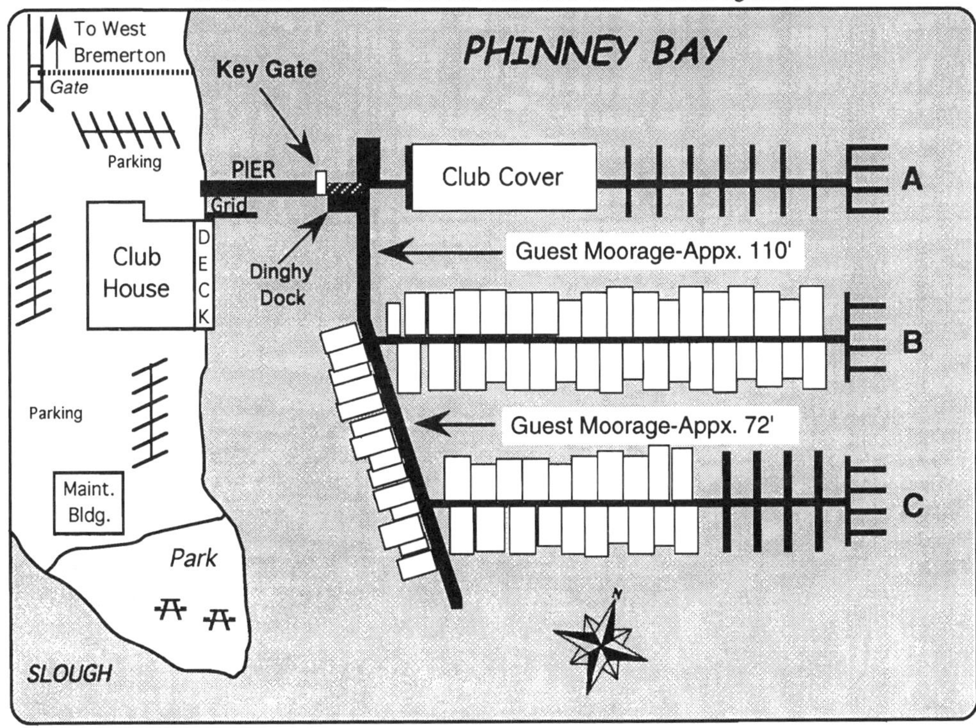

Bremerton

NAME OF PARK: *ILLAHEE STATE PARK*
ADDRESS: 3540 N.E. Bahia Vista Bremerton, WA 98310
TELEPHONE: 360-478-6460 **MGR:** Steven Kendall
SHORT DESCRIPTION & LOCATION: www.parks.wa.gov/moorage/parks

47°36.00'-122°35.65' Located in E. Bremerton on the W. shore of Port Orchard Bay appx. 4 mi. S of Battle Pt. & 1 mi S of sm. town of Illahee. The 75 acre park contains park facilities in the wooded upland portion. Park dock open to wakes and can be roly.

GUEST BOAT CAPACITY: ....Appx. 4-6 boats
DOCKSIDE DEPTH AT ZERO TIDE:.....15-20 ft.
SEASON: ....All year
AMT W/ELECTRICITY: ....None
TOILETS: ....Yes
HOT SHOWERS: ....Yes
PICNIC AREA: ....Yes
PLAY AREA: ....Yes
BASIC STORE: ....None
DAILY RATE: Dock Charge............50¢/foot ($10.00 Minimum)
Mooring Buoys...............$10.00/night

GUEST DOCK: Appx. 350 ft. TTL
MOORING BUOYS: ....5 Buoys
WATER: ....Yes
PAY PHONES: ....Yes
BOAT RAMP: ....Yes
PICNIC SHELTER: ....Yes
BBQ: ....Yes
PUMP OUT STATION: ......None
PET FRIENDLY: ....Excellent
OTHER: Campsites, fishing pier, hiking trails, horseshoes, softball, swimming, beach combing & shell fishing in season.

CAUTION! This chartlet not intended for use in navigation.

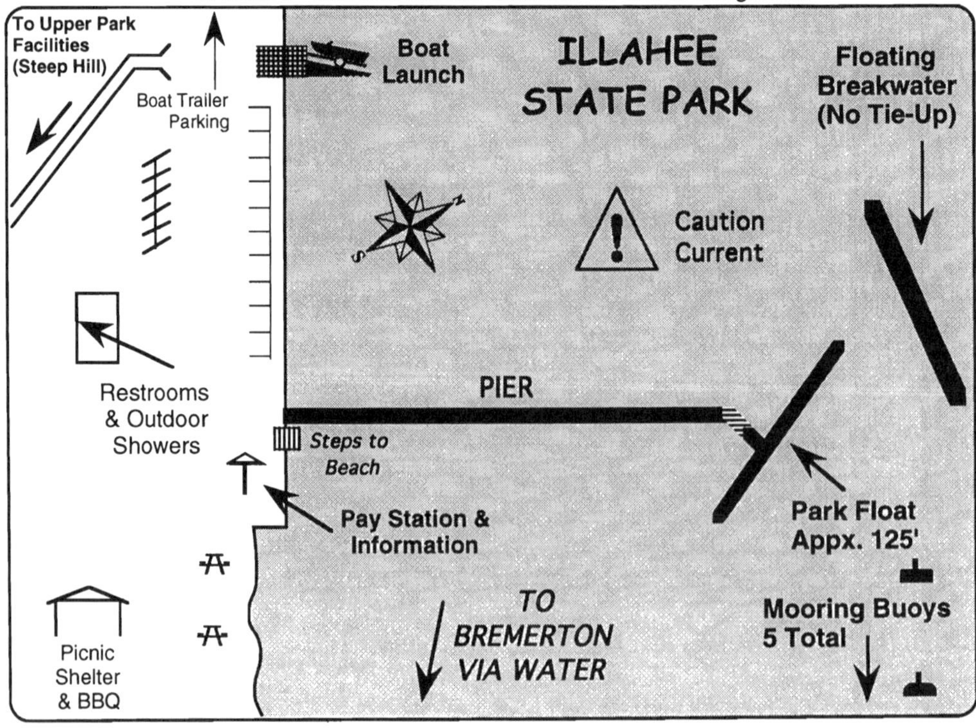

Silverdale

NAME OF MARINA: ***PORT OF SILVERDALE*** RADIO: None
TELEPHONE: **360-698-4918** MGR: Theresa Haaland
E-MAIL: portsilv@tscnet.com FAX: 360-698-2402 (Fax)
ADDRESS: P.O. Box 310 Silverdale WA 98383
SHORT DESCRIPTION & LOCATION: **www.portofsilverdale.com**

47°38.50' - 122°41.70' Located at the head of Dyes Inlet on the W. side about 4.5 miles NNW of Bremerton. The marina features a waterfront park with gazebo & picnic area which is within walking distance to many shops and restaurants.

GUEST BOAT CAPACITY: ....Appx. 40-50 boats
DOCKSIDE DEPTH AT ZERO TIDE: ....10 ft.
SEASON: ....All year
RESERVATION POLICY: ....On-line for groups
AMT W/ELECTRICITY: ....All
FUEL DOCK: ....None
MARINE REPAIRS: ....None
TOILETS: ....Yes
HOT SHOWERS: ....Yes
RESTAURANT: ....Close by
PICNIC AREA: ....Yes
BASIC STORE: ....Close by
BROADBAND/WI-FI: ....None
DAILY RATE: ....Economical (Under 75¢/foot)

GUEST DOCK: Total 1300 Lin. ft.
GUEST SLIPS: ....4 large slips
WATER: ....Yes
AMPS: ....30 A
PUMP OUT STATION ....Yes
HAUL OUT: ....None
BOAT RAMP: ....Yes
LAUNDRY: ....Yes
BAR: ....Close by
POOL: ....None
GOLF: ....Close by
PET FRIENDLY: ....Good
OTHER: Fishing pier, swimming, Old Town attractions, Whaling Days festivities held last weekend in July. No Rafting.

CAUTION! This chartlet not intended for use in navigation.

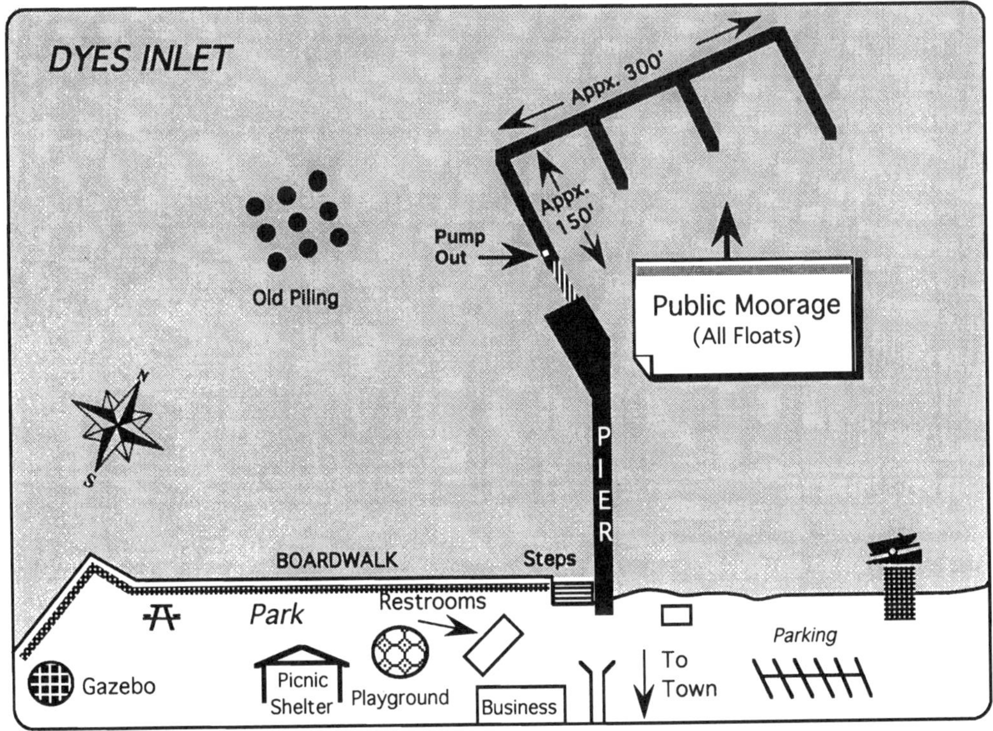

Brownsville

NAME OF MARINA: ***PORT OF BROWNSVILLE MARINA*** **RADIO:** VHF Ch.16
TELEPHONE: **360-692-5498** MGR: Jerry Rowland
E-MAIL: pob@portofbrownsville.org FAX: 360-698-8023
ADDRESS: 9790 Ogle Rd. NE Bremerton WA 98311
SHORT DESCRIPTION & LOCATION: **www.portofbrownsville.org**

47°39.10' - 122°36.65' Located on Kitsap Peninsula about 4 miles south of Agate Pass Bridge & 7 miles north of Bremerton in Port Orchard passage directly across from west side of Bainbridge Island. Full service marina in quiet neighborhood location.

GUEST BOAT CAPACITY: ....Appx. 50 boats
DOCKSIDE DEPTH AT ZERO TIDE: ....22 ft.
SEASON: ....All year
RESERVATION POLICY:..........None
AMT W/ELECTRICITY: ....All
FUEL DOCK: ....Gas, Dsl, & LP
MARINE REPAIRS: ....Close by
TOILETS: ....Yes
HOT SHOWERS: ....Yes
RESTAURANT: ....Cafe/Deli in Store
PICNIC AREA: ....Yes
BASIC STORE: ....Store & Deli
BROADBAND/WI-FI: ....BroadbandXpress
DAILY RATE:..........Economical (Under 75¢/foot)

GUEST DOCK: ....700' plus slips
GUEST SLIPS: ....Appx. 40 slips
WATER: ....Yes
AMPS: ....20-30 A
PUMP OUT STATION ....Yes
HAUL OUT: ....Grid
BOAT RAMP: ....Yes
LAUNDRY: ....Yes
BAR: ....Wine & Beer in Deli
POOL: ....None
GOLF: ....Close by
PET FRIENDLY: ....Excellent
OTHER: Gifts & fishing supplies in Store, Public shell-fish beaches close by, great group picnic area & waterfront pavilion.

NAME OF YACHT CLUB: ***BROWNSVILLE YACHT CLUB***
CLUB ADDRESS: 9756A Ogle Rd. N.E., Bremerton, WA 98311
CLUB TELEPHONE: None PERSON IN CHARGE: Commodore
LOCATION & SPECIAL NOTES: **www.portofbrownsville.org**

BVYC provides reciprocal moorage for 2 boats per night up to 90' total on the Brownsville guest dock. Rafting not permitted. Check in with the harbormaster. Must show your current YC membership card. Authorizes 2 nights max/month, 4 nights/year. If your club reciprocal & BVYC's monthly limit not reached, you will be granted reciprocal privileges.

RECIPROCAL BOAT CAPACITY:..............2 Boats
DOCKSIDE DEPTH AT ZERO TIDE: As above
RECIPROCAL SEASON:..........All year
RESERVATION POLICY: ....None
TOILETS: ....At Marina
HOT SHOWERS: ....At Marina
RESTAURANT: ....Deli at Marina
DAILY RATE: ....$3.00 electricity per day

RECIPROCAL DOCK: Guest dock
RECIPROCAL SLIPS: ...Dock only
WATER: ....Yes
AMT W/ELECTRICITY: ....All
AMPS: ....20-30 A
BAR: ....Deli w/wine & beer
OTHER: Clubhouse rental available. Info on website.

***NOTE:* THIS IS PRIVATE MOORAGE *ONLY* AVAILABLE TO MEMBERS OF RECIPROCAL YACHT CLUBS! YOUR CLUB *MUST* HAVE RECIPROCAL PRIVILEGES AND YOU MUST FLY YOUR BURGEE!**

Brownsville

CAUTION! This chartlet not intended for use in navigation.

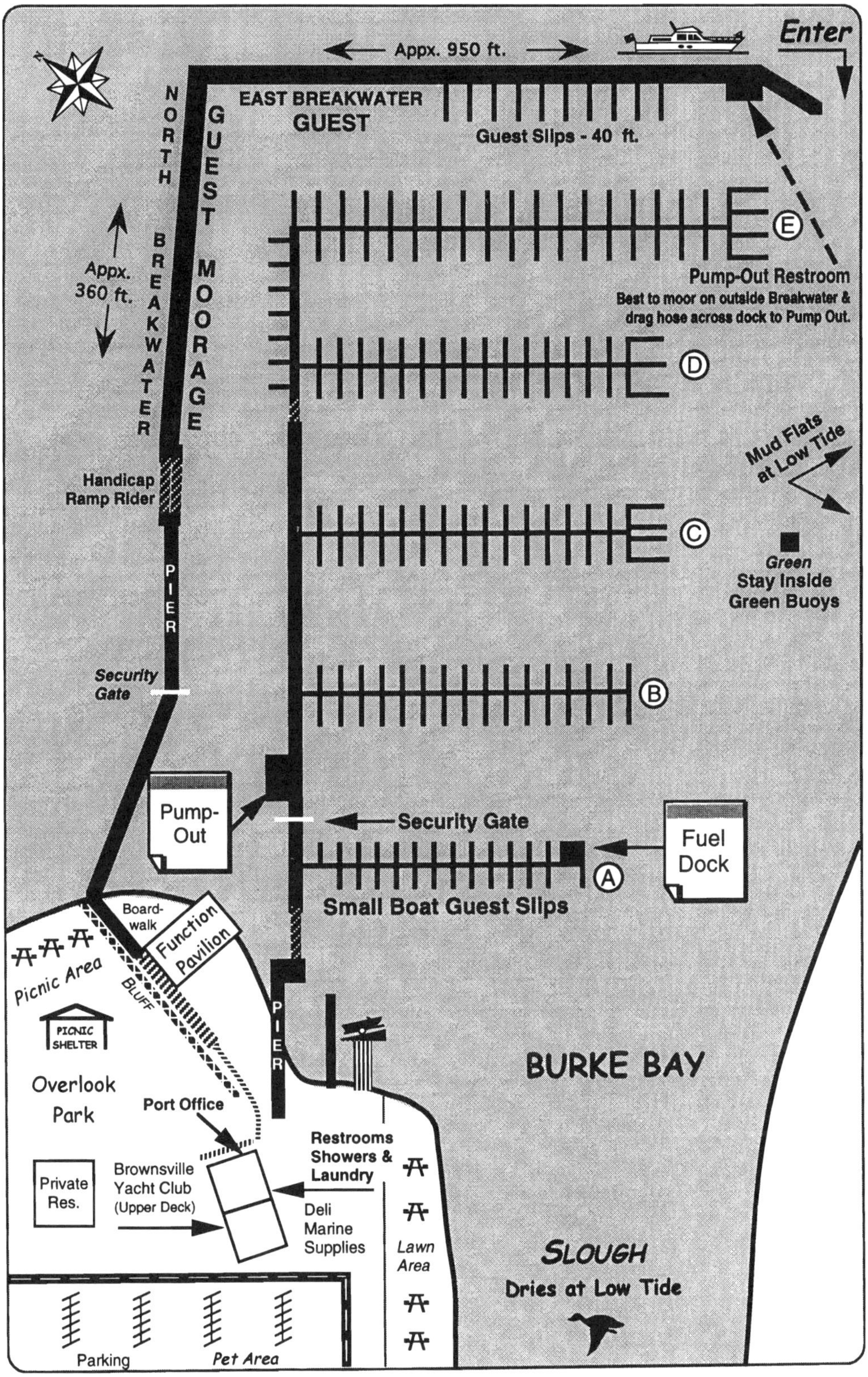

Enter
Appx. 950 ft.
EAST BREAKWATER
GUEST
Guest Slips - 40 ft.
NORTH BREAKWATER
GUEST MOORAGE
Appx. 360 ft.
Pump-Out Restroom
Best to moor on outside Breakwater & drag hose across dock to Pump Out.
E
D
Mud Flats at Low Tide
Handicap Ramp Rider
C
Green
Stay Inside Green Buoys
PIER
Security Gate
B
Pump-Out
Security Gate
Fuel Dock
A
Small Boat Guest Slips
Board-walk
Function Pavilion
Picnic Area
BLUFF
PICNIC SHELTER
PIER
BURKE BAY
Overlook Park
Port Office
Restrooms Showers & Laundry
Private Res.
Brownsville Yacht Club (Upper Deck)
Deli Marine Supplies
Lawn Area
SLOUGH
Dries at Low Tide
Parking
Pet Area

Brownsville

NAME OF YACHT CLUB: ***SEABACS - Boeing Employee Boat Club***
ADDRESS: M/S 4H-58, POB 3707 Seattle, WA 98124
TELEPHONE:
PERSON IN CHARGE: Vice Commodore

SHORT DESCRIPTION & LOCATION: **www.seabacs.org**
SEABACS offers up to 2 nights reciprocal moorage for boats up to 40 ft. at the **Brownsville Marina** Guest Dock. Visiting reciprocal yacht club members are requested to pay the moorage charges to the harbormaster & mail the moorage receipt along with a copy of your club membership card to SEABACS (Attn: Treasurer) for reimbursement.

RECIPROCAL BOAT CAPACITY ......Varies
DOCKSIDE DEPTH AT ZERO TIDE: ......22 ft.
SEASON: ......All year
RESERVATION POLICY: ......None
TOILETS: ......At Brownsville Marina
HOT SHOWERS: ......At Brownsville Marina
RESTAURANT: ......Cafe/Deli in Store
BROADBAND/WI-FI: ......BroadbandXpress
DAILY RATE: ......2 Days Free moorage via reimbursement. No reimbursement for electricity - $3.00/day

RECIPROCAL DOCK: ....At marina
RECIPROCAL SLIPS: ...At marina
WATER: ......Yes
AMT W/ELECTRICITY: ......All
AMPS: ......30 A
BAR: ......Wine & Beer in Deli
PET FRIENDLY: ......Excellent
OTHER: Services same as Brownsville Marina listing on Page 66 & 67.

***NOTE:* THIS IS PRIVATE MOORAGE AND ONLY AVAILABLE TO MEMBERS OF RECIPROCAL YACHT CLUBS. YOUR CLUB *MUST* HAVE RECIPROCAL PRIVILEGES AND YOU MUST FLY YOUR BURGEE.**

SEE DATA & CHARTLET ON PAGE: 66 & 67

NOTES

Brownsville

NAME OF YACHT CLUB: *WEST SEATTLE YACHT CLUB*

ADDRESS: P.O. Box 16095 Seattle, WA 98116

TELEPHONE: PERSON IN CHARGE: Recip. Chair

SHORT DESCRIPTION & LOCATION: **www.westseattleyachtclub.com**

WSYC provides two slips per night located at the **Brownsville Marina** guest dock. Tie up on guest dock & check in with harbormaster & request West Seattle Y.C. reciprocal. If your club is reciprocal, and WSYC's monthly budget cap is not reached yet, you will be granted reciprocal privileges. Please show current reciprocal yacht club card upon check in.

RECIPROCAL BOAT CAPACITY Appx. 4 per Mo.

DOCKSIDE DEPTH AT ZERO TIDE: ....22 ft.

SEASON: ....All year

RESERVATION POLICY: First come first served

TOILETS: ....At Brownsville Marina

HOT SHOWERS: ....At Brownsville Marina

RESTAURANT: ....Cafe/Deli in Store

BROADBAND/WI-FI: ....BroadbandXpress

DAILY RATE: No Charge for two nights moorage Electricity $3.00/day. Maximum per yacht club is 4 boats per month.

RECIPROCAL DOCK: .....At marina

RECIPROCAL SLIPS: ...At marina

WATER: ....Yes

AMT W/ELECTRICITY: ....All

AMPS: ....30 A

BAR: ....Wine & Beer in Deli

PET FRIENDLY: ....Excellent

OTHER: Services same as Brownsville Marina listing on Page 66 & 67.

***NOTE:* THIS IS PRIVATE MOORAGE AND ONLY AVAILABLE TO MEMBERS OF RECIPROCAL YACHT CLUBS. YOUR CLUB *MUST* HAVE RECIPROCAL PRIVILEGES AND YOU MUST FLY YOUR BURGEE.**

SEE DATA & CHARTLET ON PAGE: 66 & 67

NOTES

Poulsbo

NAME OF YACHT CLUB: *POULSBO YACHT CLUB*

ADDRESS: 18129 Fjord Dr. NE #T Poulsbo WA 98370

TELEPHONE: 360-779-3116 PERSON IN CHARGE: Reciprocal Chair

SHORT DESCRIPTION & LOCATION: **www.poulsboyc.org**

47°43.60' - 122°38.30' This is the 2nd marina on starboard upon entering Liberty Bay. Modern facility and clubhouse located appx. 3/4 mile SE of downtown Poulsbo, a picturesque town with a Norwegian theme and heritage. Reciprocal moorage inside of breakwater float. See instructions at Kiosk for marina gate and restroom access.

RECIPROCAL BOAT CAPACITY: ....Appx 5-7

DOCKSIDE DEPTH AT ZERO TIDE: ....18 ft.

SEASON: ....All year

RESERVATION POLICY: First come, first serve

TOILETS: ....Yes

HOT SHOWERS: ....Yes

RESTAURANT: ....None

BROADBAND/WI-FI: ....BroadbandXpress

DAILY RATE: Electricity $3.00 per day. Moorage N/C for 48 hours, with a maximum of 4 days in a 3 month period.

RECIPROCAL DOCK: ....225 ft.

RECIPROCAL SLIPS: Dock only

WATER: ....Yes

AMT W/ELECTRICITY: ....All

AMPS: ....30 A

BAR: ....None

PET FRIENDLY: ....Good

OTHER: Groups of 3 or more need advance notice. Pump out, Ice & Laundry.

***NOTE:* THIS IS PRIVATE MOORAGE AND ONLY AVAILABLE TO MEMBERS OF RECIPROCAL YACHT CLUBS. YOUR CLUB MUST HAVE RECIPROCAL PRIVILEGES AND YOU MUST FLY YOUR BURGEE.**

CAUTION! This chartlet not intended for use in navigation.

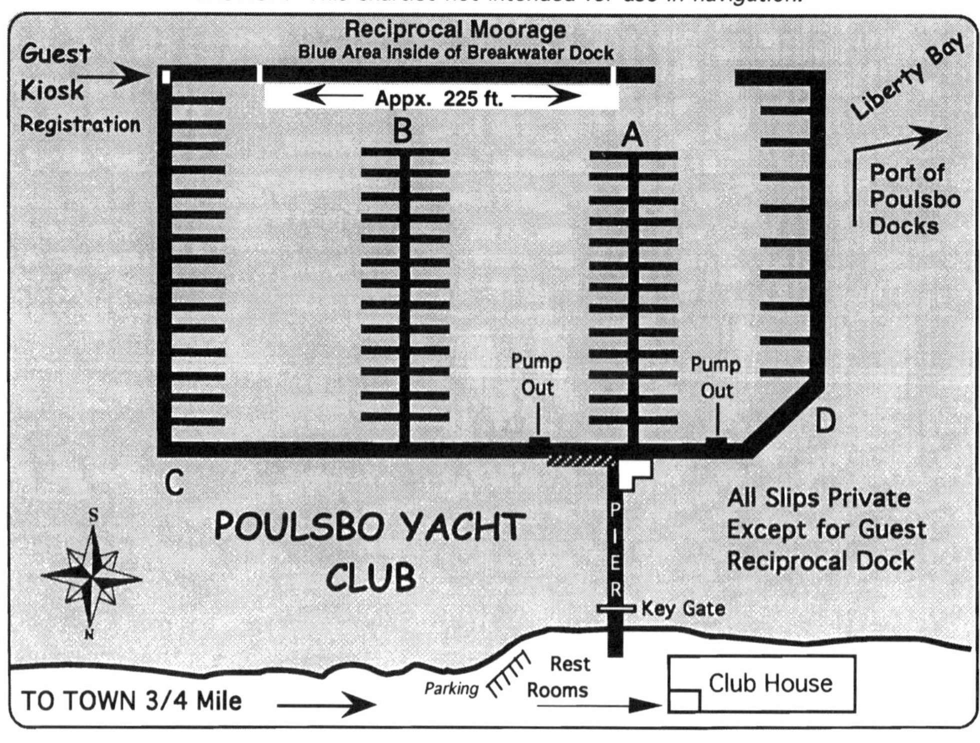

Poulsbo

NAME OF MARINA: ***PORT OF POULSBO MARINA*** **RADIO:** VHF 66A
TELEPHONE: **360-779-3505/9905** **MGR:** Kirk Stickels VHF Weekdays only
E-MAIL: portofpoulsbo@yahoo.com **FAX:** 360-779-8090 (Fax)
ADDRESS: P.O. Box 732 Poulsbo, WA 98370
SHORT DESCRIPTION & LOCATION: **http://poulsbo.net/portofpoulsbo/**

47°44.00' - 122°38.60' Located at head of Liberty Bay on north side of bay in heart of downtown Poulsbo. Picturesque town with Norwegian theme and heritage. Full boating amenities including speciality shops and restaurants. Many seasonal festivals & rendezvous's.

GUEST BOAT CAPACITY: ....Appx. 140 boats
DOCKSIDE DEPTH AT ZERO TIDE: ....12 ft.
SEASON: ....All year
RESERVATION POLICY: ....Groups only
AMT W/ELECTRICITY: ....All
FUEL DOCK: ....Gas & Diesel
MARINE REPAIRS: ....Limited - Close by
TOILETS: ....Yes
HOT SHOWERS: ....Yes
RESTAURANT: ....Close by
PICNIC AREA: ....Yes
BASIC STORE: ....Close by
BROADBAND/WI-FI: ....BroadbandXpress
DAILY RATE:..........Economical (Under 75¢/foot)
CHECK IN BEFORE 2000 HOURS TO GET COMBO FOR NICE RESTROOMS.

GUEST DOCK: ......76 ft. plus slips
GUEST SLIPS: 140
WATER: ....Yes
AMPS: ....30 A
PUMP OUT STATION ....Yes
HAUL OUT: ....None
BOAT RAMP: ....Yes
LAUNDRY: ....Yes
BAR: ....Close by
POOL: ....None
GOLF: ....Close by
PET FRIENDLY: ....Good
OTHER: Multi purpose function room for rent. Shops, famous bakery, walking trails, banks, post office, & many sights to see.

CAUTION! This chartlet not intended for use in navigation.

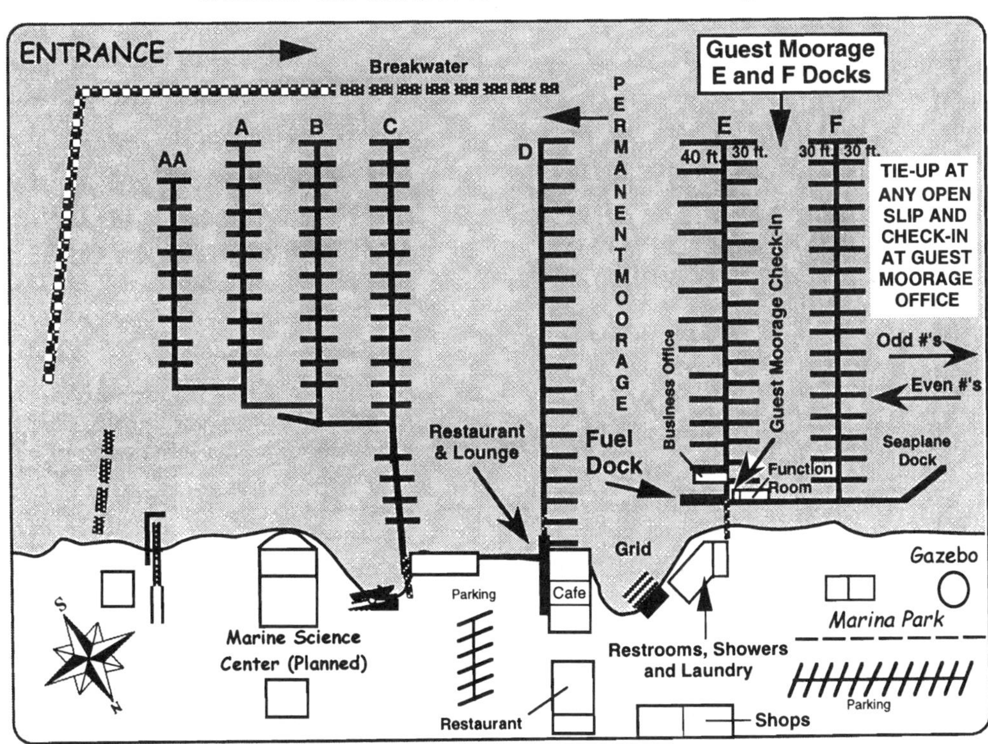

Bainbridge Island

NAME OF MARINA: ***BAINBRIDGE ISLAND MARINA*** RADIO: None
TELEPHONE: 206-842-9292/953-6767 MGR: Darrell McNabb, Owner
E-MAIL: None FAX: 206-352-3933 (Fax)
ADDRESS: P.O. Box 10325 Bainbridge Island, WA 98110

SHORT DESCRIPTION & LOCATION:

47°37.05' - 122°30.65' Located on the S. side of Eagle Harbor which indents on the E. shore of Bainbridge Is. directly across from Ferry Terminal. Caters to lg. yachts up to 160 ft. Unique marina recreating the flavor of 1890's. Ample room for yacht club cruises.

GUEST BOAT CAPACITY: ....Appx. 100 boats
DOCKSIDE DEPTH AT ZERO TIDE: ....11 ft.
SEASON: ....All year
RESERVATION POLICY: ....Accepts
AMT W/ELECTRICITY: ....All
FUEL DOCK: ....None
MARINE REPAIRS: ....None
TOILETS: ....Yes
HOT SHOWERS: ....Yes
RESTAURANT: ....None
PICNIC AREA: ....Yes
BASIC STORE: ....None
BROADBAND/WI-FI: ....None
DAILY RATE: ....Moderate (75¢-$1.25/foot)

GUEST DOCK: ....2500 ft. total
GUEST SLIPS: 30
WATER: ....Yes
AMPS: ....30-50 A
PUMP OUT STATION ....Yes
HAUL OUT: ....None
BOAT RAMP: ....None
LAUNDRY: ....None
BAR: ....None
POOL: ....None
GOLF: ....Close by
PET FRIENDLY: ....Good
OTHER: No restaurants or stores in the immediate area - must drive or take short dinghy ride across harbor to Winslow.

CAUTION! This chartlet not intended for use in navigation.

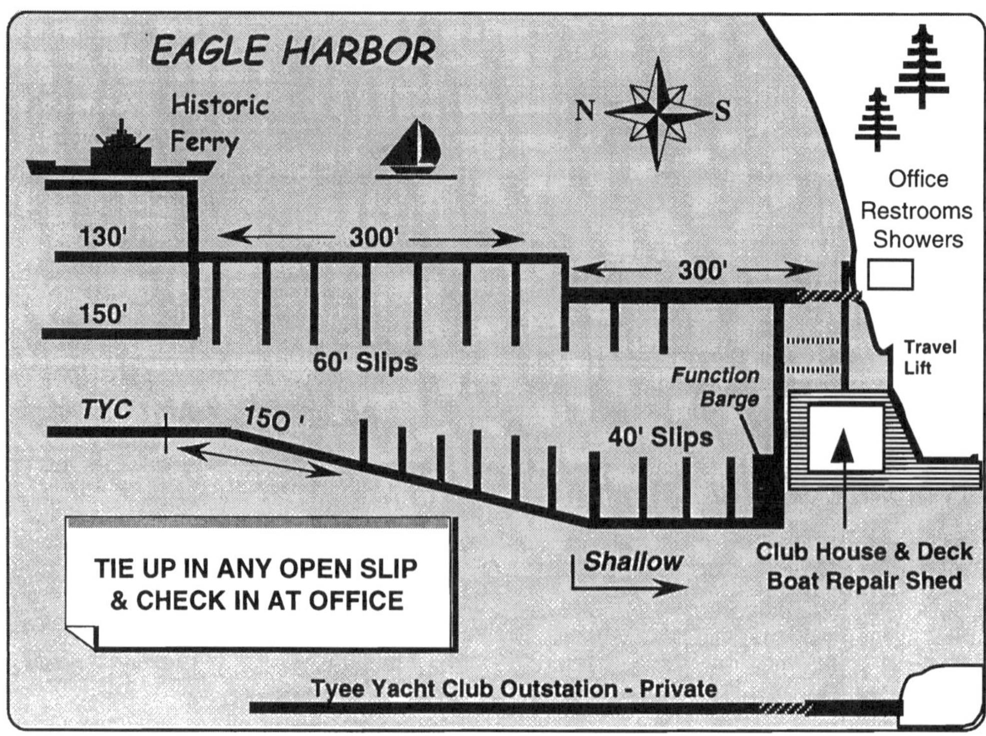

Bainbridge Island

NAME OF PARK:	***EAGLE HARBOR WATERFRONT PARK***
ADDRESS:	7305 HiddenCove Rd NE, Bainbridge Island WA 98110
TELEPHONE:	206-842-1212 MGR: Chris StRomain

SHORT DESCRIPTION & LOCATION:

47°37.00' - 122°31.00' Eagle Harbor indents the E. shore of Bainbridge Is., opposite Elliott Bay. The park float is located just W. of the State Ferry Shipyard in a typical park setting that is convenient to supermarket, shops, and restaurants in the town of Winslow.

GUEST BOAT CAPACITY:	Appx. 6-8 boats	GUEST DOCK:	Appx. 225 ft.
DOCKSIDE DEPTH AT ZERO TIDE:	0-5 ft.	MOORING BUOYS:	Linear
SEASON:	All year	WATER:	None
AMT W/ELECTRICITY:	None	PAY PHONES:	Yes
TOILETS:	Yes	BOAT RAMP:	Yes
HOT SHOWERS:	None	PICNIC SHELTER:	None
PICNIC AREA:	Yes	BBQ:	None
PLAY AREA:	Yes	PUMP OUT STATION:	Yes
BASIC STORE:	Close by	PET FRIENDLY:	Excellent
DAILY RATE:	Economical (Under 75¢/foot)	OTHER:	Great play area for kids, tennis courts, rafting OK - maximum 2 deep, walking distance to all stores & Seattle ferry.

48 HOUR LIMIT

CAUTION! This chartlet not intended for use in navigation.

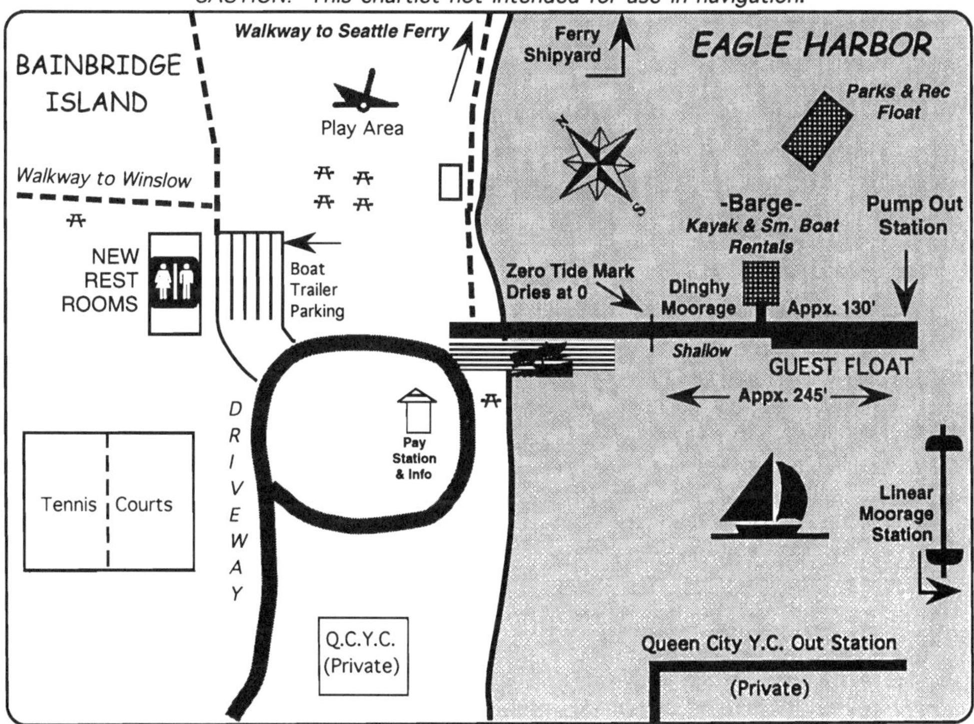

Bainbridge Island

NAME OF MARINA: ***WINSLOW WHARF MARINA*** RADIO: VHF Ch. 09
TELEPHONE: **206-842-4202** MGR: David LaFave
E-MAIL: wwmcoa@seanet.com FAX: 206-842-7785 (Fax)
ADDRESS: P.O. Box 10297 Bainbridge Island WA 98110

SHORT DESCRIPTION & LOCATION:

47°37.40' - 122°31.20' Located on the N. side of Eagle Harbor which indents on the E. shore of Bainbridge Is. appx. 1/2 mile W. of the State Ferry Shipyard. Guest moorage is available by using open slips of condo moorage tenants. Best to call ahead for slip.

GUEST BOAT CAPACITY: ....Varies- up to 15
DOCKSIDE DEPTH AT ZERO TIDE: ....20 ft.
SEASON: ....All year
RESERVATION POLICY: ....Accepts
AMT W/ELECTRICITY: ....All
FUEL DOCK: ....None
MARINE REPAIRS: ....None
TOILETS: ....Yes
HOT SHOWERS: ....Yes
RESTAURANT: ....Yes
PICNIC AREA: ....Close by
BASIC STORE: ....Close by
BROADBAND/WI-FI: ....BroadbandXpress
DAILY RATE: ....Moderate (75¢-$1.25/foot)
Maximum boat size 50 feet

GUEST DOCK: ....Slips only
GUEST SLIPS: ....Varies
WATER: ....Yes
AMPS: ....20-30 A
PUMP OUT STATION Portable
HAUL OUT: ....None
BOAT RAMP: ....Close by
LAUNDRY: ....Yes
BAR: ....Close by
POOL: ....None
GOLF: ....Close by
PET FRIENDLY: ....Good
OTHER: Walking distance to ferry, shopping center, 5 screen theatre, shops, restaurants, and marine chandlery.

NAME OF YACHT CLUB: ***EAGLE HARBOR YACHT CLUB***
CLUB ADDRESS: P.O. Box 10905, Bainbridge Island, WA 98110
CLUB TELEPHONE: None PERSON IN CHARGE: Recip. Chairman
LOCATION & SPECIAL NOTES: **www.bicomnet.com/ehyc/**

EHYC offers 1 slip per night at the Winslow Wharf Marina for members in good standing of reciprocal clubs. Check in with the Winslow Wharf harbormaster upon arrival. First come, first serve.

RECIPROCAL BOAT CAPACITY: ....1 Boat
DOCKSIDE DEPTH AT ZERO TIDE: ....20 ft.
RECIPROCAL SEASON: ....All year
RESERVATION POLICY: ....None
TOILETS: ....At Marina
HOT SHOWERS: ....At Marina
RESTAURANT: ....Close by
DAILY RATE: ....No charge

RECIPROCAL DOCK: ....Slips only
RECIPROCAL SLIPS: ....1 Siip
WATER: ....Yes
AMT W/ELECTRICITY: ....All
AMPS: ....20-30 A
BAR: ....Close by
OTHER: ....Same as marina listing

***NOTE:* THIS IS PRIVATE MOORAGE *ONLY* AVAILABLE TO MEMBERS OF RECIPROCAL YACHT CLUBS! YOUR CLUB *MUST* HAVE RECIPROCAL PRIVILEGES AND YOU MUST FLY YOUR BURGEE!**

Bainbridge Island

CAUTION! This chartlet not intended for use in navigation.

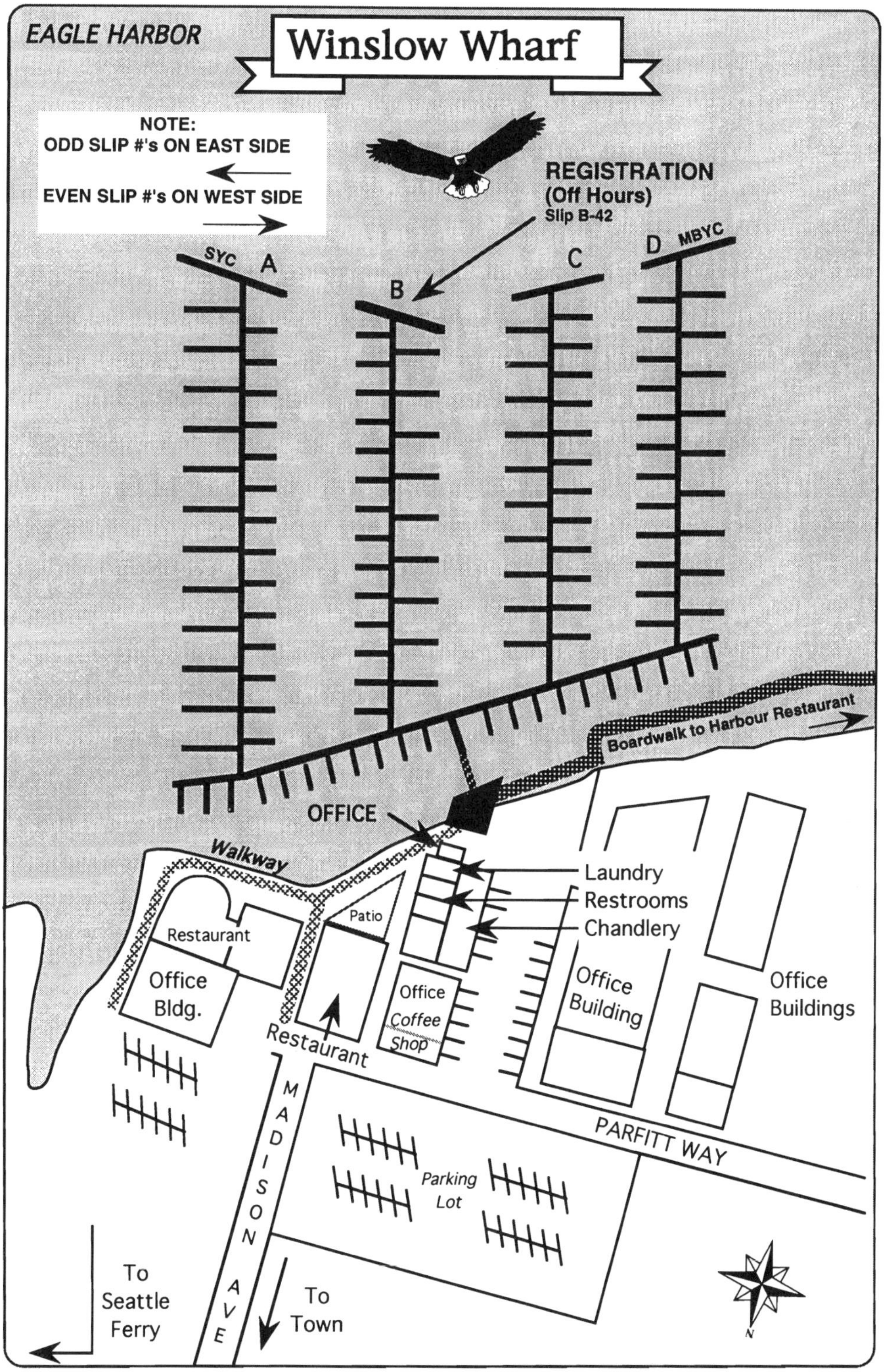

EAGLE HARBOR

Winslow Wharf

NOTE:
ODD SLIP #'s ON EAST SIDE
EVEN SLIP #'s ON WEST SIDE

REGISTRATION
(Off Hours)
Slip B-42

SYC A
B
C
D MBYC

Boardwalk to Harbour Restaurant

OFFICE

Walkway

Laundry
Restrooms
Chandlery

Patio

Restaurant

Office Bldg.

Restaurant

Office
Coffee Shop

Office Building

Office Buildings

PARFITT WAY

MADISON AVE

Parking Lot

To Seattle Ferry

To Town

N

Bainbridge Island

NAME OF MARINA: ***HARBOUR MARINA & PUB*** RADIO: None
TELEPHONE: **206-842-6502** MGR: Dockmaster 206-550-5340
E-MAIL: info@harbourpub.com FAX: 206-842-5047 (Fax)
ADDRESS: 233 Parfitt Way Bainbridge Island, WA 98110
SHORT DESCRIPTION & LOCATION: **www.harbourpub.com**

47°37.20' - 122°31.25' Located on the N. side of Eagle Harbor which indents on the E. shore of Bainbridge Is. appx. 1 miles W. of State Ferry Shipyard. Overnight guest moorage is available by using open slips of permanent tenants. Call ahead for slip.

GUEST BOAT CAPACITY: .... Varies
DOCKSIDE DEPTH AT ZERO TIDE: ...... 8-10 ft.
SEASON: .... All year
RESERVATION POLICY: .... Accepts
AMT W/ELECTRICITY: .... All
FUEL DOCK: .... None
MARINE REPAIRS: .... None
TOILETS: .... Yes
HOT SHOWERS: .... Yes
RESTAURANT: .... Yes
PICNIC AREA: .... Yes
BASIC STORE: .... Close by
BROADBAND/WI-FI: .... BroadbandXpress
DAILY RATE: .... Moderate (75¢-$1.25/foot)

GUEST DOCK: ...... 60 ft. plus slips
GUEST SLIPS: .... Varies
WATER: .... Yes
AMPS: .... 30 A
PUMP OUT STATION .... Yes
HAUL OUT: .... None
BOAT RAMP: .... Close by
LAUNDRY: .... Yes
BAR: .... Yes
POOL: .... None
GOLF: .... Close by
PET FRIENDLY: .... Good
OTHER: Directly below Harbour Public House, an English style pub with good food and spirits. <u>Minors not allowed in pub.</u>

CAUTION! This chartlet not intended for use in navigation.

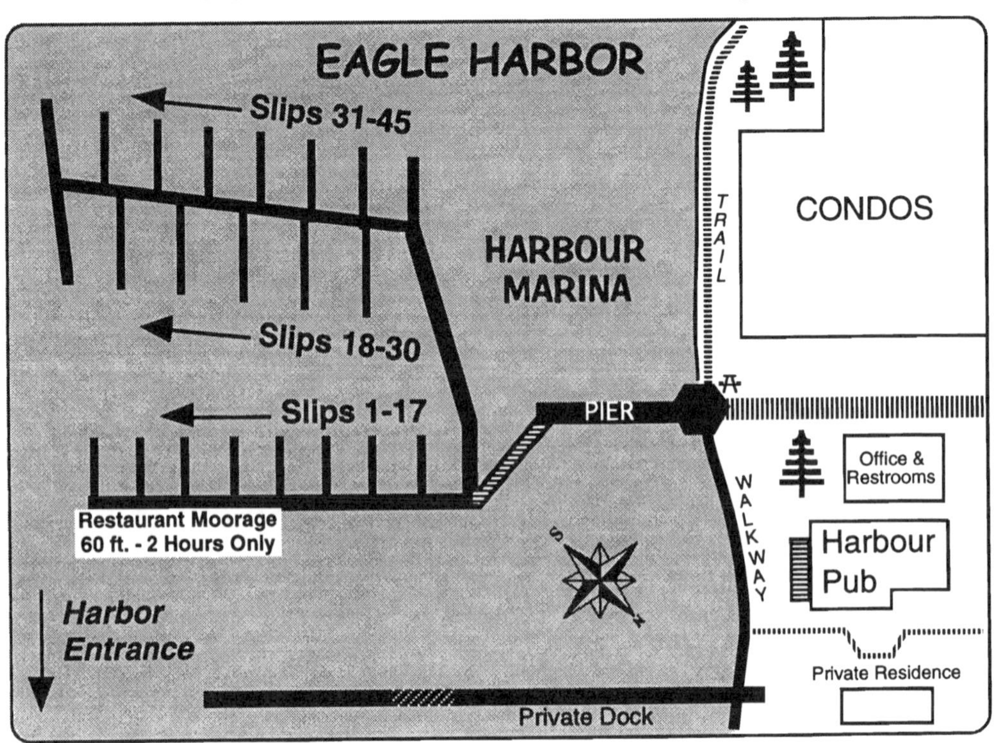

Bainbridge Island

NAME OF YACHT CLUB: *PORT MADISON YACHT CLUB*

ADDRESS: P.O. Box 10002 Bainbridge Island, WA 98110

TELEPHONE: 206-842-2102/3484 PERSON IN CHARGE: Reciprocal Chair

SHORT DESCRIPTION & LOCATION: **http://portmadisonyc.org**

47°42.30' - 122°31.50' Located on N end of Bainbridge Is. inside of Port Madison harbor appx. half way in the bay immediately west of Seattle Y.C. outstation. To approach guest dock go around west end of docks to long dock inside finger piers. Please register in the clubhouse. **The electrical outlets are not adequate for shore power.**

RECIPROCAL BOAT CAPACITY: ....Appx. 2 or 3

DOCKSIDE DEPTH AT ZERO TIDE: ....6 ft.

SEASON: ....All year

RESERVATION POLICY: ....None

TOILETS: ....Yes

HOT SHOWERS: ....None

RESTAURANT: ....None

BROADBAND/WI-FI: ....None

DAILY RATE: ....No Charge for 48 hours.

**Electricity day time use of power tools only. Kitchen in Club.

RECIPROCAL DOCK: ..Appx. 100'

RECIPROCAL SLIPS: ...Dock only

WATER: ....Yes

AMT W/ELECTRICITY: **Limited

AMPS: ....None

BAR: ....None

PET FRIENDLY: ....Good

OTHER: Closest services are Rolling Bay 4 miles & Winslow 8 miles.

***NOTE:* THIS IS PRIVATE MOORAGE AND ONLY AVAILABLE TO MEMBERS OF RECIPROCAL YACHT CLUBS. YOUR CLUB MUST HAVE RECIPROCAL PRIVILEGES AND YOU MUST FLY YOUR BURGEE.**

CAUTION! This chartlet not intended for use in navigation.

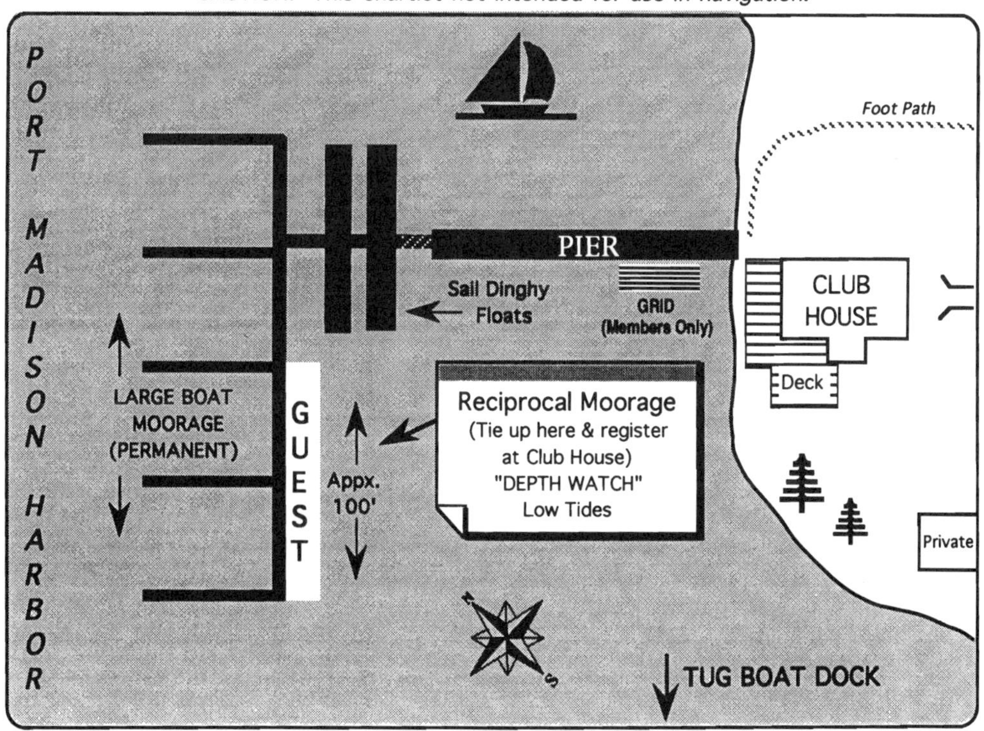

Seattle

NAME OF YACHT CLUB: *DUWAMISH YACHT CLUB*

ADDRESS: 1801 South 93rd St. Seattle, WA 98108

TELEPHONE: 206-767-9330 PERSON IN CHARGE: Marina Manager

SHORT DESCRIPTION & LOCATION:

47°31.02' - 122°18.50' Located in the Duwamish River 40 minutes cruising time from Elliott Bay, 1/2 mile past the 14th Ave. bridge. Vacant member slips are available for boats **up to 50 ft**. Small shopping center w/in walking distance. Best to call ahead for moorage arrangements to club office 206-767-9330.

RECIPROCAL BOAT CAPACITY: ....Appx . 2-4

DOCKSIDE DEPTH AT ZERO TIDE: ....10 ft.

SEASON: ....All year

RESERVATION POLICY: ....Recommended

TOILETS: ....Yes

HOT SHOWERS: ....Yes

RESTAURANT: ....Close by

BROADBAND/WI-FI: ....None

DAILY RATE: ....No charge for 3 nights in a 30 day period. **Power is $3.00 per night.**

RECIPROCAL DOCK: ......45-50 ft.

RECIPROCAL SLIPS: ....Varies

WATER: ....Yes

AMT W/ELECTRICITY: ....All

AMPS: ....30 A

BAR: ....Close by

PET FRIENDLY: ....Good

OTHER: Clubhouse w/showers & laundry, pump out station.

***NOTE:* THIS IS PRIVATE MOORAGE AND ONLY AVAILABLE TO MEMBERS OF RECIPROCAL YACHT CLUBS. YOUR CLUB *MUST* HAVE RECIPROCAL PRIVILEGES AND YOU MUST FLY YOUR BURGEE.**

CAUTION! This chartlet not intended for use in navigation.

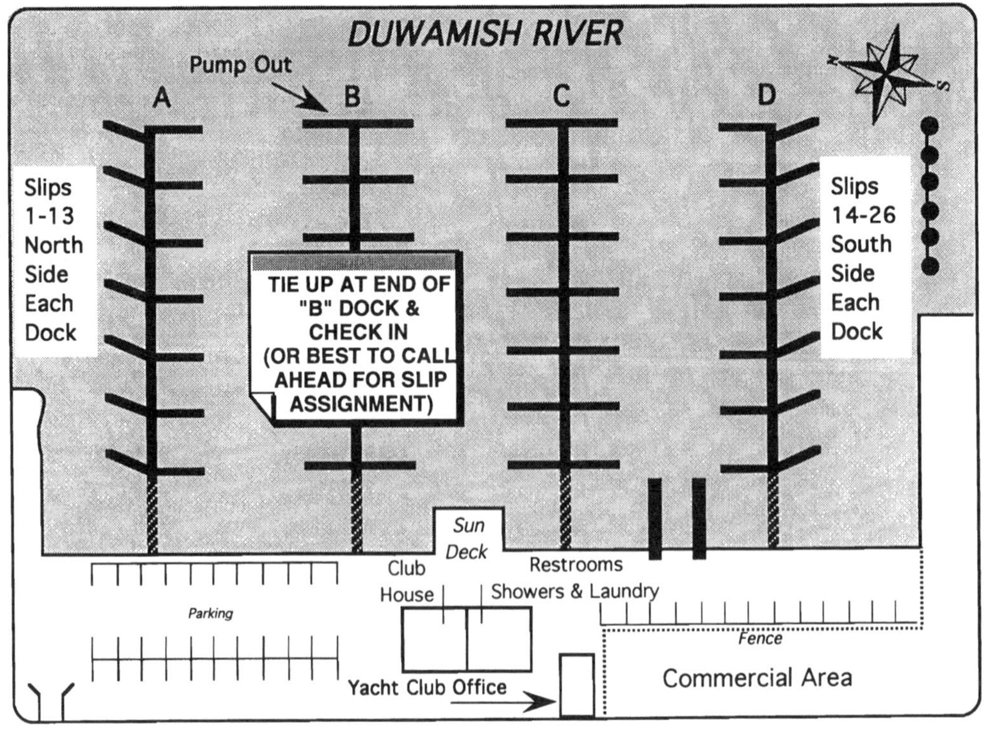

Seattle

NAME OF MARINA: *BELL HARBOR MARINA* RADIO: VHF Ch.66A
TELEPHONE: 206-615-3952 MGR: Sharon Briggs
E-MAIL: bhm@portseattle.org FAX: 206-615-3965 (Fax)
ADDRESS: Port of Seattle, POB 1209, Seattle, WA 98111
SHORT DESCRIPTION & LOCATION: **www.portseattle.org**

47°36.6'-122°20.9' The modern Bell Harbor Marina makes cruising to downtown Seattle a great destination. The marina accents & complements an entire waterfront community w/ new Cruise Ship Terminal & Conference Center. Friendly & professional staff 24 hrs/day.

GUEST BOAT CAPACITY: ....Max. 70 boats
DOCKSIDE DEPTH AT ZERO TIDE: 15-30 ft.
SEASON: ....All year
RESERVATION POLICY: ....Accepts
AMT W/ELECTRICITY: ....All
FUEL DOCK: ....Close by
MARINE REPAIRS: ....Close by
TOILETS: ....Yes
HOT SHOWERS: ....Yes
RESTAURANT: ....Yes
PICNIC AREA: ....Yes
BASIC STORE: ....Yes
BROADBAND/WI-FI: ....None
DAILY RATE:................Premium (Over $1.25/foot)
CHECK W/MARINA FOR WINTER RATES

GUEST DOCK: 125 ft. plus slips
GUEST SLIPS: ....Appx. 35
WATER: ....Yes
AMPS: ....30-50-100 A
PUMP OUT STATION ....Yes
HAUL OUT: ....None
BOAT RAMP: ....None
LAUNDRY: ....None
BAR: ....Yes
POOL: ....None
GOLF: ....None
PET FRIENDLY: ....Fair
OTHER: 24 Hr. security, Odyssey Maritime Discovery Center, walking distance to Aquarium, Pike Place Mkt., & downtown Seattle.

CAUTION! This chartlet not intended for use in navigation.

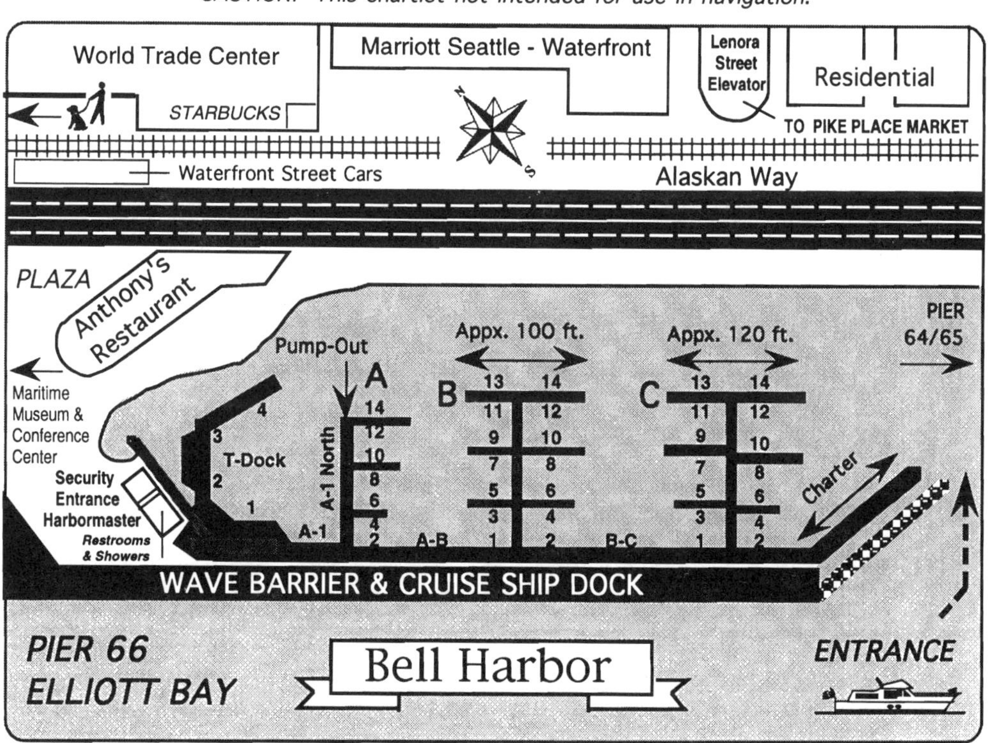

Seattle

NAME OF MARINA: ***ELLIOTT BAY MARINA*** **RADIO:** VHF 78A
TELEPHONE: **206-285-4817** **MGR:** Dan Park
E-MAIL: info@elliottbaymarina.net **FAX:** 206-282-0626 (Fax)
ADDRESS: 2601 West Marina Place Seattle, WA 98199
SHORT DESCRIPTION & LOCATION: **www.elliottbaymarina.net**

47°37.80' - 122°23.80' One of Seattle's newest marinas offers spacious guest moorage w/complete amenities. Located on N side of Elliott Bay at base of Magnolia Bluff about 3 mi. N of downtown Seattle. *Call ahead via VHF or land line for slip assignment.*

GUEST BOAT CAPACITY: ....Appx. 100 boats
DOCKSIDE DEPTH AT ZERO TIDE: ....30 ft.
SEASON: ....All year
RESERVATION POLICY: ....Accepts
AMT W/ELECTRICITY: ....All
FUEL DOCK: ....Gas & Diesel
MARINE REPAIRS: ....On premises
TOILETS: ....Yes
HOT SHOWERS: ....Yes (Free)
RESTAURANT: ....Yes - 3
PICNIC AREA: ....Yes
BASIC STORE: ....Yes
BROADBAND/WI-FI: ....BroadbandXpress
DAILY RATE:..........Moderate (75¢-$1.25/foot)
-Premium $1.50/ft for boats 65 ft. +

GUEST DOCK: ....400' plus slips
GUEST SLIPS: ....Appx. 60
WATER: ....Yes
AMPS: ....30-50 A & 3 Phase
PUMP OUT STATION ....Yes
HAUL OUT: ....Close by
BOAT RAMP: ....None
LAUNDRY: ....Yes
BAR: ....Yes
POOL: ....Close-by
GOLF: ....Close by
PET FRIENDLY: ....Excellent
OTHER: 24 Hr. security, cable TV, diving services, concierge, line handlers upon request. Walking & biking paths, park-like setting.

CAUTION! This chartlet not intended for use in navigation.

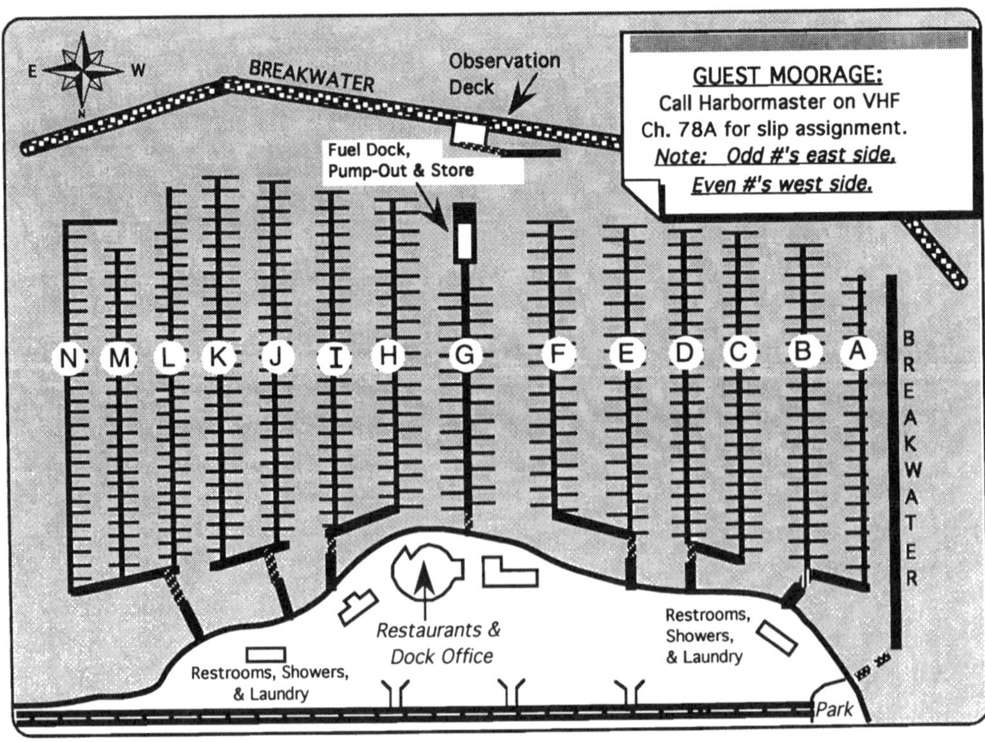

Seattle

NAME OF YACHT CLUB: *SHILSHOLE BAY YACHT CLUB*

ADDRESS: 2442 NW Market St.Box 98 Seattle, WA 98107

TELEPHONE: PERSON IN CHARGE: Commodore

SHORT DESCRIPTION & LOCATION: **www.shilsholebayyc.org**

SBYC offers reciprocal moorage located at the **Shilshole Bay Marina**. Visiting yacht club members should call ahead to marina office for berth assginment. Upon registration, request SBYC reciprocal moorage form & pay the moorage fee. Please mail forms to SBYC for reimbursement w/copy of current yacht club membership card & moorage receipt.

RECIPROCAL BOAT CAPACITY....................Varies

DOCKSIDE DEPTH AT ZERO TIDE: Per Marina

SEASON: ....All year

RESERVATION POLICY: ....None

TOILETS: ....At Shilshole Bay Marina

HOT SHOWERS: ....At Shilshole Bay Marina

RESTAURANT: ....Close by

BROADBAND/WI-FI: ....BroadbandXpress

DAILY RATE: Electricity charge per Marina Rate. 1 Night stay per boat per Calendar Year.

RECIPROCAL DOCK: ....At marina

RECIPROCAL SLIPS: ...At marina

WATER: ....Yes

AMT W/ELECTRICITY: ....All

AMPS: ....30 A

BAR: ....Close by

PET FRIENDLY: ....Excellent

OTHER: Services same as Shilshole Bay Marina listing on Page 82 & 83.

***NOTE:* THIS IS PRIVATE MOORAGE AND ONLY AVAILABLE TO MEMBERS OF RECIPROCAL YACHT CLUBS. YOUR CLUB *MUST* HAVE RECIPROCAL PRIVILEGES AND YOU MUST FLY YOUR BURGEE.**

SEE DATA & CHARTLET ON PAGE: 82 & 83

NOTES

Seattle

NAME OF MARINA: ***SHILSHOLE BAY MARINA*** **RADIO:** VHF Ch.17
TELEPHONE: **206-601-4089** or 728-3006 **MGR:** Kathy Goodman/Sharon Briggs
E-MAIL: sbm@portseattle.org **FAX:** 206-728-3391 (Fax)
ADDRESS: 7001 Seaview Ave. N.W. Seattle, WA 98117
SHORT DESCRIPTION & LOCATION: **www.portseattle.org/seaport/marinas**
47°42.00' - 122°24.25' Shilshole is one of the largest marinas on the Coast & located between Meadow Pt. & West Pt. just N. of where Ballard Locks are entered. Full services abound for boaters. During 2006-2008 re-construction period, guest moorage is limited.

GUEST BOAT CAPACITY: ....Varies
DOCKSIDE DEPTH AT ZERO TIDE: ....15 ft.
SEASON: ....All year
RESERVATION POLICY: ....Accepts
AMT W/ELECTRICITY: ....All
FUEL DOCK: ....Gas & Diesel
MARINE REPAIRS: ....Yes
TOILETS: ....Yes
HOT SHOWERS: ....Yes
RESTAURANT: ....Close by
PICNIC AREA: ....Close by
BASIC STORE: ....Yes - At Fuel Dock
BROADBAND/WI-FI: ....BroadbandXpress
DAILY RATE: Moderate (75¢-$1.25/foot-Winter)
Premium (Over $1.25/foot-Summer)
Touch & Go - 25¢/foot

GUEST DOCK: ....Varies
GUEST SLIPS: ....Varies
WATER: ....Yes
AMPS: ....20-100 A
PUMP OUT STATION ....Yes
HAUL OUT: ....Yes
BOAT RAMP: ....Yes
LAUNDRY: ....Yes
BAR: ....Close by
POOL: ....None
GOLF: ....Close by
PET FRIENDLY: ....Excellent
OTHER: Restaurants & night life close by, marine chandlery, boatyard. Garbage, recycling, haz. waste & oil stations. Sailing schools.

NAME OF YACHT CLUB: ***CORINTHIAN YACHT CLUB OF SEATTLE*** (B)
CLUB ADDRESS: 7755 Seaview Ave. NW, Seattle, WA 98117
CLUB TELEPHONE: 206-789-1919 **PERSON IN CHARGE:** Chas. Fawcett, Mgr
LOCATION & SPECIAL NOTES: **http://cycseattle.org**
CYC will reimburse reciprocal club members for guest moorage at Shilshole Bay Marina for up to 2 days moorage. Hail the Harbormaster & ask for guest berth. Request CYC Reciprocal Form and submit for reimbursement per instructions. Drop off reimbursement envelope at club office or mail. Guest moorage limited during marina re-construction.

RECIPROCAL BOAT CAPACITY: ....Appx. 2-3
DOCKSIDE DEPTH AT ZERO TIDE: ....15 ft.
RECIPROCAL SEASON: ....All year
RESERVATION POLICY: ....None
TOILETS: ....At Marina
HOT SHOWERS: ....At Marina
RESTAURANT: ....Close by
DAILY RATE: ....No charge
48 HOUR LIMIT

RECIPROCAL DOCK: ....Varies
RECIPROCAL SLIPS: ....Varies
WATER: ....Yes
AMT W/ELECTRICITY: ....All
AMPS: ....30 A
BAR: ....Close by
OTHER: ....Same as marina listing

***NOTE:* THIS IS PRIVATE MOORAGE *ONLY* AVAILABLE TO MEMBERS OF RECIPROCAL YACHT CLUBS! YOUR CLUB *MUST* HAVE RECIPROCAL PRIVILEGES AND YOU MUST FLY YOUR BURGEE!**

Seattle

CAUTION! This chartlet not intended for use in navigation.

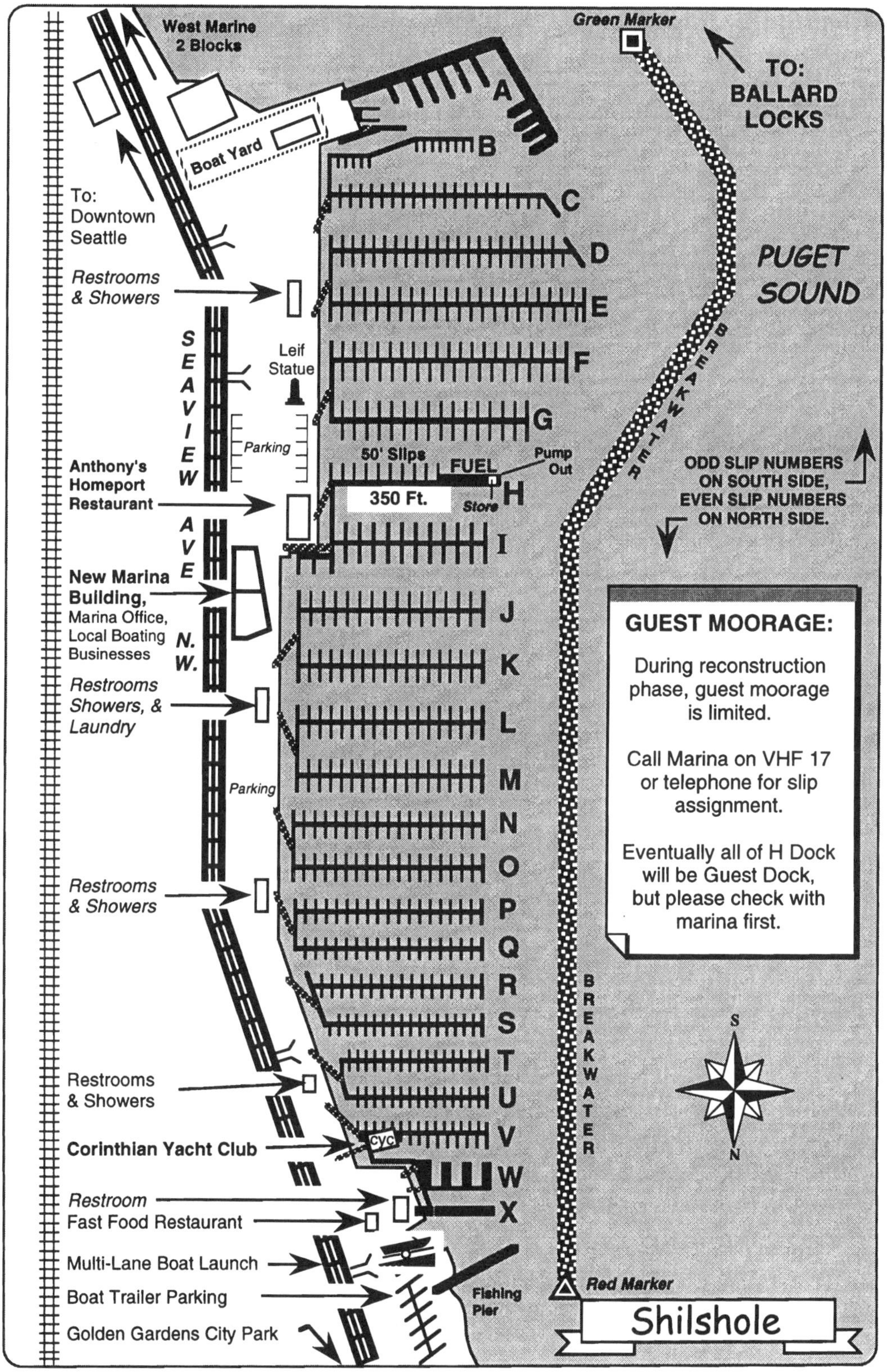

West Marine
2 Blocks
Green Marker
TO:
BALLARD
LOCKS
Boat Yard
A
B
C
D
E
F
G
H
I
J
K
L
M
N
O
P
Q
R
S
T
U
V
W
X
To:
Downtown
Seattle
Restrooms
& Showers
SEAVIEW AVE N.W.
Leif
Statue
Parking
50' Slips
FUEL
Pump
Out
350 Ft.
Store
PUGET
SOUND
BREAKWATER
ODD SLIP NUMBERS
ON SOUTH SIDE,
EVEN SLIP NUMBERS
ON NORTH SIDE.
Anthony's
Homeport
Restaurant
New Marina
Building,
Marina Office,
Local Boating
Businesses
Restrooms
Showers, &
Laundry
Parking
Restrooms
& Showers
Restrooms
& Showers
Corinthian Yacht Club
CYC
Restroom
Fast Food Restaurant
Multi-Lane Boat Launch
Boat Trailer Parking
Golden Gardens City Park
Fishing
Pier
Red Marker
S
N
Shilshole

GUEST MOORAGE:

During reconstruction phase, guest moorage is limited.

Call Marina on VHF 17 or telephone for slip assignment.

Eventually all of H Dock will be Guest Dock, but please check with marina first.

Seattle

NAME OF MARINA: ***FISHERMEN'S TERMINAL*** **RADIO:** VHF Ch. 17
TELEPHONE: **206-728-3395** **MGR:** Kenneth Lyles/Ross Perry
E-MAIL: ft@portseattle.org **FAX:** 206-728-3393 (Fax)
ADDRESS: 3919 18th Ave. W. Seattle WA 98199
SHORT DESCRIPTION & LOCATION: **www.portseattle.org/seaport**

47°39.40' - 122°22.80' Located in Salmon Bay near Ballard, this is home to N. Pacific fishing fleet. Pleasure boats are welcome. Finger piers under constructions. Boats may have to either bow or stern-in to sawtooth slip & tie off piling. Abundant services in area.

GUEST BOAT CAPACITY: ....Appx 25 boats
DOCKSIDE DEPTH: ....10 ft. plus
SEASON: ....All year
RESERVATION POLICY: ....None - Usually room
AMT W/ELECTRICITY: ....All
FUEL DOCK: ....Close by
MARINE REPAIRS: ....On premises
TOILETS: ....Yes
HOT SHOWERS: ....Yes
RESTAURANT: ....3 Restaurants
PICNIC AREA: ....None
BASIC STORE: ....Yes
BROADBAND/WI-FI: ....None
DAILY RATE: ....Moderate (75¢-$1.25/foot)

GUEST DOCK: ....Varies
GUEST SLIPS: ....Varies
WATER: ....Yes
AMPS: ....15-20 A
PUMP OUT STATION ....Yes
HAUL OUT: ....Close by
BOAT RAMP: ....Close by
LAUNDRY: ....Yes
BAR: ....Highliner Pub
POOL: ....None
GOLF: ....Close by
PET FRIENDLY: ....Fair
OTHER: Major repairs & supplies on site or nearby, public transp. to Seattle city amenities. Home to Fishermen's Memorial.

CAUTION! This chartlet not intended for use in navigation.

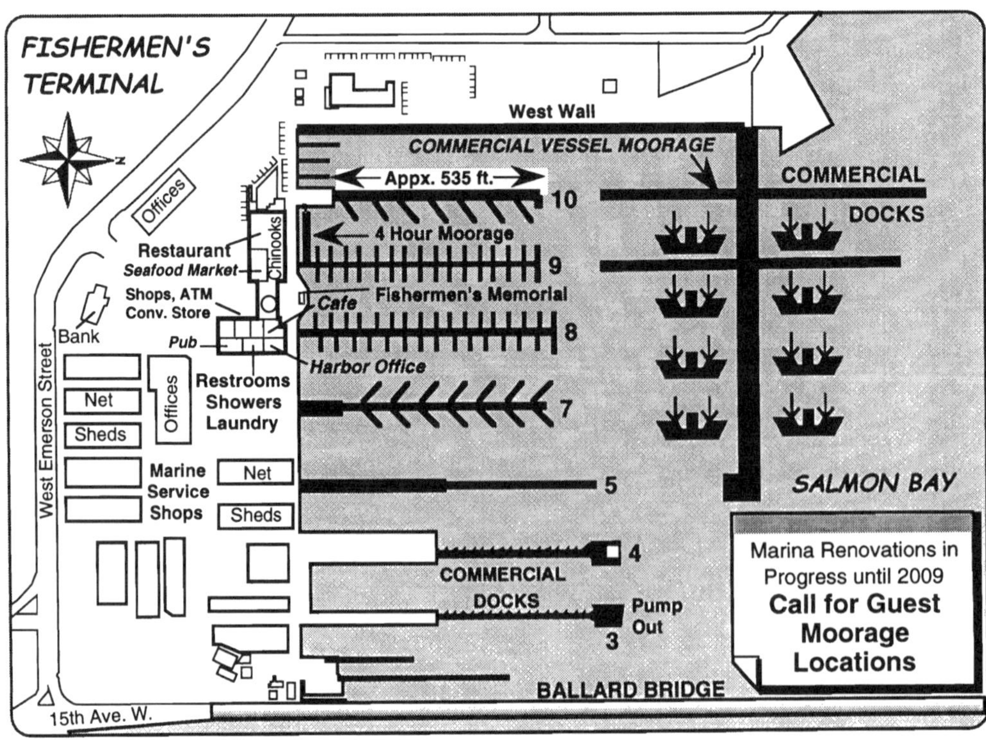

Seattle

NAME OF YACHT CLUB: *HIDDEN HARBOR YACHT CLUB*

ADDRESS: 3190 Nickerson St. #194 Seattle, WA 98109

TELEPHONE: 206-216-3440 PERSON IN CHARGE: Laurie Way

SHORT DESCRIPTION & LOCATION:

47°37.85' - 122°20.42' Reciprocal moorage located on SW side of Lake Union next to the Rock Salt Steak House (1232 Westlake Ave. N.) Dock space consists of a 65 ft. dock directly north of restaurant. Upon arrival, please register at restaurant. No gate key required. Please be considerate so more than one boat can tie up.

RECIPROCAL BOAT CAPACITY ....1 or 2 boats

DOCKSIDE DEPTH: ....Appx. 10 ft.

SEASON: ....All year

RESERVATION POLICY: ....None

TOILETS: ....None

HOT SHOWERS: ....None

RESTAURANT: ....Yes

BROADBAND/WI-FI: ....None

DAILY RATE: ....No Charge for 24 Hours
CHECK OUT TIME IS 12 NOON

RECIPROCAL DOCK: Appx. 65 Ft.

RECIPROCAL SLIPS: ...Dock only

WATER: ....No

AMT W/ELECTRICITY: ....None

AMPS: ....None

BAR: ....Yes

PET FRIENDLY: ....Challenging

OTHER: Walk to several restaurants & shops. Public transp. to Seattle sights.

***NOTE:* THIS IS PRIVATE MOORAGE AND ONLY AVAILABLE TO MEMBERS OF RECIPROCAL YACHT CLUBS. YOUR CLUB *MUST* HAVE RECIPROCAL PRIVILEGES AND YOU MUST FLY YOUR BURGEE.**

CAUTION! This chartlet not intended for use in navigation.

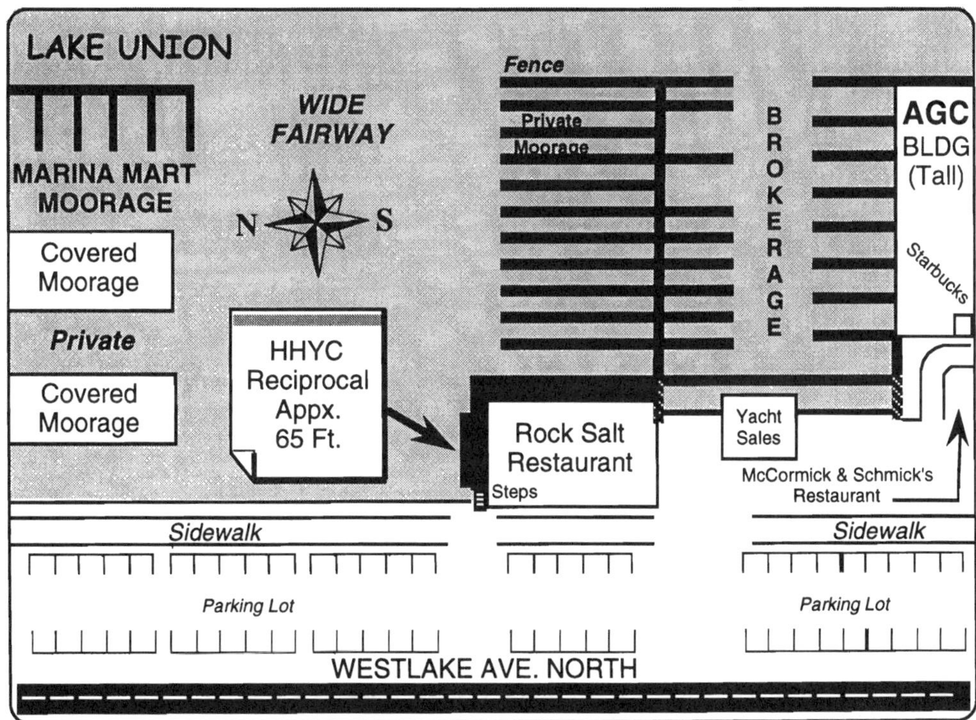

Seattle

NAME OF MARINA: ***FAIRVIEW MARINA*** **RADIO:** None
TELEPHONE: 1-888-673-1118 MGR: Steven Agnew
E-MAIL: seattleboatslips@hotmail.com FAX: None
ADDRESS: 1109 Fairview Ave. N., POB 3955 Seattle, WA 98124

SHORT DESCRIPTION & LOCATION:

47°37.90' - 122°19.85' Located on SE side of Lake Union about 1/4 mi. NE of Chandler's Cove, in between Yale St. & YachtFish Boat Yard. Marina has no designated guest moorage, but may have a vacant slip for short term moorage. Call ahead to check.

GUEST BOAT CAPACITY: ....Varies
DOCKSIDE DEPTH: ....Appx. 15-20 ft.
SEASON: ....All year
RESERVATION POLICY: ....Mandatory
AMT W/ELECTRICITY: ....All
FUEL DOCK: ....Close by
MARINE REPAIRS: ....Close by
TOILETS: ....Yes
HOT SHOWERS: ....Yes
RESTAURANT: ....Close by
PICNIC AREA: ....None
BASIC STORE: ....Close by
BROADBAND/WI-FI: ....None
DAILY RATE: ....Moderate (75¢-$1.25/foot)

GUEST DOCK: ....None
GUEST SLIPS: ....Varies
WATER: ....Yes
AMPS: ....30-50 A
PUMP OUT STATION ...Close by
HAUL OUT: Travel-Lift next door
BOAT RAMP: ....None
LAUNDRY: ....None
BAR: ....Close by
POOL: ....None
GOLF: ....None
PET FRIENDLY: ....Good
OTHER: Walking distance to Chandler's Cove amenities, shops, restaurants, chandleries.

CAUTION! This chartlet not intended for use in navigation.

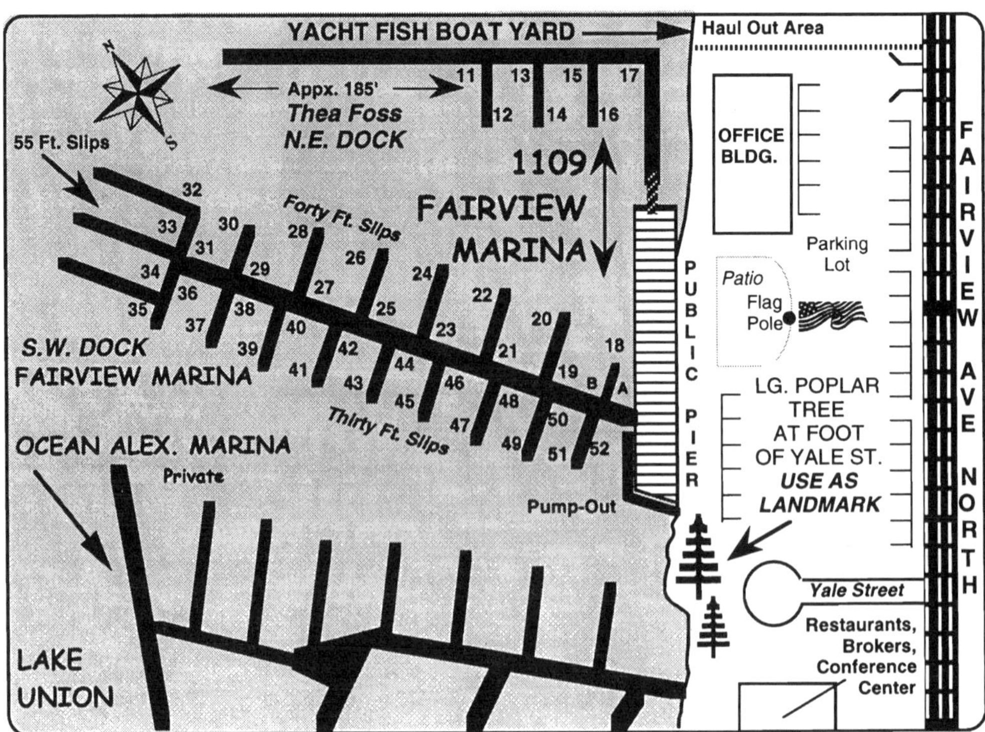

Seattle

NAME OF YACHT CLUB: *TYEE YACHT CLUB*

ADDRESS: P.O. Box 17036 Seattle WA 98127

TELEPHONE: PERSON IN CHARGE: Commodore

SHORT DESCRIPTION & LOCATION: **www.tyeeyachtclub.org**

47°39.10' - 122°19.30' Nice yacht club facility conveniently located on E. side of Lake Union almost directly under I-5 bridge. Close to restaurants and marine services. Gates to street are locked at all times. Gate keys located in locked box at foot of stairs. Please see your home club reciprocal agreement letter for combination to the lock box.

RECIPROCAL BOAT CAPACITY............Appx 6-10

DOCKSIDE DEPTH:..................................Appx. 10 ft.

SEASON: ....All year

RESERVATION POLICY: ....None

TOILETS: ....None

HOT SHOWERS: ....None

RESTAURANT: ....Close by

BROADBAND/WI-FI: ....None

DAILY RATE: ....No Charge

THREE DAY LIMIT

RECIPROCAL DOCK: Appx. 100'

RECIPROCAL SLIPS: Dock only

WATER: ....Yes

AMT W/ELECTRICITY: ....All

AMPS: ....15-30 A

BAR: ....Close by

PET FRIENDLY: ....Good

OTHER: Rafting permitted. Several excellent restaurants within 2 blocks. U District shopping within a mile.

***NOTE:* THIS IS PRIVATE MOORAGE AND ONLY AVAILABLE TO MEMBERS OF RECIPROCAL YACHT CLUBS. YOUR CLUB *MUST* HAVE RECIPROCAL PRIVILEGES AND YOU MUST FLY YOUR BURGEE.**

CAUTION! This chartlet not intended for use in navigation.

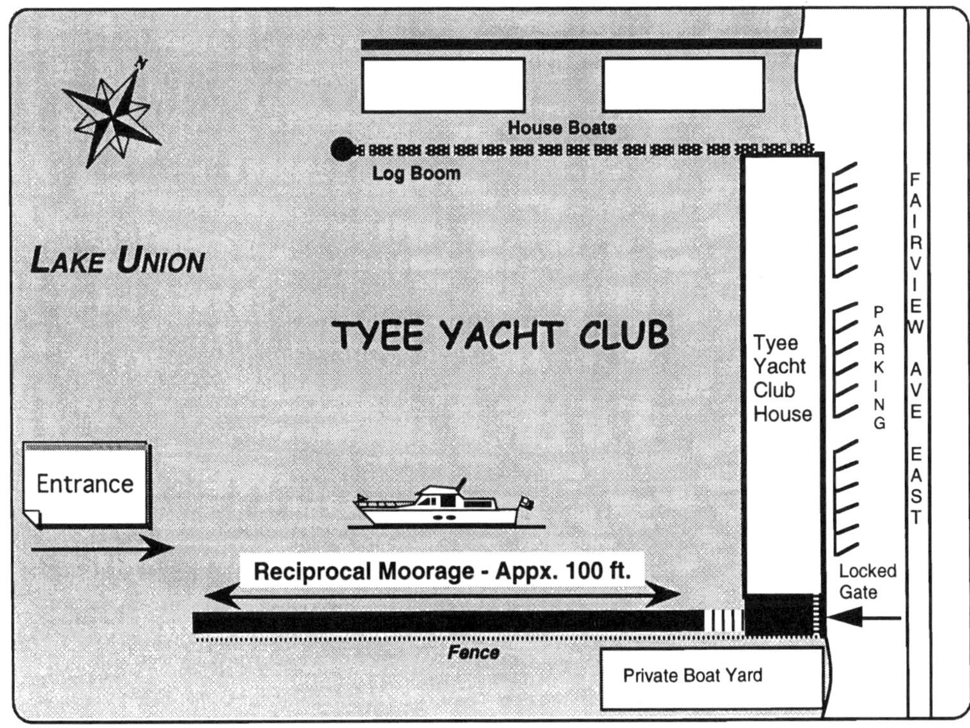

Seattle

NAME OF YACHT CLUB: *PUGET SOUND YACHT CLUB*

ADDRESS: 2321 N Northlake Way Seattle, WA 98103

TELEPHONE: 206-634-3733 **PERSON IN CHARGE:** Reciprocal Chair

SHORT DESCRIPTION & LOCATION: **www.pugetsoundyc.org**
47°39.00' -122°19.85' PSYC facility located on N. shore of Lake Union, E. of Aurora bridge, W. of the I-5 bridge & next to Dunato's Boat Yard. Reciprocal moorage is on west end of pier, marked by red cap rail. No rafting. Register at Kiosk next to flag pole for gate code. Add'l moorage (if avail) assigned by Dockmaster.

RECIPROCAL BOAT CAPACITY.....Appx. 2 boats
DOCKSIDE DEPTH:.....30 Ft.
SEASON: All yr. except holidays & opening day
RESERVATION POLICY:.....None
TOILETS:.....Yes
HOT SHOWERS:.....Yes
RESTAURANT:.....Close by
BROADBAND/WI-FI:.....None
DAILY RATE: $3/day Elec. charge. 2 Days N/C. Max 2 boats visiting from same YC. Limit 4 days/4 mos. - 12 days/yr.

RECIPROCAL DOCK:.....60 Ft.
RECIPROCAL SLIPS: ...Dock only
WATER:.....Yes
AMT W/ELECTRICITY:.....All
AMPS:.....30 A
BAR:.....Close by
PET FRIENDLY:.....Good
OTHER: Close to many marine services and chandleries on N. side of Lake Union.

***NOTE:* THIS IS PRIVATE MOORAGE AND ONLY AVAILABLE TO MEMBERS OF RECIPROCAL YACHT CLUBS. YOUR CLUB *MUST* HAVE RECIPROCAL PRIVILEGES AND YOU MUST FLY YOUR BURGEE.**

CAUTION! This chartlet not intended for use in navigation.

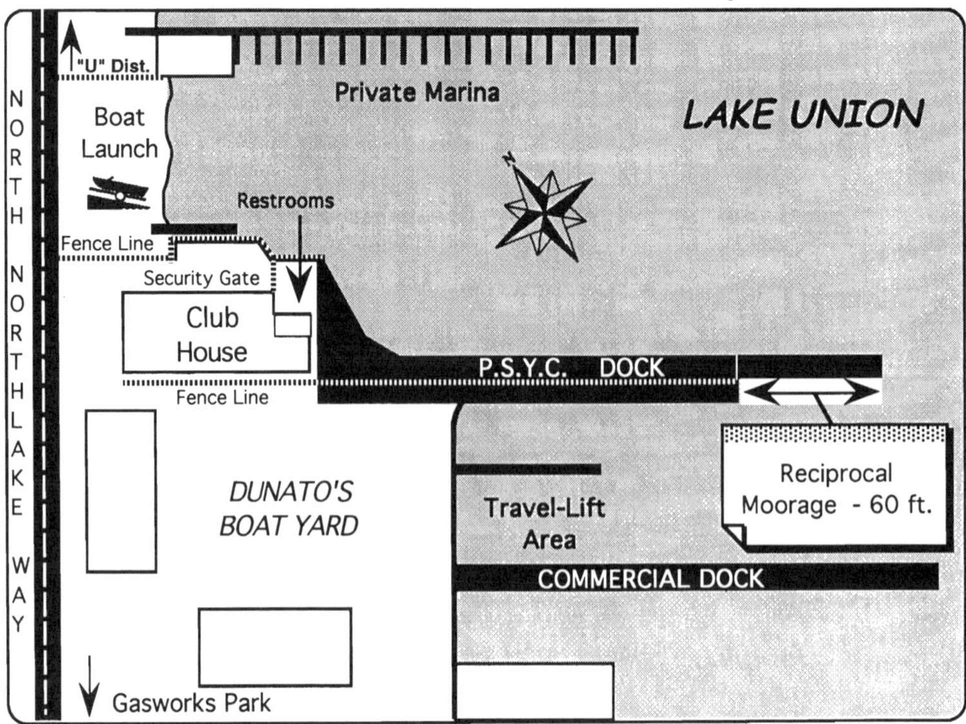

Seattle

NAME OF YACHT CLUB: *QUEEN CITY YACHT CLUB*

ADDRESS: 2608 Boyer Ave. E. Seattle WA 98102

TELEPHONE: 206-323-9602 PERSON IN CHARGE: Caretaker on duty

SHORT DESCRIPTION & LOCATION: **www.queencity.org**

47°38.60' - 122°18.90' QCYC is located on the S. end of the W. shore of Portage Bay. The reciprocal moorage is on the S.E. side of Dock 3 at the outer end. 5 Day reciprocal moorage maximum & skipper (or family) must be on board overnight. Clubhouse is open 0700-1800 daily and frequently in the evening for scheduled events.

RECIPROCAL BOAT CAPACITY: Varies
DOCKSIDE DEPTH: Appx. 4-6 ft.
SEASON: All year
RESERVATION POLICY: None
TOILETS: Yes
HOT SHOWERS: Yes
RESTAURANT: Close by
BROADBAND/WI-FI: BroadbandXpress
DAILY RATE: $3.00/Day - Electrical Hookup. Moorage no charge/5 days max per month. Get gate code from Club.

RECIPROCAL DOCK: 150 ft.
RECIPROCAL SLIPS: Varies
WATER: Yes
AMT W/ELECTRICITY: All
AMPS: 30 A
BAR: Close by
PET FRIENDLY: Fair
OTHER: Small grocery store 3 blocks west, city bus service at front door.

***NOTE:* THIS IS PRIVATE MOORAGE AND ONLY AVAILABLE TO MEMBERS OF RECIPROCAL YACHT CLUBS. YOUR CLUB *MUST* HAVE RECIPROCAL PRIVILEGES AND YOU MUST FLY YOUR BURGEE.**

CAUTION! This chartlet not intended for use in navigation.

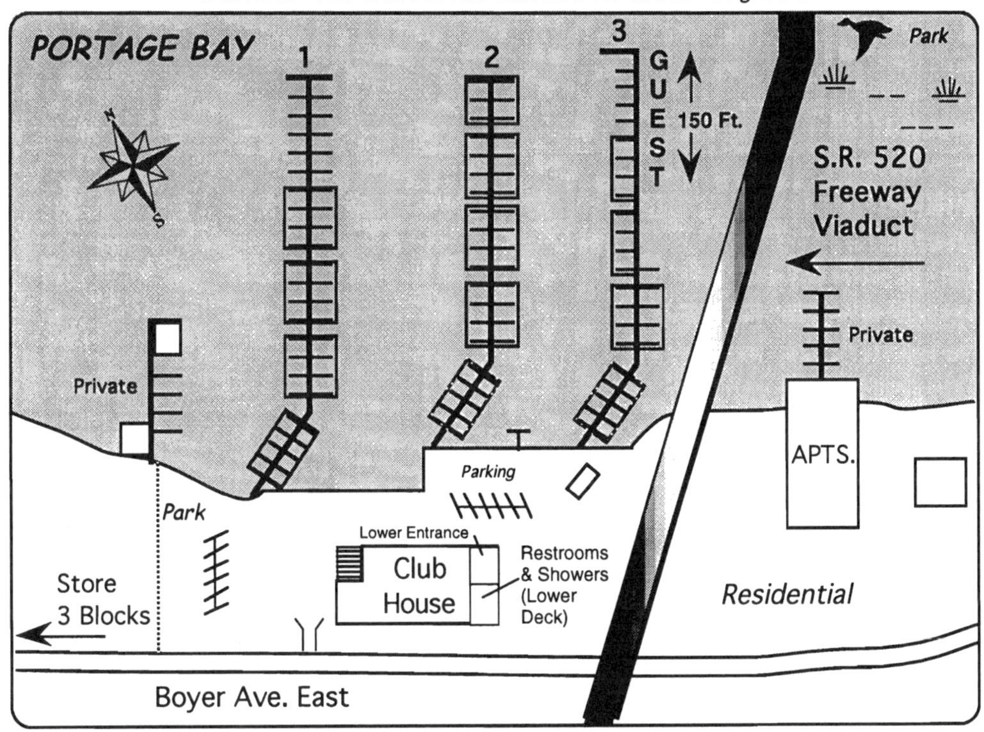

Seattle

NAME OF YACHT CLUB: *SEATTLE YACHT CLUB*
ADDRESS: 1807 E. Hamlin Seattle WA 98112
TELEPHONE: 206-325-1000 PERSON IN CHARGE: Manager
SHORT DESCRIPTION & LOCATION: **www.seattleyachtclub.org**
47°38.73' - 122°18.43' SYC is located on the west. side of Portage Bay near the entrance to Portage Cut. This *historic and stately facility* offers reciprocal amenities, including both fine & casual dining. Reciprocal moorage is on Dock (Pier) 2. Check in at front desk on 2nd floor of Clubhouse. Guests must show current club membership card.

RECIPROCAL BOAT CAPACITY: Appx. 4 boats
DOCKSIDE DEPTH AT ZERO TIDE:)px. 8-12 ft.
SEASON: .... All year
RESERVATION POLICY: First come first serve
TOILETS: .... Yes
HOT SHOWERS: .... Yes
RESTAURANT: .... Yes
BROADBAND/WI-FI: .... BroadbandXpress
DAILY RATE: Recip.guests receive moorage free for 48 hrs. Add'l nights are available w/fee for up to 14 days/yr.

RECIPROCAL DOCK: Appx. 220'
RECIPROCAL SLIPS: Dock only
WATER: .... Yes
AMT W/ELECTRICITY: .... All
AMPS: .... 30 A
BAR: .... Yes
PET FRIENDLY: .... Excellent
OTHER: Park area nearby. SYC sponsors the annual N.W. Opening Day Boat Parade. Fuel & groceries close by.

***NOTE:* THIS IS PRIVATE MOORAGE AND ONLY AVAILABLE TO MEMBERS OF RECIPROCAL YACHT CLUBS. YOUR CLUB *MUST* HAVE RECIPROCAL PRIVILEGES AND YOU MUST FLY YOUR BURGEE.**

CAUTION! This chartlet not intended for use in navigation.

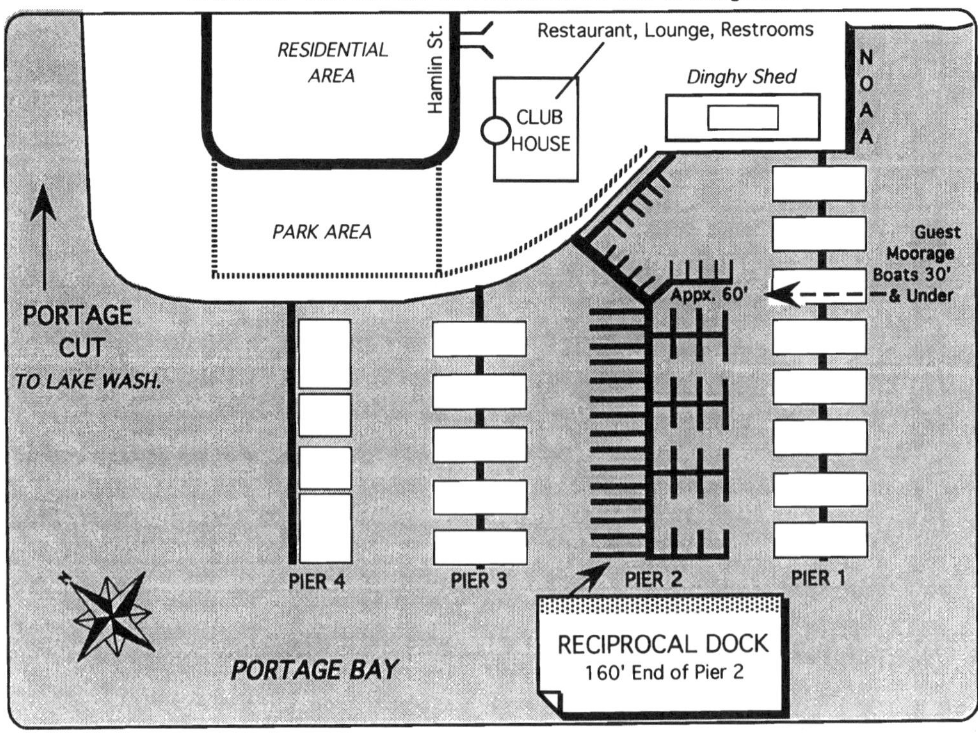

Seattle

NAME OF MARINA: ***LAKEWOOD MOORAGE*** **RADIO:** None
TELEPHONE: **206-722-3887** **MGR:** Kathleen Schober
E-MAIL: None **FAX:** None
ADDRESS: 4500 Lake Wash. Blvd S POB 18403, Seattle, WA 98118
SHORT DESCRIPTION & LOCATION: **www.seattle.gov/parks/boats/lakewood**

47°33.80' - 122°15.80' Small older marina in quiet park like setting located on W. side of Lake Washington about a mile N. of Seward Park close to the entrance of Andrews Bay. Guest moorage is available on a limited basis. Call first between 10AM-5PM.

GUEST BOAT CAPACITY: ....Appx. 2 boats
DOCKSIDE DEPTH: ....15 ft.
SEASON: ....All year
RESERVATION POLICY: ....Preferred
AMT W/ELECTRICITY: ....Limited
FUEL DOCK: ....None
MARINE REPAIRS: ....None
TOILETS: ....Yes
HOT SHOWERS: ....None
RESTAURANT: ....Close by
PICNIC AREA: ....Yes
BASIC STORE: ....Yes
BROADBAND/WI-FI: ....None
DAILY RATE: ....Moderate (75¢-$1.25/foot)

GUEST DOCK: ....60 ft.
GUEST SLIPS: ....Varies
WATER: ....Yes
AMPS: ....20 A
PUMP OUT STATION ....None
HAUL OUT: ....None
BOAT RAMP: ....None
LAUNDRY: ....Yes
BAR: ....Close by
POOL: ....None
GOLF: ....Close by
PET FRIENDLY: ....Excellent
OTHER: Bus service on street in front, waterfowl, close to Seward Park, gift & convenience store.

CAUTION! This chartlet not intended for use in navigation.

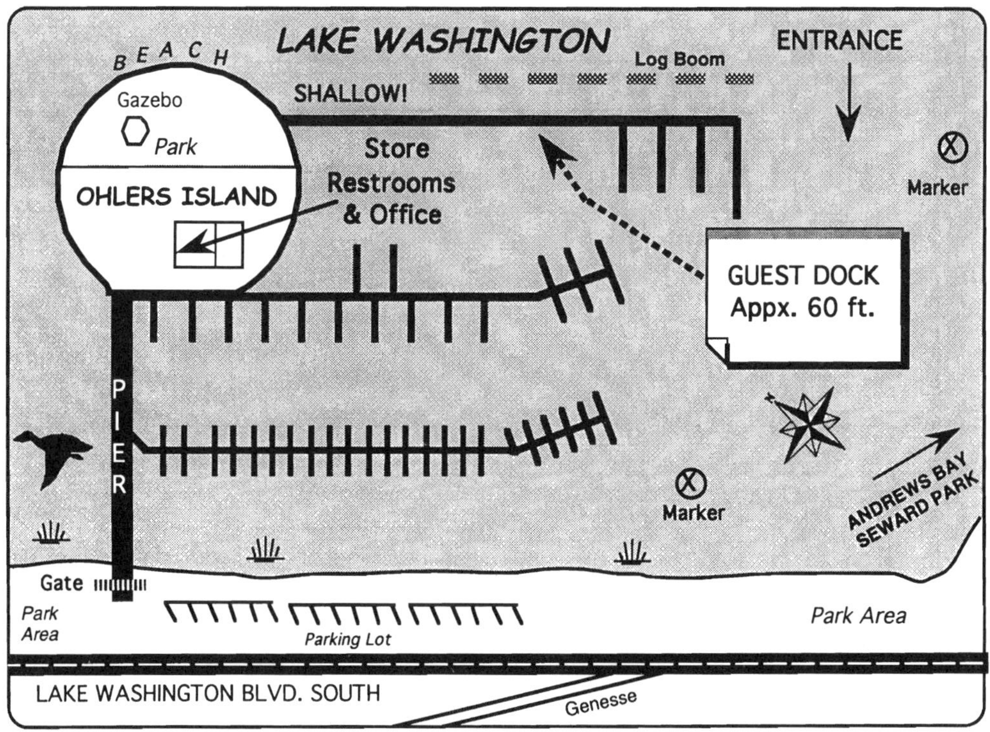

Seattle

NAME OF YACHT CLUB: *RAINIER YACHT CLUB*

ADDRESS: 9094 Seward Park Ave. S. Seattle, WA 98118

TELEPHONE: 206-722-9576 PERSON IN CHARGE: Recip. Chair

SHORT DESCRIPTION & LOCATION: **www.rainieryachtclub.com**

47°31.40' - 122°15.60' RYC offers 65 ft. slip at end of E Dock at Parkshore Marina on W. side of Lk. Washington next to Seward Park in Rainier Beach area. Moorage offered for reciprocal yacht club members only. To leave marina, a gate key is available in a lock box. Check with your home club's reciprocal information for combination & location of lock box.

RECIPROCAL BOAT CAPACITY............2-3 boats

DOCKSIDE DEPTH:..........................Appx. 10-15 ft.

SEASON: ....All year

RESERVATION POLICY: ....None

TOILETS: ....At Marina

HOT SHOWERS: ....At Marina

RESTAURANT: ....Close by

BROADBAND/WI-FI: ....None

DAILY RATE: No Charge. 7 Day limit unless other boaters are waiting.

RECIPROCAL DOCK: .....Slips only

RECIPROCAL SLIPS: ....2 Slips

WATER: ....Yes

AMT W/ELECTRICITY: ....All

AMPS: ....15-30 A

BAR: ....Close by

PET FRIENDLY: ....Good

OTHER: Close to Seward Park, groceries, bus service, pumpout. Gate key works for Marina facilities too.

NOTE: ***THIS IS PRIVATE MOORAGE AND ONLY AVAILABLE TO MEMBERS OF RECIPROCAL YACHT CLUBS. YOUR CLUB MUST HAVE RECIPROCAL PRIVILEGES AND YOU MUST FLY YOUR BURGEE.***

NOTES

NOTE: Parkshore Marina does not offer guest moorage any longer except for Rainier Yacht Club reciprocal visitors.

CAUTION! This chartlet not intended for use in navigation.

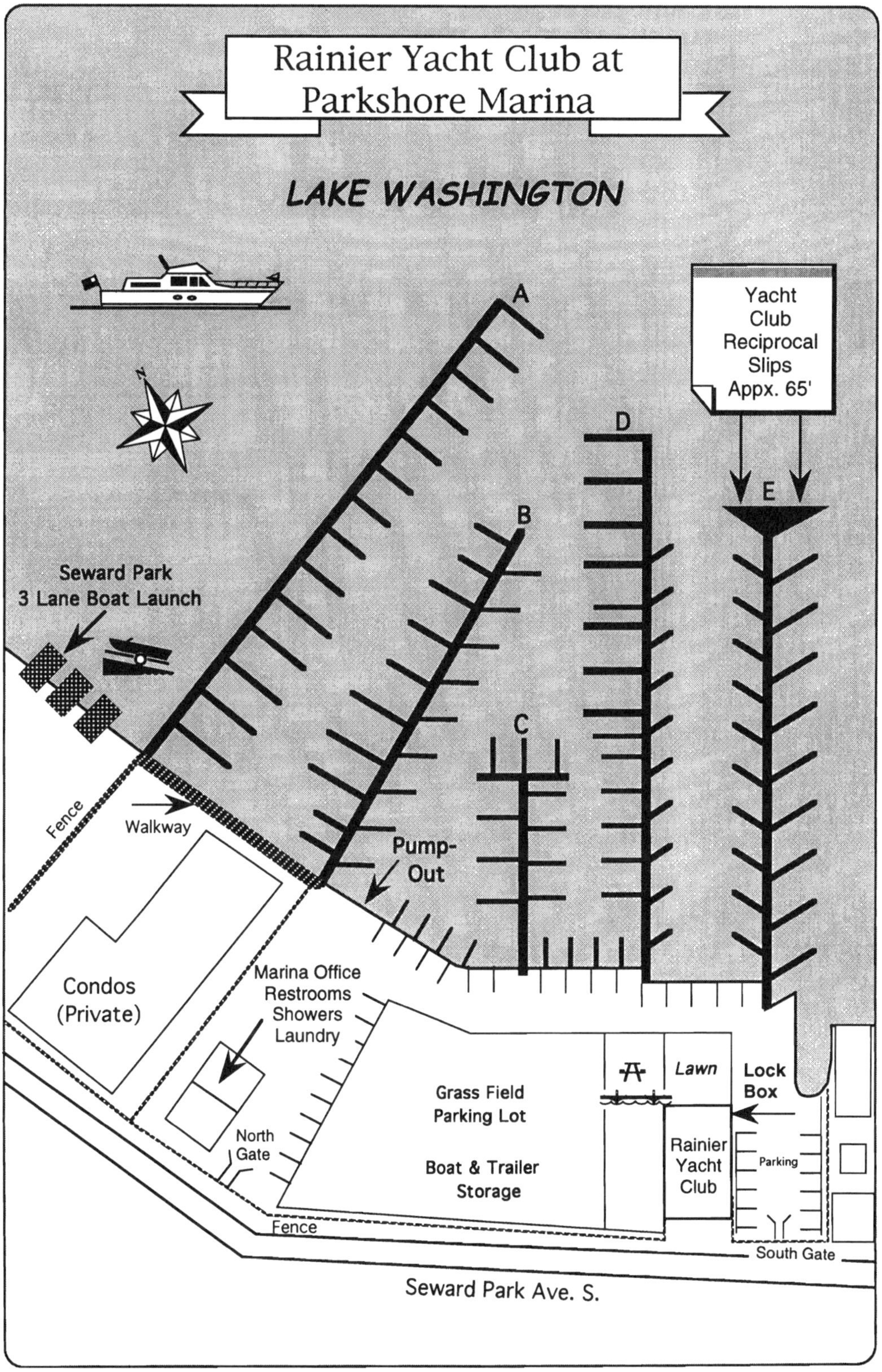

Rainier Yacht Club at Parkshore Marina

LAKE WASHINGTON

Yacht Club Reciprocal Slips Appx. 65'

A

B

C

D

E

Seward Park
3 Lane Boat Launch

Fence

Walkway

Pump-Out

Condos (Private)

Marina Office
Restrooms
Showers
Laundry

North Gate

Grass Field Parking Lot

Boat & Trailer Storage

Lawn

Lock Box

Rainier Yacht Club

Parking

Fence

South Gate

Seward Park Ave. S.

Bellevue

NAME OF YACHT CLUB: *NEWPORT YACHT CLUB*
ADDRESS: 81 Skagit Key Bellevue, WA 98006
TELEPHONE: 425-747-3291 PERSON IN CHARGE: Club Manager

SHORT DESCRIPTION & LOCATION:
47°34.50' - 122°11.30' NYC is located on the east shore, second marina south of the East Channel I-90 Bridge on Lake Washington. Upon arriving, tie up & check in with the Club Manager during working hours to obtain a slip and a gate key (deposit required). Must have key for dock exit and entrance.

RECIPROCAL BOAT CAPACITY...........Appx. 6-10
DOCKSIDE DEPTH:...6 ft. +
SEASON: ....All year
RESERVATION POLICY: ....Accepts
TOILETS: ....Yes-during office hours
HOT SHOWERS: ....Yes-during office hours
RESTAURANT: ....None
BROADBAND/WI-FI: ....None
DAILY RATE: ....Power - $2.00/Day

RECIPROCAL DOCK: ....Slips only
RECIPROCAL SLIPS: ....Varies
WATER: ....Yes
AMT W/ELECTRICITY: ....All
AMPS: ....20-30 A
BAR: ....None
PET FRIENDLY: ....Good
OTHER: Quiet neighborhood location, close to Factoria Mall. Guests have access to pool & facilities.

***NOTE:* THIS IS PRIVATE MOORAGE AND ONLY AVAILABLE TO MEMBERS OF RECIPROCAL YACHT CLUBS. YOUR CLUB *MUST* HAVE RECIPROCAL PRIVILEGES AND YOU MUST FLY YOUR BURGEE.**

CAUTION! This chartlet not intended for use in navigation.

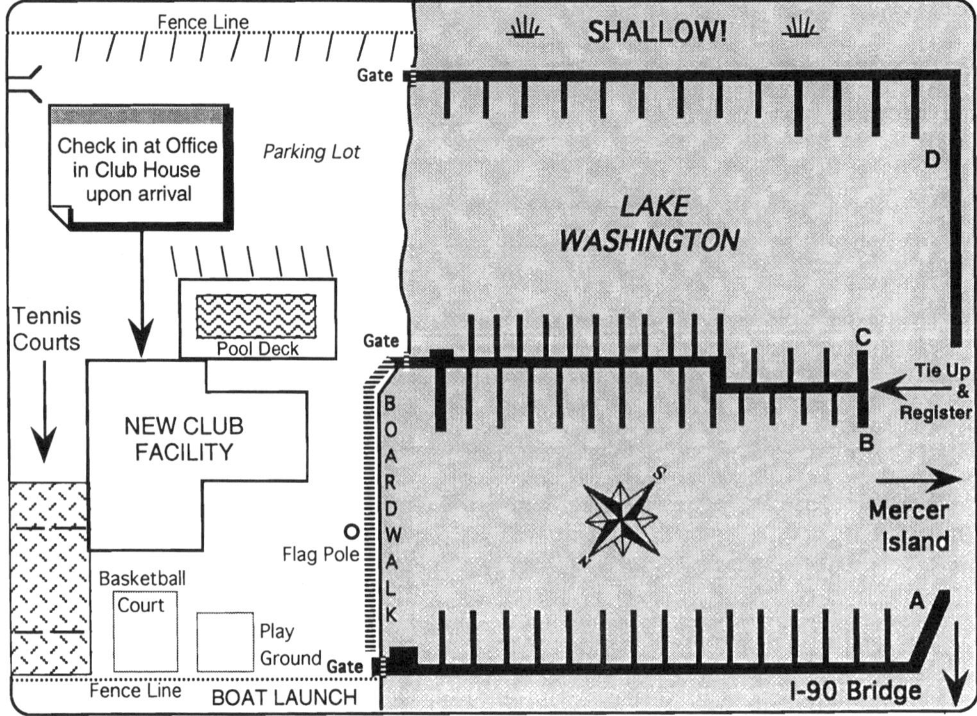

Bellevue

NAME OF YACHT CLUB: *BELLEVUE YACHT CLUB*

ADDRESS: P.O. Box 1175 Bellevue, WA 98009

TELEPHONE: 425-747-2628 PERSON IN CHARGE: Rear Commodore

SHORT DESCRIPTION & LOCATION: **www.bellevueyachtclub.com**

47°34.55' - 122°11.35' For clubs holding a current reciprocal agreement, BYC offers moorage at their home port, Newport Yacht Basin on Lake Washington just south of I-90 Bridge in Bellevue. For reservations, gate key, & directions to reciprocal slips, call Eastside Marine Brokerage @425-747-BOAT between 11 AM & 5 PM daily(except holidays).

RECIPROCAL BOAT CAPACITY....................Varies

DOCKSIDE DEPTH:....................................Appx. 6 ft.

SEASON: ....All year

RESERVATION POLICY: ....Call first

TOILETS: ....Yes

HOT SHOWERS: ....None

RESTAURANT: ....Walking distance

BROADBAND/WI-FI: ....None

DAILY RATE: First- come-first-serve with a maximum of two nights stay.

RECIPROCAL DOCK: ....45 ft.

RECIPROCAL SLIPS: ....Varies

WATER: ....Yes

AMT W/ELECTRICITY: ....All

AMPS: ....30 A

BAR: ....Walking distance

PET FRIENDLY: ....Good

OTHER: Covered dock BBQ area, fuel dock, haul out and repair facilities.

***NOTE:* THIS IS PRIVATE MOORAGE AND ONLY AVAILABLE TO MEMBERS OF RECIPROCAL YACHT CLUBS. YOUR CLUB *MUST* HAVE RECIPROCAL PRIVILEGES AND YOU MUST FLY YOUR BURGEE.**

CAUTION! This chartlet not intended for use in navigation.

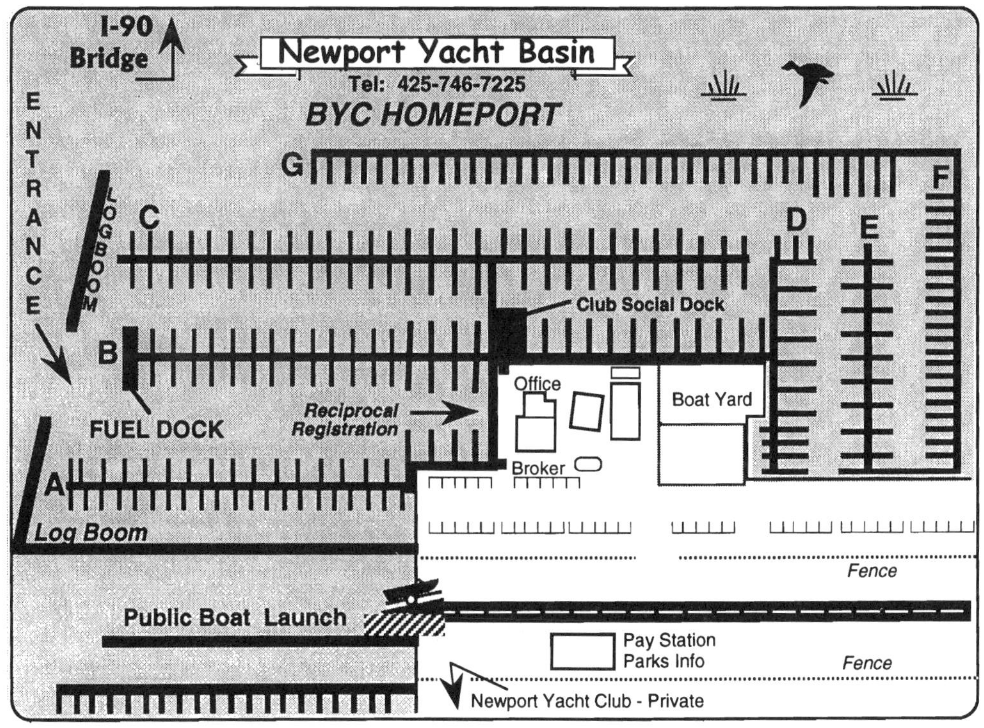

Bellevue

NAME OF YACHT CLUB: *MEYDENBAUER BAY YACHT CLUB*

ADDRESS: P.O. Box 863 Bellevue WA 98009

TELEPHONE: 425-454-8880 PERSON IN CHARGE: Club Manager

SHORT DESCRIPTION & LOCATION: **www.mbycwa.org**

47°36.50'-122°12.40' MBYC invites reciprocal club members to use their moorage at far E. end of Meydenbauer Bay in the City of Bellevue, WA on Lake Washington. The reciprocal moorage is at the end of Dock Two. Rafting is encouraged. Check in w/club manager upon arriving.

RECIPROCAL BOAT CAPACITY: Appx 4-6

DOCKSIDE DEPTH: 6 ft.

SEASON: All year

RESERVATION POLICY: First come first served

TOILETS: Yes

HOT SHOWERS: None

RESTAURANT: Close by

BROADBAND/WI-FI: None

DAILY RATE: Complimentary moorage, Power $2.00/Day. 3 Day limit per visit, per calendar month.

RECIPROCAL DOCK: 60 Ft.

RECIPROCAL SLIPS: Dock only

WATER: Yes

AMT W/ELECTRICITY: All

AMPS: Electricity - 30 A

BAR: Close by

PET FRIENDLY: Good

OTHER: Restaurants, shops, & Bellevue Square close by. Rafting OK up to 4 boats.

***NOTE:* THIS IS PRIVATE MOORAGE AND ONLY AVAILABLE TO MEMBERS OF RECIPROCAL YACHT CLUBS. YOUR CLUB *MUST* HAVE RECIPROCAL PRIVILEGES AND YOU MUST FLY YOUR BURGEE.**

CAUTION! This chartlet not intended for use in navigation.

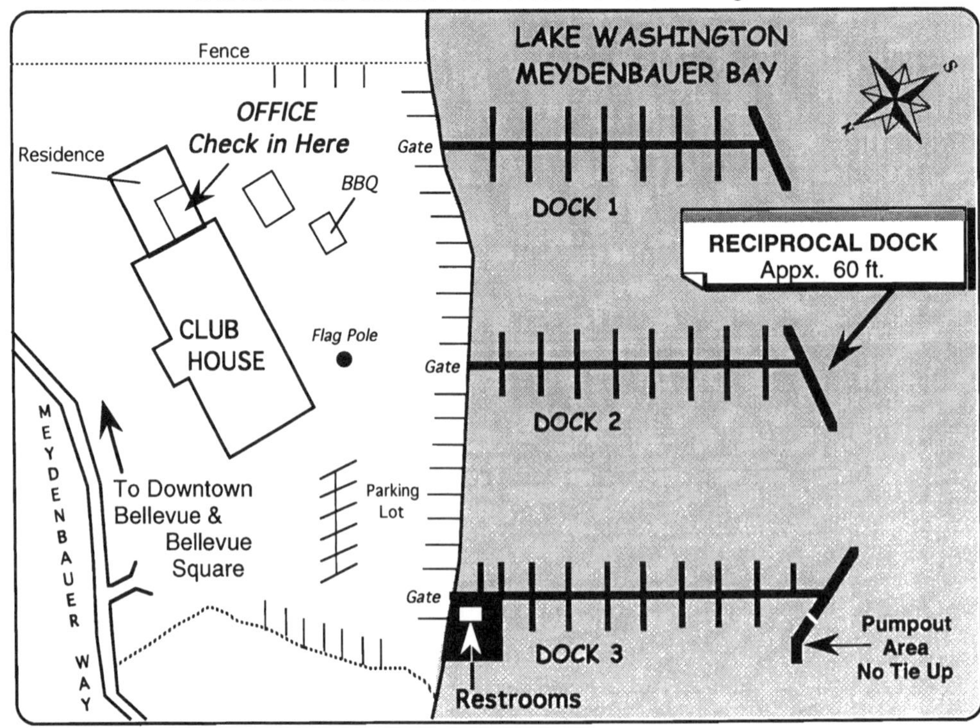

Kirkland

NAME OF PARK: ***KIRKLAND MARINA PARK - Kirkland Parks Dept.***
ADDRESS: 25 Lakeshore Plaza Dr. Kirkland, WA 98033
TELEPHONE: 425-587-3340 **MGR:** Marina Manager
SHORT DESCRIPTION & LOCATION: **www.ci.kirkland.wa.us**
47°40.50' - 122°12.50' Located in downtown Kirkland on E. side of Lake Washington. Very convenient to shops, groceries, good restaurants, art galleries, and local festivals. **There is no breakwater & the busy marina can be unsettled.**

GUEST BOAT CAPACITY: ....Appx. 60 boats
DOCKSIDE DEPTH: ....3-15 ft.
SEASON: ....All year
AMT W/ELECTRICITY: ....None
TOILETS: ....Yes
HOT SHOWERS: ....None
PICNIC AREA: ....Yes
PLAY AREA: ....None
BASIC STORE: ....Close by
DAILY RATE: ....Economical (Under 75¢/foot)
3 DAY MAXIMUM STAY
NO BOAT FOR SALE SIGNS ON DOCKS

GUEST DOCK: Appx 500' + slips
MOORING BUOYS: ....None
WATER: ....No
PAY PHONES: ....None
BOAT RAMP: ....Yes-(fee)
PICNIC SHELTER: ....None
BBQ: ....None
PUMP OUT STATION: .....None
OTHER: **NO RAFTING,** gazebo pavilion in park for entertainment, waterfowl, fishing, excellent beach.

CAUTION! This chartlet not intended for use in navigation.

Edmonds

NAME OF MARINA: ***PORT OF EDMONDS MARINA*** RADIO: VHF Ch.69
TELEPHONE: **425-775-4588** MGR: Marla Kempf
E-MAIL: info@portofedmonds.org FAX: 425-670-0583 (Fax)
ADDRESS: 336 Admiral Way Edmonds, WA 98020
SHORT DESCRIPTION & LOCATION: **www.portofedmonds.org**
47°48.60' - 122°23.50' The Port of Edmonds is located north of Seattle on the east side of Puget Sound just NE of Point Edwards. Appx. 700 lineal ft. of guest moorage is available, however a Loan-A-Slip program provides expanded guest slips. Large work yard.

GUEST BOAT CAPACITY: ....Appx. 90 boats
DOCKSIDE DEPTH AT ZERO TIDE: ....15 ft.
SEASON: ....All year
RESERVATION POLICY: ....Accepts
AMT W/ELECTRICITY: ....All
FUEL DOCK: ....Gas & Diesel
MARINE REPAIRS: ....On premises
TOILETS: ....Yes
HOT SHOWERS: ....Yes
RESTAURANT: ....Several in area
PICNIC AREA: ....Yes
BASIC STORE: ....Close by
BROADBAND/WI-FI: ....BroadbandXpress
DAILY RATE: ....Moderate
(75¢-$1.25/foot)
First 4 hours are complementary
BOATING CLUBS WELCOME

GUEST DOCK: Appx. 700' + slips
GUEST SLIPS: ....Varies
WATER: ....Yes
AMPS: ....30-50 A
PUMP OUT STATION ....Yes
HAUL OUT: ....Travel-Lift
BOAT RAMP: ....Double sling
LAUNDRY: ....Yes
BAR: ....Yes
POOL: ....None
GOLF: ....Close by
PET FRIENDLY: ....Good
OTHER: Waterfront shops & parks, fishing pier, beaches, several restaurants close by, & city museum nearby. West Marine - 4 blocks.

NAME OF YACHT CLUB: ***EDMONDS YACHT CLUB***
CLUB ADDRESS: Box 356, Edmonds, WA 98020
CLUB TELEPHONE: 425-778-5499 PERSON IN CHARGE: Rear Commo.
LOCATION & SPECIAL NOTES: **www.edmondsyachtclub.com**
Reciprocal moorage is located in the marina at the base of "H" Dock on north side. **Boats must be tied stern to dock - no exceptions.** There is a 48 foot boat length restriction & 48 hour limit. Please register at the red registration box on the dock.

RECIPROCAL BOAT CAPACITY: ....Appx. 2-3
DOCKSIDE DEPTH AT ZERO TIDE: ....15 ft.
RECIPROCAL SEASON: ....All year
RESERVATION POLICY: ....None
TOILETS: ....At Marina
HOT SHOWERS: ....At Marina
RESTAURANT: ....Close by
DAILY RATE: ....$3.00 per day for power
..........Moorage free for 48 hours

RECIPROCAL DOCK: Apx 120 ft
RECIPROCAL SLIPS: Stern tie
WATER: ....Yes
AMT W/ELECTRICITY: ....All
AMPS: ....30 A
BAR: ....Close by
OTHER: ......Same as marina listing
Club rental available.

***NOTE:* THIS IS PRIVATE MOORAGE *ONLY* AVAILABLE TO MEMBERS OF RECIPROCAL YACHT CLUBS! YOUR CLUB *MUST* HAVE RECIPROCAL PRIVILEGES AND YOU MUST FLY YOUR BURGEE!**

Edmonds

CAUTION! This chartlet not intended for use in navigation.

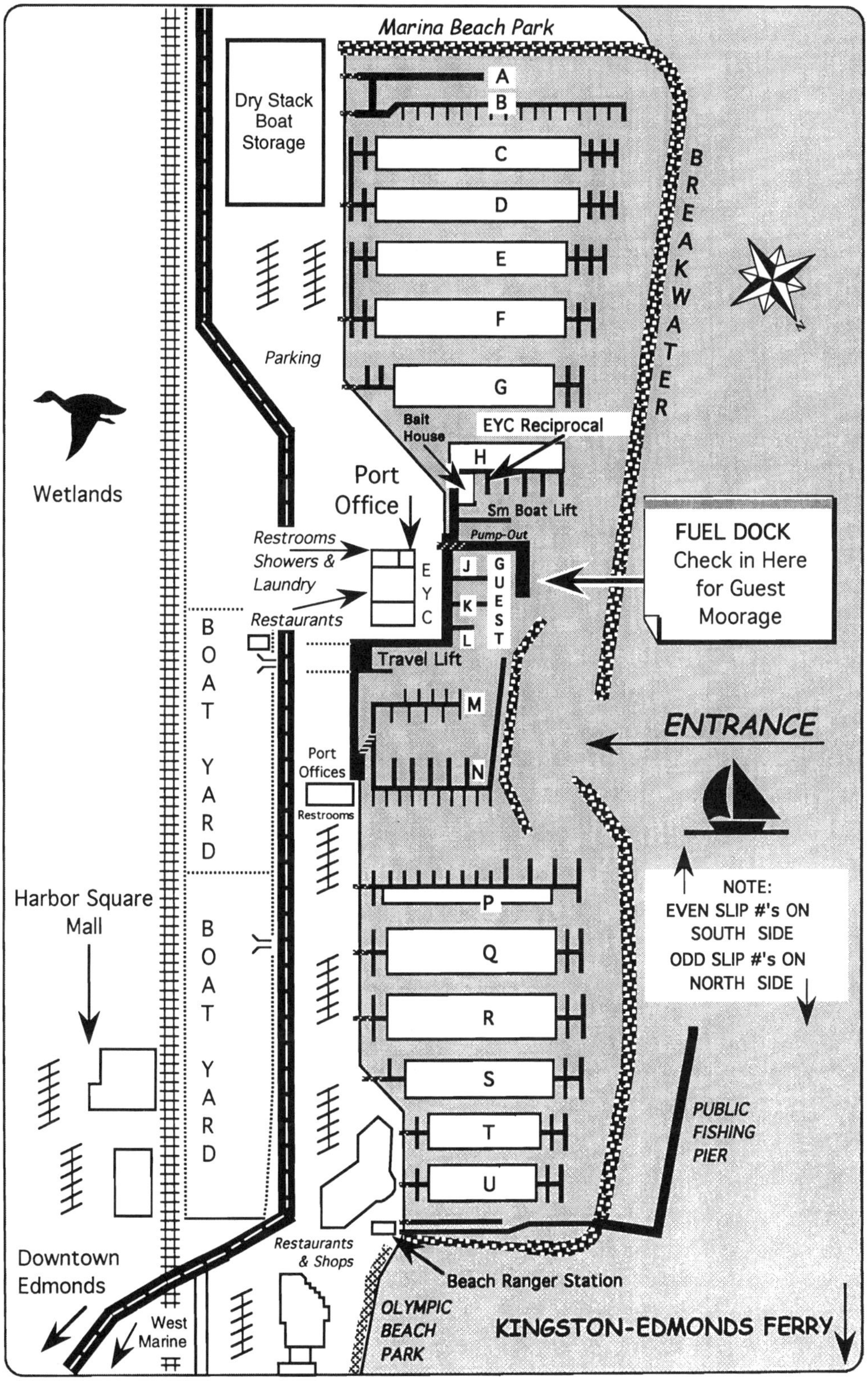

Marina Beach Park

Dry Stack Boat Storage

A
B
C
D
E
F
G

BREAKWATER

Parking

Wetlands

Bait House

EYC Reciprocal

H

Port Office

Sm Boat Lift

Pump-Out

Restrooms

Showers & Laundry

Restaurants

EYC

J
K
L

GUEST

FUEL DOCK
Check in Here
for Guest
Moorage

Travel Lift

BOAT YARD

M

ENTRANCE

Port Offices

N

Restrooms

P

Harbor Square Mall

BOAT YARD

NOTE:
EVEN SLIP #'s ON SOUTH SIDE
ODD SLIP #'s ON NORTH SIDE

Q
R
S
T
U

PUBLIC FISHING PIER

Downtown Edmonds

Restaurants & Shops

Beach Ranger Station

West Marine

OLYMPIC BEACH PARK

KINGSTON-EDMONDS FERRY

Kingston

NAME OF MARINA: ***PORT OF KINGSTON*** **RADIO:** VHF 65A
TELEPHONE: **360-297-3545** **MGR:** Tom Berry, Harbormaster
E-MAIL: ptkingston@aol.com **FAX:** 360-297-2945 (Fax)
ADDRESS: P.O. Box 559 Kingston, WA 98346
SHORT DESCRIPTION & LOCATION: **http://www.portofkingston.org**

47°46.60' - 122°29.80' Located on the N. side of Appletree Cove, about 1.5 mi. S. of Apple Cove Pt. adjacent to the Kingston-Edmonds Ferry Terminal. The modern & large marina is protected by a 340 yd. breakwater marked by a red light. **See below note.

GUEST BOAT CAPACITY: ....Appx. 50 boats
DOCKSIDE DEPTH AT ZERO TIDE: ....12 ft.
SEASON: ....All year
RESERVATION POLICY: Accepts -3 day notice
AMT W/ELECTRICITY: ....All
FUEL DOCK: ....Gas, Dsl, & LP
MARINE REPAIRS: ....None
TOILETS: ....Yes
HOT SHOWERS: ....Yes
RESTAURANT: ....Close by
PICNIC AREA: ....Yes
BASIC STORE: ....Close by
BROADBAND/WI-FI: ....BroadbandXpress
DAILY RATE: ....Economical (Under 75¢/foot)

**Note: Saturday Farmers Market in Marina Park during summer months.

GUEST DOCK: Appx. 85' + slips
GUEST SLIPS: ....Appx. 49
WATER: ....Yes
AMPS: ....30 A
PUMP OUT STATION ....Yes
HAUL OUT: ....None
BOAT RAMP: ....Yes
LAUNDRY: ....Yes
BAR: ....Close by
POOL: ....None
GOLF: ....New Course close by
PET FRIENDLY: ....Excellent
OTHER: Adjacent to shops & restaurants, shopping center 1/2 mile, park, fishing & crabbing. **See note on left.

B

NAME OF YACHT CLUB: ***KINGSTON COVE YACHT CLUB***
CLUB ADDRESS: P.O. Box 81, Kingston, WA 98346
CLUB TELEPHONE: 360-297-3371 PERSON IN CHARGE: Reciprocal Chmn
LOCATION & SPECIAL NOTES: **www.kcyc.org**

The KCYC reciprocal dock is located in the Kingston Marina at the head of the guest dock. Two 50 ft. slips are available on a first come first serve basis. After docking, register at Port Office & show your current Y.C. membership card. After hours registration available. Note: Clubhouse available for clubs to rent for boating club functions. Call for info.

RECIPROCAL BOAT CAPACITY: ....2 Boats
DOCKSIDE DEPTH AT ZERO TIDE: ....17 ft.
RECIPROCAL SEASON: ....All year
RESERVATION POLICY: ....None
TOILETS: ....At Marina
HOT SHOWERS: ....At Marina
RESTAURANT: ....Close by
DAILY RATE: ....1 Night $4.00. Extra nights: 30¢/ft. + $ 4 Utility Charge. Limit: 3 nights per visit, 1 visit per mo.

RECIPROCAL DOCK: 100' Total
RECIPROCAL SLIPS: 2-50' Slips
WATER: ....Yes
AMT W/ELECTRICITY: ....All
AMPS: ....30 A
BAR: ....Friday nights
OTHER: ....Club open for Happy Hour Friday nights

***NOTE:* THIS IS PRIVATE MOORAGE *ONLY* AVAILABLE TO MEMBERS OF RECIPROCAL YACHT CLUBS! YOUR CLUB *MUST* HAVE RECIPROCAL PRIVILEGES AND YOU MUST FLY YOUR BURGEE!**

Kingston

CAUTION! This chartlet not intended for use in navigation.

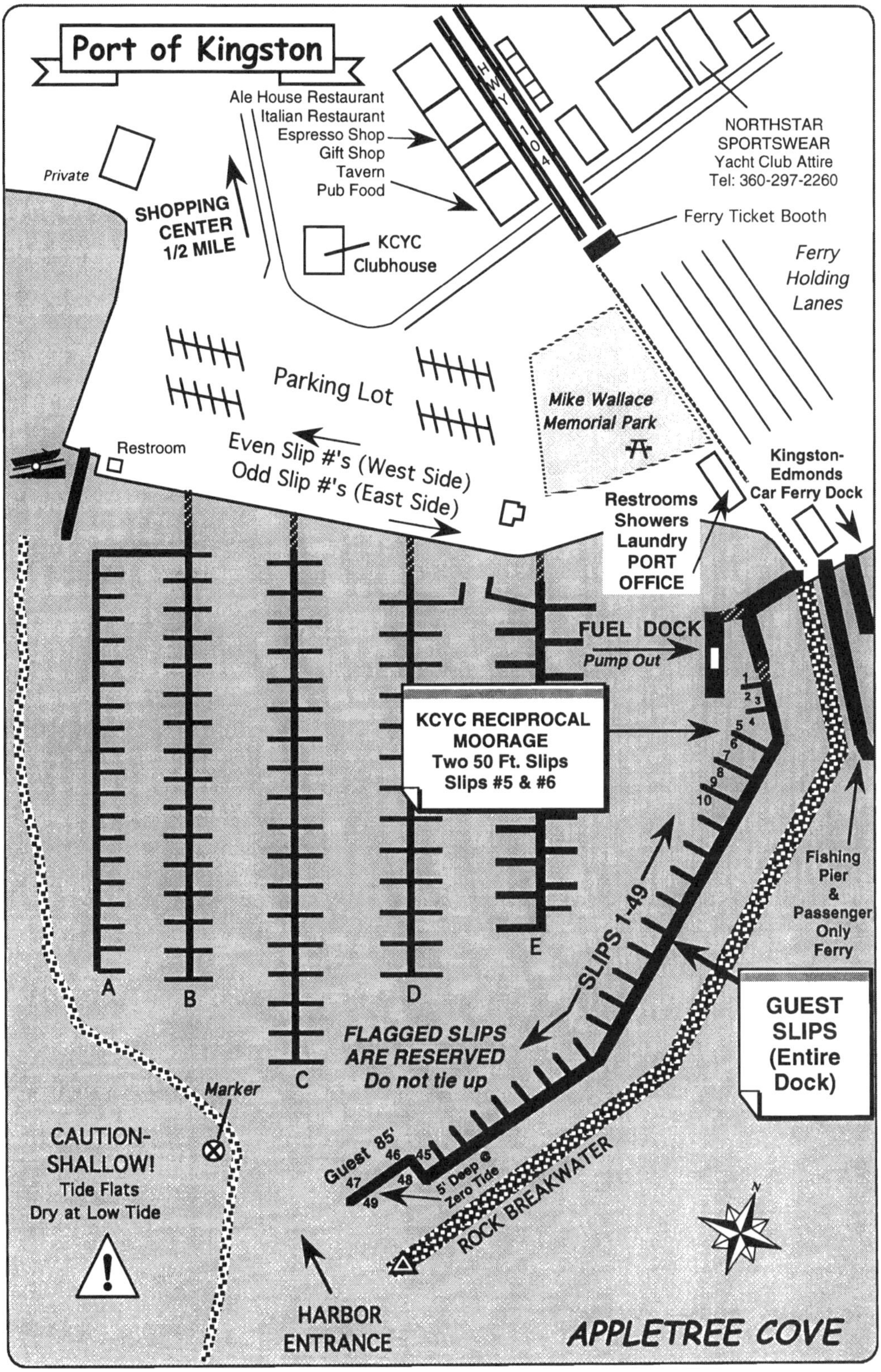

Port of Kingston

Ale House Restaurant
Italian Restaurant
Espresso Shop
Gift Shop
Tavern
Pub Food

HWY 104

NORTHSTAR
SPORTSWEAR
Yacht Club Attire
Tel: 360-297-2260

Private

SHOPPING
CENTER
1/2 MILE

KCYC
Clubhouse

Ferry Ticket Booth

*Ferry
Holding
Lanes*

Parking Lot

*Mike Wallace
Memorial Park*

Restroom

Even Slip #'s (West Side)
Odd Slip #'s (East Side)

Kingston-
Edmonds
Car Ferry Dock

Restrooms
Showers
Laundry
PORT
OFFICE

FUEL DOCK
Pump Out

KCYC RECIPROCAL
MOORAGE
Two 50 Ft. Slips
Slips #5 & #6

1 2 3 4 5 6 7 8 9 10

Fishing
Pier
&
Passenger
Only
Ferry

SLIPS 1-49

GUEST
SLIPS
(Entire
Dock)

A B C D E

*FLAGGED SLIPS
ARE RESERVED
Do not tie up*

Marker

CAUTION-
SHALLOW!
Tide Flats
Dry at Low Tide

Guest 85'
45 46 47 48 49

5' Deep @
Zero Tide

ROCK BREAKWATER

HARBOR
ENTRANCE

APPLETREE COVE

N

Everett

NAME OF MARINA: *PORT OF EVERETT MARINA* RADIO: VHF 16/69
TELEPHONE: 425-259-6001 MGR: Kim Buike
E-MAIL: marina@portofeverett.com FAX: 425-259-0860 (Fax)
ADDRESS: 1720 Marine View Dr. Everett, WA 98201
SHORT DESCRIPTION & LOCATION: **www.portofeverett.com**

47°59.50' - 122°13.50' Located about 1 mile above mouth of and on east side of Snohomish River. Largest marina in the State and offers complete marine facilities. Both Marina Village & Port Gardner Landing on south side offer many services and amenities.

GUEST BOAT CAPACITY: ....Appx. 50 boats
DOCKSIDE DEPTH AT ZERO TIDE: ....10 ft.
SEASON: ....All year
RESERVATION POLICY: ....Summer only
AMT W/ELECTRICITY: ....All
FUEL DOCK: ....Gas & Diesel
MARINE REPAIRS: ....On premises
TOILETS: ....Yes
HOT SHOWERS: ....Yes
RESTAURANT: ....Yes
PICNIC AREA: ....Yes
BASIC STORE: ....Yes
BROADBAND/WI-FI: ....BroadbandXpress
DAILY RATE: ....Moderate (75¢-$1.25/foot)
NOTE: Activity barge available for Clubs with 10 or more boats.

GUEST DOCK: ......1000' plus slips
GUEST SLIPS: 20
WATER: ....Yes
AMPS: ....20-50 A
PUMP OUT STATION Yes- 3 ea.
HAUL OUT: ....Travel-Lift
BOAT RAMP: ....Yes
LAUNDRY: ....Yes
BAR: ....Yes
POOL: ....None
GOLF: ....Close by
PET FRIENDLY: ....Good
OTHER: Many fine shops and restaurants on south & east side. Two fully stocked marine chandleries on premises.

NAME OF YACHT CLUB: *EVERETT YACHT CLUB*
CLUB ADDRESS: 404 14th Street Dock, Everett, WA 98201
CLUB TELEPHONE: 425-259-8178 PERSON IN CHARGE: Rear Comm.
LOCATION & SPECIAL NOTES: **www.everettyachtclub.com**

During the summer, EYC offers 100 feet of moorage located on south side of long dock in front of former yacht club building, just to north after entering Everett Marina adjacent to the fuel dock. Please have your current membership card & register & pay at dock box. 7 DAY LIMIT. **Maximum boat size allowable is 50 feet.** 10PM Noise curfew.

RECIPROCAL BOAT CAPACITY: Appx. 4 boats
DOCKSIDE DEPTH AT ZERO TIDE: ....8 ft.
RECIPROCAL SEASON: ....All year
RESERVATION POLICY: ....None
TOILETS: ....At Marina
HOT SHOWERS: ....At Marina
RESTAURANT: ....Close by
DAILY RATE: ....First 2 days no charge. After 2 days-75¢ /ft. Electricity $4/night.

RECIPROCAL DOCK: ....100 ft.
RECIPROCAL SLIPS: ....None
WATER: ....At fuel dock
AMT W/ELECTRICITY: ....All
AMPS: ....30 A
BAR: ....Club open Wed. eve
OTHER: ....Same as marina listing

NOTE:* *THIS IS PRIVATE MOORAGE ONLY AVAILABLE TO MEMBERS OF RECIPROCAL YACHT CLUBS! YOUR CLUB MUST HAVE RECIPROCAL PRIVILEGES AND YOU MUST FLY YOUR BURGEE!

Everett

CAUTION! This chartlet not intended for use in navigation.

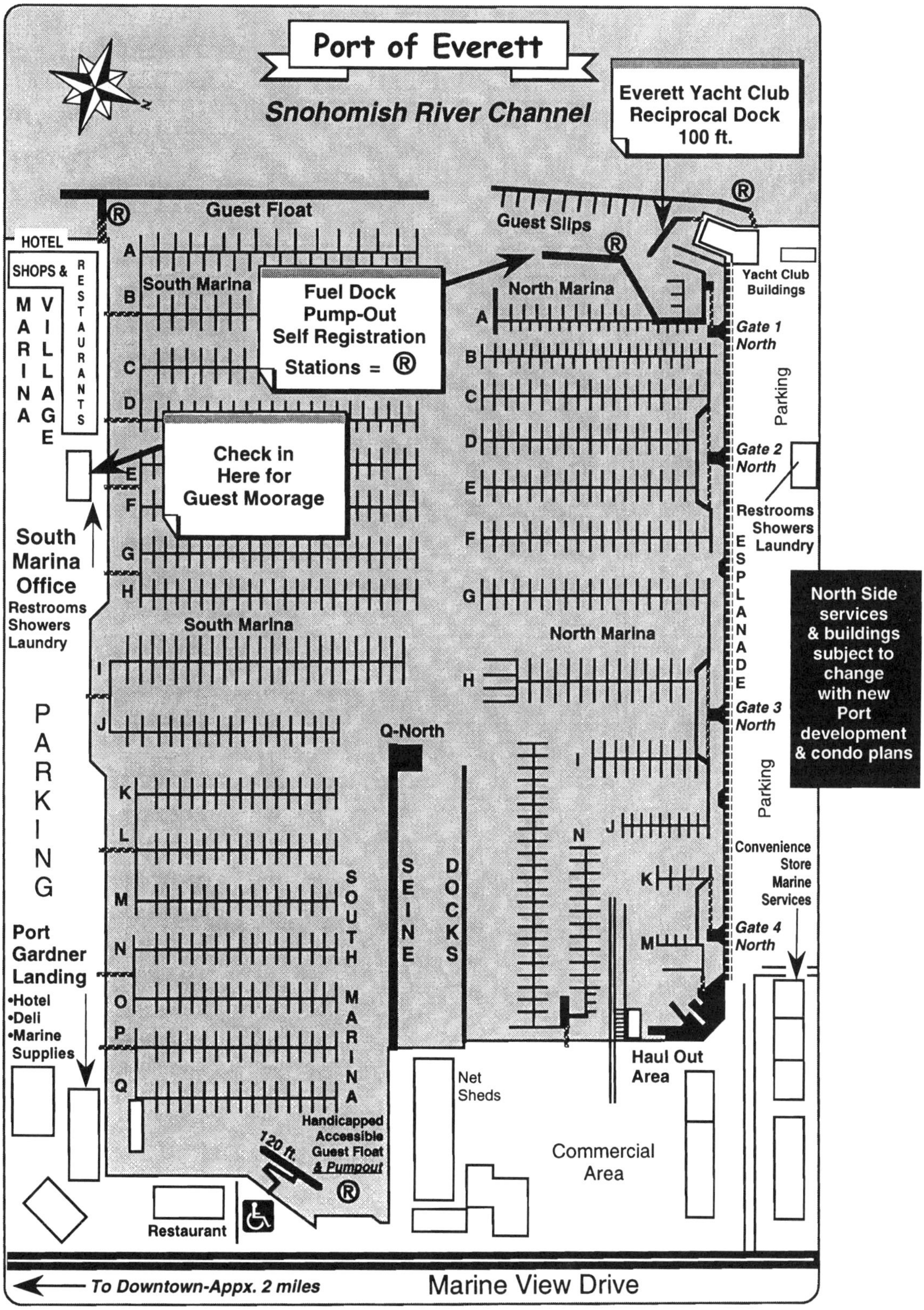

Port of Everett

Snohomish River Channel

Everett Yacht Club
Reciprocal Dock
100 ft.

Guest Float

Guest Slips

HOTEL

SHOPS &

RESTAURANTS

MARINA VILLAGE

South Marina

Fuel Dock
Pump-Out
Self Registration
Stations = ®

North Marina

Yacht Club
Buildings

Gate 1
North

Parking

Check in
Here for
Guest Moorage

Gate 2
North

Restrooms
Showers
Laundry

South
Marina
Office
Restrooms
Showers
Laundry

ESPLANADE

North Side
services
& buildings
subject to
change
with new
Port
development
& condo plans

South Marina

North Marina

PARKING

Q-North

Gate 3
North

Parking

SOUTH MARINA

SEINE DOCKS

Convenience
Store
Marine
Services

Gate 4
North

Port
Gardner
Landing
•Hotel
•Deli
•Marine
Supplies

Haul Out
Area

Net
Sheds

Handicapped
Accessible
Guest Float
& Pumpout

120 ft.

Commercial
Area

Restaurant

To Downtown-Appx. 2 miles

Marine View Drive

NOTES

Everett

NAME OF MARINA: ***12th STREET YACHT BASIN*** RADIO: VHF 16/69
TELEPHONE: 425-259-6001 MGR: Kim Buike, Port of Everett
E-MAIL: marina@portofeverett.com FAX: 425-259-0860 (Fax)
ADDRESS: 1720 Marine View Dr. Everett, WA 98201
SHORT DESCRIPTION & LOCATION: **www.portofeverett.com**

48°00.15' - 122°13.35' Located just N. of the large Port of Everett marina on E. side of Snohomish River channel. This new marina is the focal point of the North Marina Re-development project that includes new condos, retail stores, offices, & marine services.

GUEST BOAT CAPACITY: ....Appx. 70 boats
DOCKSIDE DEPTH AT ZERO TIDE: ......10 ft. +
SEASON: ....All year
RESERVATION POLICY: ....Accepts
AMT W/ELECTRICITY: ....All
FUEL DOCK: ....Gas & Diesel
MARINE REPAIRS: ....On premises
TOILETS: ....Yes
HOT SHOWERS: ....Yes
RESTAURANT: ....Yes
PICNIC AREA: ....Yes
BASIC STORE: ....Yes
BROADBAND/WI-FI: ....BroadbandXpress
DAILY RATE: ....Moderate (75¢-$1.25/foot)

NOTE: Activity barge available for Clubs with 10 or more boats.

GUEST DOCK: Appx. 1500' + slips
GUEST SLIPS: ....Appx. 35
WATER: ....Yes
AMPS: ....30-50 A
PUMP OUT STATION ....Yes
HAUL OUT: ....Travel-Lift
BOAT RAMP: ....Yes
LAUNDRY: ....Yes
BAR: ....Yes
POOL: ....None
GOLF: ....Close by
PET FRIENDLY: ....Good
OTHER: Many fine shops and restaurants in the area. Two fully stocked marine chandleries close by.

CAUTION! This chartlet not intended for use in navigation.

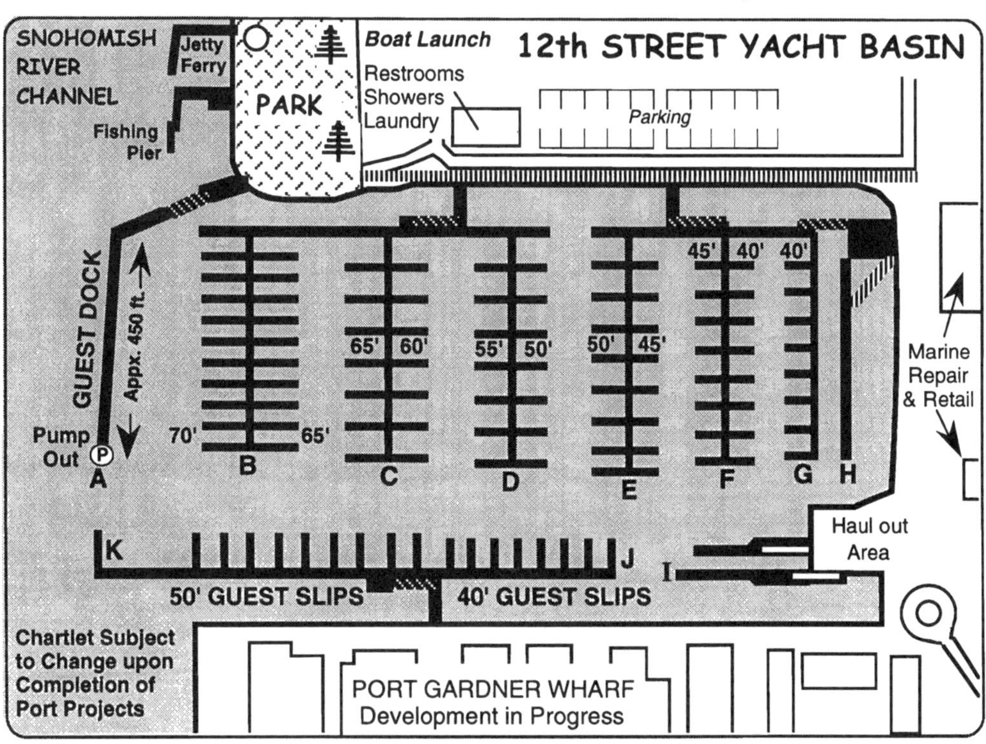

Everett

NAME OF YACHT CLUB: *MILLTOWN SAILING ASSOCIATION*

ADDRESS: P.O. BOX 2963 Everett, WA 98203

TELEPHONE: 425-259-5275 PERSON IN CHARGE: Recip. Chairman

SHORT DESCRIPTION & LOCATION: **www.ussailing.net/msa**

MSA offers reciprocal moorage consiting of two slips per per night within the **Port of Everett Marina.** Visitors must check in with Port Dock Master upon arrival for moorage assignment, or call ahead via VHF Ch. 16/69 or telephone 425-259-6001. Clubhouse located on North side of marina & open during Club functions, visitors welcome.

RECIPROCAL BOAT CAPACITY................2 Boats

DOCKSIDE DEPTH AT ZERO TIDE: ....12 ft.

SEASON: ....All year

RESERVATION POLICY: ....None

TOILETS: ....At Everett Marina

HOT SHOWERS: ....At Everett Marina

RESTAURANT: ....Close by

BROADBAND/WI-FI: ....BroadbandXpress

DAILY RATE: .... No charge for 48 hours. Limit: 1 vist per month.

RECIPROCAL DOCK: ....At marina

RECIPROCAL SLIPS: ...At marina

WATER: ....Yes

AMT W/ELECTRICITY: ....All

AMPS: ....20-30 A

BAR: ....Close by

PET FRIENDLY: ....Good

OTHER: Services same as Port of Everett Marina listing on Page 102 & 103

NOTE: ***THIS IS PRIVATE MOORAGE AND ONLY AVAILABLE TO MEMBERS OF RECIPROCAL YACHT CLUBS. YOUR CLUB MUST HAVE RECIPROCAL PRIVILEGES AND YOU MUST FLY YOUR BURGEE.***

SEE DATA & CHARTLET ON PAGE: 102-103

NOTES

Everett

NAME OF YACHT CLUB: *MUKILTEO YACHT CLUB*
ADDRESS: P.O. Box 915 Mukilteo, WA 98275
TELEPHONE: PERSON IN CHARGE: Vice Commodore
SHORT DESCRIPTION & LOCATION: **www.mukilteoyachtclub.com**
MYC extends reciprocal priveleges at the Port of Everett (POE) guest docks. A maximum of 2 slips are allowed per night. Visitors must check in with POE Dock Master upon arrival for moorage assignment. Visitors are encouraged to check in with Dock Master via VHF Ch. 16/69 or by phone 425-259-6001 prior to arrival.

RECIPROCAL BOAT CAPACITY............... 2 Boats
DOCKSIDE DEPTH AT ZERO TIDE: ....12 ft.
SEASON: ....All year
RESERVATION POLICY: .....1st Come, 1st Serve
TOILETS: ....At Everett Marina
HOT SHOWERS: ....At Everett Marina
RESTAURANT: ....Close by
BROADBAND/WI-FI: ....BroadbandXpress
DAILY RATE: 48 Hours free per visit w/power. Limit - 2 consecutive nights per month & 3 visits/year per vessel.

RECIPROCAL DOCK: ....At marina
RECIPROCAL SLIPS: 2
WATER: ....Yes
AMT W/ELECTRICITY: ....All
AMPS: ....20-30 A
BAR: ....Close by
PET FRIENDLY: ....Good
OTHER: Services same as Port of Everett Marina listing on Pages 102 & 103.

NOTE: ***THIS IS PRIVATE MOORAGE AND ONLY AVAILABLE TO MEMBERS OF RECIPROCAL YACHT CLUBS. YOUR CLUB MUST HAVE RECIPROCAL PRIVILEGES AND YOU MUST FLY YOUR BURGEE.***

SEE DATA & CHARTLET ON PAGE: 102-103

NOTES

Everett

NAME OF YACHT CLUB: *NAVY YACHT CLUB EVERETT*

ADDRESS: P.O. Box 12217 Everett, WA 98206

TELEPHONE: PERSON IN CHARGE: Reciprocal Mgr.

SHORT DESCRIPTION & LOCATION: **www.nyceinfo.org**

NYCE extends reciprocal priveleges at the Port of Everett (POE) guest docks. A maximum of 2 slips are allowed per night. Visitors must register with POE Dock Master upon arrival for moorage assignment. Visitors are encouraged to check in with Dock Master via VHF Ch. 16/69 or by phone 425-259-6001 prior to arrival.

RECIPROCAL BOAT CAPACITY................2 boats

DOCKSIDE DEPTH AT ZERO TIDE: ....12 ft.

SEASON: ....All year

RESERVATION POLICY: ....Call ahead to marina

TOILETS: ....At Everett Marina

HOT SHOWERS: ....At Everett Marina

RESTAURANT: ....Close by

BROADBAND/WI-FI: ....BroadbandXpress

DAILY RATE: Free for 48 hours except for electricity if used. Limited to 1 visit per month per vessel.

RECIPROCAL DOCK: ....At marina

RECIPROCAL SLIPS: 2

WATER: ....Yes

AMT W/ELECTRICITY: ....All

AMPS: ....20-30 A

BAR: ....Close by

PET FRIENDLY: ....Good

OTHER: Services same as Port of Everett Marina listing on Page 102 & 103.

***NOTE:* THIS IS PRIVATE MOORAGE AND ONLY AVAILABLE TO MEMBERS OF RECIPROCAL YACHT CLUBS. YOUR CLUB *MUST* HAVE RECIPROCAL PRIVILEGES AND YOU MUST FLY YOUR BURGEE.**

SEE DATA & CHARTLET ON PAGE: 102-103

NOTES

Everett

NAME OF PARK: *JETTY ISLAND MARINE PARK - Port of Everett*
ADDRESS: P.O. Box 538 Everett, WA 98206
TELEPHONE: 425-259-6001 **MGR:** Kim Buike
SHORT DESCRIPTION & LOCATION: **www.portofeverett.com**

48°00.20' - 122°13.80' Located across from Port of Everett Marina about 1 mile above mouth of and on west side of Snohomish River. Jetty Island is a low sand island in a natural river delta setting with great beach and picnic areas. No shore side camping.

GUEST BOAT CAPACITY: ....Appx. 10-15 boats
DOCKSIDE DEPTH AT ZERO TIDE:.............9 ft.
SEASON: ....All year
AMT W/ELECTRICITY: ....None
TOILETS: ....Summer only
HOT SHOWERS: ....None
PICNIC AREA: ....Yes
PLAY AREA: ....Yes
BASIC STORE: ....None
DAILY RATE: ....No Charge
Moorage limited to one night only. STAFFED IN SUMMER BY PARKS STAFF. FREE SUMMER SHUTTLE TO PORT OF EVERETT.

GUEST DOCK: 2 Docks appx 100'
MOORING BUOYS: ....None
WATER: ....None
PAY PHONES: ....None
BOAT RAMP: ....None
PICNIC SHELTER: ....None
BBQ: ....None
PUMP OUT STATION: ......None
PET FRIENDLY: NO PETS Ashore
OTHER: Clam digging, swimming beach, campfires, sensitive wildlife & bird watching area. BE CAUTIOUS OF RIVER CURRENT UPON LANDING!

CAUTION! This chartlet not intended for use in navigation.

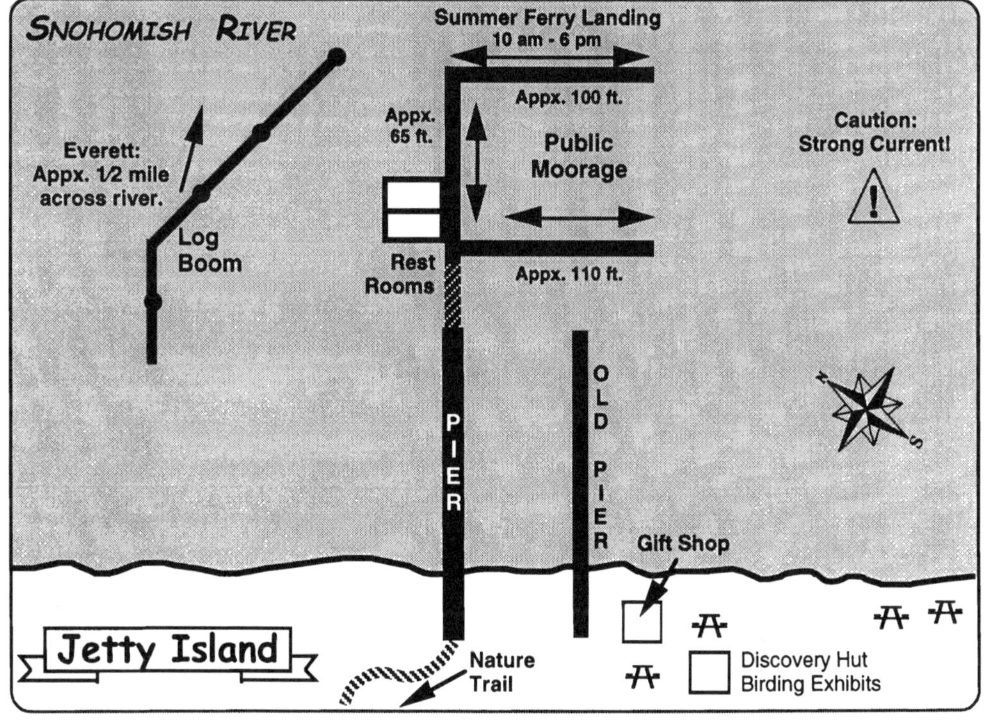

Everett

NAME OF MARINA: ***DAGMAR'S MARINA*** **RADIO:** VHF Ch.09
TELEPHONE: **425-259-6124** **MGR:** Kernan Manley
E-MAIL: Go to Website **FAX:** 425-742-5191 (Fax)
ADDRESS: 1871 Ross Ave. Everett, WA 98205
SHORT DESCRIPTION & LOCATION: **www.dagmarsmarina.com**

48°00.75' - 122°10.83' Located about 3 mi. up the Snohomish River from the Port of Everett. The marina caters primarily to it's large 800 boat dry moorage fleet, but offers guest moorage to the occasional boat that finds its way up the river.

GUEST BOAT CAPACITY: ....Appx. 20 boats
DOCKSIDE DEPTH AT ZERO TIDE: ....5 ft.
SEASON: ....All year
RESERVATION POLICY: ....None
AMT W/ELECTRICITY: ....All
FUEL DOCK: ....None
MARINE REPAIRS: ....On premises
TOILETS: ....Yes
HOT SHOWERS: ....None
RESTAURANT: ....None
PICNIC AREA: ....Yes
BASIC STORE: ....None
BROADBAND/WI-FI: ....None
DAILY RATE:............$30.00/Night - Flat Rate
Space Available - Moorage members have priority.

GUEST DOCK: ....Appx. 1000 ft.
GUEST SLIPS: ....Dock only
WATER: ....Yes
AMPS: ....30 A
PUMP OUT STATION ....None
HAUL OUT: ....Up to appx. 36 ft.
BOAT RAMP: ....Yes
LAUNDRY: ....None
BAR: ....None
POOL: ....None
GOLF: ....Close by
PET FRIENDLY: ....Good
OTHER: Nice park area, ice, snacks, phone. Closest basic services are in Marysville or Everett.

CAUTION! This chartlet not intended for use in navigation.

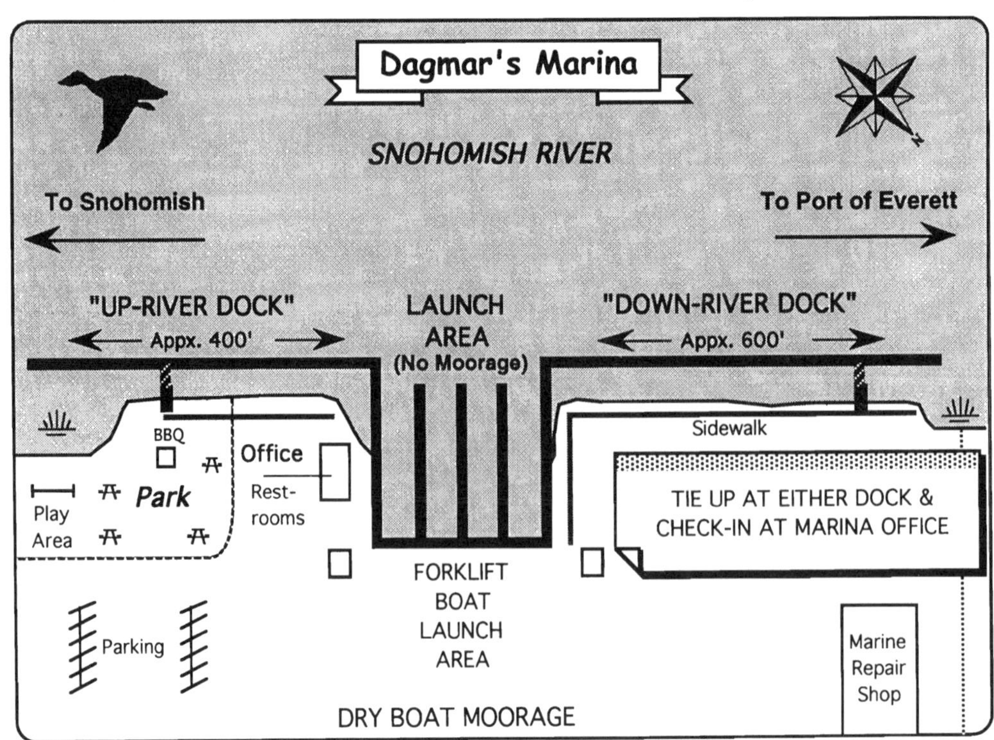

Everett

NAME OF YACHT CLUB: *DAGMAR'S YACHT CLUB*

ADDRESS: P.O. Box 1161 Everett, WA 98201

TELEPHONE: 425-745-2275 (Marina Office) PERSON IN CHARGE: Reciprocal Chair

SHORT DESCRIPTION & LOCATION: **www.dagmars.org**

48°00.75' - 122°10.83' Located on the Snohomish River, 2-1/2 miles upriver from the Port of Everett at Dagmars Marina. 1/2 mile past Railroad Bridge #37 (VHF 13). 4 ft. max draft at low tide. 150 Feet of side tie moorage along the riverfront with herons & eagles pointed directly at Mt. Rainier. Tie up and check in at Marina Office.

RECIPROCAL BOAT CAPACITY: ....3-5 boats

DOCKSIDE DEPTH AT ZERO TIDE: ....5 ft.

SEASON: ....All year

RESERVATION POLICY: ....None

TOILETS: ....Yes

HOT SHOWERS: ....None

RESTAURANT: ....None

BROADBAND/WI-FI: ....None

DAILY RATE: ......No Charge for a two night stay. FIRST COME FIRST SERVE BASIS.

RECIPROCAL DOCK: ....150 ft.

RECIPROCAL SLIPS: Dock only

WATER: ....Yes

AMT W/ELECTRICITY: ....All

AMPS: ....30 A

BAR: ....None

PET FRIENDLY: ....Good

OTHER: Ice, covered BBQ pit, childrens play area, marina services.

***NOTE:* THIS IS PRIVATE MOORAGE AND ONLY AVAILABLE TO MEMBERS OF RECIPROCAL YACHT CLUBS. YOUR CLUB *MUST* HAVE RECIPROCAL PRIVILEGES AND YOU MUST FLY YOUR BURGEE.**

CAUTION! This chartlet not intended for use in navigation.

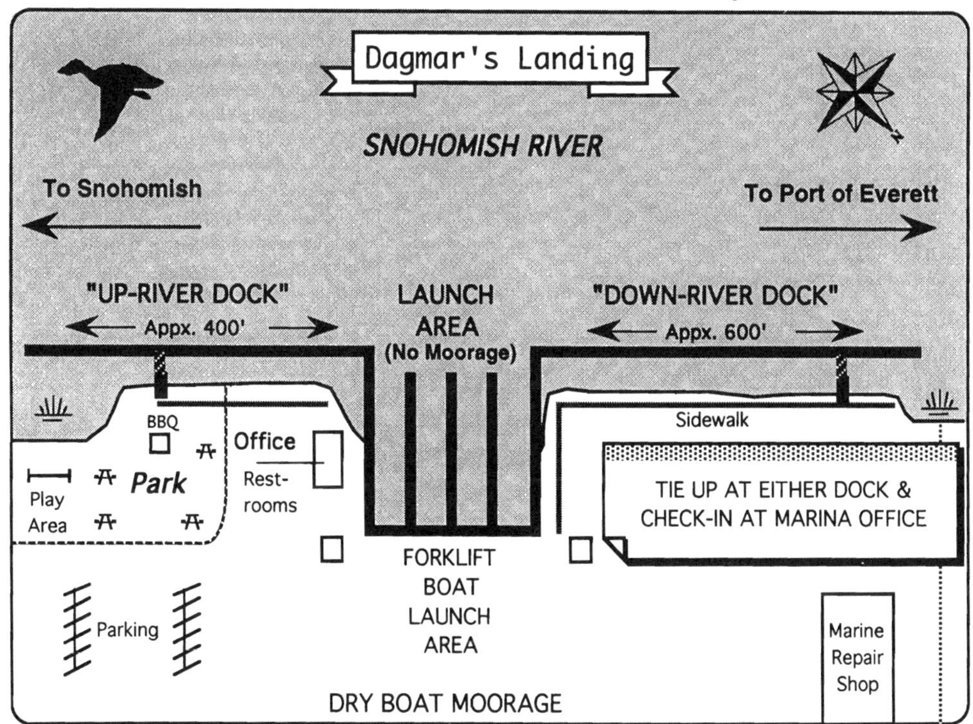

Hat (Gedney) Island

NAME OF YACHT CLUB: *HAT ISLAND YACHT & GOLF CLUB*

ADDRESS: 3616 Colby Ave. #316 Everett, WA 98201

TELEPHONE: 360-444-6611 PERSON IN CHARGE: Island Manager

SHORT DESCRIPTION & LOCATION: **www.hiyandgc.org**

48°01.20'-122°19.20' Hat Is. (Gedney Is. on chart) is located appx. 5 mi. NW of Everett. Club offers 40' reciprocal moorage on a space available basis. Entire island is private & reciprocal guests are limited to marina area. With permission, visitors can use golf course & tennis courts. Call on VHF Ch. 16, or telephone before entering harbor.

RECIPROCAL BOAT CAPACITY: ....1 per club

DOCKSIDE DEPTH AT ZERO TIDE: ....10 ft.

SEASON: .....All year except holidays not available

RESERVATION POLICY: Call in advance,VHF/Tel

TOILETS: ....Yes

HOT SHOWERS: ....Yes

RESTAURANT: ....None

BROADBAND/WI-FI: ....None

DAILY RATE: 48 Hours free annually for 40 ft. boat. Over 40 ft. = 60¢/ft. POWER CHARGE $2.00/DAY

RECIPROCAL DOCK: ....40 ft.

RECIPROCAL SLIPS: ....Varies

WATER: ....None on docks

AMT W/ELECTRICITY: ....Most

AMPS: ....30 A

BAR: ....None

PET FRIENDLY: ....Excellent

OTHER: Quiet setting, no stores or services. Golf & tennis available-see manager. No garbage facilities.

***NOTE:* THIS IS PRIVATE MOORAGE AND ONLY AVAILABLE TO MEMBERS OF RECIPROCAL YACHT CLUBS. YOUR CLUB MUST HAVE RECIPROCAL PRIVILEGES AND YOU MUST FLY YOUR BURGEE.**

CAUTION! This chartlet not intended for use in navigation.

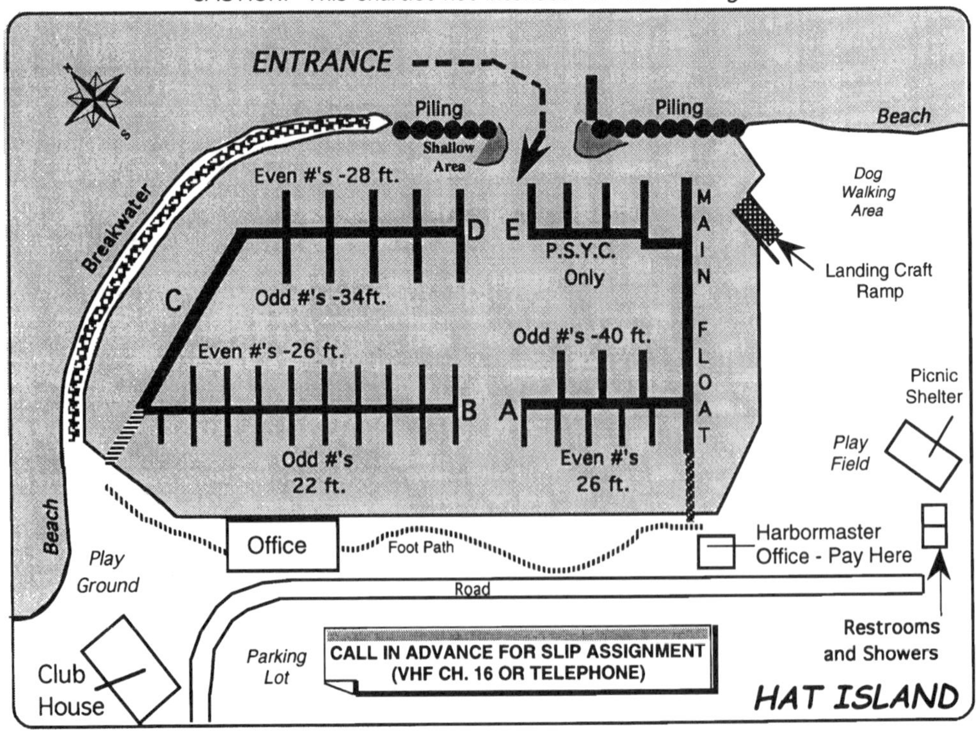

CHAPTER 3

HOOD CANAL & OLYMPIC PENINSULA

Chapter Map - Page 3

Belfair

NAME OF MARINA: ***NORTH SHORE DOCK** (Port of Allyn)* **RADIO:** None
TELEPHONE: **360-275-2430** MGR: Bonnie Knight
E-MAIL: PortofAllyn@aol.com FAX: 360-275-2455 (Fax)
ADDRESS: 4361 NE North Shore Rd. Belfair, WA 98528
SHORT DESCRIPTION & LOCATION: **www.portofallyn.com**

47°25.20' - 122°54.20' Located on the north shore of The Great Bend of Hood Canal about 1 mile W. of Belfair State Park. The facility is owned and operated by the Port of Allyn. Newly renovated pier and floats located in a quiet residential location.

GUEST BOAT CAPACITY: ....Appx. 8-10 boats
DOCKSIDE DEPTH AT ZERO TIDE: Appx. 4 ft.
SEASON: ....All year
RESERVATION POLICY: ....None
AMT W/ELECTRICITY: Call for Electrical hookup
FUEL DOCK: ....None
MARINE REPAIRS: ....None
TOILETS: ....Porta-Potty
HOT SHOWERS: ....None
RESTAURANT: ....None
PICNIC AREA: ....None
BASIC STORE: ....Close by - 1 Mile
BROADBAND/WI-FI: ....None
DAILY RATE: ....Economical (Under 75¢/foot)

GUEST DOCK: Appx 300 ft. total
GUEST SLIPS: ....Appx. 8 slips
WATER: ....None
AMPS: ....30 A
PUMP OUT STATION ....Yes
HAUL OUT: ....None
BOAT RAMP: ....Close by
LAUNDRY: ....None
BAR: ....None
POOL: ....None
GOLF: ....Close by
PET FRIENDLY: ....Excellent
OTHER: CALL 360-372-2408 FOR ELECTRICAL. Convenience store 1 mile, all other services in Belfair, 5 road miles.

CAUTION! This chartlet not intended for use in navigation.

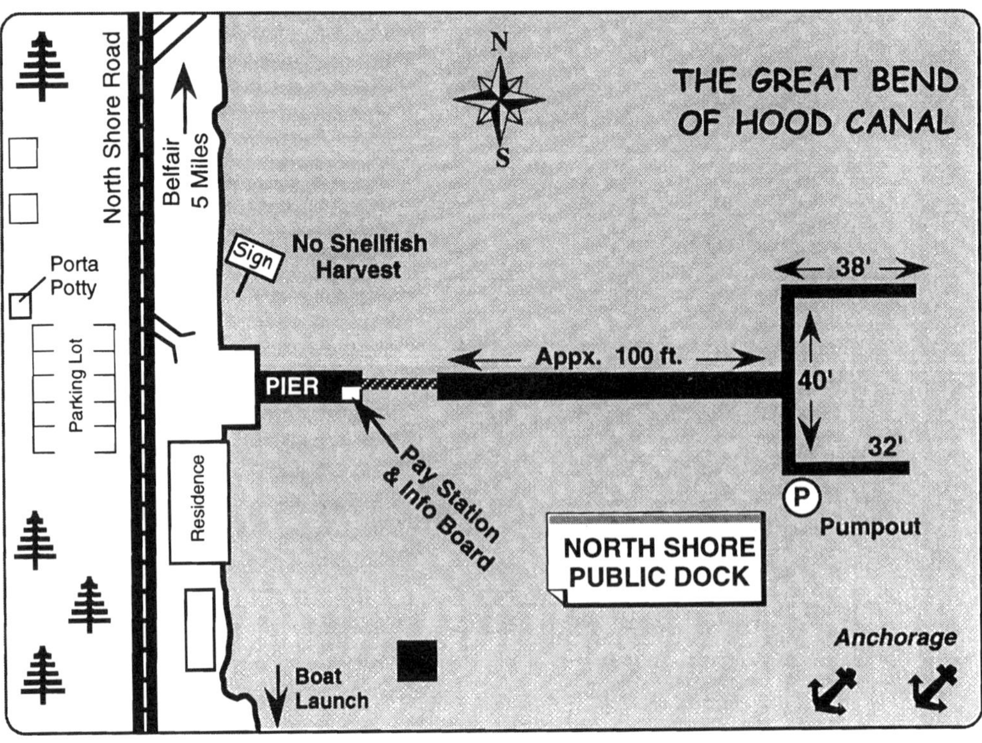

Belfair

NAME OF PARK: ***TWANOH STATE PARK***
ADDRESS: 12190 E. Hwy 106 Union, WA 98592
TELEPHONE: **360-275-2222** MGR: Larry Otto
SHORT DESCRIPTION & LOCATION: **www.parks.wa.gov/moorage/parks**

47°22.42' - 122°58.37' Located about 6 miles E. of Union on the S. shore at the bottom of Hood Canal. This full service park offers many activities for family boating and camping. Use caution at low tide on park dock. Buoys recommended for larger boats.

GUEST BOAT CAPACITY: ....Appx. 5-10 boats
DOCKSIDE DEPTH AT ZERO TIDE:..............3 ft.
SEASON: ....Closed winter
AMT W/ELECTRICITY: ....None
TOILETS: ....Yes
HOT SHOWERS: ....Yes
PICNIC AREA: ....Yes
PLAY AREA: ....Yes
BASIC STORE: ....Summer only
DAILY RATE: DockMoorage..........50¢/foot
Mooring Buoys...............$10.00/night
Minimum Charge: $10.00

GUEST DOCK: ....194 ft. total
MOORING BUOYS: ....7 buoys
WATER: ....Yes
PAY PHONES: ....Yes
BOAT RAMP: ....Yes
PICNIC SHELTER: ....Yes
BBQ: ....Yes
PUMP OUT STATION: ....Yes
PET FRIENDLY: ....Excellent
OTHER: Summer snack bar, campsites, picnicking, great beach swimming area, hiking trails, sports area, salmon spawning creek.

CAUTION! This chartlet not intended for use in navigation.

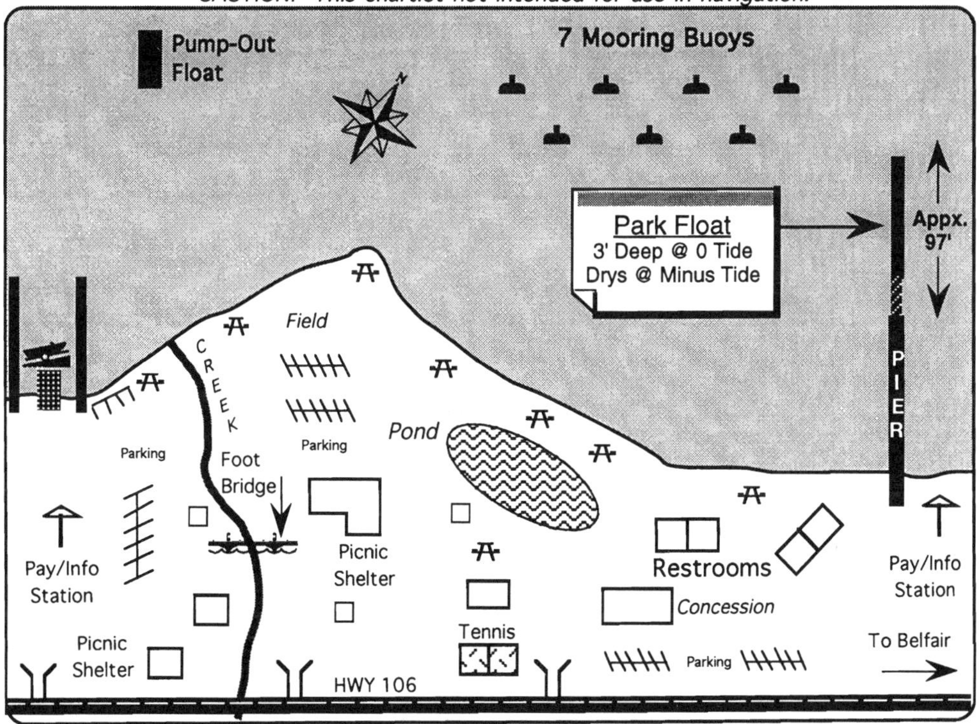

Alderbrook

NAME OF MARINA: ***ALDERBROOK RESORT MARINA*** RADIO: None
TELEPHONE: **360-898-2200** MGR: Mike Carnovale
E-MAIL: info@alderbrookresort.com FAX: 360-898-4610 (Fax)
ADDRESS: 10 Alderbrook Drive, Union, WA 98592
SHORT DESCRIPTION & LOCATION: **www.alderbrookresort.com**

47°21.97' - 123°04.16' Located in a cove on S. shore of The Great Bend of Hood Canal about 1.3 miles E. of the town of Union. Marina & Resort have been beautifully remodeled. This is again a premier golf & resort destination offering boaters all of their amenities.

GUEST BOAT CAPACITY: ....Appx. 30 boats
DOCKSIDE DEPTH AT ZERO TIDE: ....30 ft.
SEASON: ....All year
RESERVATION POLICY: ....Accepts
AMT W/ELECTRICITY: ....All
FUEL DOCK: ....None
MARINE REPAIRS: ....Close by
TOILETS: ....Yes
HOT SHOWERS: ....Yes
RESTAURANT: ...Yes-Reservations Recommended
PICNIC AREA: ....Yes
BASIC STORE: ....Close by
BROADBAND/WI-FI: ....BroadbandXpress
DAILY RATE: ....Moderate (75¢-$1.25/foot)
Moorage includes wireless internet, telephone hook up, water & power.

GUEST DOCK: ....1500 Ft. Total
GUEST SLIPS: ....Dock only
WATER: ....Yes
AMPS: ....30-50 A
PUMP OUT STATION ....Yes
HAUL OUT: ....None
BOAT RAMP: ....Close by
LAUNDRY: ....Close by
BAR: ....Yes
POOL: ....Yes
GOLF: ....Yes
PET FRIENDLY: ....Good
OTHER: Hotel and cottages, Swimming pool & whirlpool spa included in moorage. Small boat rentals, tennis, gift shop, seaplanes.

CAUTION! This chartlet not intended for use in navigation.

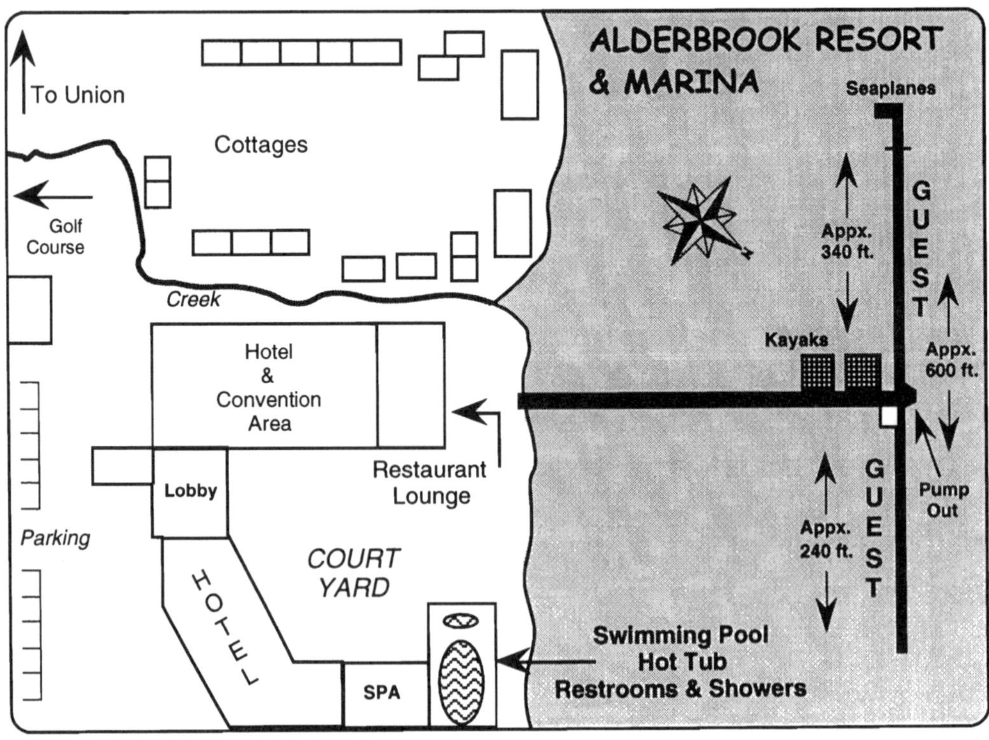

Union

NAME OF MARINA: ***HOOD CANAL MARINA*** **RADIO:** VHF Ch. 16
TELEPHONE: **360-898-2252** **MGR:** Martin Sun
E-MAIL: None **FAX:** 360-898-8888 (Fax)
ADDRESS: PO Box 86, E.5101 Hwy 106 Union, WA 98592
SHORT DESCRIPTION & LOCATION:

47°21.26' - 122°06.00' Located on the S. shore of The Great Bend of Hood Canal, the marina & town offer some services for boaters. The town of Union is a historic railway stop & the marina is protected from south winds. Rents out open slips for overnight moorage.

GUEST BOAT CAPACITY: ....Varies
DOCKSIDE DEPTH AT ZERO TIDE: ....25 ft.
SEASON: ....All year
RESERVATION POLICY: ....Accepts
AMT W/ELECTRICITY: ....All
FUEL DOCK: ....No Longer
MARINE REPAIRS: ....Yes
TOILETS: ....Yes
HOT SHOWERS: ....None
RESTAURANT: ....Close by
PICNIC AREA: ....Yes
BASIC STORE: ....Close by
BROADBAND/WI-FI: ....None
DAILY RATE: ....Moderate (75¢-$1.25/foot)

GUEST DOCK: ....Varies
GUEST SLIPS: ....Varies
WATER: ....Yes
AMPS: ....30 A
PUMP OUT STATION ....None
HAUL OUT: ....None
BOAT RAMP: ....Yes
LAUNDRY: ....None
BAR: ....Close by
POOL: ....None
GOLF: ....Close by
PET FRIENDLY: ....Good
OTHER: Oysters, clams, & crabbing close by, marine supplies, kayaking center.

CAUTION! This chartlet not intended for use in navigation.

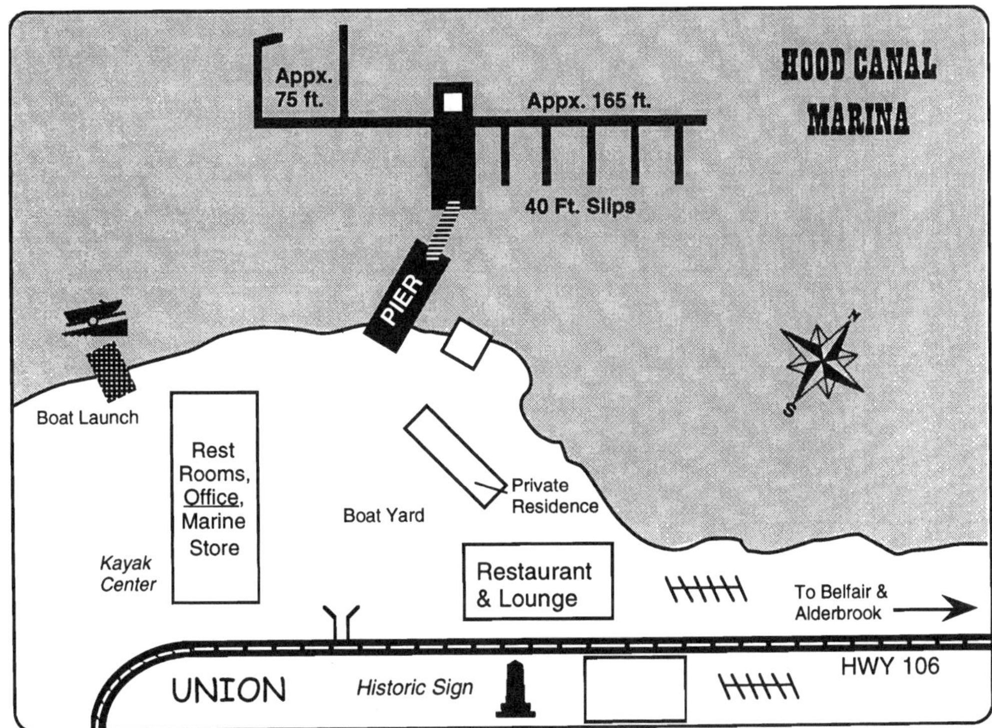

Hoodsport

NAME OF MARINA: ***PORT OF HOODSPORT*** **RADIO:** None
TELEPHONE: **360-877-9350** **MGR:** Port Commissioners
E-MAIL: portmail@hctc.com **FAX:** 360-877-9350
ADDRESS: P.O. Box 429 Hoodsport, WA 98548
SHORT DESCRIPTION & LOCATION: **www.portofhoodsport.us**

47°24.18' - 122°08.40' Located at the quaint township of Hoodsport on the west shore of Hood Canal appx. 20 mi. south of Pleasant Harbor & 4 mi. north of Union. Primarily a small boat facility but the end tie has room for a 60 ft. boat and rafting. Open to weather.

GUEST BOAT CAPACITY: ....Appx 8 boats
DOCKSIDE DEPTH AT ZERO TIDE: 6-10 ft. +
SEASON: ....All year
RESERVATION POLICY: ....None
AMT W/ELECTRICITY: ....None
FUEL DOCK: ....None
MARINE REPAIRS: ....None
TOILETS: ....Porta-Potty
HOT SHOWERS: ....None
RESTAURANT: ....Close by
PICNIC AREA: ....Yes
BASIC STORE: ....Close by
BROADBAND/WI-FI: ....Close by
DAILY RATE: ....Economical (Under 75¢/foot)

GUEST DOCK: .....60 Ft. plus slips
GUEST SLIPS: 10 sm. 25 ' slips
WATER: ....None
AMPS: ....None
PUMP OUT STATION ....None
HAUL OUT: ....None
BOAT RAMP: ....None
LAUNDRY: ....None
BAR: ....Pubs close by
POOL: ....None
GOLF: ....Close by
PET FRIENDLY: ....Good
OTHER: Close to gift shops, restaurants, groceries, public services, and excellent winery and tasting room.

CAUTION! This chartlet not intended for use in navigation.

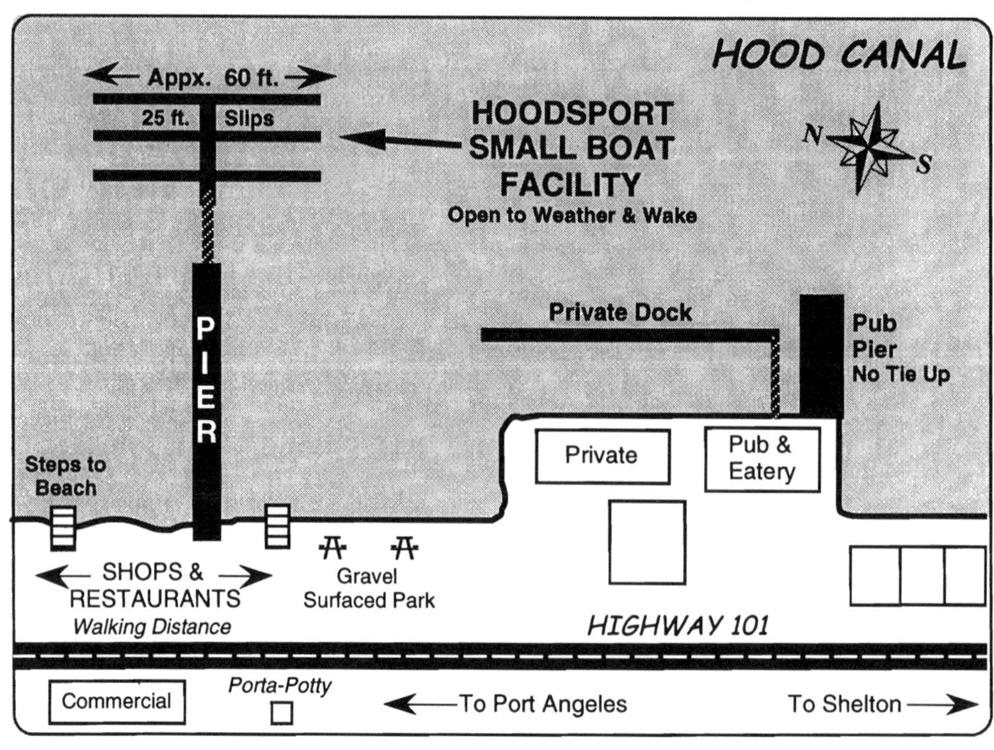

Pleasant Harbor

NAME OF PARK: *PLEASANT HARBOR STATE MARINE PARK*
ADDRESS: P.O. Box 969 Brinnon, WA 98320
TELEPHONE: **360-796-4415** MGR: Douglas Hinton
SHORT DESCRIPTION & LOCATION: **www.parks.wa.gov/moorage/parks**

47°39.44' - 122°54.60' Located just inside the Pleasant Harbor entrance behind the sand & gravel spit on the N. side of the harbor. No upland park facilities but nice wide float in beautiful & serene harbor location. Popular location during shrimping season.

GUEST BOAT CAPACITY: ....Appx. 6 boats
DOCKSIDE DEPTH AT ZERO TIDE: ....33 ft.
SEASON: ....All year
AMT W/ELECTRICITY: ....None
TOILETS: ....Vault toilet
HOT SHOWERS: ....None
PICNIC AREA: ....Yes
PLAY AREA: ....None
BASIC STORE: ....Close by
DAILY RATE: Dock Moorage........................50¢/foot
Minumum Charge: $10
RAFTING ALLOWED

GUEST DOCK: ....218 ft. total
MOORING BUOYS: ....None
WATER: ....None
PAY PHONES: ....Close by
BOAT RAMP: ....Close by
PICNIC SHELTER: ....None
BBQ: ....None
PUMP OUT STATION: Close by
PET FRIENDLY: ....Excellent
OTHER: Great dinghy area, crabbing, shrimping (in season), fishing, close to Pls. Hbr. Marina services.

CAUTION! This chartlet not intended for use in navigation.

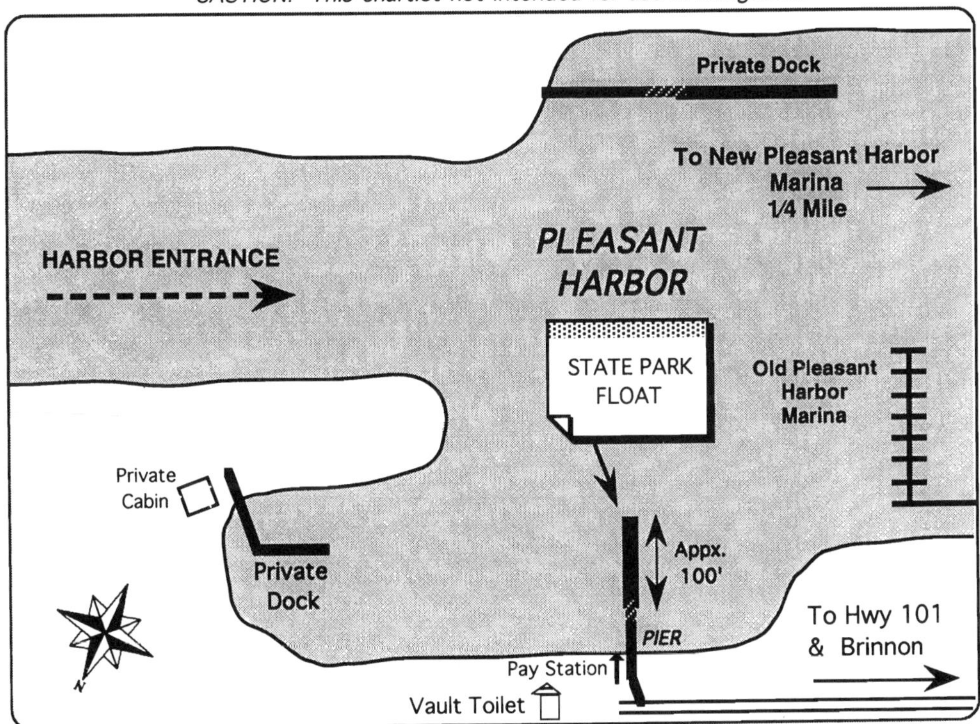

Pleasant Harbor

NAME OF MARINA: ***PLEASANT HARBOR MARINA*** RADIO: VHF Ch.16/9
TELEPHONE: **360-796-4611** MGR: Ryan Kaufman
E-MAIL: info@pleasantharbormarina.com FAX: 866-848-4612 (Fax)
ADDRESS: 308913 Highway 101 Brinnon WA 98320
SHORT DESCRIPTION & LOCATION: www.pleasantharbormarina.com
47°39.70' - 122°54.80' Family oriented marina resort located in lovely cove on Hood Canal 2 miles S. of town of Brinnon. Well protected & recently upgraded marina caters to recreational boaters & boating clubs. *Check in at fuel dock or call ahead for reservation.*

GUEST BOAT CAPACITY: ....Appx. 43 boats
DOCKSIDE DEPTH AT ZERO TIDE: ....35 ft.
SEASON: ....All year
RESERVATION POLICY: ....Accepts
AMT W/ELECTRICITY: ....All
FUEL DOCK: ....Gas & Diesel
MARINE REPAIRS: ....Close by
TOILETS: ....Yes
HOT SHOWERS: ....Yes
RESTAURANT: ....Deli & Pizza
PICNIC AREA: ....Yes
BASIC STORE: ....Yes
BROADBAND/WI-FI: ....BroadbandXpress
DAILY RATE: ....Moderate
(75¢-$1.25/foot)

GUEST DOCK: ....240' plus slips
GUEST SLIPS: 43
WATER: ....Yes
AMPS: 30-50-220A
PUMP OUT STATION ....Yes
HAUL OUT: ....None
BOAT RAMP: ....Close by
LAUNDRY: ....Yes
BAR: ....None
POOL: ....Yes
GOLF: ....Planned
PET FRIENDLY: ....Excellent
OTHER: Pizza, deli, gift shop, fishing & marine supplies, jacuzzi spa, crabbing, shrimping in season, great dinghy area in harbor.

NAME OF YACHT CLUB: ***PLEASANT HARBOR YACHT CLUB*** (B)
CLUB ADDRESS: P.O. Box 30, Brinnon, WA 98230
CLUB TELEPHONE: PERSON IN CHARGE: Vice Commodore
LOCATION & SPECIAL NOTES:
Reciprocal guests of PHYC receive moorage free of charge at the Pleasant Harbor Marina for 48 hours on a first come basis. Moorage is normally located on end of "J" Dock (end tie/outside) consisting of 86 ft. Check in at marina upon arrival.

RECIPROCAL BOAT CAPACITY: ....2-3 boats
DOCKSIDE DEPTH AT ZERO TIDE: ....35 ft.
RECIPROCAL SEASON: ....All year
RESERVATION POLICY: ....None
TOILETS: ....At Marina
HOT SHOWERS: ....At Marina
RESTAURANT: ....Deli & Pizza
DAILY RATE: ....Free for 48 Hours
Power Charge: $5/day

RECIPROCAL DOCK: ....86 ft.
RECIPROCAL SLIPS: Dock only
WATER: ....Yes
AMT W/ELECTRICITY: ....All
AMPS: ....30-50 A
BAR: ....None
OTHER: ....Same as marina listing

***NOTE:* THIS IS PRIVATE MOORAGE *ONLY* AVAILABLE TO MEMBERS OF RECIPROCAL YACHT CLUBS! YOUR CLUB *MUST* HAVE RECIPROCAL PRIVILEGES AND YOU MUST FLY YOUR BURGEE!**

Pleasant Harbor

CAUTION! This chartlet not intended for use in navigation.

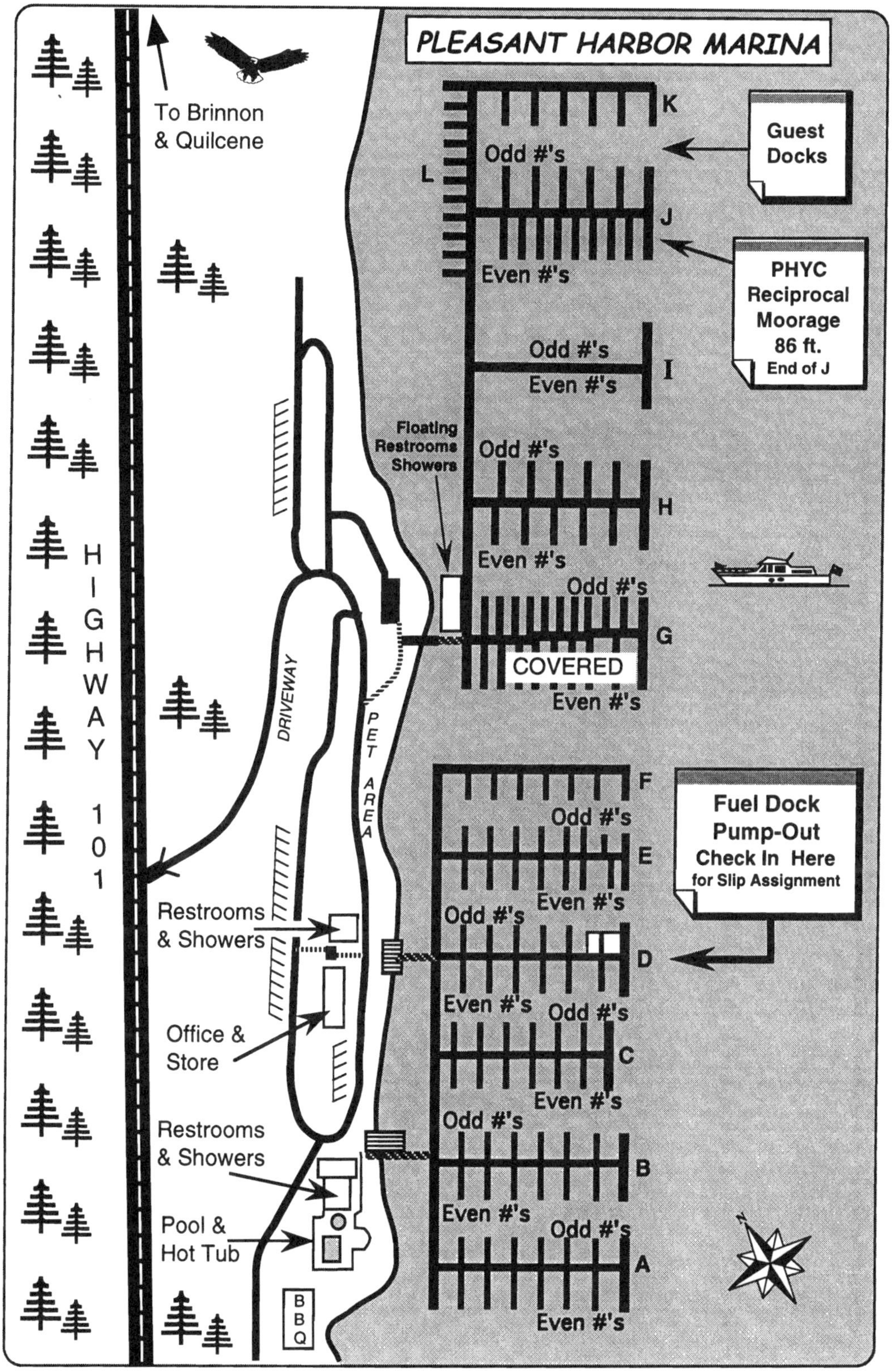

PLEASANT HARBOR MARINA

To Brinnon
& Quilcene

Guest
Docks

PHYC
Reciprocal
Moorage
86 ft.
End of J

K
L
J
I
H
G
F
E
D
C
B
A

Odd #'s
Even #'s

Floating
Restrooms
Showers

COVERED

DRIVEWAY

PET AREA

HIGHWAY 101

Fuel Dock
Pump-Out
Check In Here
for Slip Assignment

Restrooms
& Showers

Office &
Store

Restrooms
& Showers

Pool &
Hot Tub

BBQ

N

NOTES

Quilcene

NAME OF MARINA: **HERB BECK MARINA/Quilcene Boat Haven** RADIO: None
TELEPHONE: 360-765-3131 MGR: Port of Port Townsend
E-MAIL: info@portofpt.com FAX: 360-379-8205 (Fax)
ADDRESS: 1731 Linger Longer Rd. Quilcene, WA 98376
SHORT DESCRIPTION & LOCATION: **www.portofpt.com**

47°48.00' - 122°49.60' Located in Quilcene Bay, a small inlet on the W. side of Dabob Bay N. of Whitney Point. Small rustic marina is run by the Port of Port Townsend. Marina is on the W. side of Bay 1.4 mi. S. of the Town of Quilcene, famous for oysters & seafood.

GUEST BOAT CAPACITY: ....Appx 5-6 boats
DOCKSIDE DEPTH AT ZERO TIDE: ....6 ft. +
SEASON: ....All year
RESERVATION POLICY: ....None
AMT W/ELECTRICITY: ....All
FUEL DOCK: ....Gas & Diesel
MARINE REPAIRS: ....Close by
TOILETS: ....Yes
HOT SHOWERS: ....Yes
RESTAURANT: ....In Quilcene
PICNIC AREA: ....Yes
BASIC STORE: ....None
BROADBAND/WI-FI: ....None
DAILY RATE: ....Economical (Under 75¢/foot)

GUEST DOCK: 70 ft.+ open slips
GUEST SLIPS: ....Varies
WATER: ....Yes
AMPS: ....20-30 A
PUMP OUT STATION ....Yes
HAUL OUT: ....None
BOAT RAMP: ....Yes
LAUNDRY: ....Close by
BAR: ....In Quilcene
POOL: ....None
GOLF: ....None
PET FRIENDLY: ....Excellent
OTHER: Water activities, fishing clamming shrimping, swimming. RV Parking.

CAUTION! This chartlet not intended for use in navigation.

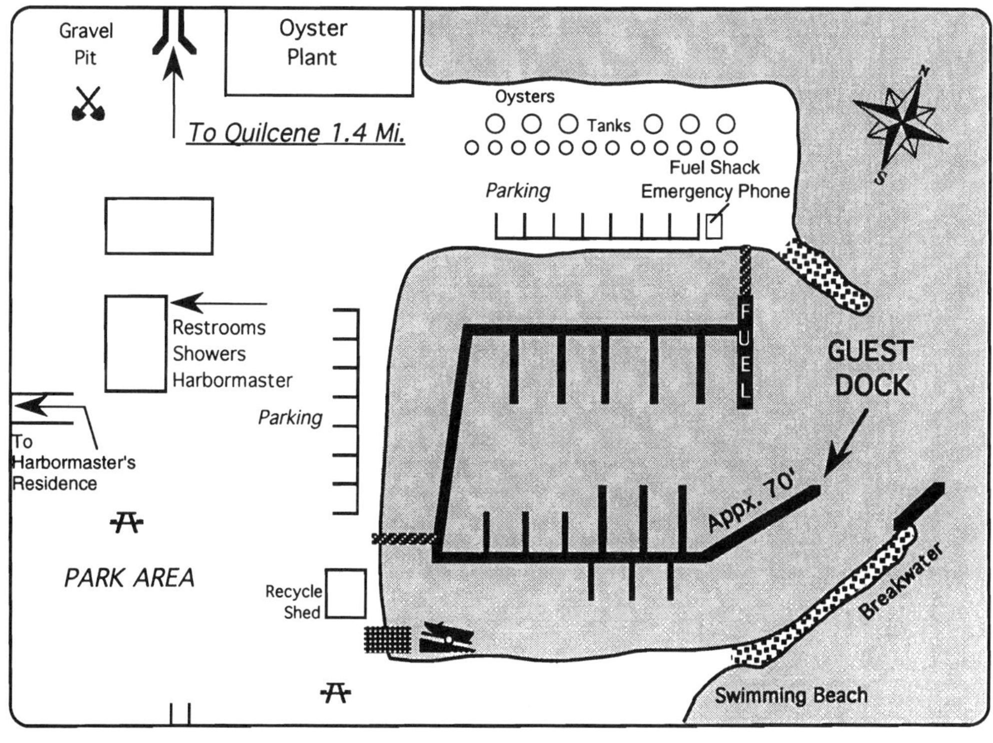

Port Ludlow

NAME OF MARINA: ***PORT LUDLOW MARINA*** RADIO: VHF 68
TELEPHONE: **360-437-0513**/800-308-7991 MGR: Kori Ward
E-MAIL: marina@portludlowresort.com FAX: 360-437-2428 (Fax)
ADDRESS: 1 Gull Drive Port Ludlow, WA 98365
SHORT DESCRIPTION & LOCATION: **www.portludlowresort.com**
47°55.30' - 122°41.10' Located about 12 mi. S. of Port Townsend on N. side of Port Ludlow Bay. The upscale resort & marina offer full marine services in a well protected harbor. There is a large guest dock plus open slips are rented out to overnighters.

GUEST BOAT CAPACITY: ....Appx. 50 boats
DOCKSIDE DEPTH AT ZERO TIDE: ....19 ft.
SEASON: ....All year
RESERVATION POLICY: ....Accepts
AMT W/ELECTRICITY: ....All
FUEL DOCK: ....Gas & Diesel
MARINE REPAIRS: ....None
TOILETS: ....Yes
HOT SHOWERS: ....Yes
RESTAURANT: ....Yes - 2 ea.
PICNIC AREA: ....Yes
BASIC STORE: ....Yes
BROADBAND/WI-FI: ....Yes
DAILY RATE: ....Moderate (90¢-$1.25/foot)

GUEST DOCK: ...Appx.300' + slips
GUEST SLIPS: ....Varies
WATER: ....Yes
AMPS: ....30-50 A
PUMP OUT STATION ....Yes
HAUL OUT: ....None
BOAT RAMP: ....None
LAUNDRY: ....Yes
BAR: ....Yes - 2 ea.
POOL: ....None
GOLF: ....Yes
PET FRIENDLY: ....Excellent
OTHER: Hotel, kayak, boat and bicycle rentals, tennis, gifts & marine supplies, free shuttle to golf course.

NAME OF YACHT CLUB: ***PORT LUDLOW YACHT CLUB***
CLUB ADDRESS: P.O. Box 65338, Port Ludlow, WA 98365
CLUB TELEPHONE: PERSON IN CHARGE: Port Captain
LOCATION & SPECIAL NOTES: **www.olympus.net/community/plyc**
PLYC offers 80 feet of reciprocal moorage located just on the inside of the first float ("A" Dock) near the fuel dock. Visiting reciprocal yachtsmen are allowed 48 hours of free moorage in a 30 day period - annual limit 5 days. Max boat size 60 ft. Please check-in with harbormaster upon arrival. Your current yacht club membership card is required.

RECIPROCAL BOAT CAPACITY: ....2-4 Boats
DOCKSIDE DEPTH AT ZERO TIDE: ....32 ft.
RECIPROCAL SEASON: ....All year
RESERVATION POLICY: ....None
TOILETS: ....At Marina
HOT SHOWERS: ....At Marina
RESTAURANT: ....Close by
DAILY RATE: ....No charge for moorage. ....Power charge at marina rates.

RECIPROCAL DOCK: ....80 ft.
RECIPROCAL SLIPS: ...Dock only
WATER: ....Yes
AMT W/ELECTRICITY: ....All
AMPS: ....30-50 A
BAR: ....Close by
OTHER: ....Same as marina listing

***NOTE:* THIS IS PRIVATE MOORAGE *ONLY* AVAILABLE TO MEMBERS OF RECIPROCAL YACHT CLUBS! YOUR CLUB *MUST* HAVE RECIPROCAL PRIVILEGES AND YOU MUST FLY YOUR BURGEE!**

Port Ludlow

CAUTION! This chartlet not intended for use in navigation.

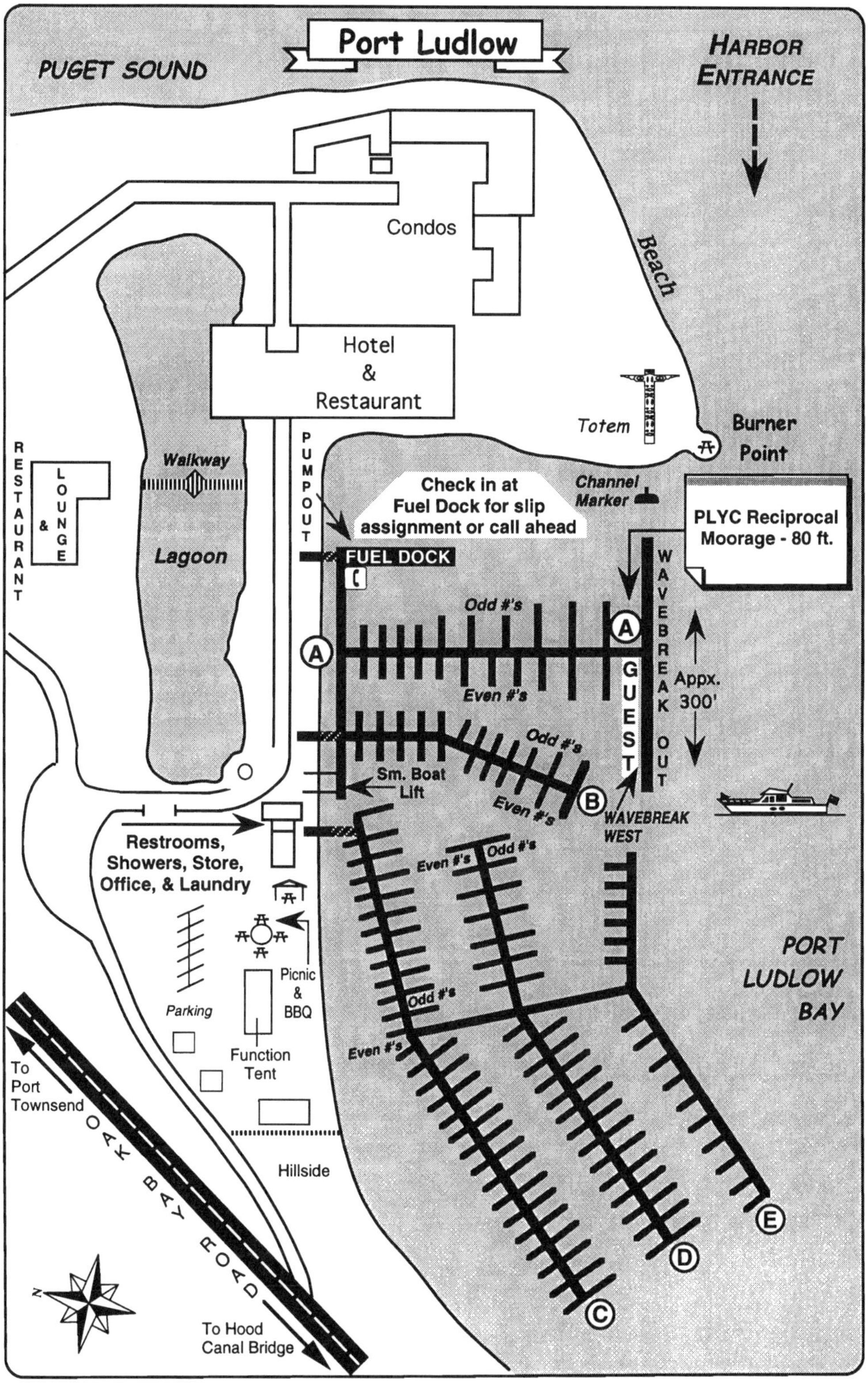

Port Ludlow

PUGET SOUND

HARBOR ENTRANCE

Condos

Beach

Hotel & Restaurant

Totem

Burner Point

RESTAURANT & LOUNGE

Walkway

Lagoon

PUMPOUT

Check in at Fuel Dock for slip assignment or call ahead

Channel Marker

PLYC Reciprocal Moorage - 80 ft.

FUEL DOCK

Odd #'s

Even #'s

A

WAVEBREAK OUT

GUEST

Appx. 300'

Odd #'s

Even #'s

B

Sm. Boat Lift

WAVEBREAK WEST

Restrooms, Showers, Store, Office, & Laundry

Even #'s

Odd #'s

Picnic & BBQ

Parking

Function Tent

Odd #'s

Even #'s

PORT LUDLOW BAY

To Port Townsend

OAK BAY ROAD

Hillside

N

To Hood Canal Bridge

C

D

E

Port Hadlock

NAME OF MARINA: *PORT HADLOCK MARINA* RADIO: VHF 16/68
TELEPHONE: **360-385-6368** MGR: Jerry Spencer
E-MAIL: phmarina@mail.com FAX: 360-385-3067 (Fax)
ADDRESS: 310 Hadlock Bay Rd, Port Hadlock, WA 98339
SHORT DESCRIPTION & LOCATION: **www.innatporthadlock.com/marina.php**

48°01.80' - 122°44.60' Located about 5 mi. S. of Port Townsend just W. of entrance to Pt. Townsend Canal. Community marina features The Inn at Port Hadlock & Restaurant. Guest moorage consists of space available slips as assigned by Harbormaster.

GUEST BOAT CAPACITY: ....Appx. 8 boats
DOCKSIDE DEPTH AT ZERO TIDE: ....13 ft.
SEASON: ....All year
RESERVATION POLICY: ......1st Come, 1st Serve
AMT W/ELECTRICITY: ....All
FUEL DOCK: ....None
MARINE REPAIRS: ....Close by
TOILETS: ....Yes
HOT SHOWERS: ....Yes
RESTAURANT: ....Yes
PICNIC AREA: ....Yes
BASIC STORE: ....Close by
BROADBAND/WI-FI: ....Yes
DAILY RATE:.................Moderate (75¢-$1.25/foot)
NOTE: Please call at least a week ahead for more than 4 boats in group.

GUEST DOCK: ....Varies
GUEST SLIPS: ....Varies
WATER: ....Yes
AMPS: ....30-50 A
PUMP OUT STATION ....Yes
HAUL OUT: ....None
BOAT RAMP: ....Close by
LAUNDRY: ....Yes
BAR: ....Yes
POOL: ....None
GOLF: ....Close by
PET FRIENDLY: ....Excellent
OTHER: Hotel & restaurant at marina. Walking distance to Port Hadlock sm. town center, supermarket, restaurants and services.

CAUTION! This chartlet not intended for use in navigation.

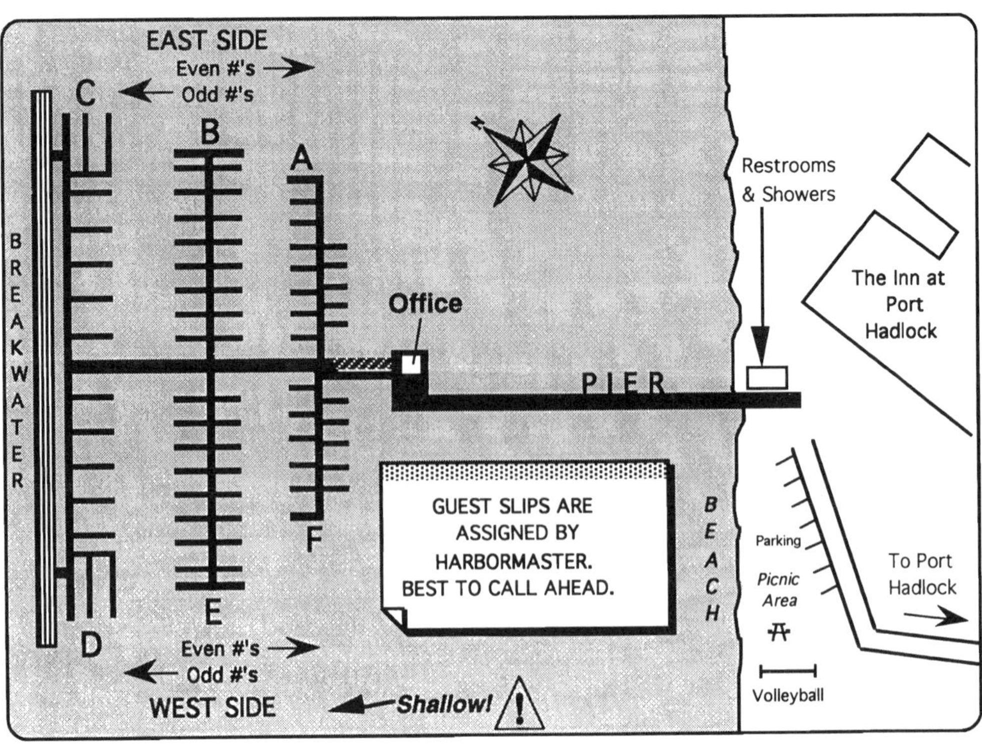

Marrowstone Is.

NAME OF PARK: *FORT FLAGLER STATE PARK*
ADDRESS: 10541 Fort Flagler Rd. Nordland WA 98358
TELEPHONE: 360-385-1259 MGR: Michael Zimmerman
SHORT DESCRIPTION & LOCATION: **www.parks.wa.gov/moorage/parks**

48°05.55' - 122°43.15' The park floats are located on the N. side of the first bend of the winding entrance to Kilisut Harbor which lies between Indian Is. on the W. and Marrowstone Is. on the E. The park is a historic former Army fort & consists of 783 acres.

GUEST BOAT CAPACITY: ....Appx. 8 boats
DOCKSIDE DEPTH AT ZERO TIDE:.............6 ft.
SEASON: ....No Docks from Oct-April
AMT W/ELECTRICITY: ....None
TOILETS: ....Yes
HOT SHOWERS: ....Yes
PICNIC AREA: ....Yes
PLAY AREA: ....Yes
BASIC STORE: ....Yes - summer only
DAILY RATE: Dock Moorage........................50¢/foot
Mooring Buoys...............$10.00/night
$10 Minimum

GUEST DOCK: ....244 ft. total
MOORING BUOYS: 7-Avail. all yr.
WATER: ....Yes
PAY PHONES: ....None
BOAT RAMP: ....Yes
PICNIC SHELTER: ....Yes
BBQ: ....Yes
PUMP OUT STATION: ......Porta
PET FRIENDLY: ....Excellent
OTHER: Lodging, campsites, marine trails, swimming, scuba diving, underwater park, fishing, porta-potty dump station.

CAUTION! This chartlet not intended for use in navigation.

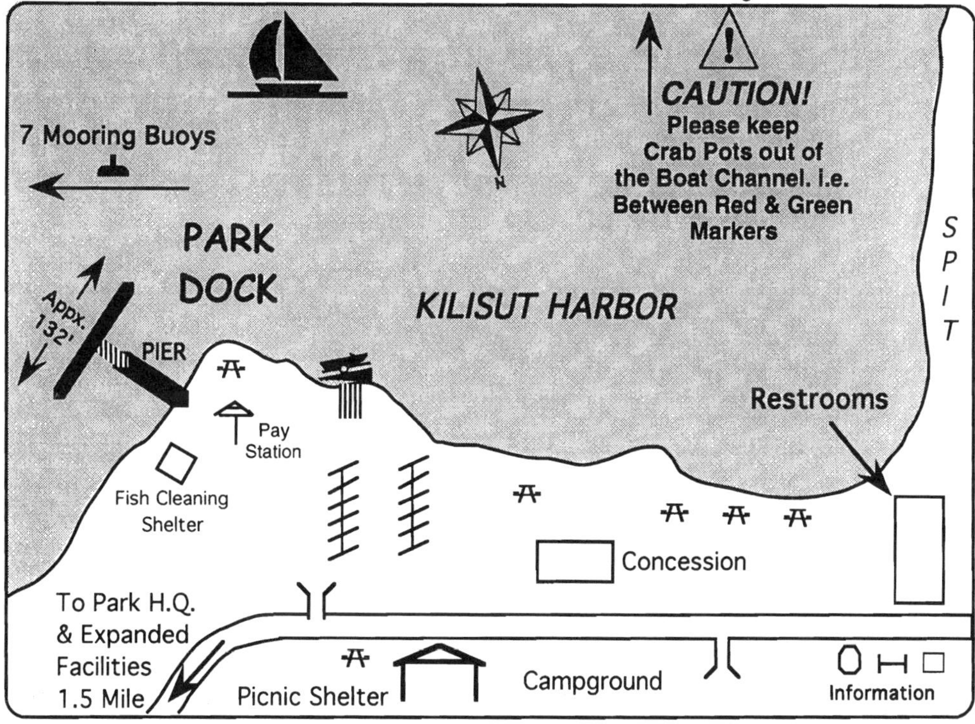

Marrowstone Is.

NAME OF PARK: *MYSTERY BAY STATE PARK*

ADDRESS: 10541 Flagler Rd. Nordland, WA 98358

TELEPHONE: 360-385-1259 MGR: Michael Zimmerman

SHORT DESCRIPTION & LOCATION: **www.parks.wa.gov/moorage/parks**

48°03.50' - 122°41.60' Located in northeastern portion of Mystery Bay in Kilisut Harbor adjacent to west side of Marrowstone Island. Ten acre day use park in quiet rural setting. The park is affiliated with Ft. Flagler State Park 3 miles north.

GUEST BOAT CAPACITY: ....Appx. 15-20 boats

DOCKSIDE DEPTH AT ZERO TIDE:..........4.5 ft.

SEASON: ....All year

AMT W/ELECTRICITY: ....None

TOILETS: ....Vault toilets

HOT SHOWERS: ....None

PICNIC AREA: ....Yes

PLAY AREA: ....Yes

BASIC STORE: ....Close by

DAILY RATE: Dock Moorage....................50¢/foot
Mooring Buoys...............$10.00/night
$10 Minimum

GUEST DOCK: ....Appx. 640 ft.

MOORING BUOYS: ....7 Buoys

WATER: ....March thru October

PAY PHONES: ....None

BOAT RAMP: ....Yes

PICNIC SHELTER: ....Yes

BBQ: ....Yes

PUMP OUT STATION: ....Yes

PET FRIENDLY: ....Excellent

OTHER: Clam beaches, unguarded swimming beach, bicycling, and great shore side picnic area. No camping.

CAUTION! This chartlet not intended for use in navigation.

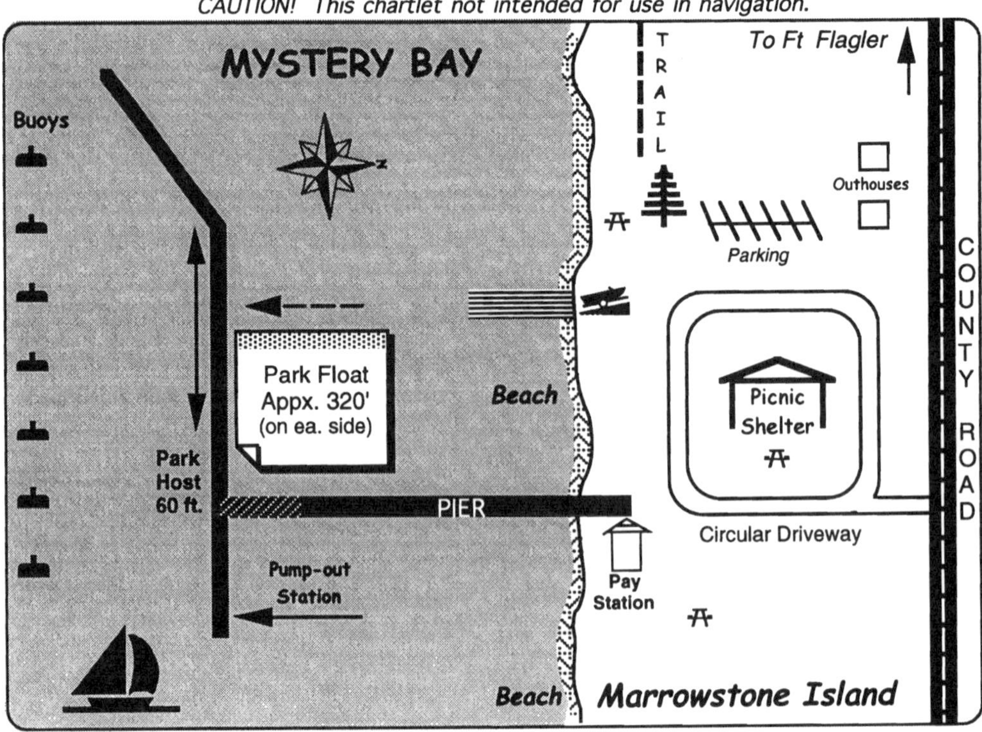

Port Townsend

NAME OF MARINA: ***HUDSON POINT MARINA*** **RADIO:** VHF Ch. 9
TELEPHONE: **800-228-2803**/360-385-2828 MGR: Chris Wenger
E-MAIL: info@portofpt.com FAX: 360-385-7331 (Fax)
ADDRESS: 103 Hudson St., POB 1180 Port Townsend, WA 98368
SHORT DESCRIPTION & LOCATION: **www.portofpt.com**

48°07.00'-122°44.90' Located in downtown Port Townsend in a protected dredged harbor, formerly a Coast Guard station. The well maintained docks have been reconstructed by the Port in 2007. Close to shops, galleries, restaurants, historic sights & Victorian bldgs.

GUEST BOAT CAPACITY: ....Appx. 100 boats
DOCKSIDE DEPTH AT ZERO TIDE: ....12 ft.
SEASON: ....All year
RESERVATION POLICY: ....Accepts
AMT W/ELECTRICITY: ....All
FUEL DOCK: ....Close by
MARINE REPAIRS: ....On premises
TOILETS: ....Yes
HOT SHOWERS: ....Yes
RESTAURANT: ....3 On premises
PICNIC AREA: ....Yes
BASIC STORE: ....Close by
BROADBAND/WI-FI: ....BroadbandXpress & S2S
DAILY RATE: ....Moderate (75¢-$1.25/foot)

GUEST DOCK: ....570' plus slips
GUEST SLIPS:. Appx. 45 + Dock
WATER: ....Yes
AMPS: ....30-50 A
PUMP OUT STATION ....Yes
HAUL OUT: ....Travel-Lift
BOAT RAMP: ....None
LAUNDRY: ....Yes
BAR: ....Close by
POOL: ....None
GOLF: ....Close by
PET FRIENDLY: ....Excellent
OTHER: RV Park, group event facility, charts, ice. Home of Wooden Boat Festival held in Sept.

CAUTION! This chartlet not intended for use in navigation.

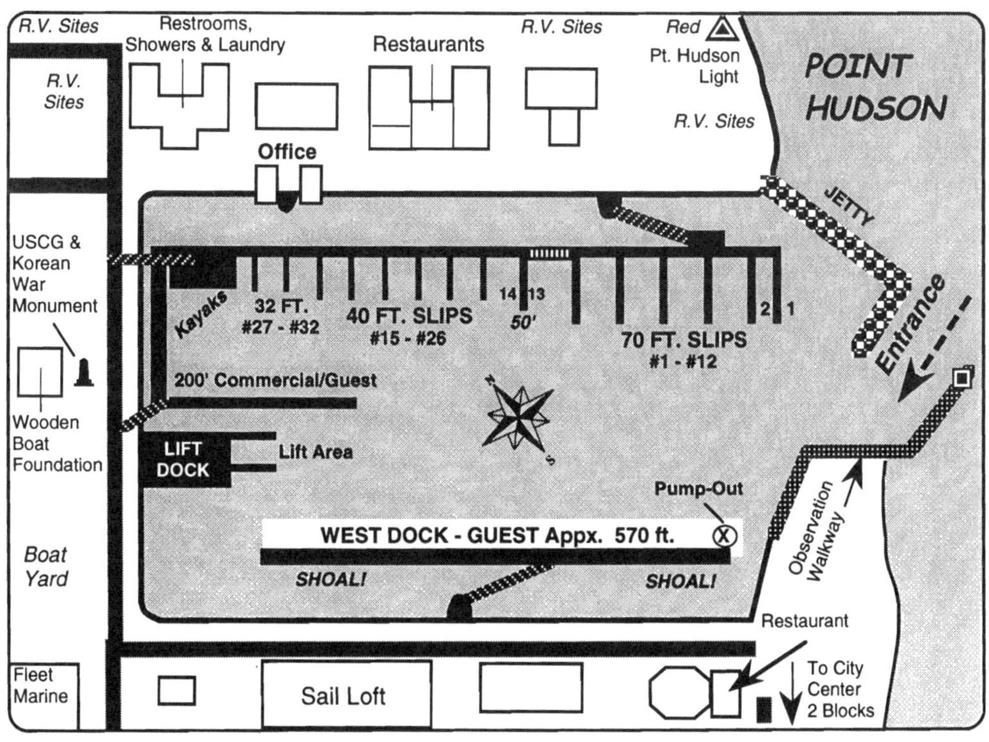

Port Townsend

NAME OF MARINA: ***PORT TOWNSEND BOAT HAVEN*** RADIO: VHF 9/66A
TELEPHONE: 800-228-2803 MGR: Ken Radon/Tami Ruby
E-MAIL: info@portofpt.com FAX: 360-379-8205 (Fax)
ADDRESS: P.O. Box 1180 Port Townsend, WA 98368
SHORT DESCRIPTION & LOCATION: www.portofpt.com
48°06.40' - 122°46.25' Located one mile SSW of Point Hudson in historic Port Townsend. Largest public boat yard on west coast. ***Tie up at Fuel Dock & register at Port Office for slip.*** After hours open slip list posted nightly at Port Office.

GUEST BOAT CAPACITY: ....Varies
DOCKSIDE DEPTH AT ZERO TIDE: ....12 ft.
SEASON: ....All year
RESERVATION POLICY: ....None
AMT W/ELECTRICITY: ....All
FUEL DOCK: ....Gas & Diesel
MARINE REPAIRS: ....On premises
TOILETS: ....Yes
HOT SHOWERS: ....Yes
RESTAURANT: ....Close by
PICNIC AREA: ....Yes
BASIC STORE: ....Yes
BROADBAND/WI-FI: ....BroadbandXpress
DAILY RATE: ....Moderate (75¢-$1.25/foot)

GUEST DOCK: ....No
GUEST SLIPS: ....Varies
WATER: ....Yes
AMPS: ....30 A
PUMP OUT STATION ....Yes
HAUL OUT: ....Travel-Lift
BOAT RAMP: ....Yes
LAUNDRY: ....Yes
BAR: ....Close by
POOL: ....None
GOLF: ....Close by
PET FRIENDLY: ....Fair
OTHER: Close to P.T. attractions & services, dry storage, large boatyard, 75 & 300 ton lift, many chandleries and skilled ship wrights.

B

NAME OF YACHT CLUB: ***PORT TOWNSEND YACHT CLUB***
CLUB ADDRESS: P.O. Box 75, Port Townsend, WA 98368
CLUB TELEPHONE: Use website PERSON IN CHARGE: Vice Commodore
LOCATION & SPECIAL NOTES: **www.porttownsendyachtclub.org**
PTYC offers slip #281 on "D" Dock for reciprocal boats up to 45 ft. all year long. An additional slip is available June thru August - be sure to check with Port Office for location. Plz call Port Office on tel. or VHF to check in, or visit moorage office before proceeding to reciprocal moorage. Current membership card required from reciprocal club.

RECIPROCAL BOAT CAPACITY: ....1-2 boats
DOCKSIDE DEPTH AT ZERO TIDE: ....12 ft.
RECIPROCAL SEASON: ....All year
RESERVATION POLICY: ....None
TOILETS: ....At Marina
HOT SHOWERS: ....At Marina
RESTAURANT: ....Close by
DAILY RATE: $3.00/Power. **Limit:** 1 night from May-Sept. & 1 night/mo. Oct-Apr.

RECIPROCAL DOCK: See above
RECIPROCAL SLIPS: ....1 or 2
WATER: ....Yes
AMT W/ELECTRICITY: ....All
AMPS: ....30 A
BAR: ....Close by
OTHER: NO RAFTING ALLOWED. CHECK OUT BY 12 NOON

***NOTE:* THIS IS PRIVATE MOORAGE *ONLY* AVAILABLE TO MEMBERS OF RECIPROCAL YACHT CLUBS! YOUR CLUB *MUST* HAVE RECIPROCAL PRIVILEGES AND YOU MUST FLY YOUR BURGEE!**

Port Townsend

CAUTION! This chartlet not intended for use in navigation.

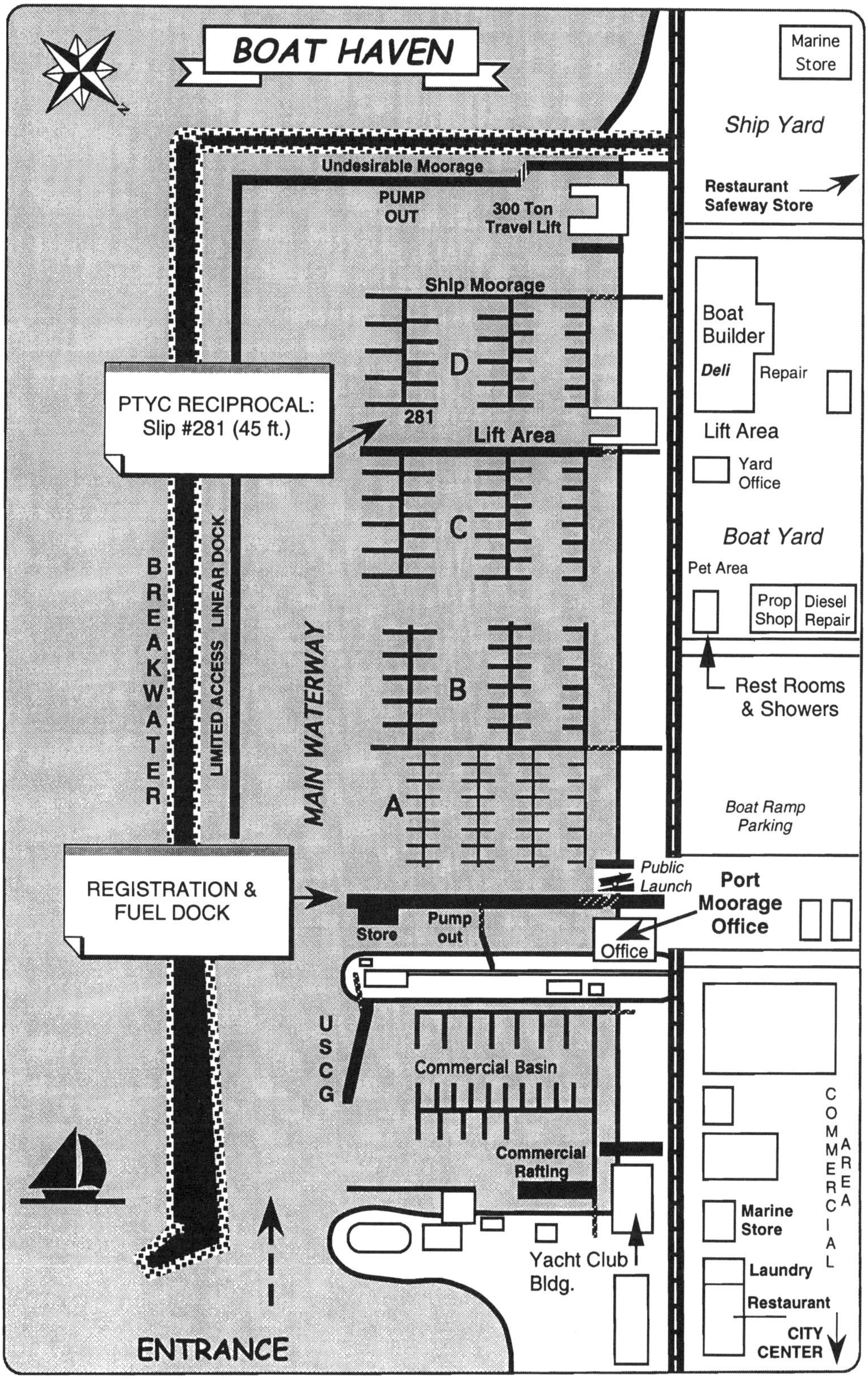

BOAT HAVEN

Marine Store

Ship Yard

Restaurant
Safeway Store

Undesirable Moorage

PUMP OUT

300 Ton Travel Lift

Ship Moorage

D

281

Lift Area

PTYC RECIPROCAL:
Slip #281 (45 ft.)

Boat Builder

Deli

Repair

Lift Area

Yard Office

C

Boat Yard

Pet Area

Prop Shop

Diesel Repair

BREAKWATER

LIMITED ACCESS LINEAR DOCK

MAIN WATERWAY

B

Rest Rooms & Showers

A

Boat Ramp Parking

Public Launch

Port Moorage Office

REGISTRATION & FUEL DOCK

Store

Pump out

Office

USCG

Commercial Basin

COMMERCIAL AREA

Commercial Rafting

Marine Store

Yacht Club Bldg.

Laundry

Restaurant

CITY CENTER

ENTRANCE

Port Townsend

NAME OF PARK: *FORT WORDEN STATE PARK & CONFERENCE CENTER*
ADDRESS: 200 Battery Way Port Townsend, WA 98368
TELEPHONE: 360-344-4400 MGR: Kate Burke/Current Park Mgr
SHORT DESCRIPTION & LOCATION: **www.parks.wa.gov/moorage/parks**

48°08.20' - 122°45.50' The park floats are located about 1.5 mi. N. of Pt. Hudson and 1/2 mi. S. of Pt. Wilson. The large 443 acre park was once a military installation and the landscaped and historic grounds offer many attractions for visitors.

GUEST BOAT CAPACITY:	Appx. 2-4 boats	GUEST DOCK:	128 ft. total
DOCKSIDE DEPTH AT ZERO TIDE:	Appx 15 ft.	MOORING BUOYS:	7 Buoys
SEASON:	All year	WATER:	Yes
AMT W/ELECTRICITY:	None	PAY PHONES:	Yes
TOILETS:	Yes	BOAT RAMP:	Yes - Fee
HOT SHOWERS:	Yes	PICNIC SHELTER:	Yes
PICNIC AREA:	Yes	BBQ:	Yes
PLAY AREA:	Yes	PUMP OUT STATION:	None
BASIC STORE: & CAFE:	All Year	PET FRIENDLY:	Excellent
DAILY RATE:	Dock Moorage 50¢/foot Mooring Buoys $10.00/night Minumum Charge: $10	OTHER:	Marine Science Center, Natural History Museum, tennis, athletic fields, diving, fishing, swimming, & marine trails.

CAUTION! This chartlet not intended for use in navigation.

"Canteen" (Store)
Restrooms & Showers
Natural History Museum
Campground
Pay Station
Kitchen Shelter
Beach
Guest Float Appx. 60 ft.
Breakwater
PIER
PORT TOWNSEND
Marine Science Center
PIER
Mooring Buoys

Sequim

NAME OF PARK: ***SEQUIM BAY STATE PARK***
ADDRESS: 269035 Hwy 101 Sequim, WA 98382
TELEPHONE: **360-683-4235** MGR: Steve Gilstrom
SHORT DESCRIPTION & LOCATION: **www.parks.wa.gov/moorage/parks**

48°02.45' -123°01.50' Located in Sequim Bay, 6 mi. S.E. of Dungeness Bay. The 3.8 mi. long bay is landlocked and the Park is at the S.W. end of the bay N. of Schoolhouse Point. At low tide the water shallows at the float & buoys recommended for large boats.

GUEST BOAT CAPACITY: ....Appx. 15-20 boats
DOCKSIDE DEPTH AT ZERO TIDE:.............7 ft.
SEASON: ....All year
AMT W/ELECTRICITY: ....None
TOILETS: ....Yes
HOT SHOWERS: ....Yes
PICNIC AREA: ....Yes
PLAY AREA: ....Yes
BASIC STORE: ....3 Miles
DAILY RATE: Dock Moorage........................50¢/foot
Mooring Buoys...............$10.00/night
Minumum Charge: $10

GUEST DOCK: Appx. 400 ft. TTL
MOORING BUOYS: ....6 Buoys
WATER: ....Yes
PAY PHONES: ....Yes
BOAT RAMP: ....Yes
PICNIC SHELTER: ....Yes
BBQ: ....Yes
PUMP OUT STATION: ......None
PET FRIENDLY: ....Excellent
OTHER: ....92 Acre park, campsites, fishing, tennis, hiking, swimming, games, scuba diving.

CAUTION! This chartlet not intended for use in navigation.

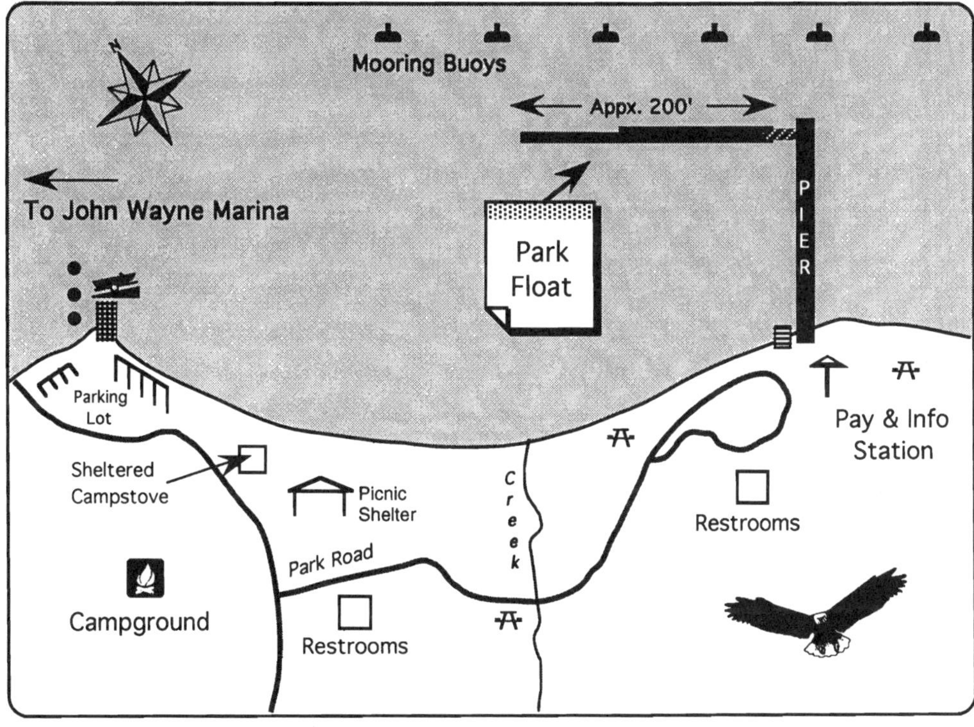

Sequim

NAME OF MARINA: ***JOHN WAYNE MARINA*** RADIO: None
TELEPHONE: **360-417-3440** MGR: Ron Amundson
E-MAIL: rona@portofpa.com FAX: 360-417-3442 (Fax)
ADDRESS: 2577 W. Sequim Bay Rd. Sequim, WA 98382
SHORT DESCRIPTION & LOCATION: **www.portofpa.com/havens.html**
48°03.99' -123°02.22' Located in Sequim Bay, 6 mi. S.E. of Dungeness Bay. The bay is accessed via a well marked channel at N.W. corner & marina is marked by nav. lights at the E. ends of the breakwaters at Pitship Pt. The modern facility offers full boater services.

GUEST BOAT CAPACITY: ....Appx. 25 boats
DOCKSIDE DEPTH AT ZERO TIDE: ....15 ft.
SEASON: ....All year
RESERVATION POLICY:..........None
AMT W/ELECTRICITY: ....All
FUEL DOCK: ....Gas & Diesel
MARINE REPAIRS: ....Close by
TOILETS: ....Yes
HOT SHOWERS: ....Yes
RESTAURANT: ....None
PICNIC AREA: ....Yes
BASIC STORE: ....Yes + marine gifts & supplies
BROADBAND/WI-FI: ....Yes
DAILY RATE: ....Moderate (75¢-$1.25/foot)

GUEST DOCK: ....160' plus slips
GUEST SLIPS: ....Varies
WATER: ....Yes
AMPS: ....20-30 A
PUMP OUT STATION ....Yes
HAUL OUT: ....None
BOAT RAMP: ....Yes
LAUNDRY: ....Yes
BAR: ....None
POOL: ....None
GOLF: ....Close by
PET FRIENDLY: ....Excellent
OTHER: Bus service to town of Sequim 3 mi. away, boat rentals, fishing, clamming, crabbing, groceries, bait & tackle.

NAME OF YACHT CLUB: ***THE SEQUIM BAY YACHT CLUB***
CLUB ADDRESS: P.O. Box 1261, Sequim, WA 98382
CLUB TELEPHONE: 360-683-1338 PERSON IN CHARGE: Reciprocal Chair
LOCATION & SPECIAL NOTES: **www.sequimbayyachtclub.com**
1 nights moorage per month in John Wayne Marina is extended to members of reciprocal clubs. A minimum of two slips are available. Guest slips will be assigned by harbormaster; after hours use sign up sheets outside harbormasters office. Limits: 2 Boats per club at one time and 1 visit per month. Please show current yacht club card to harbormaster.

RECIPROCAL BOAT CAPACITY:.............Min. of 2
DOCKSIDE DEPTH AT ZERO TIDE: ....15 ft.
RECIPROCAL SEASON:..........All year
RESERVATION POLICY: ....None
TOILETS: ....At Marina
HOT SHOWERS: ....At Marina
RESTAURANT: ....None
DAILY RATE: ....No charge (Except for electricity charge)

RECIPROCAL DOCK: ....None
RECIPROCAL SLIPS: ....Varies
WATER: ....Yes
AMT W/ELECTRICITY: .....Varies
AMPS: ....30 A
BAR: ....None
OTHER: ....Same as marina listing

***NOTE:* THIS IS PRIVATE MOORAGE *ONLY* AVAILABLE TO MEMBERS OF RECIPROCAL YACHT CLUBS! YOUR CLUB *MUST* HAVE RECIPROCAL PRIVILEGES AND YOU MUST FLY YOUR BURGEE!**

Sequim

CAUTION! This chartlet not intended for use in navigation.

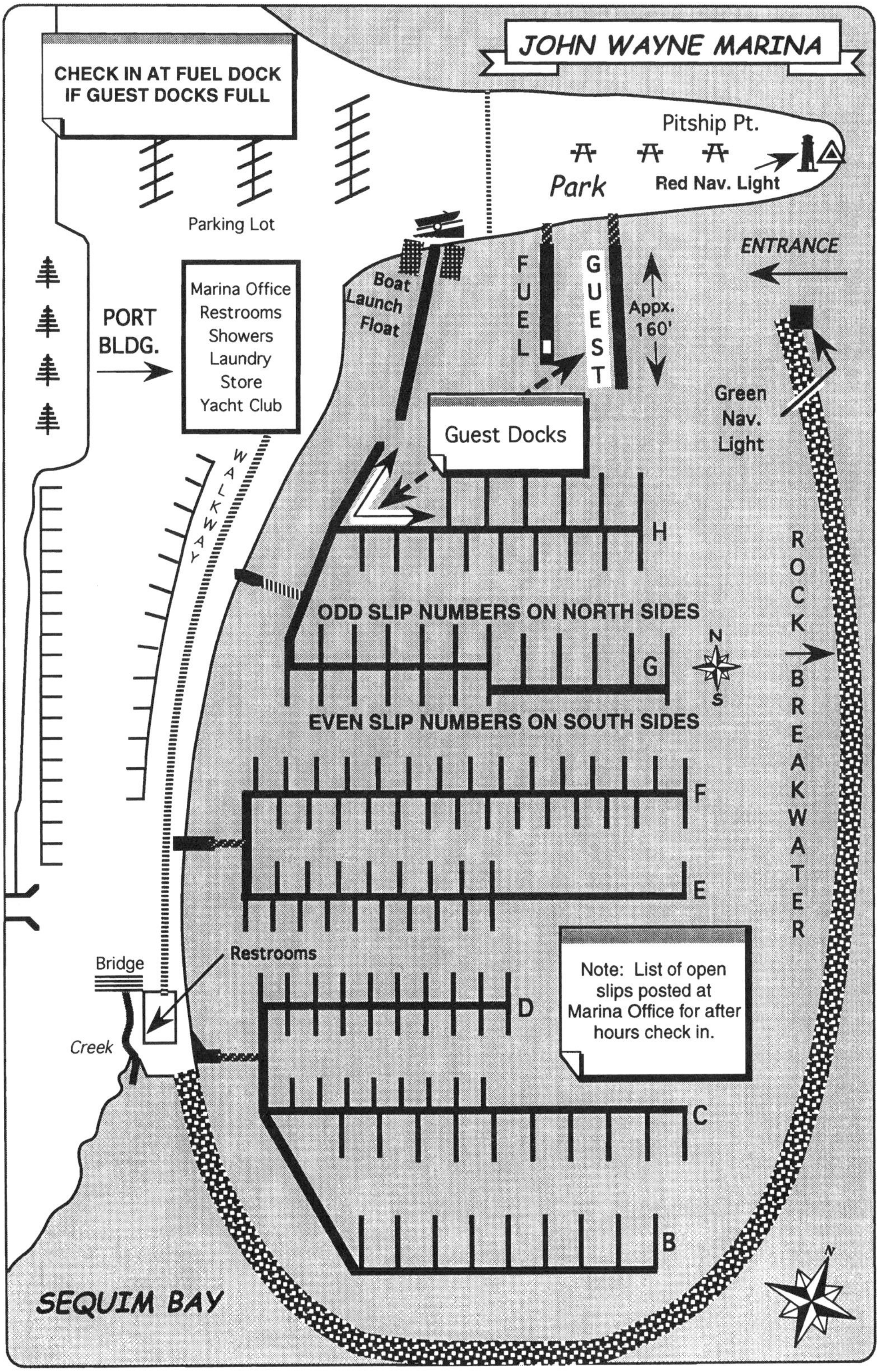

JOHN WAYNE MARINA

CHECK IN AT FUEL DOCK
IF GUEST DOCKS FULL

Pitship Pt.

Park

Red Nav. Light

Parking Lot

ENTRANCE

PORT
BLDG.

Marina Office
Restrooms
Showers
Laundry
Store
Yacht Club

Boat
Launch
Float

FUEL

GUEST

Appx.
160'

Green
Nav.
Light

Guest Docks

WALKWAY

H

ROCK BREAKWATER

ODD SLIP NUMBERS ON NORTH SIDES

G

N

S

EVEN SLIP NUMBERS ON SOUTH SIDES

F

E

Bridge

Restrooms

Creek

D

Note: List of open
slips posted at
Marina Office for after
hours check in.

C

B

N

SEQUIM BAY

Port Angeles

NAME OF MARINA: ***PORT OF PORT ANGELES** Boat Haven* **RADIO:** None
TELEPHONE: **360-457-4505** MGR: Chuck Faires
E-MAIL: pamarina@olypen.com FAX: 360-452-4921 (Fax)
ADDRESS: 832 Boathaven Drive Port Angeles, WA 98363
SHORT DESCRIPTION & LOCATION: **www.portofpa.com**
48°07.60' - 123°27.08' Marina is located at southwest corner of Port Angeles Harbor just west of downtown Port Angeles. The well protected facility offers full services for the boater. Additional moorage is located at the Municipal Pier, east of the ferry dock.

GUEST BOAT CAPACITY: ....Appx. 20-30 boats
DOCKSIDE DEPTH AT ZERO TIDE: ....15 ft.
SEASON: ....All year
RESERVATION POLICY:..........None
AMT W/ELECTRICITY: ....All
FUEL DOCK: ....Gas & Diesel
MARINE REPAIRS: ....On premises
TOILETS: ....Yes
HOT SHOWERS: ....Yes
RESTAURANT: ....Yes
PICNIC AREA: ....None
BASIC STORE: ....Close by
BROADBAND/WI-FI: ....Yes
DAILY RATE:..........Economical (Under 75¢/foot)
MAXIMUM BOAT SIZE 130 FT.

GUEST DOCK: ....700 feet
GUEST SLIPS: ....Guest dock only
WATER: ....Yes
AMPS: ....20 A
PUMP OUT STATION ....Yes
HAUL OUT: ....Travel-Lift
BOAT RAMP: ....Yes
LAUNDRY: ....Close by
BAR: ....Close by
POOL: ....None
GOLF: ....Close by
PET FRIENDLY: ....Good
OTHER: Customs port-of-entry, city tours, good fishing grounds, walking distance to shops & restaurants, walking & biking trails.

NAME OF YACHT CLUB: ***PORT ANGELES YACHT CLUB***
CLUB ADDRESS: P.O. Box 692, Port Angeles, WA 98362
CLUB TELEPHONE: 360-457-4132 PERSON IN CHARGE: Reciprocal Chair
LOCATION & SPECIAL NOTES: **www.payc.org**
Club house is located on W. side of boat harbor entrance w/dinghy dock. PAYC provides reciprocal moorage on P.A. Boat Haven guest dock via reimbursing two night's moorage. Mail receipt from Boat Haven & copy of your yacht club card. Limit 2 boats per night - 3 nights per year. Max of $25 per night reimbursed per boat.

RECIPROCAL BOAT CAPACITY:.....2 Boats max.
DOCKSIDE DEPTH AT ZERO TIDE: ....15 ft.
RECIPROCAL SEASON:..........All year
RESERVATION POLICY: ....None
TOILETS: ....At Marina
HOT SHOWERS: ....At Marina
RESTAURANT: ....Close by
DAILY RATE: ....No Charge except shore power Two nights reimbursed up to $25 per night per boat.

RECIPROCAL DOCK: Guest dock
RECIPROCAL SLIPS: ...Dock only
WATER: ....Yes
AMT W/ELECTRICITY: ....All
AMPS: ....20-30 A
BAR: ....Close by
OTHER: Officers phone numbers posted at club house.

***NOTE:* THIS IS PRIVATE MOORAGE *ONLY* AVAILABLE TO MEMBERS OF RECIPROCAL YACHT CLUBS! YOUR CLUB *MUST* HAVE RECIPROCAL PRIVILEGES AND YOU MUST FLY YOUR BURGEE!**

Port Angeles

CAUTION! This chartlet not intended for use in navigation.

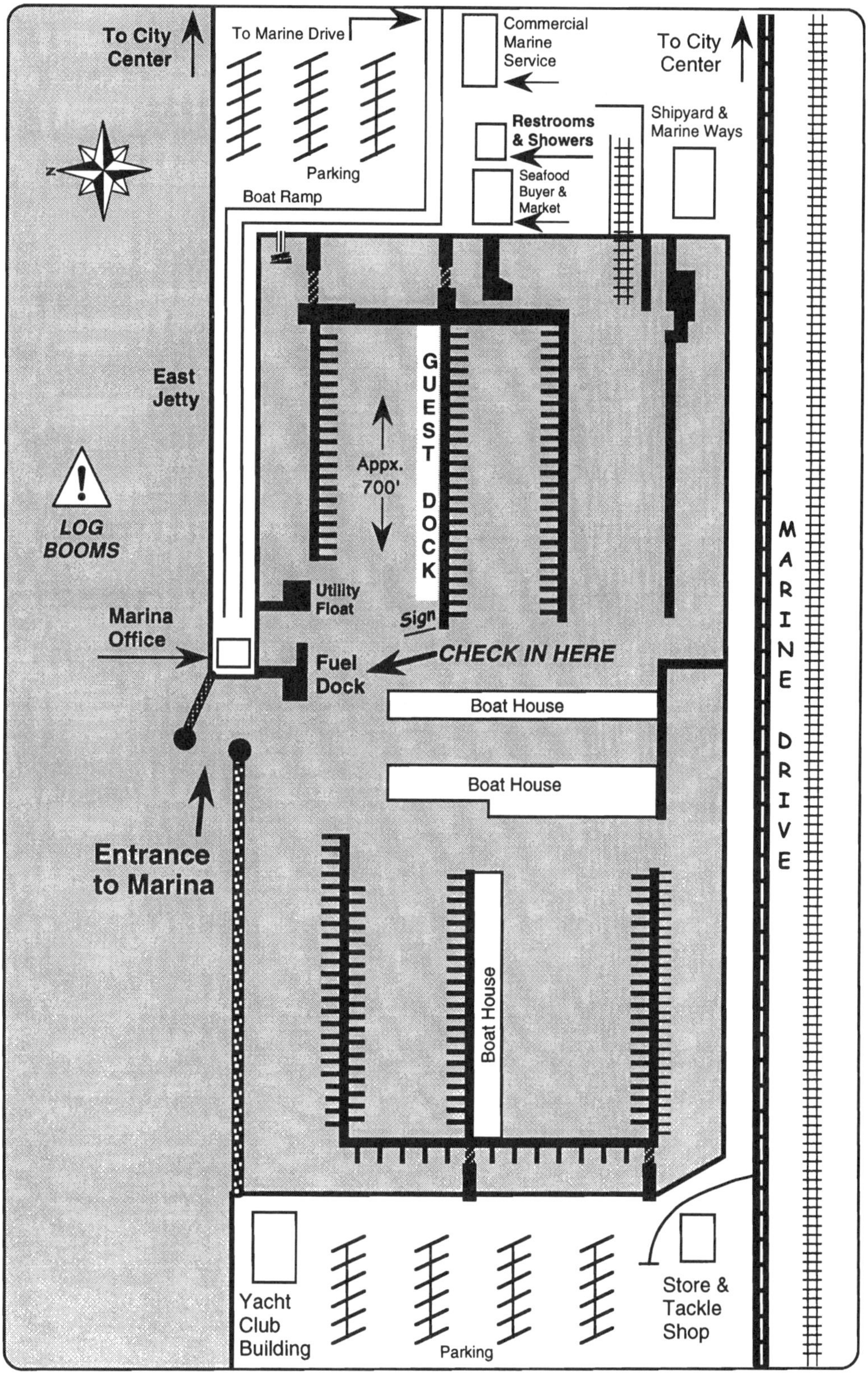

To City Center

To Marine Drive

Commercial Marine Service

To City Center

Restrooms & Showers

Shipyard & Marine Ways

Parking

Seafood Buyer & Market

Boat Ramp

East Jetty

GUEST DOCK

Appx. 700'

LOG BOOMS

Utility Float

Marina Office

Sign

CHECK IN HERE

Fuel Dock

Boat House

Boat House

MARINE DRIVE

Entrance to Marina

Boat House

Yacht Club Building

Parking

Store & Tackle Shop

Port Angeles

NAME OF PARK: *PORT ANGELES MUNICIPAL PARK & CITY PIER*
ADDRESS: P.O. Box 1150, City Hall Port Angeles, WA 98362
TELEPHONE: **360-417-4550** **MGR:** City of Port Angeles Parks Dept
SHORT DESCRIPTION & LOCATION: **www.ci.port-angeles.wa.us**

48°07.30' - 123°25.65' Located in downtown Port Angeles in a 2 acre waterfront park. A 50' tall observation tower provides a good landmark. Full services for boaters can be found as well as historic and tourist related activities. Maximum boat size is 40 feet.

GUEST BOAT CAPACITY:	Appx. 20 boats	GUEST DOCK:	4/Appx. 80 ft.
DOCKSIDE DEPTH AT ZERO TIDE:	Appx 20 ft	MOORING BUOYS:	None
SEASON:	Open Memorial Day to Labor Day	WATER:	None
AMT W/ELECTRICITY:	None	PAY PHONES:	None
TOILETS:	Yes	BOAT RAMP:	None
HOT SHOWERS:	None	PICNIC SHELTER:	Yes
PICNIC AREA:	Yes	BBQ:	None
PLAY AREA:	Yes	PUMP OUT STATION:	None
BASIC STORE:	Close by	PET FRIENDLY:	Good
DAILY RATE:	$10.00 per 24 hours	OTHER:	Marine laboratory on Pier open to public, sandy beach, fishing close by, walking paths, playground, special events, restaurants & shops close by.

NOTE: WATCH DEPTHS AT HEAD OF FLOATS

CAUTION! This chartlet not intended for use in navigation.

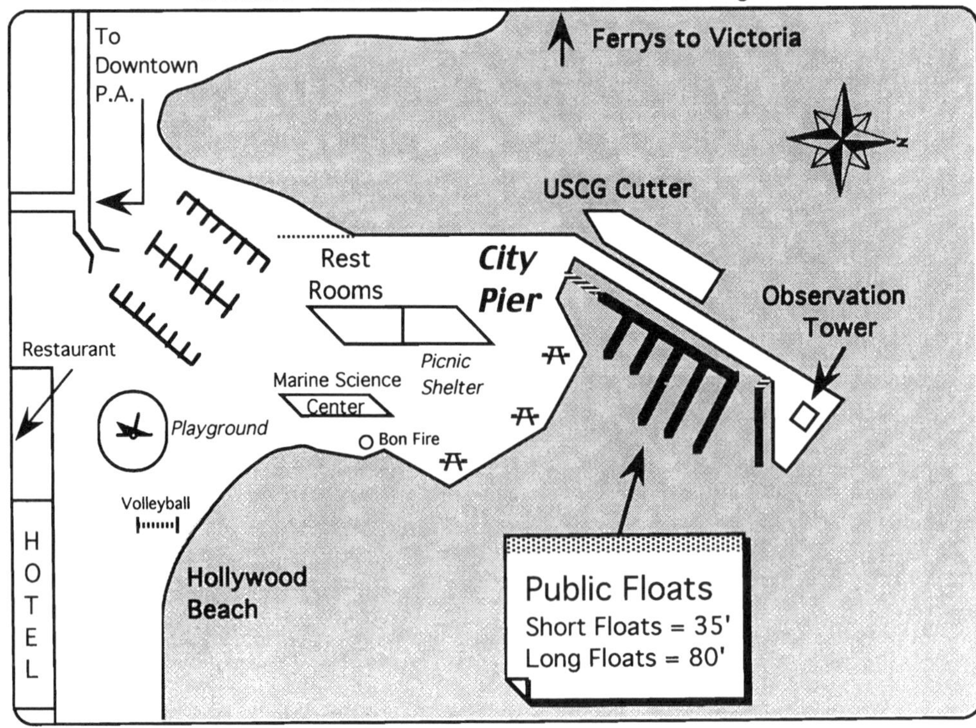

CHAPTER 4

WHIDBEY ISLAND & SKAGIT COUNTY

Chapter Map - Page 3

Langley

NAME OF MARINA: *LANGLEY BOAT HARBOR* **RADIO:** None
TELEPHONE: 360-221-2611 **MGR:** Harbormaster, Cell 360-914-1739
E-MAIL: None **FAX:** 360-221-4265 (Fax)
ADDRESS: Langley City Hall, P.O. Box 366 Langley, WA 98260
SHORT DESCRIPTION & LOCATION: **www.langleywa.org**

48°02.30' - 122°24.15' Located on E. side of Whidbey Island on Saratoga Passage about 1 mile N. of Sandy Point. Small friendly marina is walking distance to picturesque little town full of tourist oriented amenities. Creative docking & rafting often necessary.

GUEST BOAT CAPACITY: ....Appx. 40 boats
DOCKSIDE DEPTH AT ZERO TIDE: ......26 ft. +
SEASON: ....All year
RESERVATION POLICY: First come, first served
AMT W/ELECTRICITY: ....Appx. 32
FUEL DOCK: ....None
MARINE REPAIRS: ....Close by
TOILETS: ....Yes
HOT SHOWERS: ....Yes
RESTAURANT: ....Close by
PICNIC AREA: ....Yes
BASIC STORE: ....Close by
BROADBAND/WI-FI: ....None
DAILY RATE:.............Economical (Under 75¢/foot)
•**NOTE:** In winter about 50% of space is rented for permanent moorage. Summer - all guest.

GUEST DOCK: 100 ft. plus slips
GUEST SLIPS: Appx. 35 summer
WATER: ....Yes
AMPS: ....20-30 A
PUMP OUT STATION ....Yes
HAUL OUT: ....None
BOAT RAMP: ....Yes
LAUNDRY: ....Close by
BAR: ....Close by
POOL: ....None
GOLF: ....Close by
PET FRIENDLY: ....Good
OTHER: Fishing pier, restaurants, groceries, shops, antiques, & many services. **Marina Motto: "We'll try to fit you in."**

CAUTION! This chartlet not intended for use in navigation.

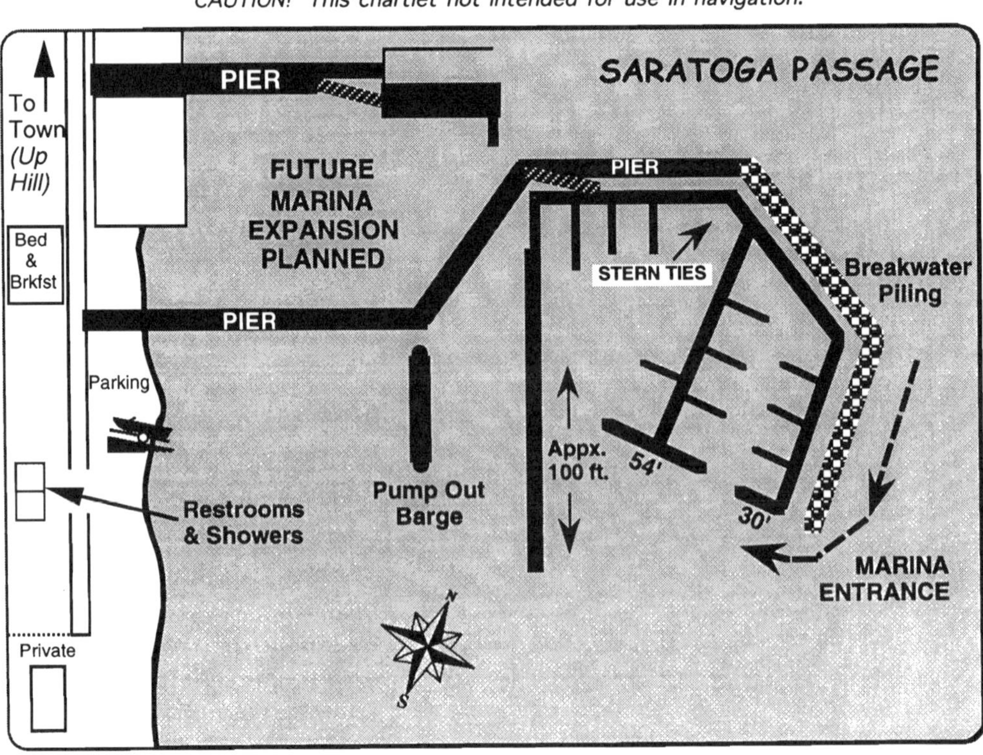

Coupeville

NAME OF MARINA: *PORT OF COUPEVILLE FLOATS* RADIO: None
TELEPHONE: 360-678-5020/3625 MGR: Long Bechard, Harbormaster
E-MAIL: FAX: 360-678-7424 (Fax)
ADDRESS: P.O. Box 577 Coupeville, WA 98239

SHORT DESCRIPTION & LOCATION:

48°13.30' - 122°41.10' Located on E. Whidbey Is. on the S. shore of Penn Cove, about 2 miles from head of Cove. Moorage is on the floats adjacent to the port's historic wharf. Coupeville's waterfront Front St. offers picturesque shops, restaurants & museum.

GUEST BOAT CAPACITY: ....Appx. 8-15 boats
DOCKSIDE DEPTH AT ZERO TIDE: ....3.5 ft.
SEASON: ....All year
RESERVATION POLICY: ....None
AMT W/ELECTRICITY: ....None
FUEL DOCK: ....Gas & Diesel
MARINE REPAIRS: ....Close by
TOILETS: ....Yes
HOT SHOWERS: ....Yes
RESTAURANT: ....Yes
PICNIC AREA: ....Yes
BASIC STORE: ....Yes
BROADBAND/WI-FI: ....None
DAILY RATE: ....Economical (Under 75¢/foot)

GUEST DOCK: Appx. 500 ft. TTL
GUEST SLIPS: ....Dock only
WATER: ....None
AMPS: ....None
PUMP OUT STATION ....None
HAUL OUT: ....None
BOAT RAMP: ....Close by
LAUNDRY: ....None
BAR: ....Close by
POOL: ....None
GOLF: ....None
PET FRIENDLY: ....Excellent
OTHER: Liquor store, chandlery, Post Office, bank, Sat. Farmers Market, Victorian homes. Shuttle to supermarket - call 360-678-5611

CAUTION! This chartlet not intended for use in navigation.

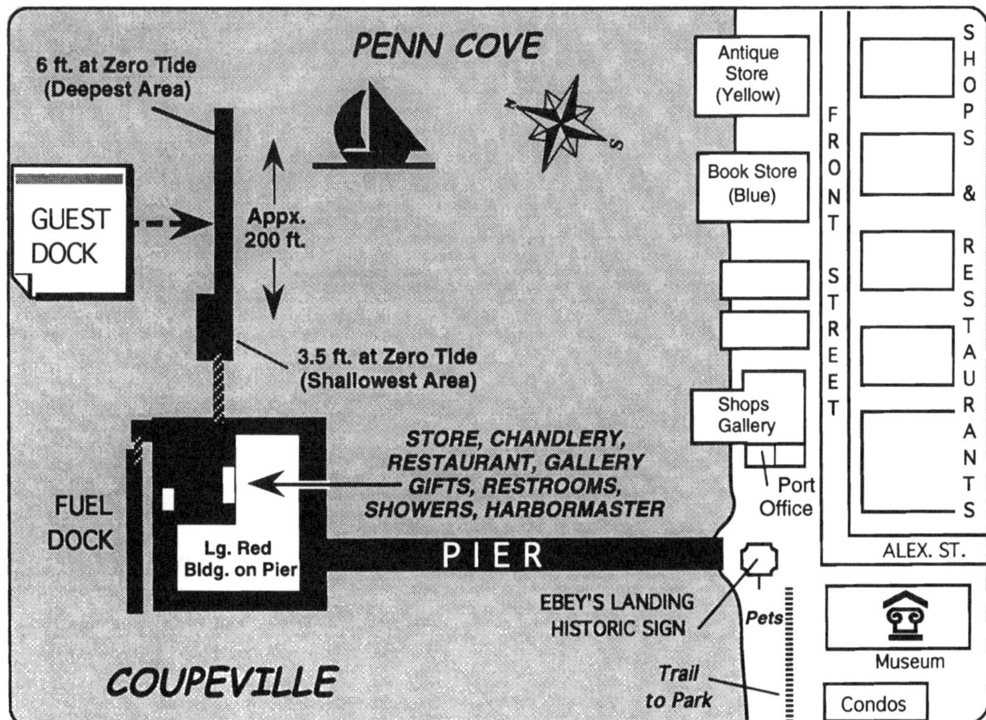

Oak Harbor

NAME OF MARINA: ***OAK HARBOR MARINA*** RADIO: VHF Ch.16
TELEPHONE: **360-679-2628** MGR: Dave Williams
E-MAIL: ohmarina@whidbey.net FAX: 360-240-0603 (Fax)
ADDRESS: 865 S.E. Barrington Dr. Oak Harbor, WA 98277
SHORT DESCRIPTION & LOCATION: **www.whidbey.com/ohmarina**

48°17.10' - 122°38.10' Marina is located at the head of Oak Harbor Bay which is situated at the northern end of Saratoga passage on east side of Whidbey Island. The facilities are modern & friendly within a 3/4 mile walk or dinghy ride from Town services.

GUEST BOAT CAPACITY: ....Appx. 100 boats
DOCKSIDE DEPTH AT ZERO TIDE: ....12-14 ft.
SEASON: ....All year
RESERVATION POLICY:....................Groups only
AMT W/ELECTRICITY: ....All
FUEL DOCK: ....Gas, Dsl, & LP
MARINE REPAIRS: ....Close by
TOILETS: ....Yes
HOT SHOWERS: ....Yes
RESTAURANT: ....Close by
PICNIC AREA: ....Yes
BASIC STORE: ....Appx. 2 miles away
BROADBAND/WI-FI: ....BroadbandXpress
DAILY RATE: ....Moderate (75¢-$1.25/foot)

GUEST DOCK: ....900' plus slips
GUEST SLIPS: 52
WATER: ....Yes
AMPS: ....30 A
PUMP OUT STATION ....Yes
HAUL OUT: ....Close by
BOAT RAMP: ....Yes
LAUNDRY: ....Yes
BAR: ....Close by
POOL: ....None
GOLF: ....Close by
PET FRIENDLY: ....Excellent
OTHER: Shopping nearby, volleyball, horseshoes, marine chandlery & Navy Exchange/ Commissary.

NAME OF YACHT CLUB: ***OAK HARBOR YACHT CLUB***
CLUB ADDRESS: P.O. Box 121, Oak Harbor, WA 98277
CLUB TELEPHONE: 360-675-1314 PERSON IN CHARGE: Bob Mitchell
LOCATION & SPECIAL NOTES: www.ohyc.org

OHYC provides four 40' slips on the end of "F" Dock in the Oak Harbor Marina on a first come basis except July Race Week when no free space is available. Enter the marina via the south gate. The first 4 slips to your port are posted reciprocal. Upon arrival, check in with Harbormaster to attain gate key. The club bar & lounge are open Wed. thru Sun.

RECIPROCAL BOAT CAPACITY:..............4 Boats
DOCKSIDE DEPTH AT ZERO TIDE: ....14 ft.
RECIPROCAL SEASON:....................All year
RESERVATION POLICY: ....None
TOILETS: ....At Marina
HOT SHOWERS: ....At Marina
RESTAURANT: ....Close by
DAILY RATE: ....No charge
....................**48 Hour Limit**

RECIPROCAL DOCK: ....Slips only
RECIPROCAL SLIPS: ..4-40' Slips
WATER: ....Yes
AMT W/ELECTRICITY: ....All
AMPS: ....30 A
BAR: ....Wed. thru Sun.
OTHER: Fri. Night is Burger Night, bring a potluck dish.

NOTE: THIS IS PRIVATE MOORAGE ONLY AVAILABLE TO MEMBERS OF RECIPROCAL YACHT CLUBS! YOUR CLUB MUST HAVE RECIPROCAL PRIVILEGES AND YOU MUST FLY YOUR BURGEE!

Oak Harbor

CAUTION! This chartlet not intended for use in navigation.

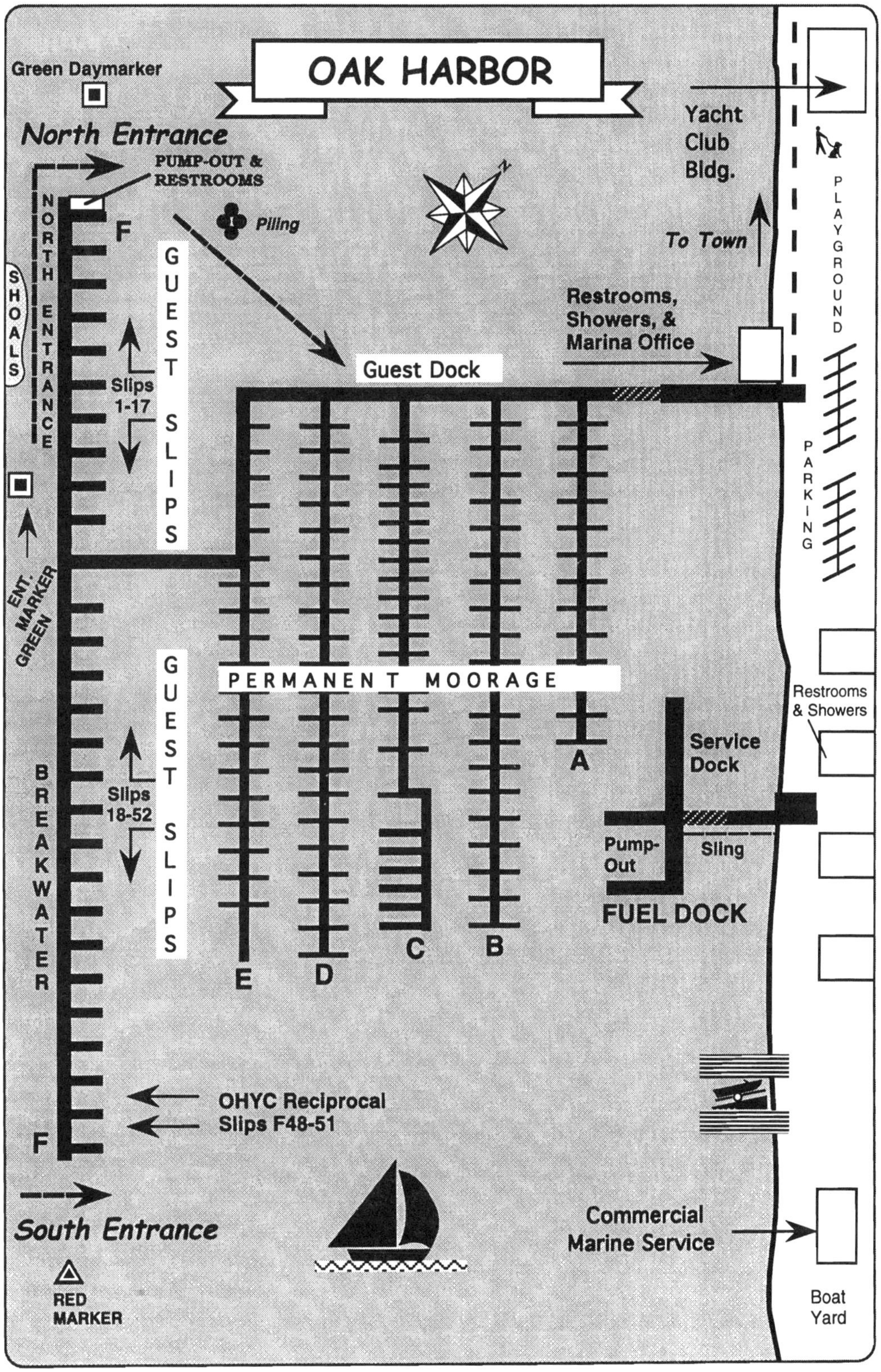

OAK HARBOR

Green Daymarker

North Entrance

PUMP-OUT & RESTROOMS

Piling

Yacht Club Bldg.

PLAYGROUND

To Town

Restrooms, Showers, & Marina Office

Guest Dock

SHOALS

NORTH ENTRANCE

F

GUEST SLIPS

Slips 1-17

PARKING

ENT. MARKER GREEN

PERMANENT MOORAGE

GUEST SLIPS

Slips 18-52

BREAKWATER

A

Service Dock

Restrooms & Showers

Pump-Out

Sling

FUEL DOCK

E

D

C

B

OHYC Reciprocal Slips F48-51

F

South Entrance

Commercial Marine Service

RED MARKER

Boat Yard

Deception Pass

NAME OF PARK: *DECEPTION PASS STATE PARK at Cornet Bay*
ADDRESS: 41229 State Hwy 20 Oak Harbor, WA 98277
TELEPHONE: 360-675-2417 MGR: W.M. Overby
SHORT DESCRIPTION & LOCATION: **www.parks.wa.gov/moorage/parks**

48°24.10' - 122°37.40' Located in Cornet Bay about 1/4 mile east of Deception Pass Marina on N. end of Whidbey off S.E. side of Deception Pass. Large boat launching area and a very convenient stop for boaters cruising to & from the San Juan Islands.

GUEST BOAT CAPACITY:	Appx. 30-40	GUEST DOCK:	1140 ft. total
DOCKSIDE DEPTH AT ZERO TIDE:	Appx. 5 ft.	MOORING BUOYS:	11 Buoys
SEASON:	All year	WATER:	Shoreside
AMT W/ELECTRICITY:	None	PAY PHONES:	Close by
TOILETS:	Yes	BOAT RAMP:	Yes
HOT SHOWERS:	Yes	PICNIC SHELTER:	None
PICNIC AREA:	Yes	BBQ:	Yes
PLAY AREA:	Yes	PUMP OUT STATION:	Yes
BASIC STORE:	Close by	PET FRIENDLY:	Excellent
DAILY RATE:	Dock Moorage 50¢/foot; Mooring Buoys $10.00/night; Minumum Charge: $10	OTHER:	Scenic walking trail in the Hoypus point area of the park, fishing, beachcombing, crabbing.

CAUTION! This chartlet not intended for use in navigation.

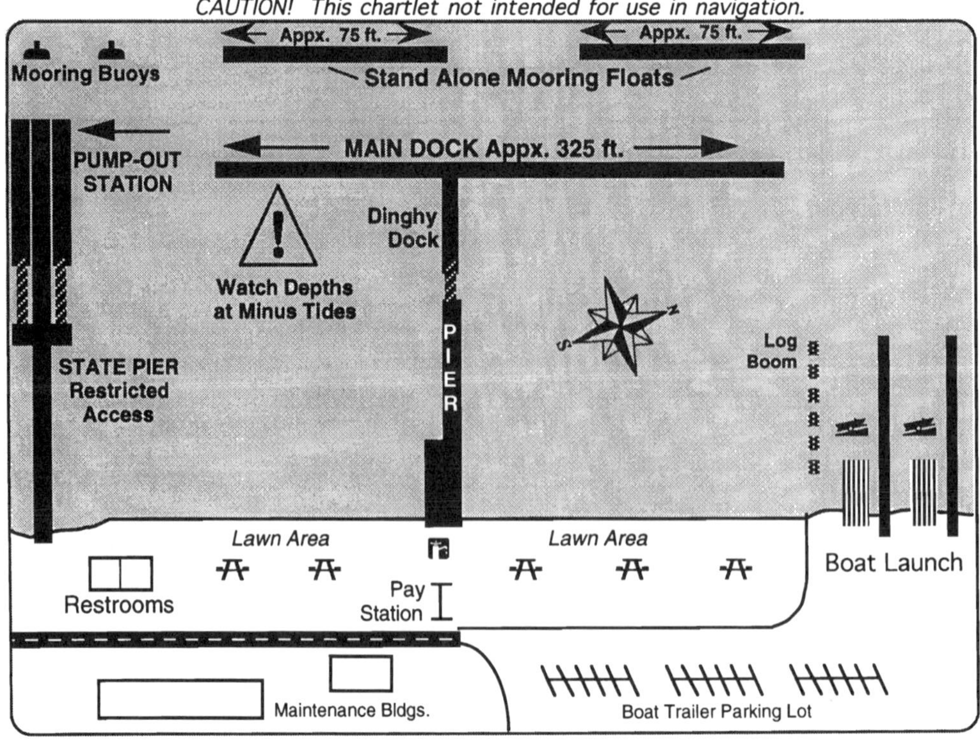

Deception Pass

NAME OF MARINA: ***DECEPTION PASS MARINA*** RADIO: VHF Ch.16
TELEPHONE: 360-675-5411 MGR: Dundee Woods
E-MAIL: None FAX: None
ADDRESS: 200 West Cornet Bay Rd. Oak Harbor, WA 98277

SHORT DESCRIPTION & LOCATION:

48°23.90' - 122°37.60' Located in Cornet Bay, which indents into the N. end of Whidbey Is. off the S.E. side of Deception Pass. Guest moorage consists of open slips of permanent marina tenants. Usually can accommodate 5-10 boats.

GUEST BOAT CAPACITY: ....Varies
DOCKSIDE DEPTH AT ZERO TIDE: ....6-7 ft.
SEASON: ....All year
RESERVATION POLICY: ....None
AMT W/ELECTRICITY: 90%
FUEL DOCK: ....Gas, Dsl, & LP
MARINE REPAIRS: ....Close by
TOILETS: ....Yes
HOT SHOWERS: ....Close by
RESTAURANT: ....Close by
PICNIC AREA: ....Yes
BASIC STORE: ....Yes
BROADBAND/WI-FI: ....None
DAILY RATE: ....Moderate (75¢-$1.25/foot)

GUEST DOCK: ....As available
GUEST SLIPS: ....Varies
WATER: ....Yes
AMPS: ....30 A
PUMP OUT STATION Close by
HAUL OUT: ....Close-by
BOAT RAMP: ....Close by
LAUNDRY: ....Close by
BAR: ....None
POOL: ....None
GOLF: ....Close by
PET FRIENDLY: ....Excellent
OTHER: Next door to Deception Pass State Park, well stocked store - propane, marine & fishing supplies.

CAUTION! This chartlet not intended for use in navigation.

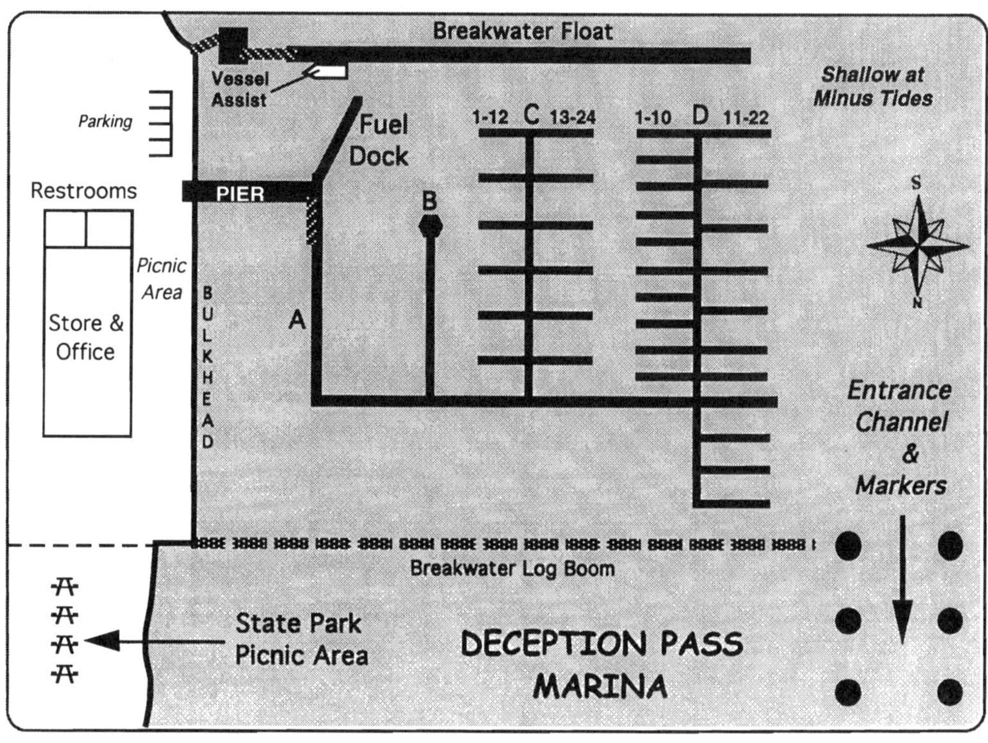

Deception Pass

NAME OF YACHT CLUB: *EMERALD ROSE YACHT CLUB*
ADDRESS: P.O. Box 18292 Seattle, WA 98118
TELEPHONE: PERSON IN CHARGE: Commodore

SHORT DESCRIPTION & LOCATION: **www.emeraldroseyachtclub.org**
Emerald Rose Yacht club offers reciprocal moorage located at the DECEPTION PASS MARINA for boats up to 45 feet. Visiting reciprocal club members are requested to pay the moorage charges to the harbormaster & mail the moorage receipt along with a copy of current club membership card to ERYC (Attn: Treasurer) for reimbursement.

RECIPROCAL BOAT CAPACITY: ....2 Boats
DOCKSIDE DEPTH AT ZERO TIDE: ....6-7 ft.
SEASON: ....All year
RESERVATION POLICY: ....None
TOILETS: ....At Deception Pass Marina
HOT SHOWERS: ....At Deception Pass Marina
RESTAURANT: ....None
BROADBAND/WI-FI: ....None
DAILY RATE: ....2 Days free per year via reimbursement. Max 2 boats from same club on same night.

RECIPROCAL DOCK: As available
RECIPROCAL SLIPS: ....Varies
WATER: ....Yes
AMT W/ELECTRICITY: 90%
AMPS: ....30 A
BAR: ....None
PET FRIENDLY: ....Excellent
OTHER: Services same as Deception Pass Marina listing on Page 145.

***NOTE:* THIS IS PRIVATE MOORAGE AND ONLY AVAILABLE TO MEMBERS OF RECIPROCAL YACHT CLUBS. YOUR CLUB MUST HAVE RECIPROCAL PRIVILEGES AND YOU MUST FLY YOUR BURGEE.**

CAUTION! This chartlet not intended for use in navigation.

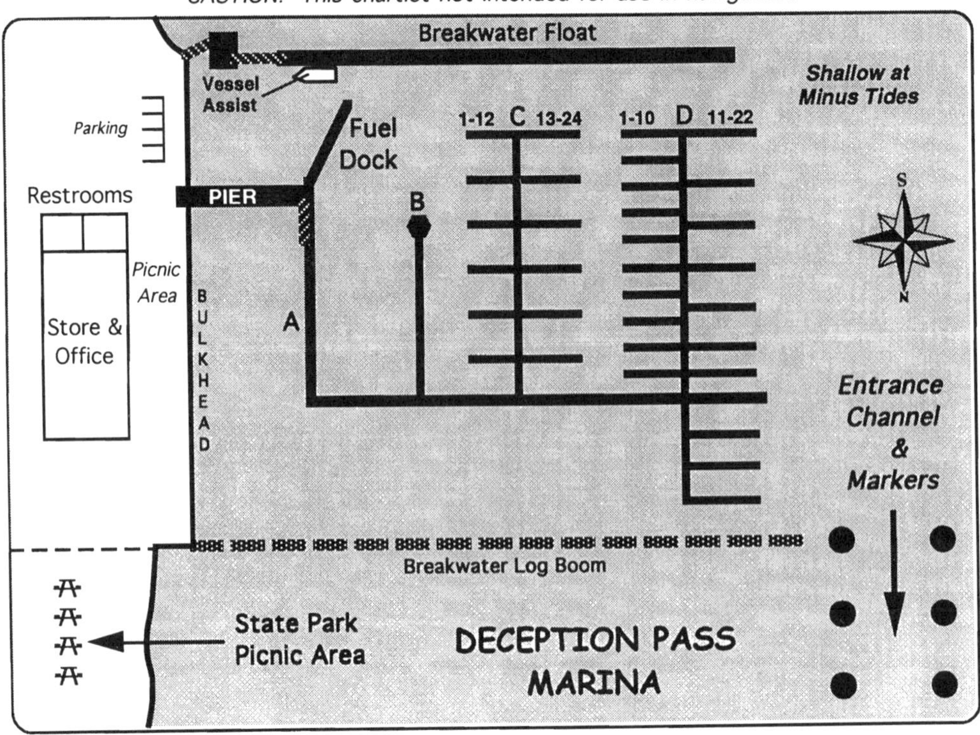

La Conner

NAME OF YACHT CLUB: *SHELTER BAY YACHT CLUB*

ADDRESS: 1000 Shoshone Dr. La Conner, WA 98257

TELEPHONE: 360-466-2680 PERSON IN CHARGE: Reciprocal Chair

SHORT DESCRIPTION & LOCATION:

48°22.93' - 122°30.50' Shelter Bay is located south of the Rainbow Bridge on W. side of Swinomish Channel in La Conner. Enter waterway & continue to the 5 large docks A thru E. Must call 360-466-2680 in advance to hear list of open slips. Max. boat size 45'. Register at Marina Restroom Building near head of C Dock. Residential private community.

RECIPROCAL BOAT CAPACITY: ....Varies

DOCKSIDE DEPTH AT ZERO TIDE: ....6-7 ft.

SEASON: ....All year

RESERVATION POLICY: ....None

TOILETS: ....Yes

HOT SHOWERS: ....Yes

RESTAURANT: ....None

BROADBAND/WI-FI: ....None

DAILY RATE: Two days free within a 10 day period. $3/day utility fee for each day.

RECIPROCAL DOCK: .....Up to 45'

RECIPROCAL SLIPS: ....Varies

WATER: ....Yes

AMT W/ELECTRICITY: ....All

AMPS: ....30 A

BAR: ....None

PET FRIENDLY: ....Excellent

OTHER: Unique community developed around marina. Facilities beyond marina common area are private.

***NOTE:* THIS IS PRIVATE MOORAGE AND ONLY AVAILABLE TO MEMBERS OF RECIPROCAL YACHT CLUBS. YOUR CLUB MUST HAVE RECIPROCAL PRIVILEGES AND YOU MUST FLY YOUR BURGEE.**

CAUTION! This chartlet not intended for use in navigation.

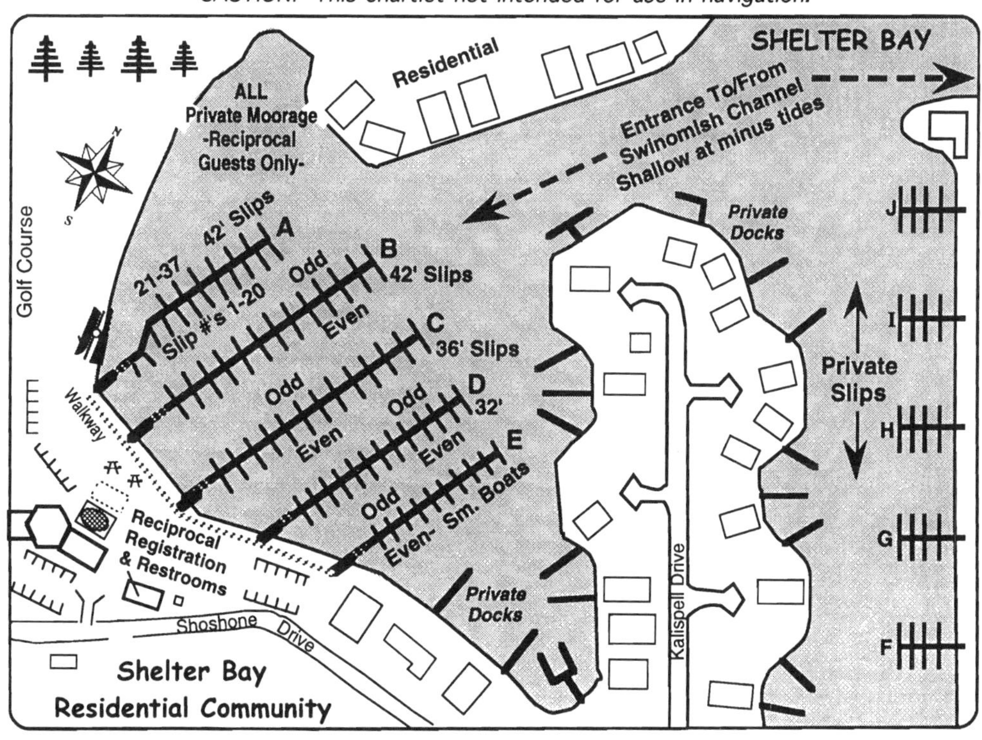

La Conner

NAME OF MARINA: ***THE LA CONNER MARINA*** **RADIO:** VHF 66A
TELEPHONE: **360-466-3118** MGR: Harbormaster
E-MAIL: posc@portofskagit.com FAX: 360-466-3119 (Fax)
ADDRESS: P.O. Box 1120 La Conner, WA 98257
SHORT DESCRIPTION & LOCATION: **www.portofskagit.com**

48°23.50' - 122°29.40' Located midway on the Swinomish Channel which connects Skagit Bay (south) and Padilla Bay (north). This marina in this historic and scenic town offers moorage in two boat basins on two guest docks as well as vacated tenant's slips.

GUEST BOAT CAPACITY: ....Appx. 50-75 boats
DOCKSIDE DEPTH AT ZERO TIDE: ....10 ft.
SEASON: ....All year
RESERVATION POLICY: ....None
AMT W/ELECTRICITY: ....All
FUEL DOCK: ....Gas & Diesel
MARINE REPAIRS: ....On premises
TOILETS: ....Yes
HOT SHOWERS: ....Yes
RESTAURANT: ....Close by
PICNIC AREA: ....Yes
BASIC STORE: ....Close by
BROADBAND/WI-FI: ....Planned
DAILY RATE: ....Moderate (75¢-$1.25/foot)

LOCAL KNOWLEDGE:
Slack water occurs approx. 3.5 hrs. after a Seattle high or low tide.

GUEST DOCK: ....2400 ft. total
GUEST SLIPS: ....Varies
WATER: ....Yes
AMPS: ....30 A
PUMP OUT STATION ....Yes
HAUL OUT: ....Travel-Lift
BOAT RAMP: ....Close by
LAUNDRY: ....Yes
BAR: ....Close by
POOL: ....None
GOLF: ....Close by
PET FRIENDLY: ....Good
OTHER: Tourist oriented town with specialty shops, restaurants, marine stores, & art galleries.

NAME OF YACHT CLUB: ***SWINOMISH YACHT CLUB***
CLUB ADDRESS: P.O. Box 602, La Conner, WA 98257
CLUB TELEPHONE: 360-466-4902 PERSON IN CHARGE: Dock Master
LOCATION & SPECIAL NOTES: **www.swinomishyachtclub.org**

Reciprocal dock is on west side "E" Dock in the South boat basin & offers appx. 75 feet. All reciprocal visitors must register at dockside registration box & have valid reciprocal club membership card. All guests must moor boat as far south on the dock as possible. The current runs swift & it is difficult for another boat to get around your boat.

RECIPROCAL BOAT CAPACITY: Appx 2 boats
DOCKSIDE DEPTH AT ZERO TIDE: ....10 ft.
RECIPROCAL SEASON: ....All year
RESERVATION POLICY: ....None
TOILETS: ....At Marina
HOT SHOWERS: ....At Marina
RESTAURANT: ....Close by
DAILY RATE: Power $3 each night. 1st Night moorage free. 2nd night charged at Port rate. MAX STAY 48 HRS.

RECIPROCAL DOCK: ....75 ft.
RECIPROCAL SLIPS: Dock only
WATER: ....Yes
AMT W/ELECTRICITY: ....All
AMPS: ....30 A
BAR: ....Close by
OTHER: ....Same as marina listing
NO BOATS OVER 50'

***NOTE:* THIS IS PRIVATE MOORAGE *ONLY* AVAILABLE TO MEMBERS OF RECIPROCAL YACHT CLUBS! YOUR CLUB *MUST* HAVE RECIPROCAL PRIVILEGES AND YOU MUST FLY YOUR BURGEE!**

La Conner

CAUTION! This chartlet not intended for use in navigation.

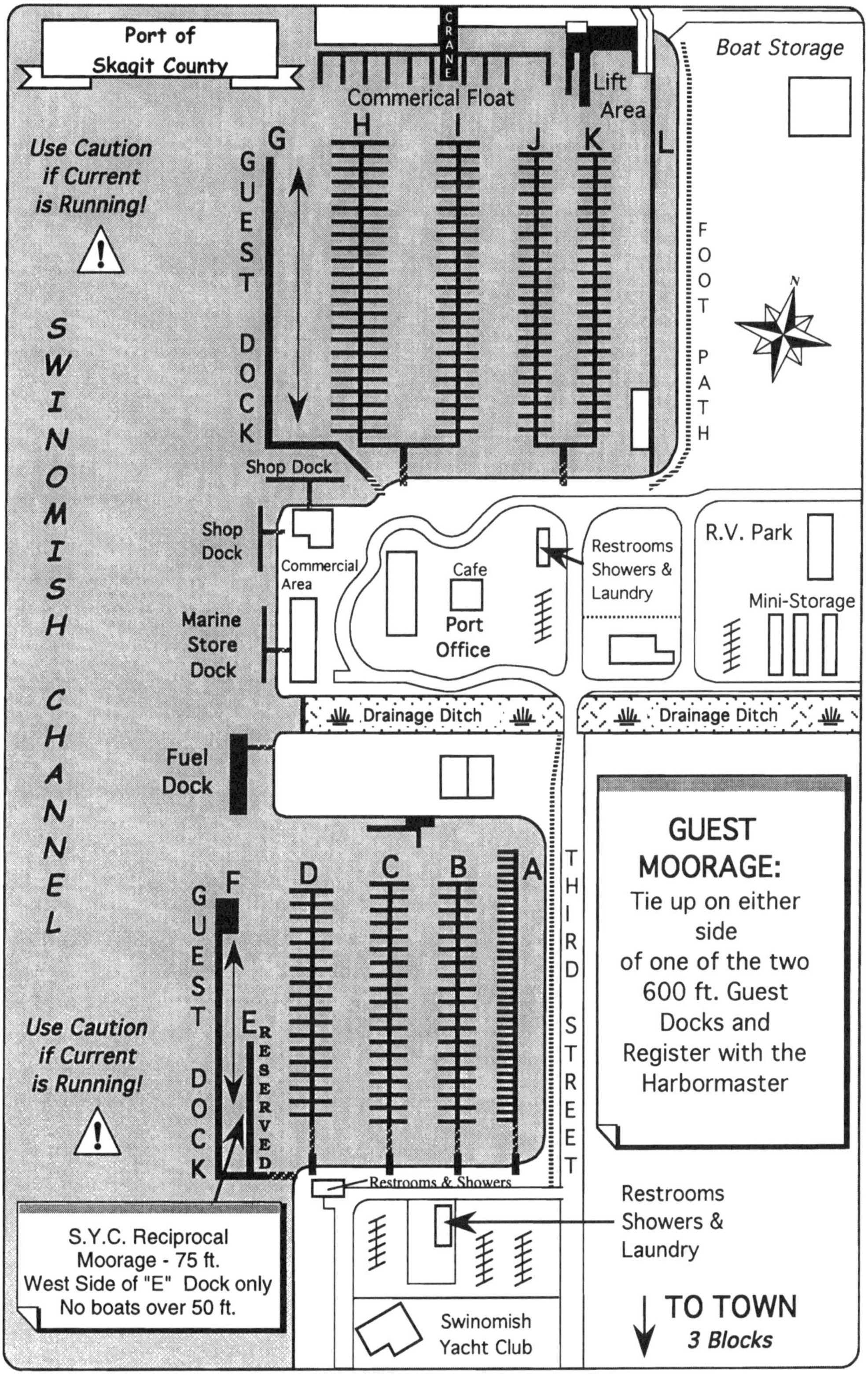

Port of Skagit County

C R A N E

Commerical Float

Lift Area

Boat Storage

G H I J K L

Use Caution if Current is Running!

GUEST DOCK

FOOT PATH

N

SWINOMISH CHANNEL

Shop Dock

Shop Dock

Commercial Area

Cafe

Port Office

Restrooms Showers & Laundry

R.V. Park

Mini-Storage

Marine Store Dock

Drainage Ditch

Drainage Ditch

Fuel Dock

F

D C B A

THIRD STREET

GUEST MOORAGE:

Tie up on either side of one of the two 600 ft. Guest Docks and Register with the Harbormaster

GUEST DOCK

E RESERVED

Use Caution if Current is Running!

Restrooms & Showers

Restrooms Showers & Laundry

S.Y.C. Reciprocal Moorage - 75 ft.
West Side of "E" Dock only
No boats over 50 ft.

Swinomish Yacht Club

TO TOWN
3 Blocks

La Conner

NAME OF MARINA: *LA CONNER TOWN FLOATS* **RADIO:** None
TELEPHONE: **360-466-3933** MGR: Brian Lease
E-MAIL: publicworks@townoflaconner.org FAX: 360-466-4293 (Fax)
ADDRESS: Town of Laconner, Box 400, La Conner, WA 98257
SHORT DESCRIPTION & LOCATION: **www.townoflaconner.net**

48°23.50' - 122°30.20' Located midway on the Swinomish Channel which connects Skagit Bay (south) and Padilla Bay (north). The town floats in this historic and scenic town offer modest moorage in three locations right at the base of many tourist activities.

GUEST BOAT CAPACITY: ....Appx. 6-8 boats
DOCKSIDE DEPTH AT ZERO TIDE: ....6-8 ft.
SEASON: ....All year
RESERVATION POLICY: ....None
AMT W/ELECTRICITY: ....None
FUEL DOCK: ....Close by
MARINE REPAIRS: ....Close by
TOILETS: ....Close by
HOT SHOWERS: ....None
RESTAURANT: ....Close by
PICNIC AREA: ....Close by
BASIC STORE: ....Close by
BROADBAND/WI-FI: ....Yes
DAILY RATE:..............Economical (Under 75¢/foot)

LOCAL KNOWLEDGE:
Slack water occurs approx. 3.5 hrs. after a Seattle high or low tide.

GUEST DOCK: 40' (Max size-45')
GUEST SLIPS: ....6 - 40' berths
WATER: ....None
AMPS: ....None
PUMP OUT STATION ....None
HAUL OUT: ....Close by
BOAT RAMP: ....Close by
LAUNDRY: ....Close by
BAR: ....Close by
POOL: ....None
GOLF: ....Close by
PET FRIENDLY: ....Fair
OTHER: Tourist oriented town with specialty shops, restaurants, marine stores, & art galleries.

CAUTION! This chartlet not intended for use in navigation.

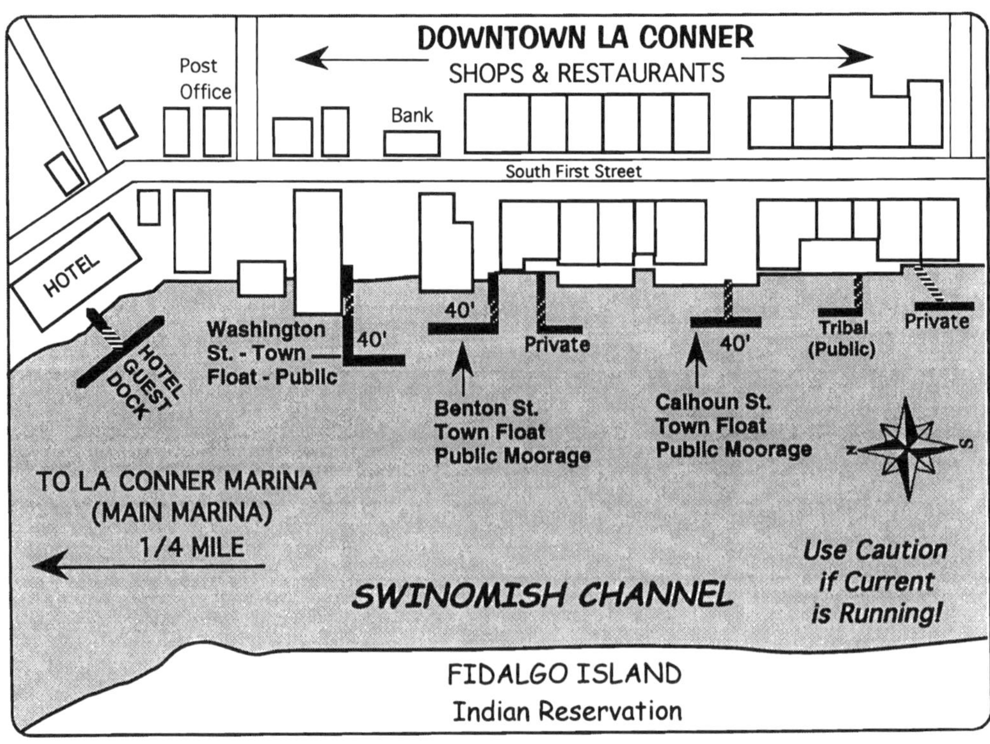

CHAPTER 5

SAN JUAN ISLANDS & NORTHERN WASHINGTON

Chapter Map - Page 3

NOTES

Anacortes

NAME OF MARINA: ***CAP SANTE BOAT HAVEN*** **RADIO:** VHF 66A
TELEPHONE: **360-293-0694** MGR: Dale Fowler
E-MAIL: marina@portofanacortes.net FAX: 360-299-0998 (Fax)
ADDRESS: P.O. Box 297 Anacortes, WA 98221
SHORT DESCRIPTION & LOCATION: **www.portofanacortes.com**

48°30.60' - 122°36.20' Located in Fidalgo Bay by Cap Sante Head, this modern & spacious marina is close to any marine service a boater could want. The town is one of the boating capitals of the N.W. **MARINA REDEVELOPMENT & CHANGES ON-GOING.**

GUEST BOAT CAPACITY: ....Appx. 150 boats
DOCKSIDE DEPTH AT ZERO TIDE: ......8-12 ft.
SEASON: ....All year
RESERVATION POLICY: Same day reservations
AMT W/ELECTRICITY: ....All
FUEL DOCK: ....Gas, Dsl, & LP
MARINE REPAIRS: ....On premises
TOILETS: ....Yes
HOT SHOWERS: ....Yes
RESTAURANT: ....Yes
PICNIC AREA: ....Yes
BASIC STORE: ....Yes
BROADBAND/WI-FI: ....BroadbandXpress
DAILY RATE: ....Moderate (75¢-$1.25/foot)
MARINA REDEVELOPMENT & CHANGES ON-GOING.

GUEST DOCK: 170 ft. plus slips
GUEST SLIPS: ....150 +
WATER: ....Yes
AMPS: ....20-50 A
PUMP OUT STATION ....Yes
HAUL OUT: ....Travel-Lift
BOAT RAMP: ....Close by
LAUNDRY: ....Yes
BAR: ....Close by
POOL: ....Close by
GOLF: ....Close by
PET FRIENDLY: ....Excellent
OTHER: Customs port-of-entry, close to many shops, provisioning, art galleries, historic steam railroad. public transit in county.

CAUTION! This chartlet not intended for use in navigation.

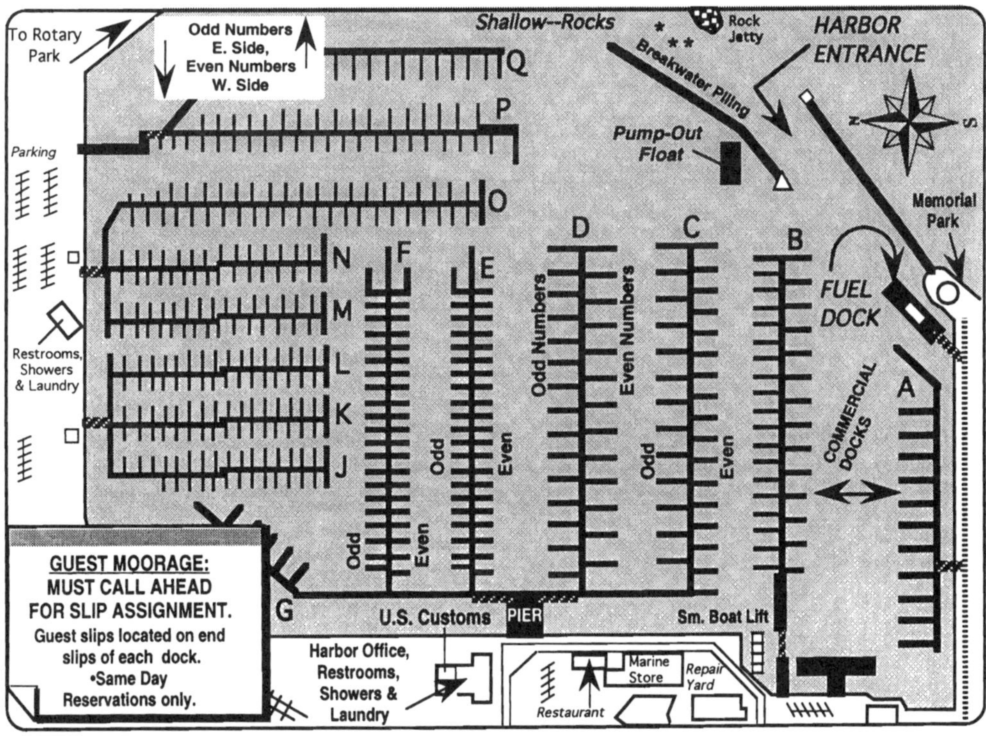

Anacortes

NAME OF YACHT CLUB: *ANACORTES YACHT CLUB*

ADDRESS: 504 7th Street Anacortes, WA 98221

TELEPHONE: 360-293-5277 PERSON IN CHARGE: Recip. Chairman

SHORT DESCRIPTION & LOCATION: **www.anacortesyachtclub.org**

48°30.60' - 122°36.20' AYC offers reciprocal moorage at Cap Sante Marina in a new location - Slips P-19 (40' max) & Q-6 (50' max) Call marina 360-293-0694/VHF 66A for availability. Register w/marina staff upon arrival. Usage is limited to 1 night a week per boat. The club is open most Fri. nights (1730-2030 hrs.), located at NE end of marina.

RECIPROCAL BOAT CAPACITY: ....2 Boats

DOCKSIDE DEPTH AT ZERO TIDE: ....5-6 ft.

SEASON: ....All year

RESERVATION POLICY: Same day - call Marina

TOILETS: ....At Marina

HOT SHOWERS: ....At Marina

RESTAURANT: ....Close by

BROADBAND/WI-FI: ....BroadbandXpress

DAILY RATE: During summer, one night free every 7 days. Electricity $3.00 each night. Check out time NOON.

RECIPROCAL DOCK: Slips only

RECIPROCAL SLIPS: ....2 Slips

WATER: ....Yes

AMT W/ELECTRICITY: ....All

AMPS: ....30 A

BAR: ....Close by

PET FRIENDLY: ....Excellent

OTHER: Services same as marina listing for Cap Sante Boat Haven on Page 153.

***NOTE:* THIS IS PRIVATE MOORAGE AND ONLY AVAILABLE TO MEMBERS OF RECIPROCAL YACHT CLUBS. YOUR CLUB MUST HAVE RECIPROCAL PRIVILEGES AND YOU MUST FLY YOUR BURGEE.**

CAUTION! This chartlet not intended for use in navigation.

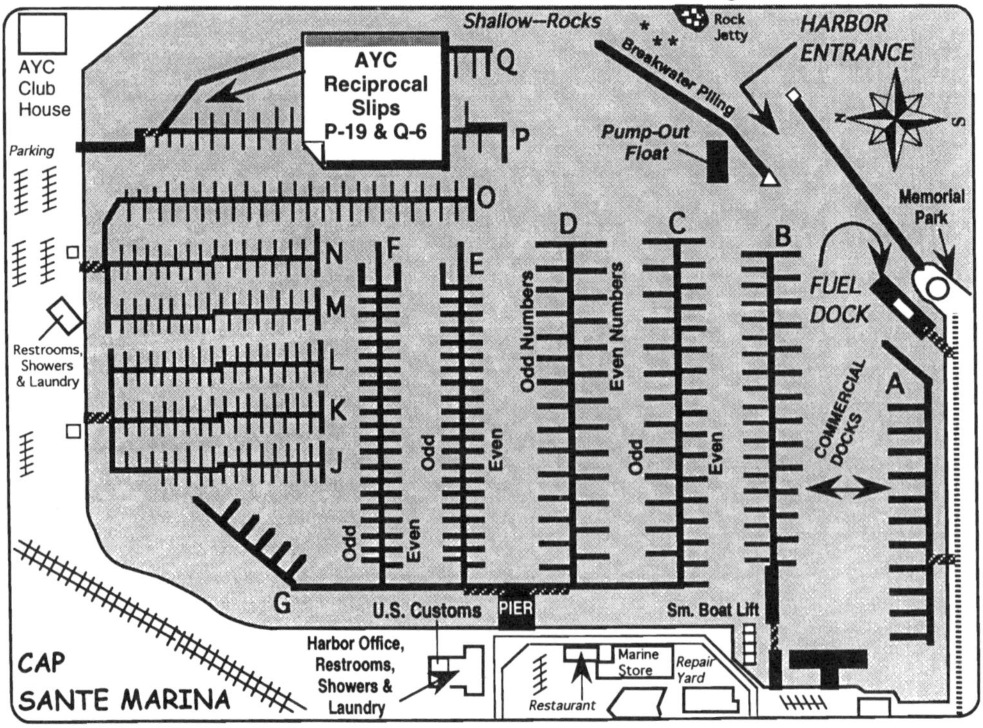

Anacortes

NAME OF YACHT CLUB: *FIDALGO YACHT CLUB*

ADDRESS: P.O. Box 1838 Anacortes, WA 98221

TELEPHONE: 360-293-0694 - Marina Office PERSON IN CHARGE: Fleet Captain

SHORT DESCRIPTION & LOCATION: **www.fidalgoyachtclub.org**

48°30.90' - 122°36.30' FYC offers reciprocal moorage at Cap Sante Marina in a new location - **Slip Q-10** (50' max). **Call marina 360-293-0694/VHF 66A for availability.** Register w/marina staff upon arrival - your current yacht club membership card will be required.

RECIPROCAL BOAT CAPACITY: ....1 Boat

DOCKSIDE DEPTH AT ZERO TIDE: ....5-6 ft.

SEASON: ....All year

RESERVATION POLICY: Same day - call Marina

TOILETS: ....At Marina

HOT SHOWERS: ....At Marina

RESTAURANT: ....Close by

BROADBAND/WI-FI: ....BroadbandXpress

DAILY RATE: During summer, one night free every 7 days. Electricity $3.00 each night. Check out time NOON.

RECIPROCAL DOCK: ....Slip only

RECIPROCAL SLIPS: ....1 Slip

WATER: ....Yes

AMT W/ELECTRICITY: ....All

AMPS: ....30 A

BAR: ....Close by

PET FRIENDLY: ....Excellent

OTHER: Services same as marina listing for Cap Sante Boat Haven on Page 153.

***NOTE:* THIS IS PRIVATE MOORAGE AND ONLY AVAILABLE TO MEMBERS OF RECIPROCAL YACHT CLUBS. YOUR CLUB *MUST* HAVE RECIPROCAL PRIVILEGES AND YOU MUST FLY YOUR BURGEE.**

CAUTION! This chartlet not intended for use in navigation.

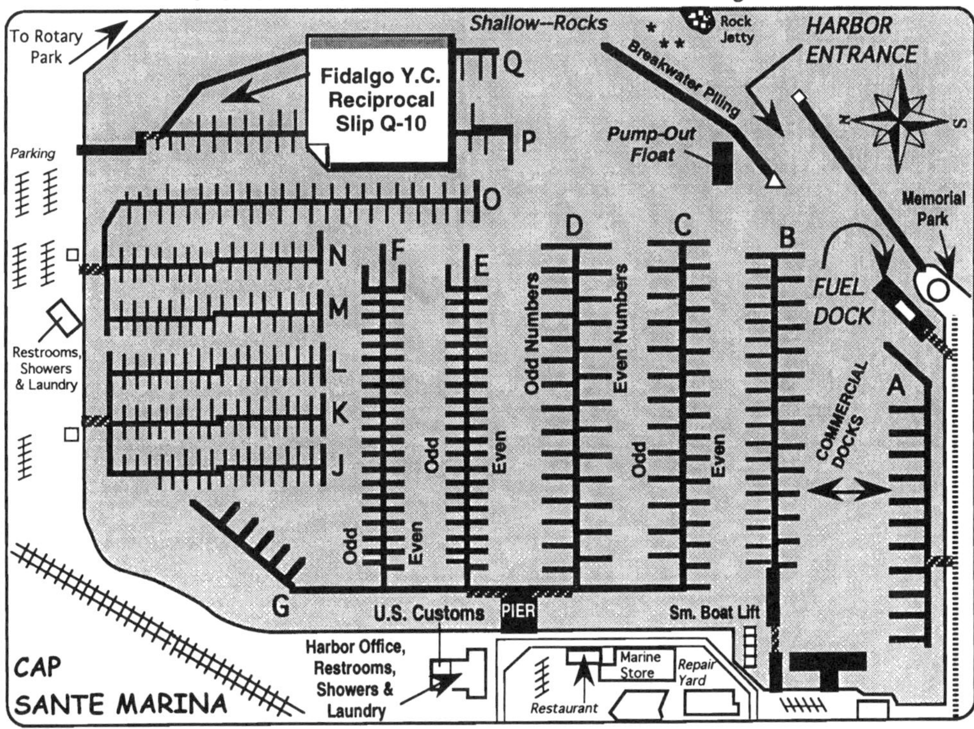

Anacortes

NAME OF MARINA: ***SKYLINE MARINA*** RADIO: None
TELEPHONE: **360-293-5134** MGR: Kelly Larkin, Manager
E-MAIL: Info@skylinemarinecenter.com FAX: 360-293-7557 (Fax)
ADDRESS: 2011 Skyline Way Anacortes, WA 98221
SHORT DESCRIPTION & LOCATION: **www.skylinemarinecenter.com**

48°29.40' - 122°40.60' Located near Fidalgo Head in Flounder Bay at the N end of Burrows Bay about 1/2 mile by land W of ferry terminal. This well protected large boat basin is situated in a marine oriented and residential area appx. 4 miles W of Anacortes.

GUEST BOAT CAPACITY: ....Varies
DOCKSIDE DEPTH AT ZERO TIDE: Appx. 6' +
SEASON: ....All year
RESERVATION POLICY: ....Accepts
AMT W/ELECTRICITY: ....All
FUEL DOCK: ....Gas, Dsl, & LP
MARINE REPAIRS: ....On premises
TOILETS: ....Yes
HOT SHOWERS: ....Yes
RESTAURANT: ....Close by
PICNIC AREA: ....Yes
BASIC STORE: ....Close by
BROADBAND/WI-FI: ....BroadbandXpress
DAILY RATE: ....Moderate (75¢-$1.25/foot)

Note: Guest moorage is available by renting unoccupied slips of permanent marina tenants.

GUEST DOCK: None designated
GUEST SLIPS: ....Varies
WATER: ....Yes
AMPS: ....30-30 A
PUMP OUT STATION ....Yes
HAUL OUT: ....Travel-Lift
BOAT RAMP: ....Sling only
LAUNDRY: ....Yes
BAR: ....Close by
POOL: ....None
GOLF: ....Close by
PET FRIENDLY: ....Good
OTHER: Water taxi, fishing supplies, charts, books, chandlery, charters.

Entrance to Harbor has 6-7 ft. of water at 0 Tide.

NAME OF YACHT CLUB: ***FLOUNDER BAY YACHT CLUB***
CLUB ADDRESS: 2400 Skyline Way, Anacortes, WA 98221
CLUB TELEPHONE: 360-299-3325 PERSON IN CHARGE: Dockmaster
LOCATION & SPECIAL NOTES: **www.fbyc.com**

The club dock, located in front of the Club building, is used for visitor check in only. The building is between "D" and "E" Docks. Follow the simple instructions given in the club building laundry room to be assigned a member's vacant slip. Usually ample slips are open. Visiting vessels may not be left unattended overnight or overhang over 4' into fairway.

RECIPROCAL BOAT CAPACITY: ....Varies
DOCKSIDE DEPTH AT ZERO TIDE: ....7 ft. +
RECIPROCAL SEASON: ....All year
RESERVATION POLICY: ....None
TOILETS: ....Yes
HOT SHOWERS: ....Yes
RESTAURANT: ....Close by
DAILY RATE: ....Power - $3.00/day every day ....First day moorage is free. Ea. add'l day is $10.00- 5 days max yr.

RECIPROCAL DOCK: ....Slips only
RECIPROCAL SLIPS: ....Varies
WATER: ....Yes
AMT W/ELECTRICITY: ....All
AMPS: ....30 A
BAR: ....Close by
OTHER: Most slips 30'-44'. Very few 50' slips available.

NOTE: THIS IS PRIVATE MOORAGE ONLY AVAILABLE TO MEMBERS OF RECIPROCAL YACHT CLUBS! YOUR CLUB MUST HAVE RECIPROCAL PRIVILEGES AND YOU MUST FLY YOUR BURGEE!

Anacortes

CAUTION! This chartlet not intended for use in navigation.

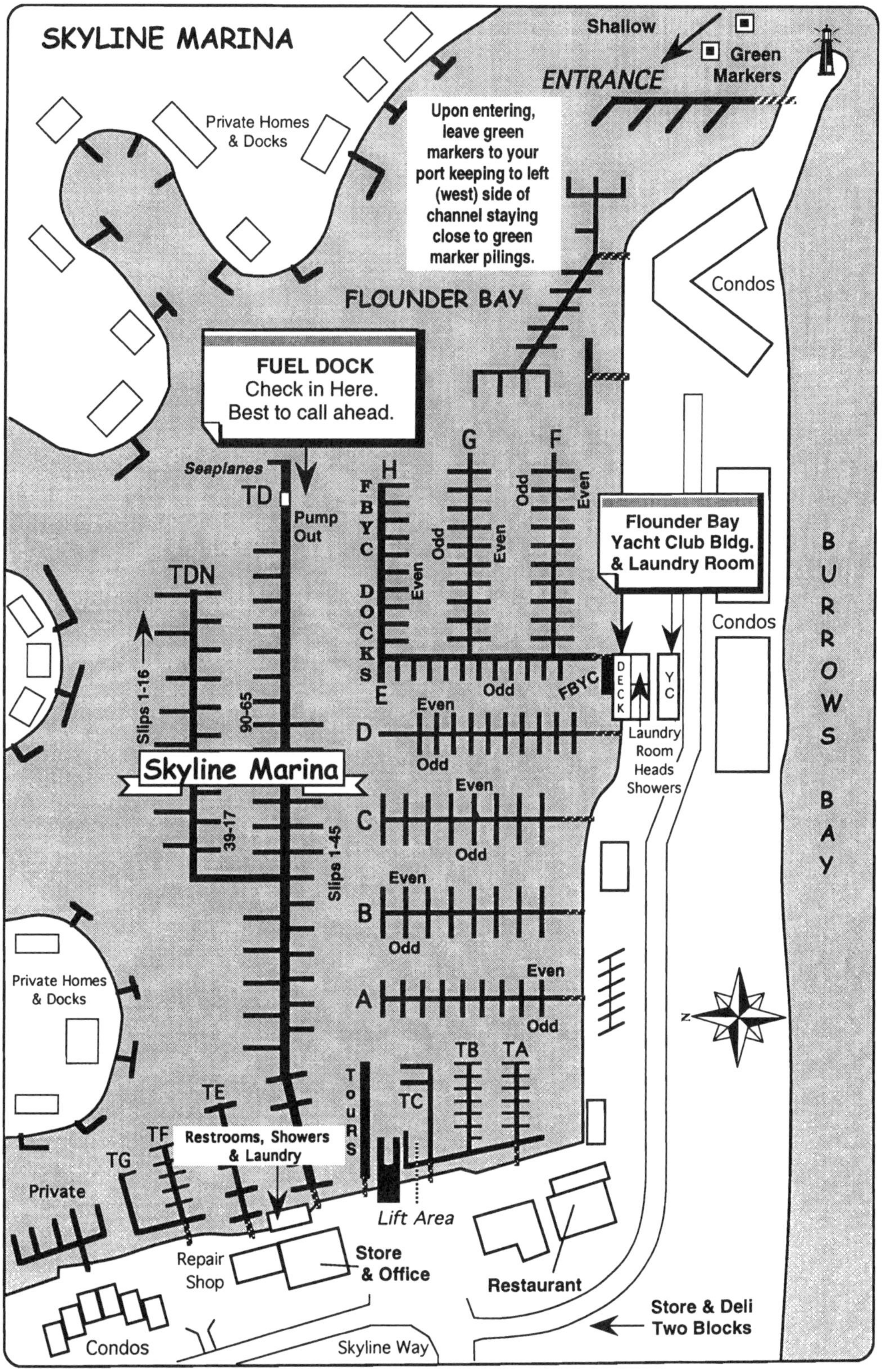

SKYLINE MARINA

Shallow

Green Markers

ENTRANCE

Private Homes & Docks

Upon entering, leave green markers to your port keeping to left (west) side of channel staying close to green marker pilings.

Condos

FLOUNDER BAY

FUEL DOCK
Check in Here.
Best to call ahead.

Seaplanes

TD

Pump Out

H G F

FBYC DOCKS

Odd Even

Flounder Bay Yacht Club Bldg. & Laundry Room

BURROWS BAY

Condos

TDN

Slips 1-16

90-65

E

FBYC

DECK

YC

D

Laundry Room Heads Showers

Skyline Marina

39-17

Slips 1-45

C

B

A

Private Homes & Docks

TE TF TG

TOURS

TB TA TC

Restrooms, Showers & Laundry

Private

Lift Area

Repair Shop

Store & Office

Restaurant

Store & Deli Two Blocks

Condos

Skyline Way

Lopez Island

NAME OF MARINA: ***ISLANDER LOPEZ MARINA RESORT*** RADIO: VHF Ch.78
TELEPHONE: **800-736-3434** MGR: Earl & Bill Diller
E-MAIL: desk@lopezislander.com FAX: 360-468-3382 (Fax)
ADDRESS: PO Box 459 Fisherman Bay Rd. Lopez Island, WA 98261
SHORT DESCRIPTION & LOCATION: **www.lopezislander.com**

48°31.00' - 122°54.80' Family oriented marina & hotel resort located in Fisherman Bay 1 mile south of Lopez Village on W. side of Lopez Is. Hotel accommodations available. Upgraded docks and close to marine services. Great fishing and crabbing in Bay.

GUEST BOAT CAPACITY: ....Appx. 60 boats
DOCKSIDE DEPTH AT ZERO TIDE: ....7 ft.
SEASON: ....All year
RESERVATION POLICY: ....Recommended
AMT W/ELECTRICITY: ....All
FUEL DOCK: ....Gas & Diesel
MARINE REPAIRS: ....Close by
TOILETS: ....Yes
HOT SHOWERS: ....Yes
RESTAURANT: ....Yes
PICNIC AREA: ....Yes
BASIC STORE: ....Yes
BROADBAND/WI-FI: ....BroadbandXpress
DAILY RATE: ....Moderate (75¢-$1.25/foot)

GUEST DOCK: ....350' plus slips
GUEST SLIPS: ....Appx. 55
WATER: ....Yes
AMPS: ....30-50 A
PUMP OUT STATION Close by
HAUL OUT: ....Close by
BOAT RAMP: ....Close by
LAUNDRY: ....Yes
BAR: ....Yes
POOL: ....Yes - Seasonal
GOLF: ....Close by
PET FRIENDLY: ....Excellent
OTHER: R/V Park & camping, volleyball, horseshoes, bike & kayak rentals, hot tub, jacuzzi, meeting & banquet facilities.

CAUTION! This chartlet not intended for use in navigation.

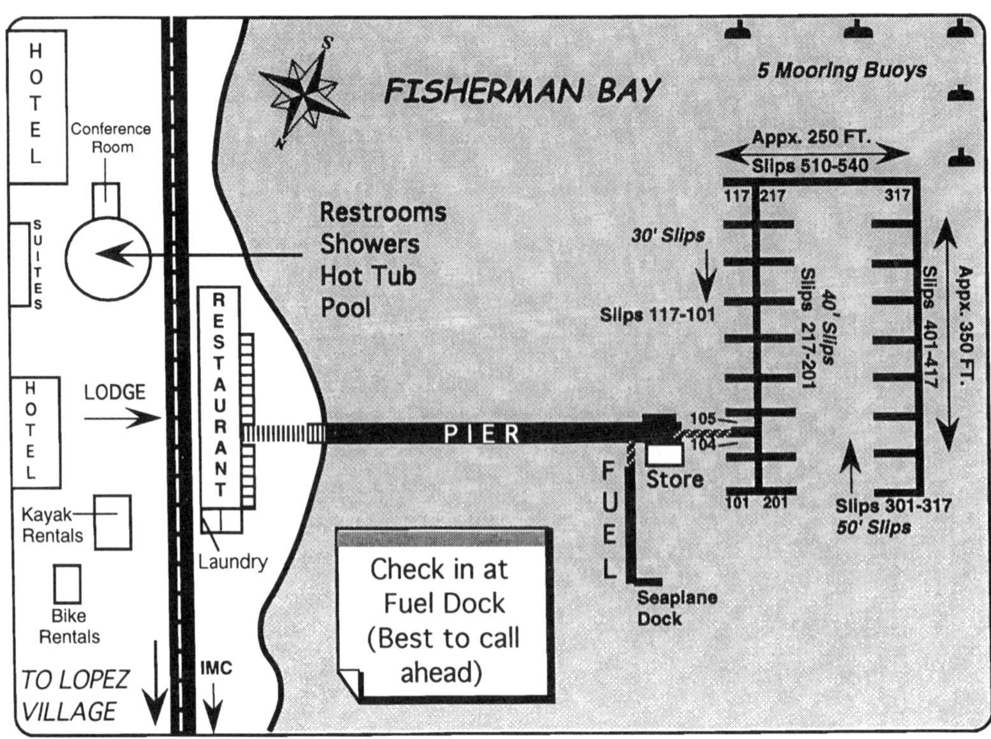

Lopez Island

NAME OF MARINA: **ISLANDS MARINE CENTER** RADIO: VHF Ch.69
TELEPHONE: 360-468-3377 MGR: Ron & Jennifer Meng
E-MAIL: imc@rockisland.com FAX: 360-468-2283 (Fax)
ADDRESS: PO Box 88 FishermanBayRd Lopez Island, WA 98261
SHORT DESCRIPTION & LOCATION: **www.islandsmarinecenter.com**

48°31.10' - 122°54.80' Modern & complete marina facility w/chandlery & auto parts store located in Fisherman Bay on W side of Lopez Is. 1/2 mile south of Lopez Village. Walking distance to bike rentals, shops, & restaurants. Full marine repair service center.

GUEST BOAT CAPACITY: ....Appx. 40 boats
DOCKSIDE DEPTH AT ZERO TIDE: Appx. 30'
SEASON: ....All year
RESERVATION POLICY: ....Accepts
AMT W/ELECTRICITY: ....All
FUEL DOCK: ....Close by
MARINE REPAIRS: ....On premises
TOILETS: ....Yes
HOT SHOWERS: ....Yes
RESTAURANT: ....Close by
PICNIC AREA: ....Yes
BASIC STORE: ....Yes
BROADBAND/WI-FI: ....BroadbandXpress
DAILY RATE: ....Moderate (75¢-$1.25/foot)

GUEST DOCK: ....1000' overall
GUEST SLIPS: ....Docks only
WATER: ....Yes
AMPS: ....30 A
PUMP OUT STATION ....Yes
HAUL OUT: ....Travel-Lift
BOAT RAMP: ....Yes
LAUNDRY: ....Close by
BAR: ....Close by
POOL: ....None
GOLF: ....Close by
PET FRIENDLY: ....Excellent
OTHER: Emergency marine repairs 6 days a week. Fishing & crabbing in Bay. Float plane service to Seattle. Guest apartment rentals.

CAUTION! This chartlet not intended for use in navigation.

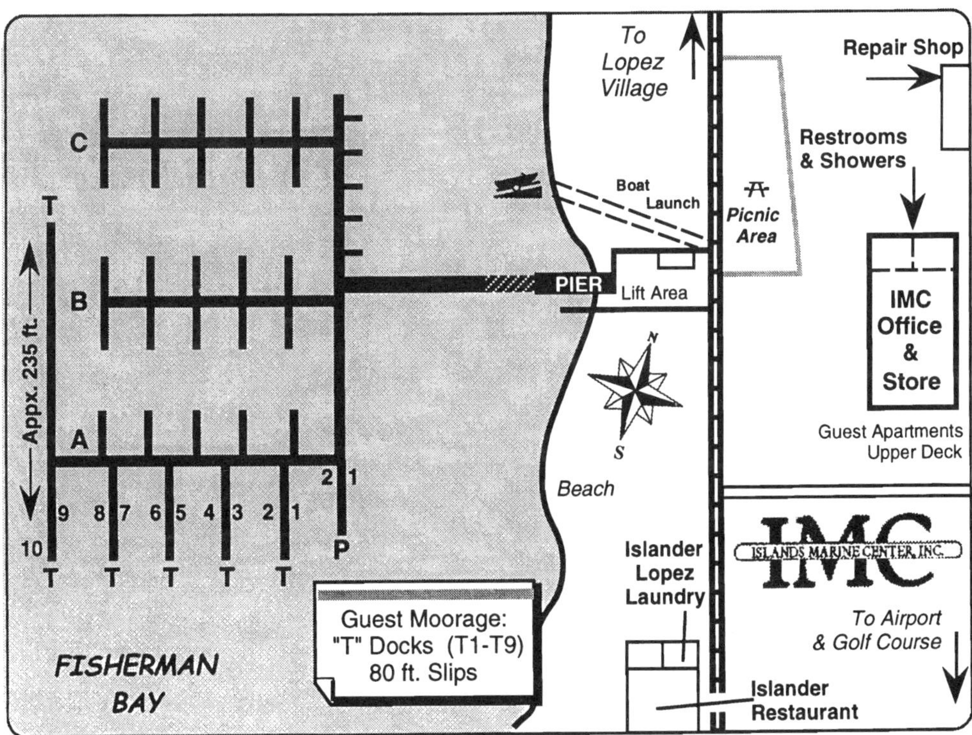

Friday Harbor

NAME OF MARINA: ***PORT OF FRIDAY HARBOR MARINA*** RADIO: VHF 66A
TELEPHONE: **360-378-2688** MGR: Tami Hayes
E-MAIL: tamih@portfridayharbor.org FAX: 360-378-6114 (Fax)
ADDRESS: P.O. Box 889 Friday Harbor, WA 98250
SHORT DESCRIPTION & LOCATION: **www.portfridayharbor.org**

48°32.12' - 123°00.88' Located on E. side in the middle of San Juan Island in a scenic harbor. Largest & busiest marina in the Islands. Services abound for boaters at the marina & in town. Call on VHF 66A for slip or breakwater dock assignment. Customs port-of-entry.

GUEST BOAT CAPACITY: ....Appx. 150 boats
DOCKSIDE DEPTH AT ZERO TIDE: ......20 ft. +
SEASON: ....All year
RESERVATION POLICY:......Accepts w/surcharge
AMT W/ELECTRICITY: ....All
FUEL DOCK: ....Gas & Diesel
MARINE REPAIRS: ....Yes
TOILETS: ....Yes
HOT SHOWERS: ....Yes
RESTAURANT: ....Close by
PICNIC AREA: ....Yes
BASIC STORE: ....Close by
BROADBAND/WI-FI: ....BroadbandXpress
DAILY RATE:..........Moderate (75¢-$1.25/foot)
Reservation surchage applies.

GUEST DOCK: 1500 ft. plus slips
GUEST SLIPS: ....Appx. 85
WATER: ....Yes
AMPS: ....30-50 A
PUMP OUT STATION ....Yes
HAUL OUT: ....Close by
BOAT RAMP: ....Close by
LAUNDRY: ....Close by
BAR: ....Close by
POOL: ....Close by
GOLF: ....Close by
PET FRIENDLY: ....Challenging
OTHER: Boats over 45' must use side-tie breakwater docks. Museums, shopping, air & water transportation to Seattle & mainland ports.

NOTES

Friday Harbor

CAUTION! This chartlet not intended for use in navigation.

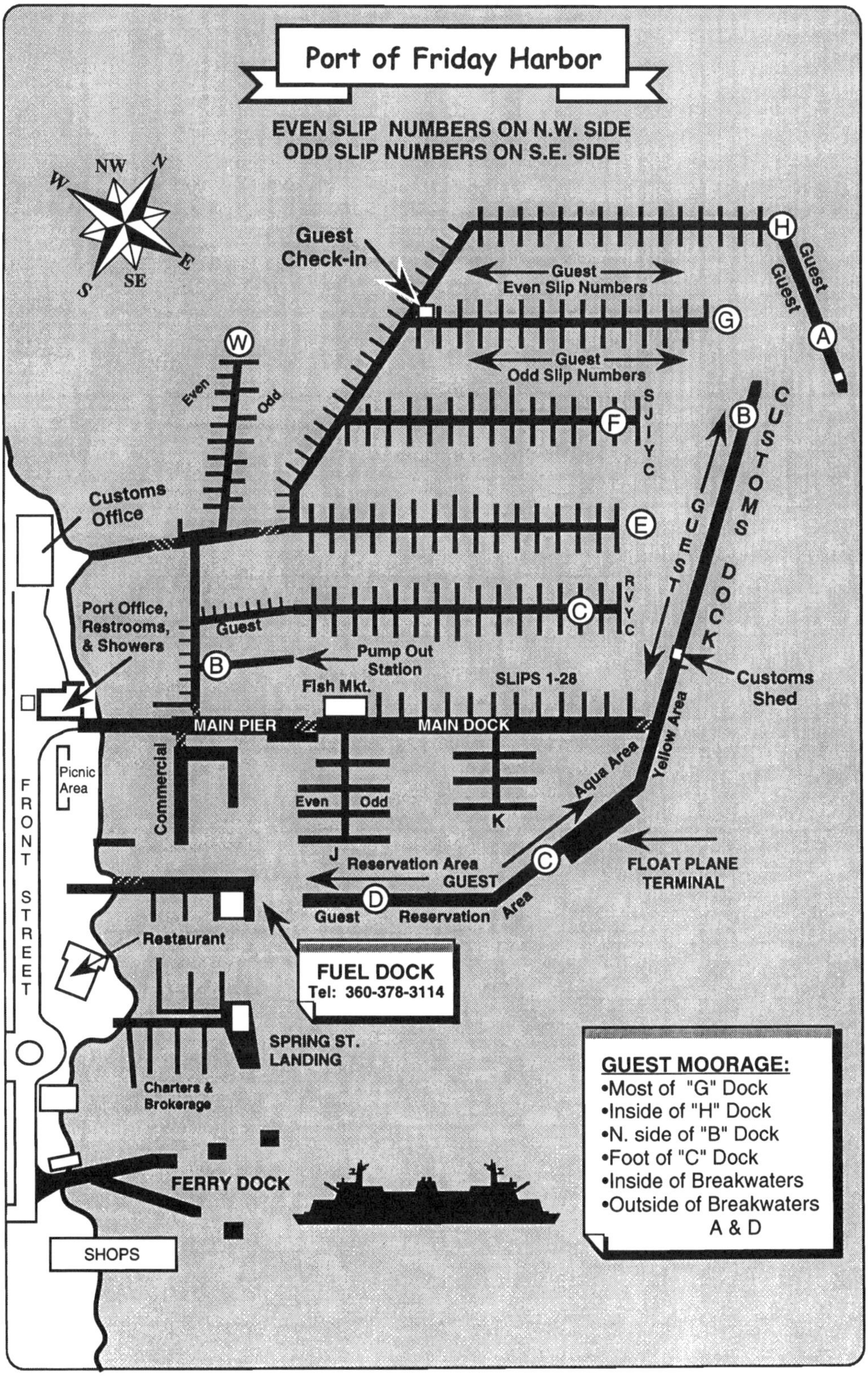

Port of Friday Harbor

EVEN SLIP NUMBERS ON N.W. SIDE
ODD SLIP NUMBERS ON S.E. SIDE

NW
N
W
E
S
SE

Guest Check-in
H
Guest
Guest
A
Guest Even Slip Numbers
G
Guest Odd Slip Numbers
W
Even
Odd
F
SJIYC
B
CUSTOMS
Customs Office
E
GUEST
DOCK
R V Y C
C
Port Office, Restrooms, & Showers
Guest
B
Pump Out Station
Customs Shed
Fish Mkt.
SLIPS 1-28
MAIN PIER
MAIN DOCK
Yellow Area
Aqua Area
Picnic Area
Commercial
Even
Odd
K
J
FRONT STREET
Reservation Area
GUEST
C
FLOAT PLANE TERMINAL
D
Guest
Reservation
Area
Restaurant

FUEL DOCK
Tel: 360-378-3114

SPRING ST. LANDING
Charters & Brokerage
FERRY DOCK
SHOPS

GUEST MOORAGE:

- Most of "G" Dock
- Inside of "H" Dock
- N. side of "B" Dock
- Foot of "C" Dock
- Inside of Breakwaters
- Outside of Breakwaters A & D

Roche Harbor

NAME OF MARINA: ***ROCHE HARBOR RESORT*** **RADIO:** VHF 78A
TELEPHONE: **1-800-586-3590** MGR: Kevin Carlton
E-MAIL: marina@rocheharbor.com FAX: 360-378-9800 (Fax)
ADDRESS: P.O. Box 4001 Roche Harbor, WA 98250
SHORT DESCRIPTION & LOCATION: **www.rocheharbor.com**

48°36.60' - 123°09.10' Historic & charming marina resort. Extensive moorage facilities & vacation amenities located on N. end of San Juan Island appx. 10 mi. N. of Friday Harbor. Slips from 20 to 150 ft. U.S. Customs Port-of-Entry. *Don't miss traditional evening colors.*

GUEST BOAT CAPACITY: ....Appx. 250 boats
DOCKSIDE DEPTH AT ZERO TIDE: ....20 ft.
SEASON: ....All year
RESERVATION POLICY: ....Accepts
AMT W/ELECTRICITY: ....All
FUEL DOCK: ....Gas, Dsl, & LP
MARINE REPAIRS: ....Yes
TOILETS: ....Yes
HOT SHOWERS: ....Yes
RESTAURANT: ....Yes
PICNIC AREA: ....Yes
BASIC STORE: ....Yes
BROADBAND/WI-FI: ....BroadbandXpress
DAILY RATE: ....Premium (Over $1.25/foot)
$10-$20 Holiday Nightly Surcharge.
(Memorial Day, July 4th, Labor Day)

GUEST DOCK: 720 ft. plus slips
GUEST SLIPS: 200
WATER: ....Yes
AMPS: ....30-50-100 A
PUMP OUT STATION ....Yes
HAUL OUT: ....None
BOAT RAMP: ....Yes
LAUNDRY: ....Yes
BAR: ....Yes
POOL: ....Yes
GOLF: ....Close by
PET FRIENDLY: ....Excellent
OTHER: Tennis courts, hotel, airport, float plane service, shops, Post Office, party barge for group events, & mobile pump-out barge.

NOTES

Roche Harbor

CAUTION! This chartlet not intended for use in navigation.

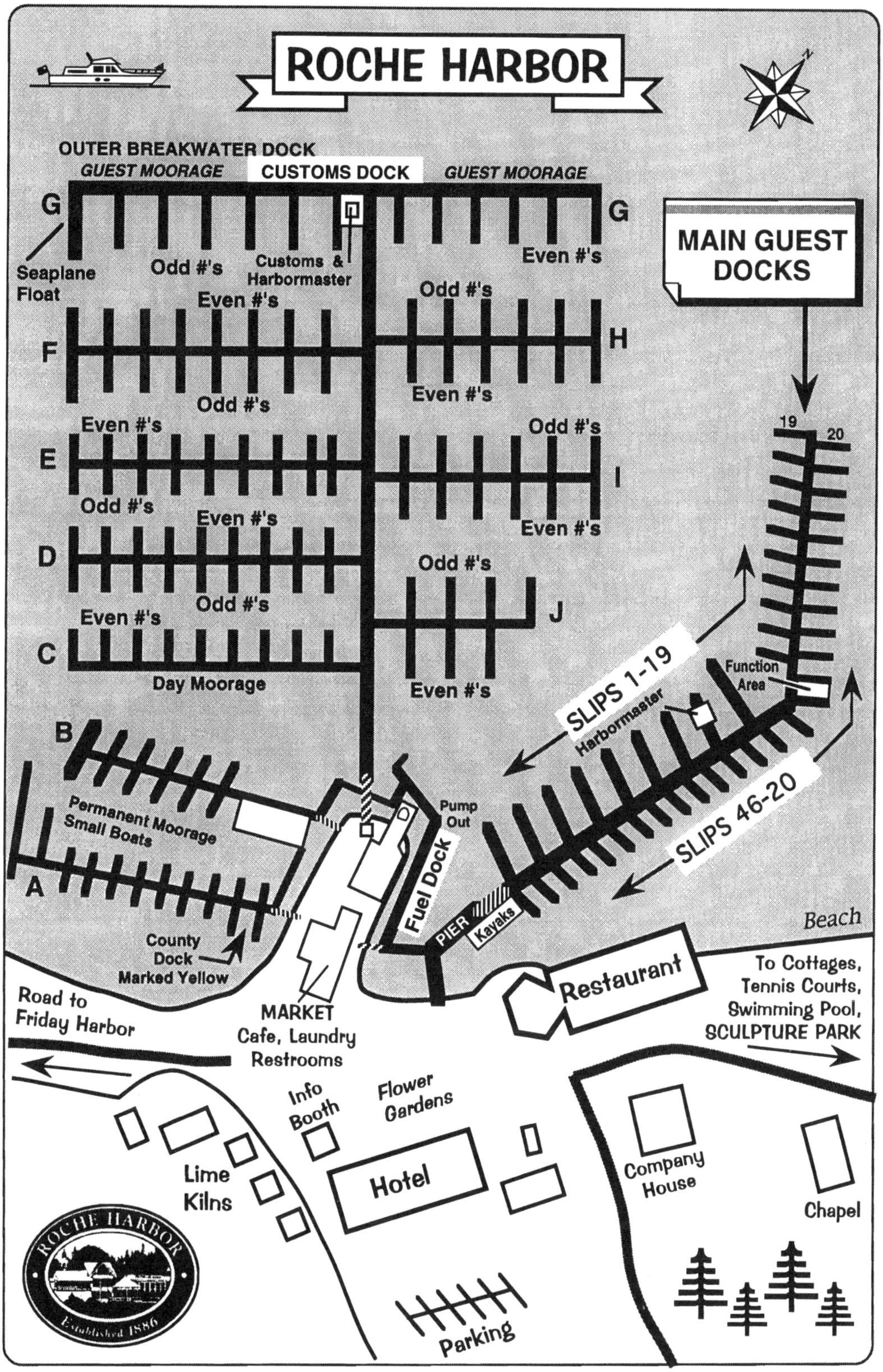

ROCHE HARBOR

OUTER BREAKWATER DOCK
GUEST MOORAGE
CUSTOMS DOCK
GUEST MOORAGE
G
G
Seaplane Float
Odd #'s
Customs & Harbormaster
Even #'s
MAIN GUEST DOCKS
Even #'s
Odd #'s
F
H
Odd #'s
Even #'s
Even #'s
Odd #'s
E
I
Odd #'s
Even #'s
Even #'s
D
Odd #'s
Odd #'s
Even #'s
J
C
Day Moorage
Even #'s
19
20
Function Area
SLIPS 1-19
Harbormaster
B
Permanent Moorage Small Boats
Pump Out
SLIPS 46-20
Fuel Dock
A
PIER
Kayaks
Beach
County Dock Marked Yellow
Restaurant
To Cottages, Tennis Courts, Swimming Pool, SCULPTURE PARK
Road to Friday Harbor
MARKET
Cafe, Laundry
Restrooms
Info Booth
Flower Gardens
Lime Kilns
Hotel
Company House
Chapel
ROCHE HARBOR
Established 1886
Parking

San Juan Island

NAME OF MARINA: ***SNUG HARBOR MARINA RESORT*** **RADIO:** None
TELEPHONE: **360-378-4762** **MGR:** Kevin Culmback
E-MAIL: sneakaway@snugresort.com **FAX:** 360-378-8859 (Fax)
ADDRESS: 1997 Mitchell Bay Rd. Friday Harbor, WA 98250
SHORT DESCRIPTION & LOCATION: **www.snugresort.com**

48°34.30' - 123°09.90' Charming and rustic marina fishing resort located In peaceful Mitchell Bay on N.E. corner of San Juan Island. Recently upgraded docks & floats. ***Note: Maximum 8 ft. working draft at zero tide upon entering Mitchell Bay.***

GUEST BOAT CAPACITY: ....Varies
DOCKSIDE DEPTH AT ZERO TIDE: ....6-8 ft.
SEASON: ....All year
RESERVATION POLICY: ....Accepts
AMT W/ELECTRICITY: 90%
FUEL DOCK: ....Gas & LP
MARINE REPAIRS: ....None
TOILETS: ....Yes
HOT SHOWERS: ....None
RESTAURANT: ....Planned
PICNIC AREA: ....Yes
BASIC STORE: ....Yes
BROADBAND/WI-FI: ....None
DAILY RATE: ....Moderate (75¢-$1.25/foot) Winter rates available.

GUEST DOCK: 77 ft. plus slips
GUEST SLIPS: ....Varies (0-10)
WATER: ....Yes
AMPS: ....20-30 A
PUMP OUT STATION ....None
HAUL OUT: ....None
BOAT RAMP: ....Yes
LAUNDRY: ....None
BAR: ....None
POOL: ....None
GOLF: ....None
PET FRIENDLY: ....Good
OTHER: Great dinghy area in Bay. Crabbing, fishing, whale watching, cabins, gift shop. Kayak, skiff & boat rentals. Crab & fish gear available.

CAUTION! This chartlet not intended for use in navigation.

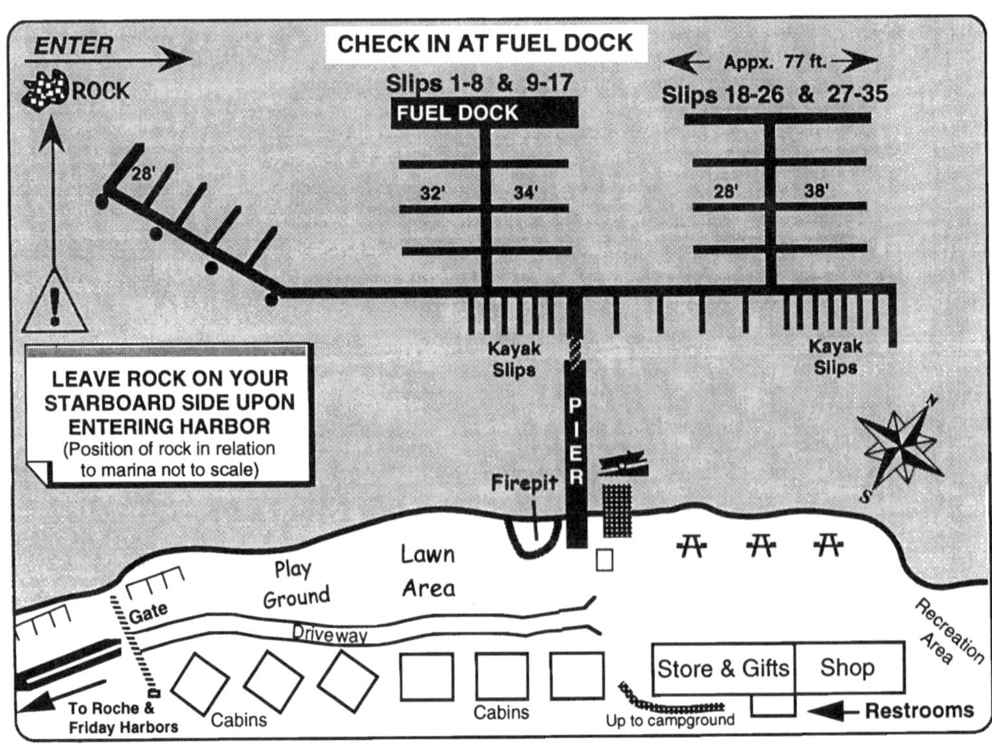

Jones Island

NAME OF PARK: ***JONES ISLAND STATE MARINE PARK***
ADDRESS: 1567 West Side Rd. Friday Harbor, WA 98250
TELEPHONE: **360-378-2044** **MGR:** Chris Guidotti
SHORT DESCRIPTION & LOCATION: **www.parks.wa.gov/moorage/parks**
48°37.30' - 123°02.80' Located 3/4 mile off the S.W. end of Orcas Is. in Spring Passage. This popular island marine park is only accessible by boat. Beauty & wildlife abound. Mooring buoys can be found on both sides of the Island - North & South.

GUEST BOAT CAPACITY: ....Appx. 8 boats
DOCKSIDE DEPTH AT ZERO TIDE:..........5 ft. +
SEASON: ....Dock removed in winter
AMT W/ELECTRICITY: ....None
TOILETS: ....Composting & pit toilets
HOT SHOWERS: ....None
PICNIC AREA: ....Yes
PLAY AREA: ....Yes
BASIC STORE: ....None
DAILY RATE: Dock Moorage.........................50¢/foot
Mooring Buoys...............$10.00/night

Garbage: Pack-it-Out.

GUEST DOCK: Appx. 260 ft. total
MOORING BUOYS: .....All year - 7
WATER: ....May thru Sept. Only
PAY PHONES: ....None
BOAT RAMP: ....None
PICNIC SHELTER: ....None
BBQ: ....None
PUMP OUT STATION: ......None
PET FRIENDLY: ....Excellent
OTHER: 4 mooring buoys on N. side of Island & 3 on S. side. Campsites, Cascadia Marine Trail site (kayaks only), wildlife reserve, hiking trails.

CAUTION! This chartlet not intended for use in navigation.

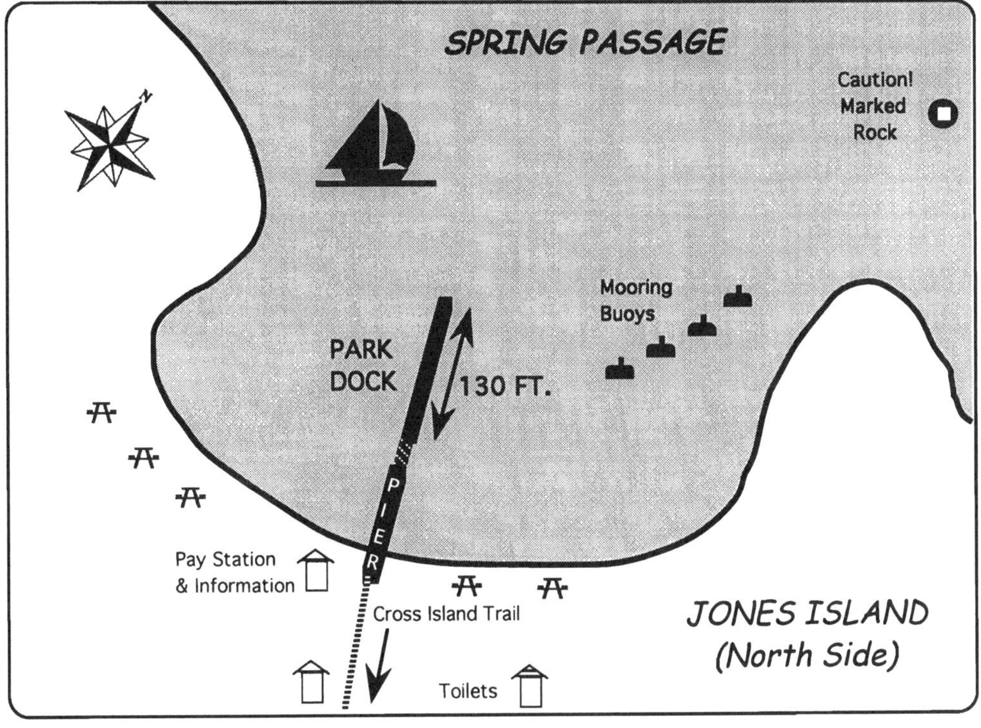

Shaw Island

NAME OF MARINA: ***SHAW GENERAL STORE*** RADIO: None
TELEPHONE: 360-468-2288 MGR: Steve & Terri Mason
E-MAIL: None FAX: None
ADDRESS: Box 455, Shaw Island, WA 98268

SHORT DESCRIPTION & LOCATION:

48°35.06' - 122°55.76' Sm. marina located on N.E. corner of Blind Bay on north side of Shaw Is. next to ferry landing. The historic store & dock are operated by the Mason family who also act as State Ferry agents. No big boats allowed overnight - maximum size 25 ft.

GUEST BOAT CAPACITY: ....Limited
DOCKSIDE DEPTH AT ZERO TIDE: ....3 ft. +
SEASON: ....All year
RESERVATION POLICY: Call ahead for Moorage
AMT W/ELECTRICITY: ....None
FUEL DOCK: ....None
MARINE REPAIRS: ....None
TOILETS: ....Yes
HOT SHOWERS: ....None
RESTAURANT: ....None
PICNIC AREA: ....None
BASIC STORE: ....Yes
BROADBAND/WI-FI: ....None
DAILY RATE: ....Economical (Under 75¢/foot)

Note: Best to anchor or take mooring buoy in Blind Bay and visit marina by dinghy.

GUEST DOCK: ....Appx. 30 ft.
GUEST SLIPS: ....None
WATER: ....None
AMPS: ....None
PUMP OUT STATION ....None
HAUL OUT: ....None
BOAT RAMP: ....None
LAUNDRY: ....None
BAR: ....None
POOL: ....None
GOLF: ....None
PET FRIENDLY: ....Good
OTHER: USPO, well stocked Store, county park, campground & museum 2.5 miles away, bicycling, & Marine Park adjacent.

CAUTION! This chartlet not intended for use in navigation.

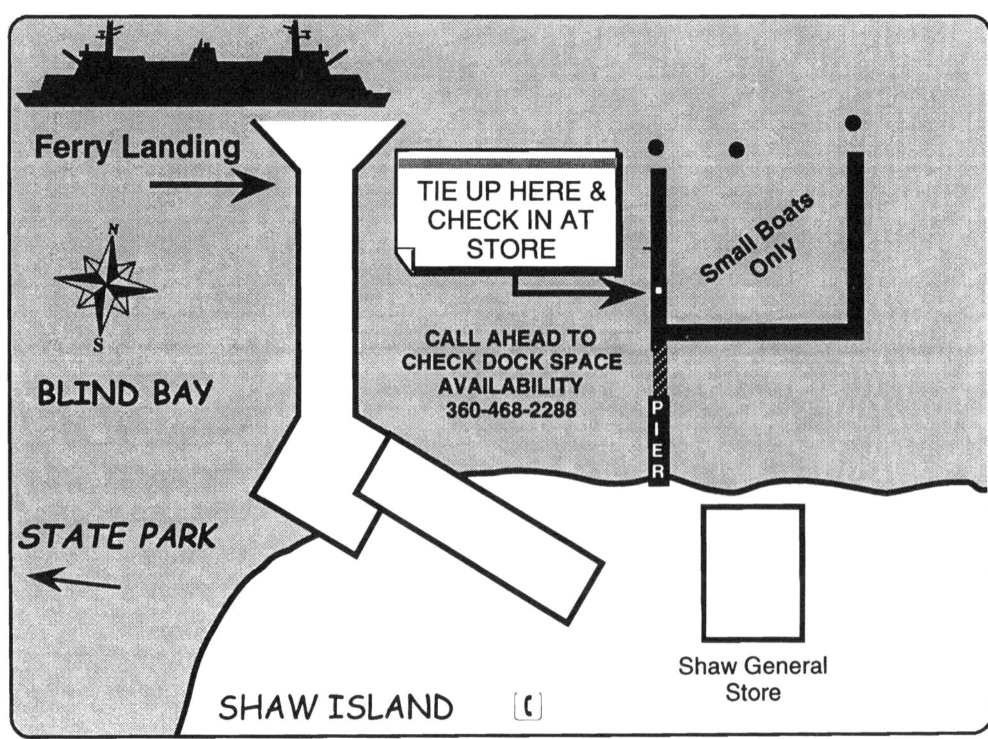

Deer Harbor

NAME OF MARINA: ***DEER HARBOR MARINA*** **RADIO:** VHF 78A

TELEPHONE: **360-376-3037** MGR: Marc Broman

E-MAIL: info@bellportgroup.com FAX: 360-376-6091 (Fax)

ADDRESS: P.O. Box 344 Deer Harbor, WA 98243

SHORT DESCRIPTION & LOCATION: **www.deerharbormarina.com**

48°37.20' - 123°00.20' Small harbor on W. side of Orcas Island between Pole Pass and Steep Point. Popular & family oriented marina & resort located on E. side of harbor in very beautiful San Juan Islands setting. Docks & slips were nicely upgraded in the late 90's.

GUEST BOAT CAPACITY: ....Appx. 100 boats
DOCKSIDE DEPTH AT ZERO TIDE: ......12 ft. +
SEASON: ....All year
RESERVATION POLICY: ....Accepts
AMT W/ELECTRICITY: ....All
FUEL DOCK: ....Gas & Diesel
MARINE REPAIRS: ....Close by
TOILETS: ....Yes
HOT SHOWERS: ....Yes
RESTAURANT: ....Close by
PICNIC AREA: ....Yes
BASIC STORE: ....Yes
BROADBAND/WI-FI: ....BroadbandXpress
DAILY RATE: ....Premium (Over $1.25/foot)

GUEST DOCK: ....Appx. 270 ft.
GUEST SLIPS: ....Appx. 80
WATER: ....Yes
AMPS: ....30 A
PUMP OUT STATION ....Yes
HAUL OUT: ....Close by
BOAT RAMP: ....Close by
LAUNDRY: ....Yes
BAR: ....Yes
POOL: ....Yes
GOLF: ....12 miles away
PET FRIENDLY: ....Excellent
OTHER: Great dinghy area, fishing, crabbing, Post Office, gift shop, bike & kayak rentals, espresso & ice cream.

CAUTION! This chartlet not intended for use in navigation.

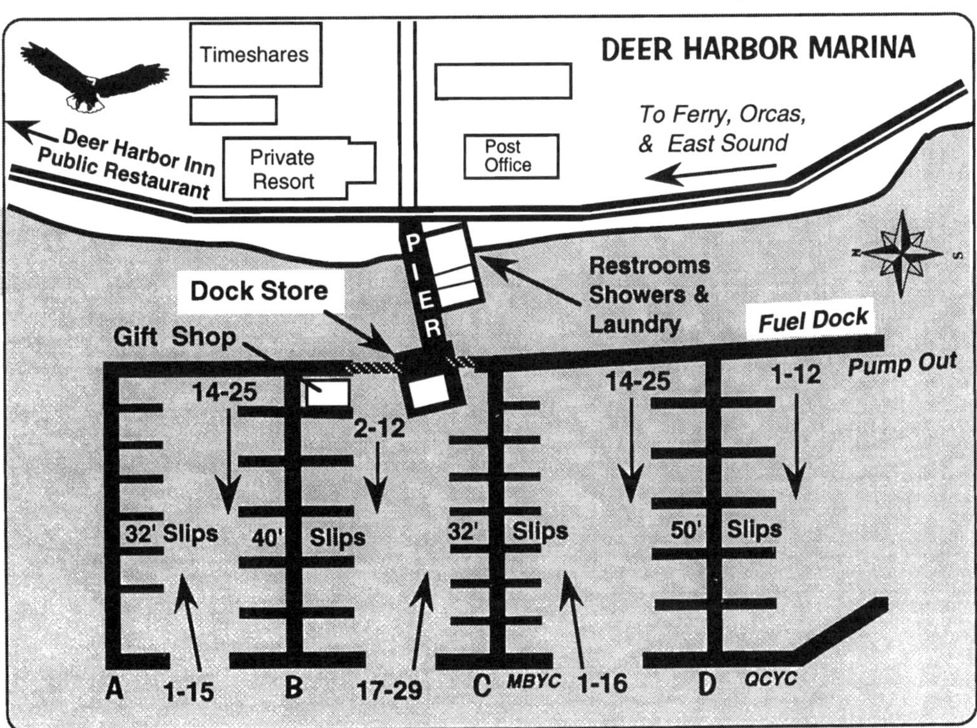

Orcas Island

NAME OF YACHT CLUB: *ORCAS ISLAND YACHT CLUB*

ADDRESS: P.O. Box 686 Eastsound WA 98245

TELEPHONE: See phone list on dock PERSON IN CHARGE: Corresp.Secretary

SHORT DESCRIPTION & LOCATION: **www.oiyc.org**

48°37.80' -122°57.50' OIYC has a small "T" float located in the N.E. corner of West Sound for Reciprocal use. Reciprocal members are invited to use dock, picnic shelter & BBQ. Yellow Zone on dock may be used when no OIYC members present. Combination to lock on facilities door is published in Reciprocal Letter sent to your Club. Log in on dock.

RECIPROCAL BOAT CAPACITY:	Appx. 6-8	RECIPROCAL DOCK:	Appx. 200'
DOCKSIDE DEPTH AT ZERO TIDE:	Appx. 8 ft.	RECIPROCAL SLIPS:	Dock only
SEASON:	Closed winter	WATER:	No
RESERVATION POLICY:	None	AMT W/ELECTRICITY:	None
TOILETS:	Yes	AMPS:	None
HOT SHOWERS:	Yes	BAR:	None
RESTAURANT:	Cafe next door	PET FRIENDLY:	Excellent
BROADBAND/WI-FI:	None	OTHER:	West Sound Marina next door with full marine services. Quiet neighborhood location.
DAILY RATE:	No Charge 48 HOUR LIMIT		

***NOTE:* THIS IS PRIVATE MOORAGE AND ONLY AVAILABLE TO MEMBERS OF RECIPROCAL YACHT CLUBS. YOUR CLUB *MUST* HAVE RECIPROCAL PRIVILEGES AND YOU MUST FLY YOUR BURGEE.**

CAUTION! This chartlet not intended for use in navigation.

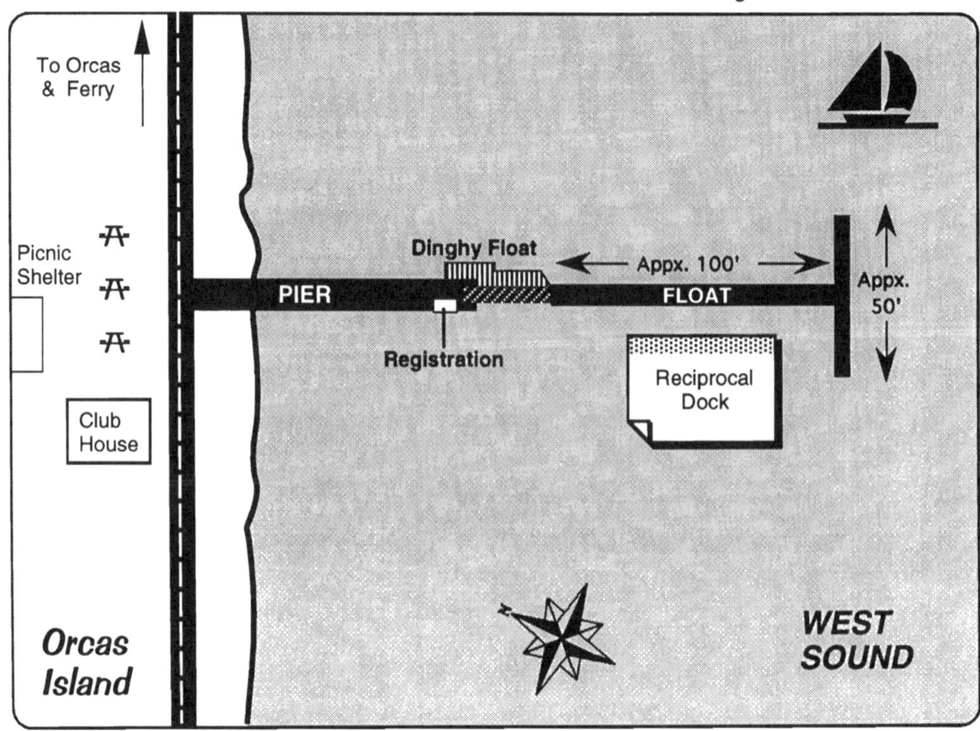

Orcas Island

NAME OF MARINA: ***WEST SOUND MARINA INC.*** **RADIO:** VHF Ch.16
TELEPHONE: **360-376-2314** MGR: Betsy Wareham
E-MAIL: info@westsoundmarina.com FAX: 360-376-4634 (Fax)
ADDRESS: P.O. Box 119 Orcas, WA 98280
SHORT DESCRIPTION & LOCATION:

48°37.80' - 122°57.40' Quiet local marina located in a picturesque cove on N.W. corner of West Sound on Orcas Island. Marina sits in the lee of Picnic Island. Marine services available. When approaching use caution as water shoals toward Picnic Island.

GUEST BOAT CAPACITY: ....Appx. 6-8 boats
DOCKSIDE DEPTH AT ZERO TIDE: ....5 ft. +
SEASON: ....All year
RESERVATION POLICY: ....Accepts
AMT W/ELECTRICITY: ....6 Outlets
FUEL DOCK: ....Gas, Dsl, & LP
MARINE REPAIRS: ....On premises
TOILETS: ....Yes
HOT SHOWERS: ....Yes
RESTAURANT: ....Close by
PICNIC AREA: ....None
BASIC STORE: ....Yes
BROADBAND/WI-FI: ....None
DAILY RATE: ....Moderate
(75¢-$1.25/foot)

GUEST DOCK: Appx. 400 ft. TTL
GUEST SLIPS: ....Dock only
WATER: ....Yes
AMPS: ....30 A
PUMP OUT STATION ....Yes
HAUL OUT: ....30 Ton Travel-Haul
BOAT RAMP: ....None
LAUNDRY: ....None
BAR: ....None
POOL: ....None
GOLF: ....3 Miles
PET FRIENDLY: ....Good
OTHER: Marine chandlery, fishing supplies. ice, & propane. Deli close by. Factory certified mechanics & techs.

CAUTION! This chartlet not intended for use in navigation.

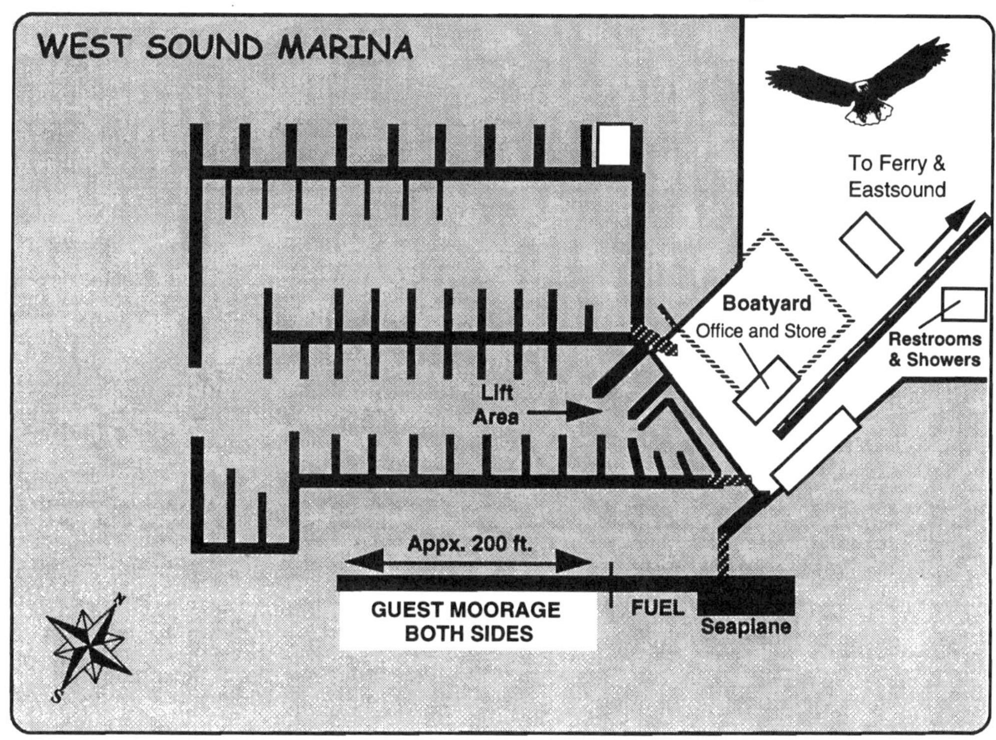

Orcas Island

NAME OF MARINA: ***OLGA COMMUNITY FLOAT*** **RADIO:** None
TELEPHONE: None **MGR:** Dockmaster
E-MAIL: None **FAX:** None
ADDRESS: Olga Community Club Olga, WA 98279

SHORT DESCRIPTION & LOCATION:

48°37.00' - 122°50.20' Olga is located on Orcas Island on the S.E. shore of East Sound. The guest float is maintained w/community labor and moorage fees. The quiet moorage is adjacent to the Village of Olga and is a pleasant walk to Orcas Island Artworks.

GUEST BOAT CAPACITY: ....Appx. 6-8 boats
DOCKSIDE DEPTH AT ZERO TIDE: ......10 ft. +
SEASON: ....May thru September
RESERVATION POLICY: ....None
AMT W/ELECTRICITY: ....None
FUEL DOCK: ....None
MARINE REPAIRS: ....None
TOILETS: ....None
HOT SHOWERS: ....None
RESTAURANT: ....Close by
PICNIC AREA: ....Yes
BASIC STORE: ....None
BROADBAND/WI-FI: ....None
DAILY RATE:..Economical
(Under 75¢/foot)

GUEST DOCK: ....90 ft.
GUEST SLIPS: ....Dock only
WATER: ....Yes
AMPS: ....None
PUMP OUT STATION ....None
HAUL OUT: ....None
BOAT RAMP: ....None
LAUNDRY: ....None
BAR: ....None
POOL: ....None
GOLF: ....None
PET FRIENDLY: ....Good
OTHER: Please side tie dinghy, all mooring buoys are private, closest stores by boat are Blakely Island & Orcas Ferry Landing.

CAUTION! This chartlet not intended for use in navigation.

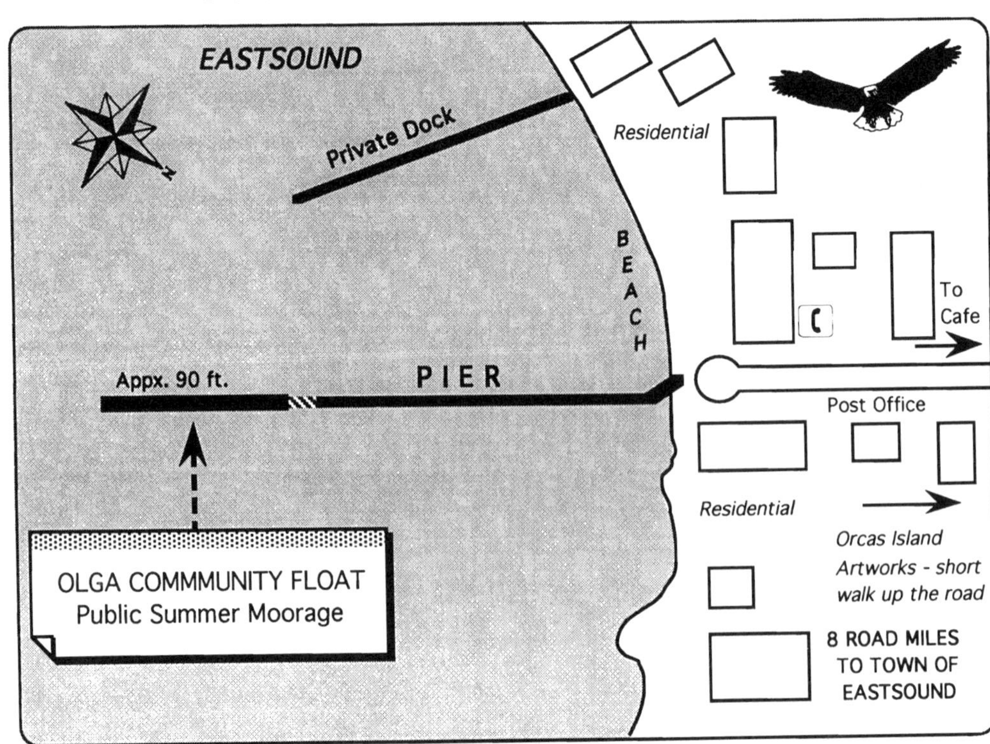

Orcas Island

NAME OF MARINA: *ROSARIO RESORT* RADIO: VHF 78A
TELEPHONE: 800-562-8820 MGR: Nelson Moulton
E-MAIL: harbormaster@rosarioresort.com FAX: 360-376-3038 (Fax)
ADDRESS: 1400 Rosario Road Eastsound, WA 98245
SHORT DESCRIPTION & LOCATION: **www.rosarioresort.com**

48°38.50' - 122°52.15' Located on Orcas Island on east side of East Sound at base of Cascade Bay. World famous & historic resort & spa in park like setting with multiple amenities for boaters. **Reservations are a must in the summer.**

GUEST BOAT CAPACITY: ....Appx. 45 boats
DOCKSIDE DEPTH AT ZERO TIDE: ....9 ft.
SEASON: ....All year
RESERVATION POLICY: ....Recommended
AMT W/ELECTRICITY: ....All
FUEL DOCK: ....Gas & Diesel
MARINE REPAIRS: ....None
TOILETS: ....Yes
HOT SHOWERS: ....Yes
RESTAURANT: ....Yes
PICNIC AREA: ....None
BASIC STORE: ....Seasonal
BROADBAND/WI-FI: ....Yes
DAILY RATE: ....Premium (Over $1.25/foot)

Anchorage: A landing fee of $25.00/day includes use of all resort facilities. Fee is required to come ashore on resort property whether using resort buoy or anchored.

GUEST DOCK: ....105' plus slips
GUEST SLIPS: ....Appx. 45
WATER: ....Yes
AMPS: ....30 A
PUMP OUT STATION ....None
HAUL OUT: ....None
BOAT RAMP: ....None
LAUNDRY: ....Yes
BAR: ....Yes
POOL: ....Yes
GOLF: ....Close by
PET FRIENDLY: ....Excellent
OTHER: 20 Mooring buoys. Moran Museum, Recreational equipment, spa, tennis, shops, car rentals. Kenmore Air to Seattle.

CAUTION! This chartlet not intended for use in navigation.

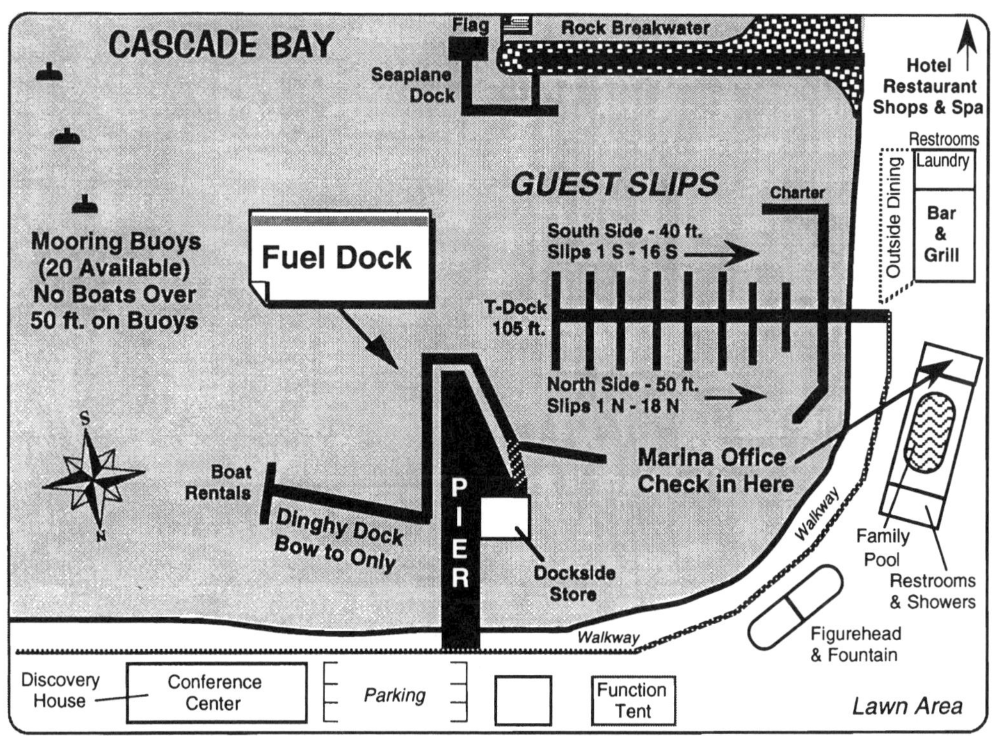

James Island

NAME OF PARK: *JAMES ISLAND STATE MARINE PARK*
ADDRESS: Mail: P.O. Box 66 Olga, WA 98279
TELEPHONE: 360-376-2073 **MGR:** San Juan Marine Area Ranger
SHORT DESCRIPTION & LOCATION: **www.parks.wa.gov/moorage/parks**

48°30.80' - 122°46.50' Located in Rosario Strait close off Decatur Head on the east end of Decatur Island, just south of Thatcher Pass. This 114 acre naturally beautiful island park offers moorage, buoys, hiking trails, gravel beaches, and campsites.

GUEST BOAT CAPACITY: 2-4 boats
DOCKSIDE DEPTH AT ZERO TIDE: 8 ft. +
SEASON: Float removed Oct. thru March
AMT W/ELECTRICITY: None
TOILETS: Compositing & pit toilets
HOT SHOWERS: None
PICNIC AREA: Yes
PLAY AREA: None
BASIC STORE: None
DAILY RATE: Dock Moorage 50¢/foot
Mooring Buoys $10.00/night

GARBAGE: Pack-it-Out.

GUEST DOCK: 80 ft. total
MOORING BUOYS: All year - 4
WATER: None
PAY PHONES: None
BOAT RAMP: None
PICNIC SHELTER: Yes
BBQ: None
PUMP OUT STATION: None
PET FRIENDLY: Excellent
OTHER: 13 Campsites including 3 Cascadia Marine Trail sites, rafting permitted, 1.5 mile network of wildlife trails, & beach combing.

CAUTION! This chartlet not intended for use in navigation.

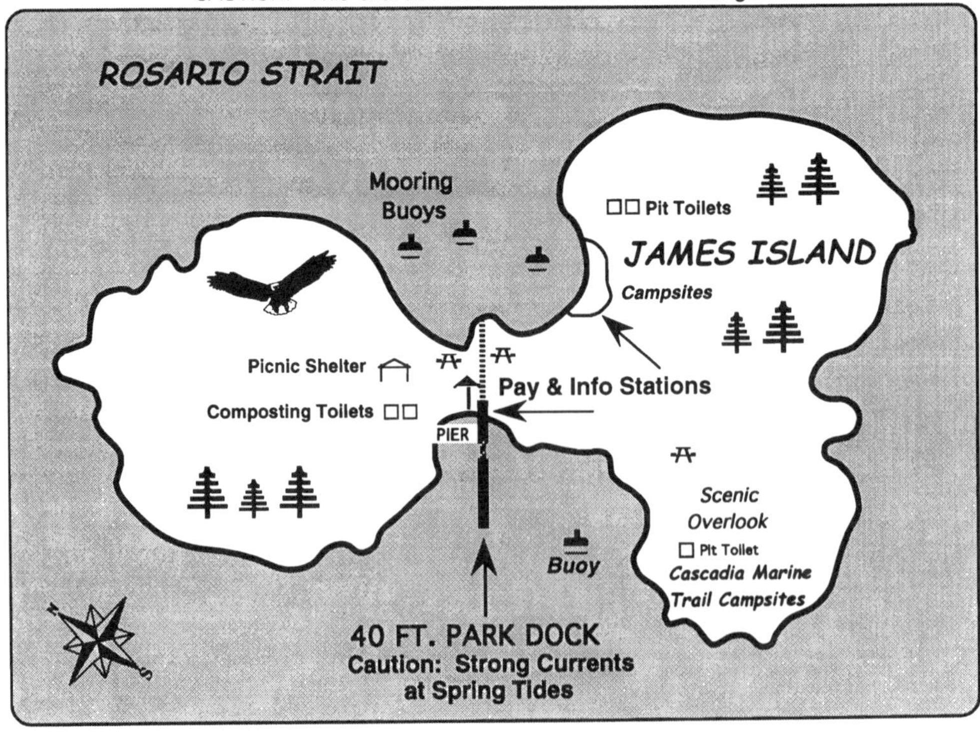

Blakely Island

NAME OF MARINA: ***BLAKELY ISLAND MARINA*** RADIO: VHF 66A
TELEPHONE: 360-375-6121 MGR: Nick & Ken Parker
E-MAIL: FAX: 360-375-6141 (Fax)
ADDRESS: #1 Marina Drive Blakely Island, WA 98222

SHORT DESCRIPTION & LOCATION:

48°35.12' - 122°48.95' Located on N. end of Blakely Island just off the W. end of Peavine Pass. Pristine & well maintained full service marina in a picturesque & quiet setting. The Island is completely private. General store serves island residents & boaters.

GUEST BOAT CAPACITY: ....Appx. 17 boats
DOCKSIDE DEPTH AT ZERO TIDE: ....8 ft.
SEASON: ....May 15 - Sept. 15
RESERVATION POLICY: ....Accepts
AMT W/ELECTRICITY: ....All
FUEL DOCK: ....Gas & Diesel
MARINE REPAIRS: ....None
TOILETS: ....Yes
HOT SHOWERS: ....Yes
RESTAURANT: ....None
PICNIC AREA: ....Yes
BASIC STORE: ....Yes
BROADBAND/WI-FI: ....Yes
DAILY RATE: ....Moderate (75¢-$1.25/foot)

GUEST DOCK: ......32' to 80' Slips
GUEST SLIPS: ....Appx. 17
WATER: ....Yes
AMPS: ....30 A
PUMP OUT STATION ....None
HAUL OUT: ....None
BOAT RAMP: ....Yes
LAUNDRY: ....Yes
BAR: ....None
POOL: ....None
GOLF: ....None
PET FRIENDLY: ....Good
OTHER: Great covered Cabana for group events. Well stocked store with food, clothes, & gifts.

CAUTION! This chartlet not intended for use in navigation.

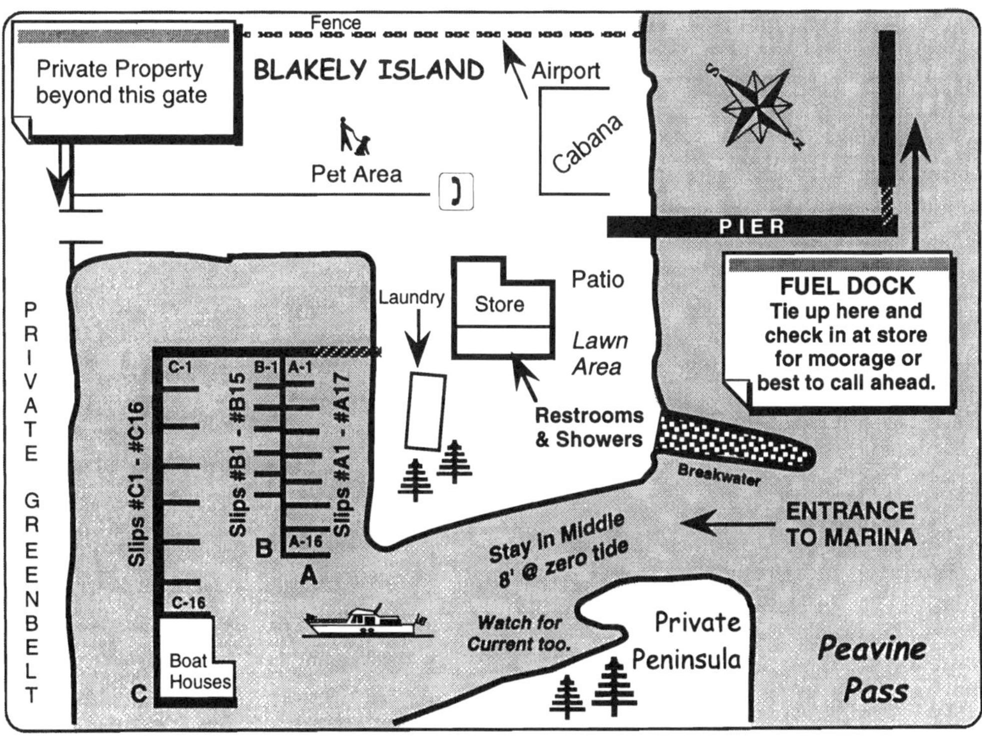

Doe Island

NAME OF PARK: *DOE ISLAND STATE MARINE PARK*
ADDRESS: P.O. Box 66 Olga, WA 98279
TELEPHONE: **360-376-2073** MGR: David Castor
SHORT DESCRIPTION & LOCATION: **www.parks.wa.gov/moorage/parks**

48°38.00' - 122°47.10' Six acre island park located off S.E. shore of Orcas Island midway between Deer and Lawrence Points. Dock on inward side of island. Trails and primitive camp sites located on island. No facilities. Adjacent mooring buoys are private.

GUEST BOAT CAPACITY: Appx. 2-4 boats	GUEST DOCK: Appx. 60 ft. total
DOCKSIDE DEPTH AT ZERO TIDE: 5-10 ft.	MOORING BUOYS: **None**
SEASON: Float removed Oct. thru March	WATER: None
AMT W/ELECTRICITY: None	PAY PHONES: None
TOILETS: Vault toilet	BOAT RAMP: None
HOT SHOWERS: None	PICNIC SHELTER: None
PICNIC AREA: Yes	BBQ: None
PLAY AREA: None	PUMP OUT STATION: None
BASIC STORE: None	PET FRIENDLY: Excellent
DAILY RATE: Dock Moorage 50¢/foot Minimum Charge: $10.00 No Mooring Buoys	OTHER: Fishing, crabbing, scuba, loop trail, 5 campsites each with table & stove, Good beach combing. GARBAGE: Pack-it-Out

CAUTION! This chartlet not intended for use in navigation.

Orcas Island
DANGER Shallow Passage!
BEST APPROACH FROM NORTH SIDE
PIER
Park Float Appx. 30' *Open to Swells*
Island Trail
Cliffs on Trail Edge
Toilet
Pay & Info Station
Doe Island
Island Trail
ROSARIO STRAIT

Matia Island

NAME OF PARK: *MATIA ISLAND STATE MARINE PARK*
ADDRESS: P.O. Box 66 Olga, WA 98279
TELEPHONE: 360-376-2073 MGR: Dave Castor
SHORT DESCRIPTION & LOCATION: www.parks.wa.gov/moorage/parks

48°44.90' - 122°50.50' Matia Island is located at the base of the Strait of Georgia, three miles north of Orcas Is. This 145 acre park is a wildlife refuge. Only the 5 acre park & dock area at Rolfe Cove, & the loop hiking trail are open to the public. No Campfires.

GUEST BOAT CAPACITY: ....Appx. 4-6 boats
DOCKSIDE DEPTH AT ZERO TIDE:.............7 ft.
SEASON: ....Float removed Oct. thru March
AMT W/ELECTRICITY: ....None
TOILETS: ....Composting toilet
HOT SHOWERS: ....None
PICNIC AREA: ....Yes
PLAY AREA: ....Yes
BASIC STORE: ....None
DAILY RATE: Dock Moorage........................50¢/foot
Mooring Buoys...............$10.00/night
Minimum Charge: $10.00
GARBAGE: Pack-it-Out

GUEST DOCK: ....128 Ft. total
MOORING BUOYS: ....2 Buoys
WATER: ....None
PAY PHONES: ....None
BOAT RAMP: ....None
PICNIC SHELTER: ....None
BBQ: ....Yes
PUMP OUT STATION: ......None
PET FRIENDLY: ....Fair
OTHER: One mile loop hiking trail, wildlife, good anchorage, fishing and beachcombing. No Pets allowed on trails but OK in campground.

CAUTION! This chartlet not intended for use in navigation.

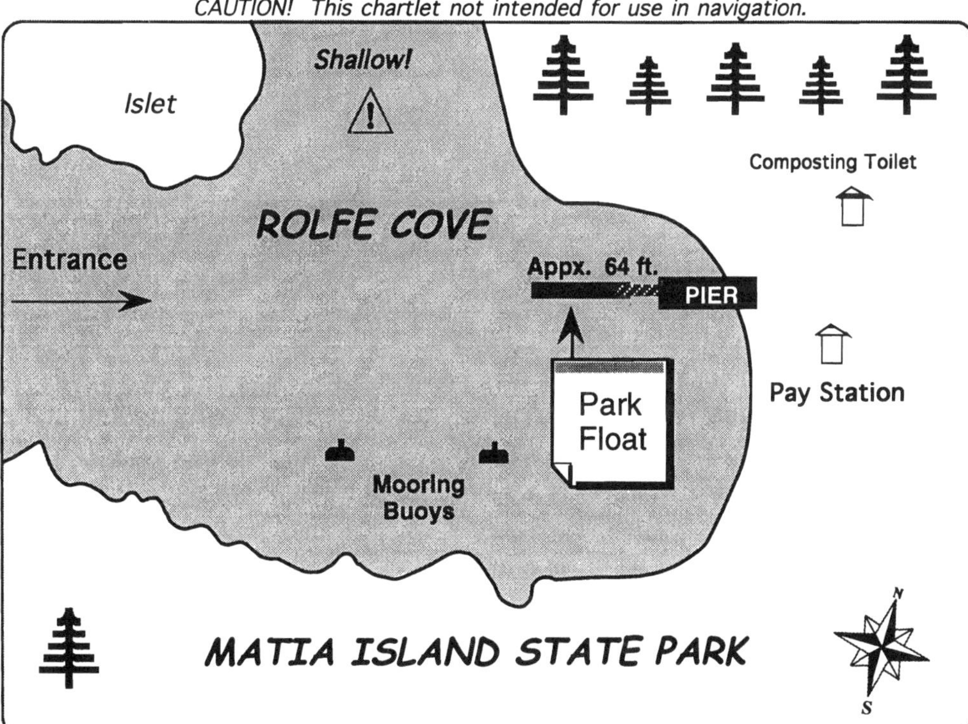

Sucia Island

NAME OF PARK: *SUCIA ISLAND STATE MARINE PARK (FOSSIL BAY)*
ADDRESS: P.O. Box 66 Olga, WA 98279
TELEPHONE: **360-376-2073** **MGR:** David Castor, SJMA Manager
SHORT DESCRIPTION & LOCATION: **www.parks.wa.gov/moorage/parks**
48°45.05' - 123°54.10' Sucia is located at the base of the Strait of Georgia 2.5 mi. N. of mid Orcas Is. This park is 562 acres & comprised of 11 islands. Several bays have park facilities & buoys, but Fossil Bay on the SW side has 2 docks & the most buoys.

GUEST BOAT CAPACITY: ....At docks: 20-25
DOCKSIDE DEPTH AT ZERO TIDE:..........4-5 ft.
SEASON: ....All year
AMT W/ELECTRICITY: ....None
TOILETS: ....Composting toilets
HOT SHOWERS: ....None
PICNIC AREA: ....Yes
PLAY AREA: ....Yes
BASIC STORE: ....None
DAILY RATE: Dock Moorage....................50¢/foot
Buoys/Linear Moorage - $10./night
Minimum Charge: $10.00
Note: 2/Ea. Linear Moorage Systems at Echo Bay - 800 ft.

GUEST DOCK: 2 lg. docks-640 ft.
MOORING BUOYS: Park Total: 48
WATER: ....April-October
PAY PHONES: ....None
BOAT RAMP: ....None
PICNIC SHELTER: ....Yes
BBQ: ....Yes
PUMP OUT STATION: ......None
PET FRIENDLY: ....Excellent
OTHER: Extensive marine trails, underwater scuba park, clams, oysters, crabbing, 55 campsites, and 2 group camping areas.

CAUTION! This chartlet not intended for use in navigation.

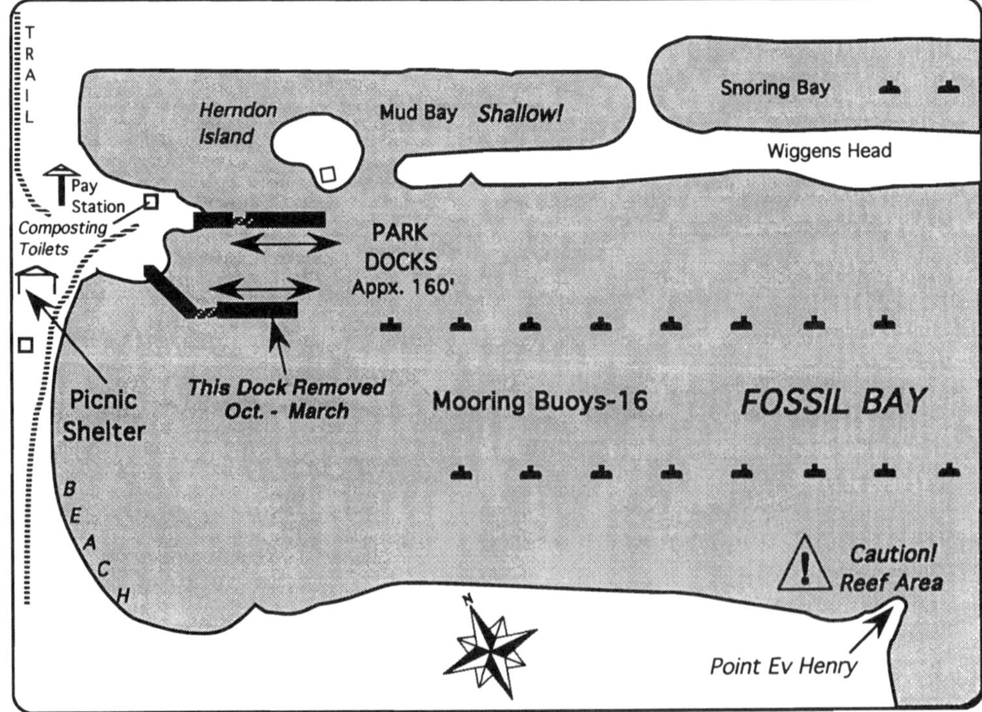

Stuart Island

NAME OF PARK: *REID HARBOR & PREVOST HARBOR MARINE PARKS*
ADDRESS: 1567 Westside Rd. Friday Harbor, WA 98250
TELEPHONE: **360-378-2044** MGR: Chris Guidotti
SHORT DESCRIPTION & LOCATION: **www.parks.wa.gov/moorage/parks**

48°41.60'-123°12' (Appx) Stuart Island is the most westerly island in the San Juans and hosts 2 marine parks with floats, buoys, and good anchorages. Reid Harbor is only 4 mi. N. of Roche Harbor while Prevost Harbor is about 5 mi. S. of Bedwell Harbour, Canada.

GUEST BOAT CAPACITY: ....Appx. 50 boats
DOCKSIDE DEPTH AT ZERO TIDE:..........4 ft. +
SEASON: ....All year
AMT W/ELECTRICITY: ....None
TOILETS: ....Yes
HOT SHOWERS: ....None
PICNIC AREA: ....Yes
PLAY AREA: ....None
BASIC STORE: ....None
DAILY RATE: Dock Moorage......................50¢/foot
Mooring Buoys...............$10.00/night
Minimum Charge: $10.00

GUEST DOCK: Reid: 620'/Pr: 250'
MOORING BUOYS: Reid: 12/Pr: 7
WATER: ....May-Sept.
PAY PHONES: ....None
BOAT RAMP: ....None
PICNIC SHELTER: ....None
BBQ: ....Yes
PUMP OUT STATION: Reid Hbr
PET FRIENDLY: ....Good
OTHER: Extensive network of trails, Cascadia Marine Trail site at head of Reid Harbor, 18 campsites, wildlife, clams, oysters nearby, fishing, and

CAUTION! This chartlet not intended for use in navigation.

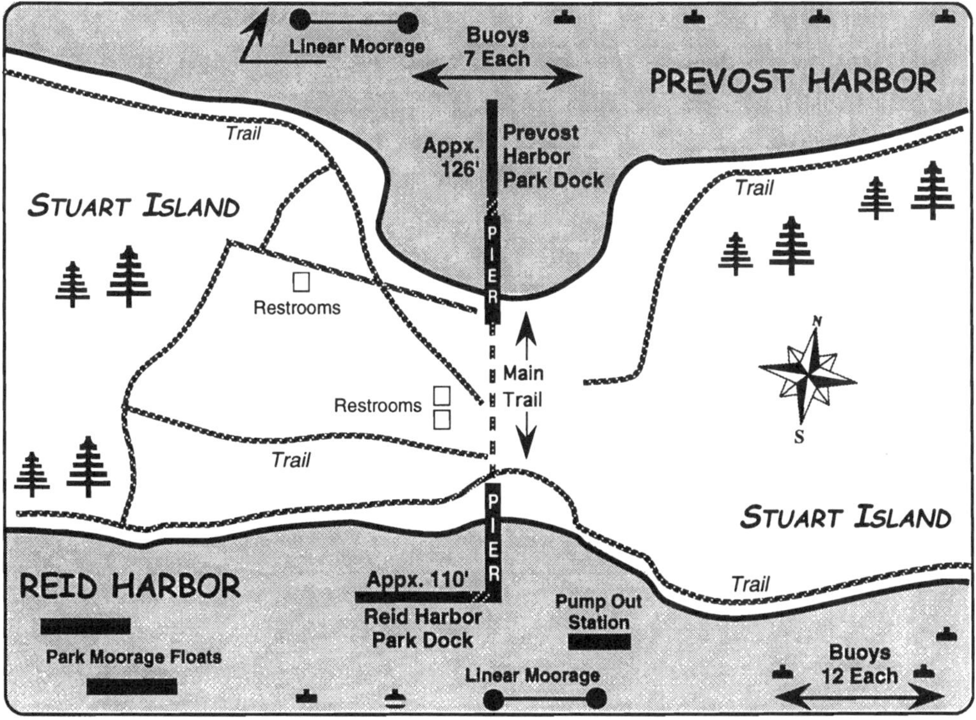

Bellingham

NAME OF MARINA: ***SQUALICUM HARBOR**-Port of Bellingham* **RADIO:** VHF Ch.16
TELEPHONE: **360-676-2542** MGR: Mike Endsley
E-MAIL: squalicum@portofbellingham.com FAX: 360-671-6149 (Fax)
ADDRESS: 722 Coho Way Bellingham, WA 98225
SHORT DESCRIPTION & LOCATION: **www.portofbellingham.com**

48°45.20' - 122°30.50' Bellingham is at the head of Bellingham Bay on the E shore & the large marina is adjacent to & SE of the Squalicum Creek Waterway. Two moorage basins are well protected by breakwaters & offer many marine-oriented services close by.

GUEST BOAT CAPACITY: ....Varies
DOCKSIDE DEPTH AT ZERO TIDE: ......7-10 ft.
SEASON: ....All year
RESERVATION POLICY: ....None
AMT W/ELECTRICITY: ....All
FUEL DOCK: ....Gas, Dsl, & LP
MARINE REPAIRS: ....On premises
TOILETS: ....Yes
HOT SHOWERS: ....Yes
RESTAURANT: ....Yes
PICNIC AREA: ....Yes
BASIC STORE: ....Close by
BROADBAND/WI-FI: ....BroadbandXpress
DAILY RATE: ....Economical (Under 75¢/foot)

COURTESY SHUTTLE - CONTACT OFFICE FOR AVAILABILITY

GUEST DOCK: ....Total 1500 ft.
GUEST SLIPS: ....Varies
WATER: ....Yes
AMPS: ....20-30 A
PUMP OUT STATION ....Yes
HAUL OUT: ....Travel-Lift
BOAT RAMP: ....Yes
LAUNDRY: ....Yes
BAR: ....Yes
POOL: ....None
GOLF: ....Close by
PET FRIENDLY: ....Fair
OTHER: Customs port-of-entry, shopping mall, shipyard, chandlery, public transit, tourist attractions. Eco charter excursions.

NAME OF YACHT CLUB: ***BELLINGHAM YACHT CLUB***
CLUB ADDRESS: 2625 Harbor Loop, Bellingham, WA 98225
CLUB TELEPHONE: 360-733-7390 PERSON IN CHARGE: Recip. Chairman
LOCATION & SPECIAL NOTES: **www.byc.org**

The BYC reciprocal dock is directly in front of the club house & accommodates 2-3 guest boats. Enter on W. entrance of breakwater, proceed E. past Shell fuel dock then N. to the Club's dock. About 100' of dock is available for reciprocal. The largest boat that can be accomodated by fairways turning channel is 65'. Tie up close to boat ahead or behind you.

RECIPROCAL BOAT CAPACITY: ....Varies
DOCKSIDE DEPTH AT ZERO TIDE: ....7 ft.
RECIPROCAL SEASON: ....All year
RESERVATION POLICY: ....None
TOILETS: ....Yes
HOT SHOWERS: ....Yes
RESTAURANT: ....Close by
DAILY RATE: $3/day power charge each day. First Day Free Moorage - 50¢/ft. thereafter. **3 day limit.**

RECIPROCAL DOCK: Appx. 100'
RECIPROCAL SLIPS: ....Varies
WATER: ....Yes
AMT W/ELECTRICITY: ....All
AMPS: ....30 A
BAR: ....Yes
OTHER: Club lounge open most weekends for beverages. Hours vary during summer & winter.

NOTE: THIS IS PRIVATE MOORAGE ONLY AVAILABLE TO MEMBERS OF RECIPROCAL YACHT CLUBS! YOUR CLUB MUST HAVE RECIPROCAL PRIVILEGES AND YOU MUST FLY YOUR BURGEE!

Bellingham

CAUTION! This chartlet not intended for use in navigation.

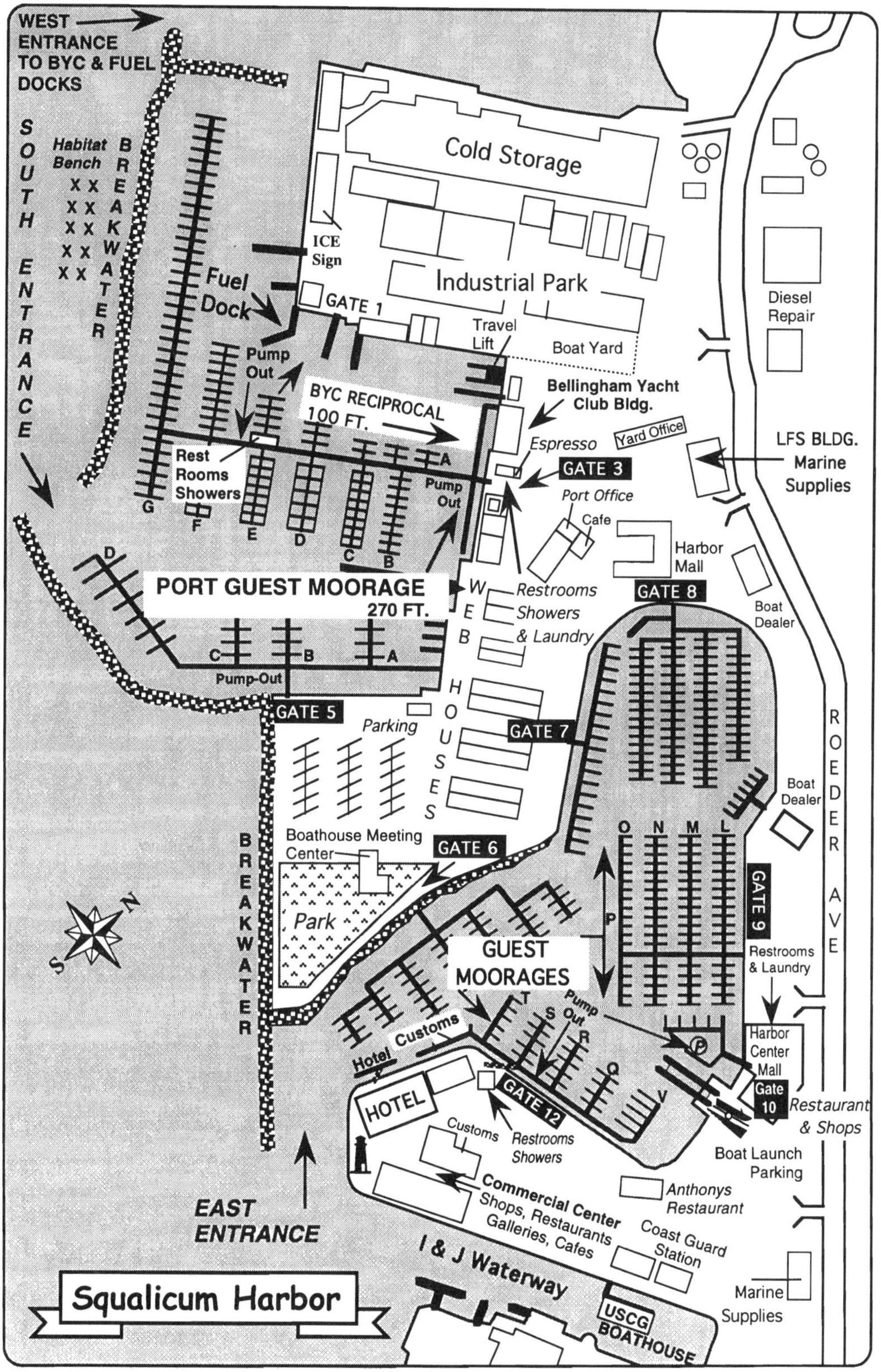

WEST ENTRANCE TO BYC & FUEL DOCKS
SOUTH ENTRANCE
Habitat Bench
BREAKWATER
Cold Storage
ICE Sign
Industrial Park
Fuel Dock
GATE 1
Travel Lift
Boat Yard
Diesel Repair
Pump Out
BYC RECIPROCAL 100 FT.
Bellingham Yacht Club Bldg.
Espresso
Yard Office
LFS BLDG. Marine Supplies
GATE 3
Port Office
Cafe
Harbor Mall
Rest Rooms Showers
G F E D C B A
Pump Out
PORT GUEST MOORAGE 270 FT.
WEB HOUSES
Restrooms Showers & Laundry
GATE 8
Boat Dealer
D C B A
Pump-Out
GATE 5
Parking
GATE 7
ROEDER AVE
Boat Dealer
O N M L
Boathouse Meeting Center
GATE 6
GATE 9
Park
BREAKWATER
P
GUEST MOORAGES
Restrooms & Laundry
Pump Out
T S R Q V
Customs
Hotel
HOTEL
GATE 12
Harbor Center Mall
Gate 10
Restaurant & Shops
Customs
Restrooms Showers
Boat Launch Parking
Commercial Center Shops, Restaurants Galleries, Cafes
Anthonys Restaurant
Coast Guard Station
EAST ENTRANCE
I & J Waterway
USCG BOATHOUSE
Marine Supplies
S N
Squalicum Harbor

NOTES

Birch Bay

NAME OF YACHT CLUB: ***BIRCH BAY VILLAGE YACHT CLUB***

ADDRESS: 8055 Cowichan Rd. Birch Bay, WA 98230

TELEPHONE: 360-371-7744 Marina Office PERSON IN CHARGE: Harbormaster

SHORT DESCRIPTION & LOCATION: **www.bbvcc.com**

48°55.90' - 122°47.21' BBVYC marina entrance located just south of Blaine, and Birch Pt. on NW corner of Birch Bay. Reciprocal guest dock on starboard side of entrance channel at S. end of fuel dock. Upon arrival follow instructions posted on fuel dock bldg. Give shoreline wide berth before turning between channel markers - 8 ft. depth @ 0 tide. **Office Open Weekdays Only.**

RECIPROCAL BOAT CAPACITY: ....2-4 boats

DOCKSIDE DEPTH AT ZERO TIDE: 8 ft. plus

SEASON: ....All year

RESERVATION POLICY: ....Accepts

TOILETS: ....Yes

HOT SHOWERS: ....None

RESTAURANT: ....None

BROADBAND/WI-FI: ....None

DAILY RATE: ....No Charge for 72 hours

NOTE: No commercial services. Moorage in quiet residential area.

RECIPROCAL DOCK: ....150 ft.

RECIPROCAL SLIPS: Dock only

WATER: ....Yes

AMT W/ELECTRICITY: ....All

AMPS: ....30 A

BAR: ....None

PET FRIENDLY: ....Excellent

OTHER: Pumpout, kids play area, BBQ, Yacht Club volunteer transportation & assistance-see office.

***NOTE:* THIS IS PRIVATE MOORAGE AND ONLY AVAILABLE TO MEMBERS OF RECIPROCAL YACHT CLUBS. YOUR CLUB *MUST* HAVE RECIPROCAL PRIVILEGES AND YOU MUST FLY YOUR BURGEE.**

CAUTION! This chartlet not intended for use in navigation.

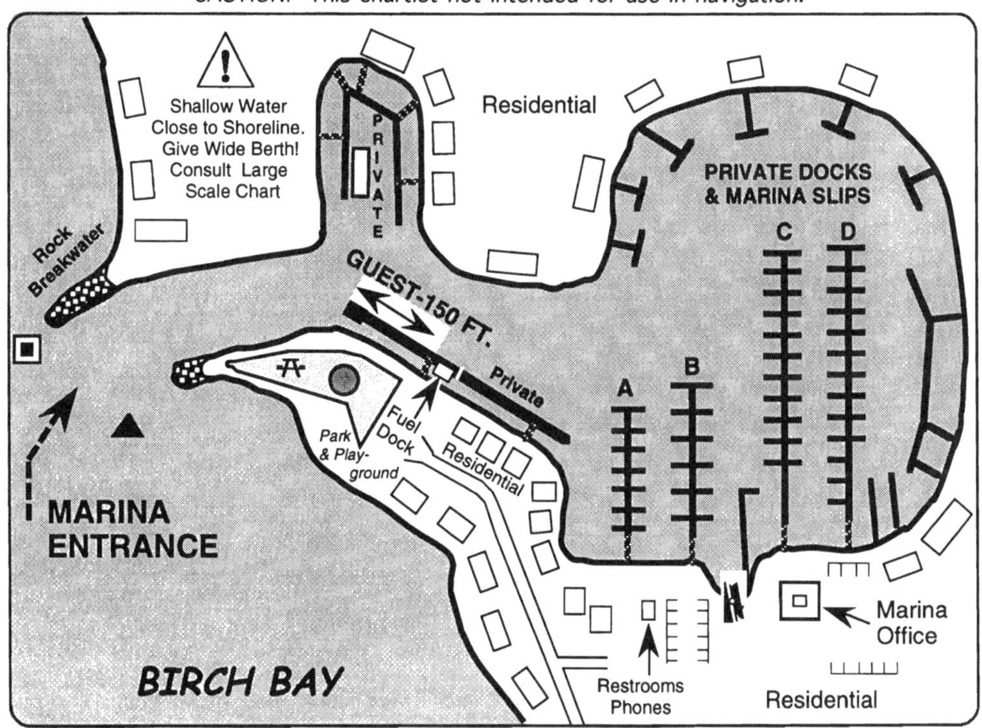

Blaine

NAME OF MARINA: ***SEMIAHMOO MARINA*** **RADIO:** VHF Ch. 68
TELEPHONE: **360-371-0440** MGR: Rick Greenhow
E-MAIL: semimarina@bbxmail.net FAX: 360-371-0200 (Fax)
ADDRESS: 9540 Semiahmoo Pkwy. Blaine, WA 98230
SHORT DESCRIPTION & LOCATION: **www.semiahmoomarina.com**

48°59.41' - 122°47.64' Located across from Blaine on the S. side of the entrance to Drayton Harbor on the N. end of the sand spit that forms the S.W. side of the harbor. The modern & well maintained marina offers all the amenities of a first-class resort.

GUEST BOAT CAPACITY: ....Ample
DOCKSIDE DEPTH AT ZERO TIDE: ....12 ft.
SEASON: ....All year
RESERVATION POLICY: ....Accepts
AMT W/ELECTRICITY: ....All
FUEL DOCK: ....Gas, Dsl, & LP
MARINE REPAIRS: ....Locally available
TOILETS: ....Yes
HOT SHOWERS: ....Yes-FREE
RESTAURANT: ....Yes - 3
PICNIC AREA: ....Yes
BASIC STORE: ....Check with Marina
BROADBAND/WI-FI: ....BroadbandXpress
DAILY RATE: ....Moderate
(75¢-$1.25/foot)

GUEST DOCK: ....Slips only
GUEST SLIPS: ....Varies
WATER: ....Yes
AMPS: ....30-50 A
PUMP OUT STATION ....Yes
HAUL OUT: ....None
BOAT RAMP: ....None
LAUNDRY: ....Yes
BAR: ....Yes
POOL: ....Yes
GOLF: ....Yes
PET FRIENDLY: ....Excellent
OTHER: Arnold Palmer designed golf course, health club, hiking & cycling trails, boutique, fishing tackle, park area nearby.

NAME OF YACHT CLUB: ***SEMIAHMOO YACHT CLUB***
CLUB ADDRESS: 9540 Semiahmoo Parkway #117, Blaine, WA 98230
CLUB TELEPHONE: 360-371-0440 PERSON IN CHARGE: Port Captain
LOCATION & SPECIAL NOTES: **www.semiahmooyachtclub.com**

Please tie up on the main dock or fuel dock & check with the Marina Managers office for availability. Best to call ahead. Reciprocal moorage is offered to members of reciprocal clubs in good standing for 2 nights at no charge. 45 Feet marked in blue on Main Dock. After 2 days, check at marina for regular guest slip at marina rates.

RECIPROCAL BOAT CAPACITY: ....Varies
DOCKSIDE DEPTH AT ZERO TIDE: ....12 ft.
RECIPROCAL SEASON: ....All year
RESERVATION POLICY: ....Call for availability
TOILETS: ....At Marina
HOT SHOWERS: ....At Marina
RESTAURANT: ....At the Inn
DAILY RATE: ....2 Nights Free

RECIPROCAL DOCK: ....Yes
RECIPROCAL SLIPS: Varies
WATER: ....Yes
AMT W/ELECTRICITY: ....All
AMPS: ....30 A
BAR: ....At the Inn
OTHER: Friday night BBQ open to reciprocal. BYOB & meat.

***NOTE:* THIS IS PRIVATE MOORAGE *ONLY* AVAILABLE TO MEMBERS OF RECIPROCAL YACHT CLUBS! YOUR CLUB *MUST* HAVE RECIPROCAL PRIVILEGES AND YOU MUST FLY YOUR BURGEE!**

Blaine

CAUTION! This chartlet not intended for use in navigation.

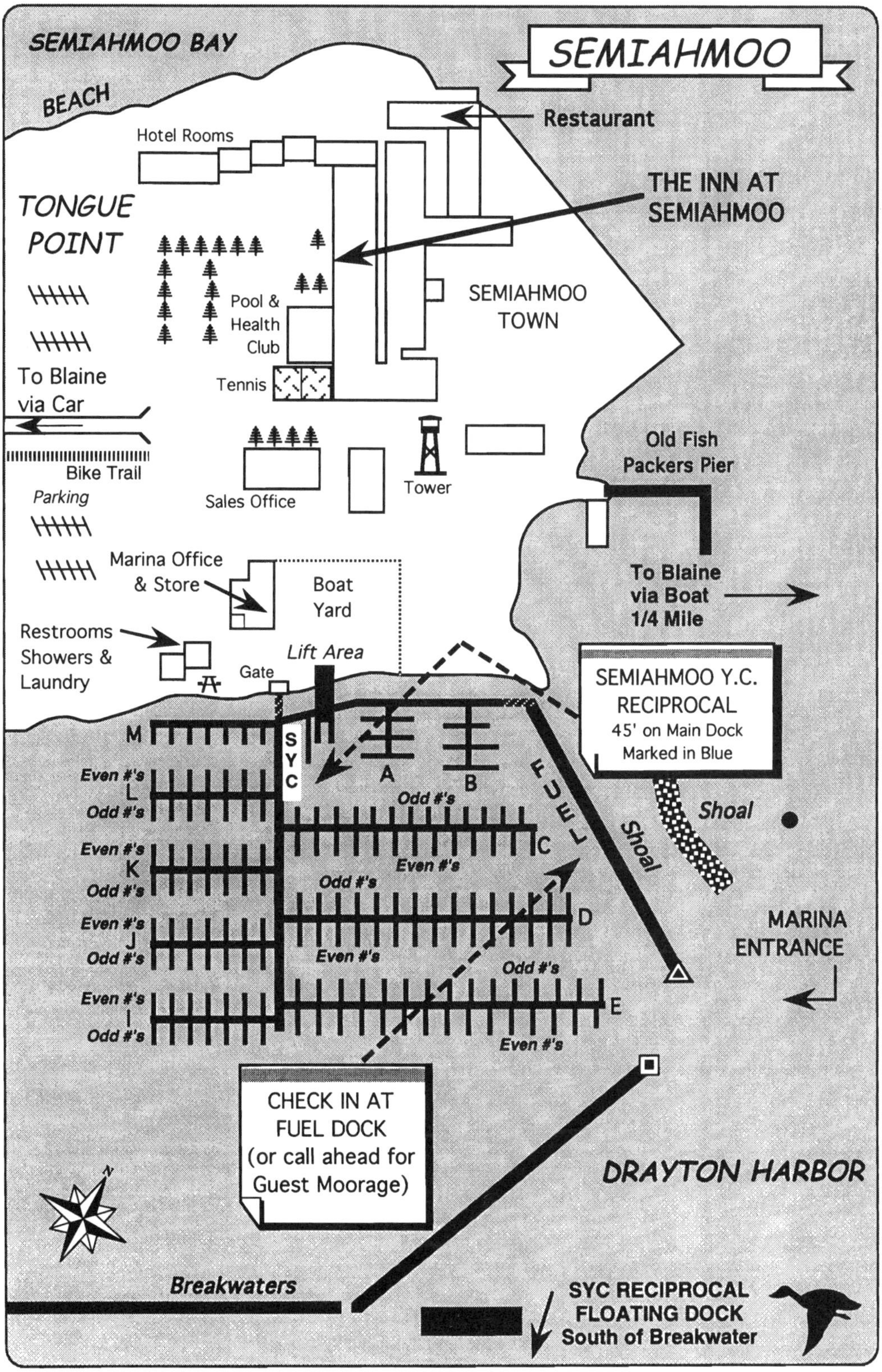

SEMIAHMOO

SEMIAHMOO BAY
BEACH
Restaurant
Hotel Rooms
THE INN AT SEMIAHMOO
TONGUE POINT
Pool & Health Club
SEMIAHMOO TOWN
To Blaine via Car
Tennis
Bike Trail
Parking
Sales Office
Tower
Old Fish Packers Pier
Marina Office & Store
Boat Yard
To Blaine via Boat 1/4 Mile
Restrooms Showers & Laundry
Lift Area
Gate
SEMIAHMOO Y.C. RECIPROCAL
45' on Main Dock
Marked in Blue
M
SYC
A
B
Even #'s
L
Odd #'s
Odd #'s
FUEL
Shoal
Shoal
Even #'s
K
Odd #'s
C
Even #'s
Odd #'s
Even #'s
J
Odd #'s
D
MARINA ENTRANCE
Even #'s
Even #'s
I
Odd #'s
Odd #'s
E
Even #'s
CHECK IN AT FUEL DOCK (or call ahead for Guest Moorage)
DRAYTON HARBOR
N
Breakwaters
SYC RECIPROCAL FLOATING DOCK
South of Breakwater

Blaine

NAME OF MARINA: ***BLAINE HARBOR**-Port of Bellingham* **RADIO:** VHF Ch. 16
TELEPHONE: **360-647-6176** MGR: Pam Taft
E-MAIL: blaineharbor@portofbellingham.com FAX: 360-332-1043 (Fax)
ADDRESS: PO Box 1245, 235 Marine Dr. Blaine, WA 98230
SHORT DESCRIPTION & LOCATION: **www.portofbellingham.com**
48°59.50' - 122°45.90' Located near the entrance to Drayton Harbor on the N. shore of Blaine. The large & very top notch marina is an active fishing center and offers many marine services for recreational boaters. Restaurants and parks close by.

GUEST BOAT CAPACITY: ....Appx. 20-30 boats
DOCKSIDE DEPTH AT ZERO TIDE: ....12 ft.
SEASON: ....All year
RESERVATION POLICY: ....None
AMT W/ELECTRICITY: ....Main guest dock only
FUEL DOCK: ....Gas & Diesel
MARINE REPAIRS: ....On premises
TOILETS: ....Yes
HOT SHOWERS: ....Yes
RESTAURANT: ....Close by
PICNIC AREA: ....Yes
BASIC STORE: ....Close by
BROADBAND/WI-FI: ....BroadbandXpress
DAILY RATE: ....Economical (Under 75¢/foot)

GUEST DOCK: ....760 ft. total
GUEST SLIPS: ....Varies
WATER: ....Yes
AMPS: ....30 A
PUMP OUT STATION ....Yes
HAUL OUT: ....Travel-Lift
BOAT RAMP: ....Yes
LAUNDRY: ....Yes
BAR: ....Close by
POOL: ....None
GOLF: ....Close by
PET FRIENDLY: ....Excellent
OTHER: Customs port-of-entry by prior telephone appt. Pump-out, charter fishing, close to Town and many service facilities.

NAME OF YACHT CLUB: *INTERNATIONAL YACHT CLUB*
CLUB ADDRESS: P.O. Box 75339, White Rock, B.C. V4B 5L5
CLUB TELEPHONE: 604-535-4699 PERSON IN CHARGE: Commodore
LOCATION & SPECIAL NOTES: **www.interyachtclub.com**
International Yacht Club of White Rock, B.C. offers reciprocal moorage in the Blaine Harbor Marina on the guest dock (appx. 700 ft.) on a first-come basis. One night complimentary moorage is available. Upon entering the marina proceed to the guest dock below the marina office and register with your reciprocal yacht club card with the harbormaster.

RECIPROCAL BOAT CAPACITY:...............2 Boats
DOCKSIDE DEPTH AT ZERO TIDE:...........12 ft.
RECIPROCAL SEASON:........All year
RESERVATION POLICY: ....None
TOILETS: ....At Marina
HOT SHOWERS: ....At Marina
RESTAURANT: ....Close by
DAILY RATE: ....ONE NIGHT FREE MOORAGE RECIP. MAX 2 BOATS PER NIGHT. 2 NIGHTS PER YEAR PER BOAT.

RECIPROCAL DOCK: ....700 ft.
RECIPROCAL SLIPS: Dock only
WATER: ....Yes
AMT W/ELECTRICITY: ....All
AMPS: ....30 A
BAR: ....Close by
OTHER: ......Same as marina listing

***NOTE:* THIS IS PRIVATE MOORAGE *ONLY* AVAILABLE TO MEMBERS OF RECIPROCAL YACHT CLUBS! YOUR CLUB *MUST* HAVE RECIPROCAL PRIVILEGES AND YOU MUST FLY YOUR BURGEE!**

Blaine

CAUTION! This chartlet not intended for use in navigation.

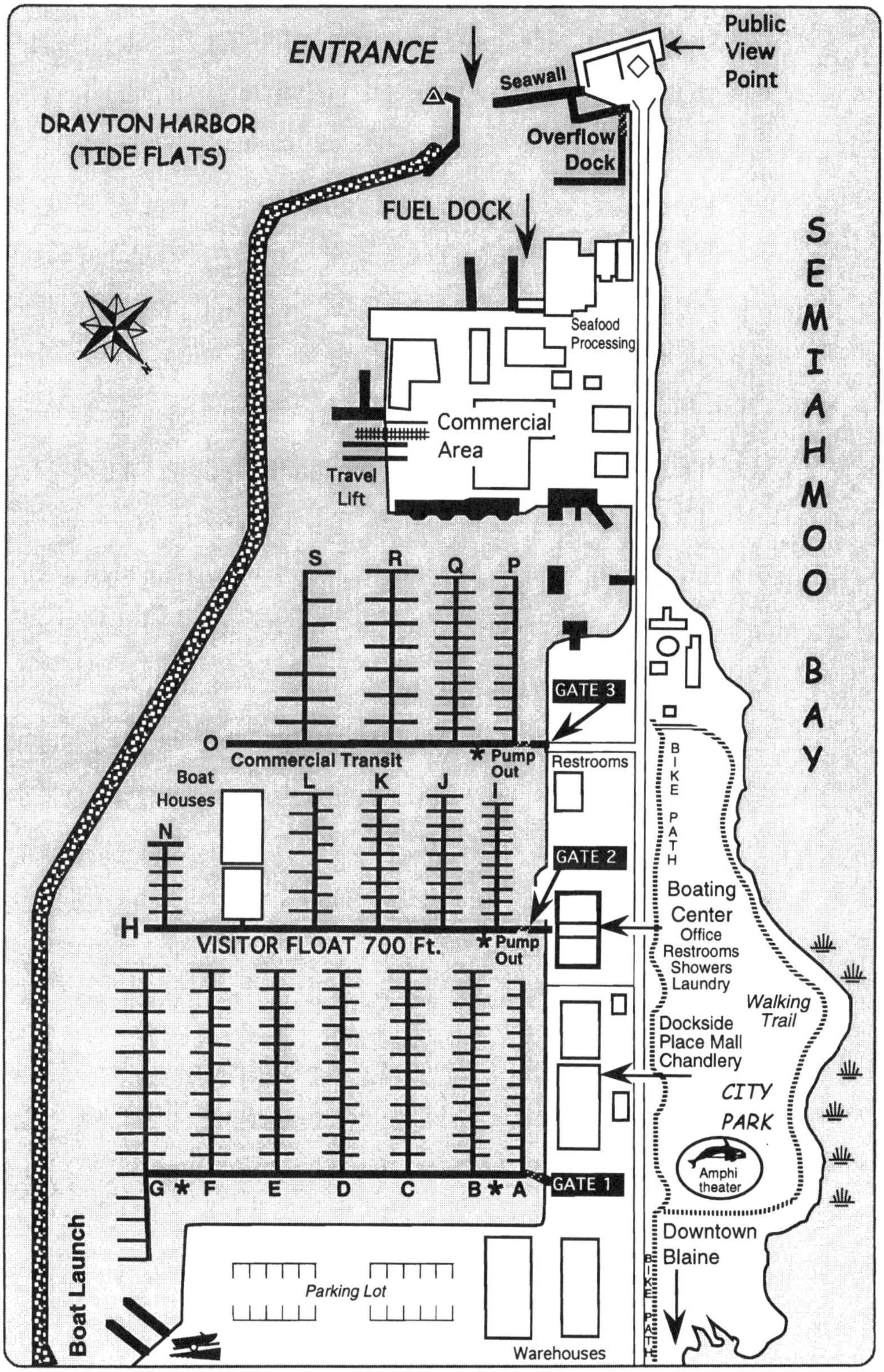

ENTRANCE
Public View Point
Seawall
DRAYTON HARBOR (TIDE FLATS)
Overflow Dock
FUEL DOCK
SEMIAHMOO BAY
Seafood Processing
Commercial Area
Travel Lift
S R Q P
GATE 3
O
Commercial Transit
Pump Out
Restrooms
Boat Houses
L K J I
N
BIKE PATH
GATE 2
Boating Center
Office
Restrooms
Showers
Laundry
H
VISITOR FLOAT 700 Ft.
Pump Out
Walking Trail
Dockside Place Mall
Chandlery
CITY PARK
Amphi theater
G * F E D C B * A
GATE 1
Downtown Blaine
Boat Launch
Parking Lot
BIKE PATH
Warehouses

Point Roberts

NAME OF MARINA: *POINT ROBERTS MARINA* RADIO: VHF 16/66A
TELEPHONE: 360-945-2255 MGR: Allan Sharp
E-MAIL: prmarina@pointrobertsmarina.com FAX: 360-945-0927 (Fax)
ADDRESS: 713 Simundson Dr. Point Roberts, WA 98281
SHORT DESCRIPTION & LOCATION: www.pointrobertsmarina.com

48°58.23' - 123°03.50' Located at the S.W. end of a peninsula that is an unconnected portion of the USA on the W. shore of Boundary Bay. The tear dropped shaped boat basin & marina offer a pleasant quiet setting with complete marine services.

GUEST BOAT CAPACITY: ....Ample
DOCKSIDE DEPTH AT ZERO TIDE: ....9 ft.
SEASON: ....All year
RESERVATION POLICY: ....None
AMT W/ELECTRICITY: ....All
FUEL DOCK: ....Gas, Dsl, & LP
MARINE REPAIRS: ....On premises
TOILETS: ....Yes
HOT SHOWERS: ....Yes
RESTAURANT: ....Yes
PICNIC AREA: ....Yes
BASIC STORE: ....Yes
BROADBAND/WI-FI: ....BroadbandXpress
DAILY RATE: ....Moderate (75¢-$1.25/foot)

GUEST DOCK: ....Up to 125 ft.
GUEST SLIPS: ....Varies
WATER: ....Yes
AMPS: ....30-100 A
PUMP OUT STATION ....Yes
HAUL OUT: ....Travel-Lift
BOAT RAMP: ....Monorail
LAUNDRY: ....Yes
BAR: ....Close by
POOL: ....None
GOLF: ....Close by
PET FRIENDLY: ....Excellent
OTHER: Customs port-of-entry, provisions, marine chandlery. Lighthouse Marine Park - nature & wildlife walks.

CAUTION! This chartlet not intended for use in navigation.

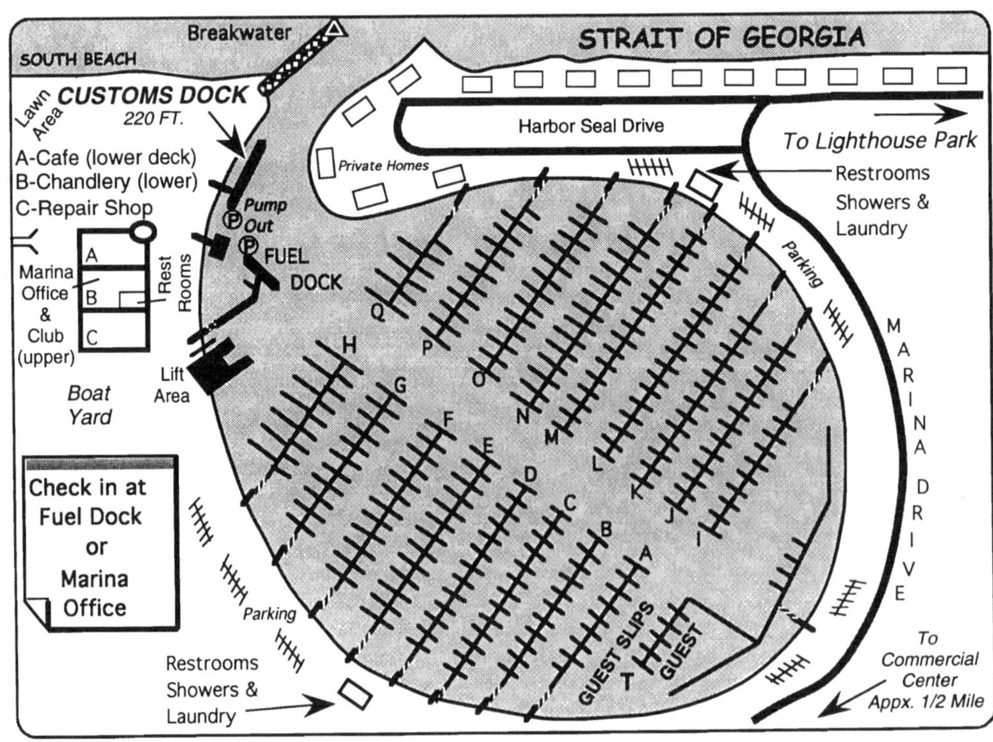

Point Roberts

NAME OF YACHT CLUB: *POINT ROBERTS YACHT CLUB*

ADDRESS: P.O. Box 889 Point Roberts, WA 98281

TELEPHONE: 360-945-2255/2166 PERSON IN CHARGE: Commodore

SHORT DESCRIPTION & LOCATION: **www.pryc.net**

P.R.Y.C. offers 2 days reciprocal moorage at the **Point Roberts Marina.** Prior to arriving contact the Marina on VHF & request reciprocal moorage. After slip assignment & docking, register at Marina Office & present your current membership card. Customs Port-of-entry. After hours arrival, call Security at 945-2166.

RECIPROCAL BOAT CAPACITY: ....2 boats

DOCKSIDE DEPTH AT ZERO TIDE: ....9 ft.

SEASON: ....All year

RESERVATION POLICY: ....None

TOILETS: ....At Point Roberts Marina

HOT SHOWERS: ....At Point Roberts Marina

RESTAURANT: ....Close by

BROADBAND/WI-FI: ....BroadbandXpress

DAILY RATE: Electricity charge - $5.00/ day
48 hours free limited to 2 boats
at 1 time & 1 visit per month.

RECIPROCAL DOCK: At marina

RECIPROCAL SLIPS: At marina

WATER: ....Yes

AMT W/ELECTRICITY: ....All

AMPS: 30A

BAR: ....Close by

PET FRIENDLY: ....Excellent

OTHER: Services same as Point Roberts Marina listing on Page 186.

***NOTE:* THIS IS PRIVATE MOORAGE AND ONLY AVAILABLE TO MEMBERS OF RECIPROCAL YACHT CLUBS. YOUR CLUB *MUST* HAVE RECIPROCAL PRIVILEGES AND YOU MUST FLY YOUR BURGEE.**

CAUTION! This chartlet not intended for use in navigation.

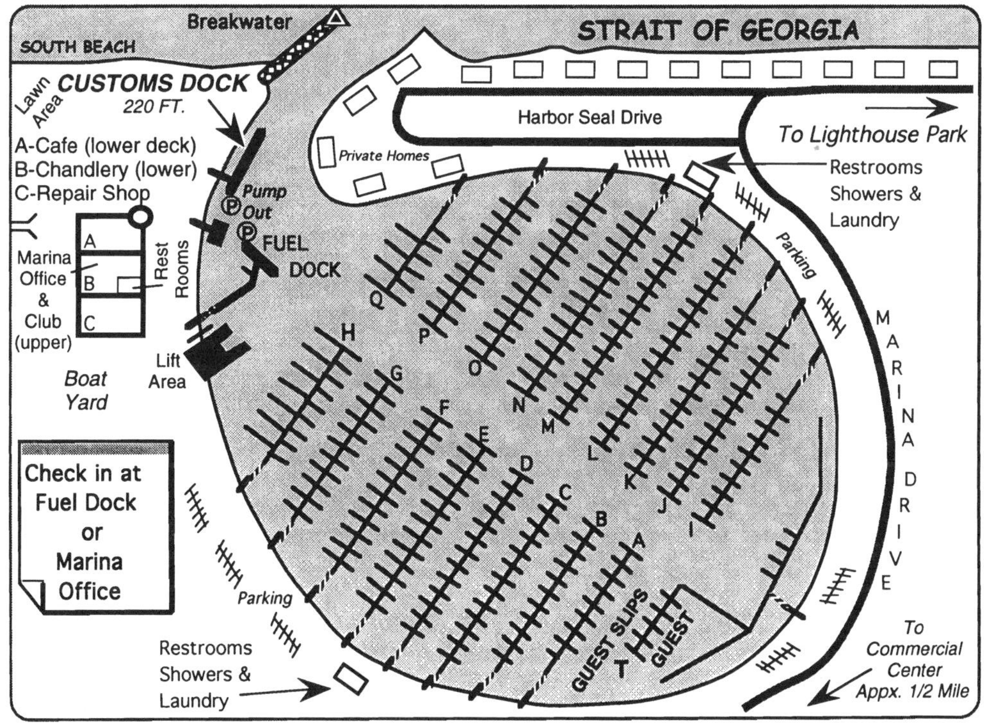

CHAPTER 6

GULF ISLANDS CANADA

Chapter Map - Page 3

Bedwell Harbour

NAME OF MARINA: *POETS COVE MARINA AT BEDWELL HBR.* RADIO: VHF 66A
TELEPHONE: **866-888-2683** MGR: Paul Jackson, Asst. Mgr.
E-MAIL: marina@poetscove.com FAX: 250-629-3202
ADDRESS: 9801 Spalding Rd. So. Pender Island, B.C. Canada VON2M3
SHORT DESCRIPTION & LOCATION: **www.poetscove.com**

48°44.80' - 123°13.60' Nestled in the tranquil Cove of Bedwell Harbour, Poets Cove is one of B.C.'s newest luxury Seaside Resorts. The marina has served boaters for years & continues to offer many family amenities & fine resort dining. Canada Customs Port-of-Entry.

GUEST BOAT CAPACITY: ....Appx 110 boats
DOCKSIDE DEPTH AT ZERO TIDE: ......8-55 ft.
SEASON: ....All year
RESERVATION POLICY: ....Accepts
AMT W/ELECTRICITY: All except Breakwater
FUEL DOCK: ....Gas & Diesel
MARINE REPAIRS: ...."On Call"
TOILETS: ....Yes
HOT SHOWERS: ....Yes
RESTAURANT: ....Pub & Fine Dining
PICNIC AREA: ....Yes
BASIC STORE: ....Moorings Market
BROADBAND/WI-FI: ....Yes
DAILY RATE: ....Premium
(Over $1.25/foot)
Winter discounts available

GUEST DOCK: ....110' plus slips
GUEST SLIPS: 110
WATER: ....Yes
AMPS: ....15 &30 A
PUMP OUT STATION Planned
HAUL OUT: ....None
BOAT RAMP: ....None
LAUNDRY: ....Yes
BAR: ....Yes
POOL: ....Yes
GOLF: ....On Island
PET FRIENDLY: ....Fair
OTHER: Provincial park nearby, Bicycling, tennis courts, lodge, conference center, BBQ, spa. ice, rentals and well stocked Gen. Store.

CAUTION! This chartlet not intended for use in navigation.

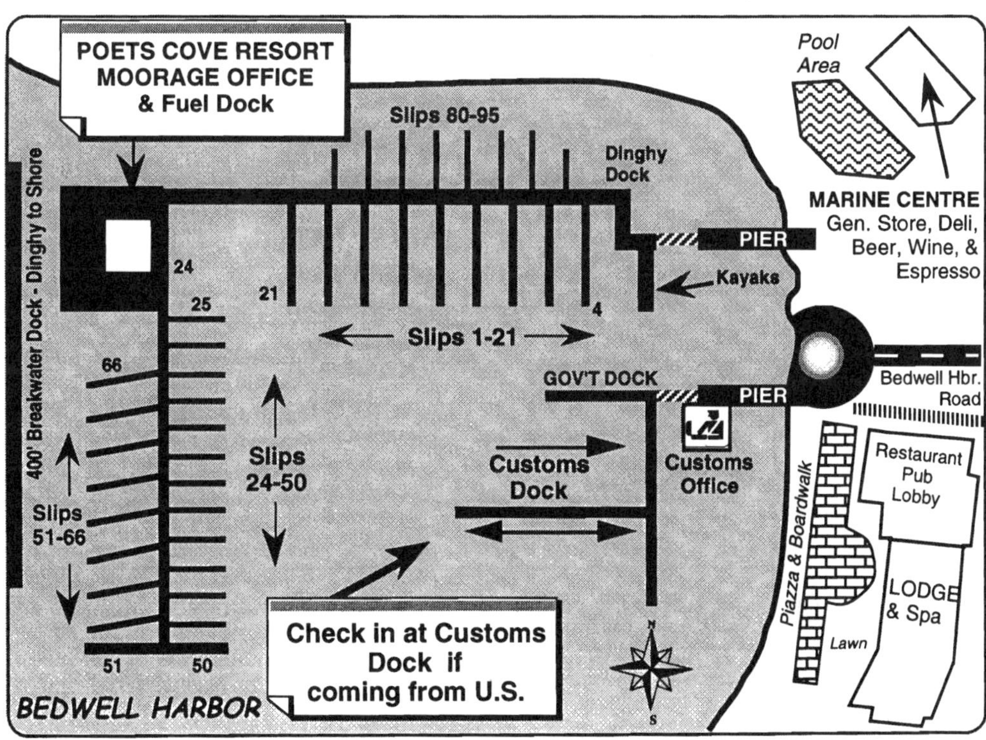

North Pender Is.

NAME OF MARINA: ***PORT BROWNING MARINA RESORT*** RADIO: VHF 66A
TELEPHONE: 250-629-3493 MGR: Lou Henshaw
E-MAIL: portbrowning@cablelan.net FAX: 250-629-3945 (Fax)
ADDRESS: P.O. Box 126 Pender Island, B.C., Canada V0N 2M1
SHORT DESCRIPTION & LOCATION: **www.portbrowning.com**

48°46.60' - 123°16.30' Situated on east side of North Pender Island at head of Port Browning. Relaxed country atmosphere with lots to do for entire family. Walking distance to Driftwood Shopping Centre w/groceries, liquor store, gift shops, and gas station.

GUEST BOAT CAPACITY: ....Appx. 110 boats
DOCKSIDE DEPTH AT ZERO TIDE: ......8-29 ft.
SEASON: ....All year
RESERVATION POLICY: ....Accepts
AMT W/ELECTRICITY: ....All
FUEL DOCK: ....None
MARINE REPAIRS: ....Close by
TOILETS: ....Yes
HOT SHOWERS: ....Yes
RESTAURANT: ....Yes
PICNIC AREA: ....Yes
BASIC STORE: ....Yes
BROADBAND/WI-FI: ....Yes
DAILY RATE: ....Moderate (75¢-$1.25/foot)

GUEST DOCK: ....3000 ft. total
GUEST SLIPS: ....46 large slips
WATER: ....Limited
AMPS: ....15 A
PUMP OUT STATION ....None
HAUL OUT: ....None
BOAT RAMP: ....Close by
LAUNDRY: ....Yes
BAR: ....Yes
POOL: ....Yes
GOLF: ....Close by
PET FRIENDLY: ....Excellent
OTHER: Cold beer & wine store, campground, cabins, tennis, convenience store, swimming beach, eagle watching.

CAUTION! This chartlet not intended for use in navigation.

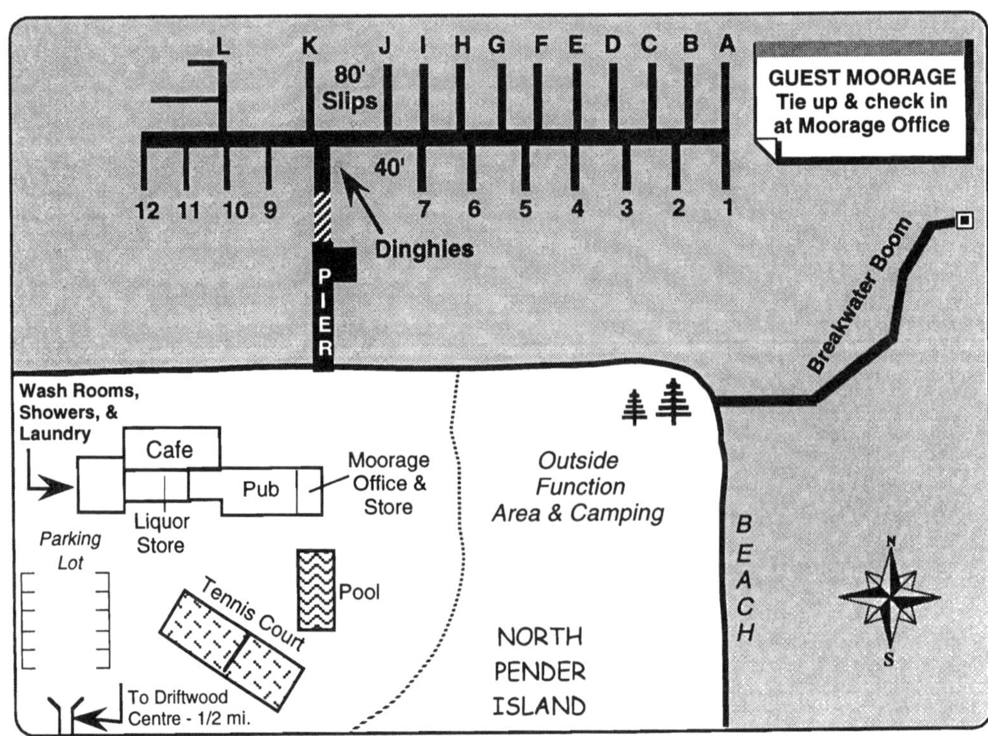

North Pender Is.

NAME OF MARINA: *HOPE BAY GOVERNMENT WHARF* RADIO: None
TELEPHONE: 250-629-9990 MGR: P. Binner, Wharfinger/Goldsmith
E-MAIL: pbinner@cablelan.net FAX: 250-629-3168 (Realty Fax)
ADDRESS: The Goldsmith Shop, Hope Bay, Pender Is, B.C. Canada V0N 2M1
SHORT DESCRIPTION & LOCATION: **www.hopebayrising.com**

48°48.20' - 123°16.40' Fairly large gov't dock situated in pretty Hope Bay on E. side of N. Pender Is. behind Fane Island. Overnight moorage possible but can be open to swells. The old store has been redeveloped to a vibrant little centre with cafe, shops, artisans, etc.

GUEST BOAT CAPACITY: ....Appx 9-12 boats
DOCKSIDE DEPTH AT ZERO TIDE: ....16 ft.
SEASON: ....All year
RESERVATION POLICY: ....None
AMT W/ELECTRICITY: ....None
FUEL DOCK: ....None
MARINE REPAIRS: ....Close by
TOILETS: ....None
HOT SHOWERS: ....None
RESTAURANT: ....Cafe: Breakfast, Lunch, Dinner
PICNIC AREA: ....None
BASIC STORE: ....None
BROADBAND/WI-FI: ....Internet at Business Center
DAILY RATE:..............Economical (Under 75¢/foot)
OTHER: Parks Canada Headquarters, Dockside Realty, Business Center, Clothing Store.

GUEST DOCK: ....226 ft. Total
GUEST SLIPS: ....Dock only
WATER: ....None
AMPS: ....None
PUMP OUT STATION ....None
HAUL OUT: ....None
BOAT RAMP: ....None
LAUNDRY: ....None
BAR: ....Beer & Wine
POOL: ....None
GOLF: ....Close by
PET FRIENDLY: ....Excellent
OTHER: Gallery, artisans & local crafts, home decor shop, Goldsmith jewelry shop, Hope Bay Farm: eggs & produce 5 min. walk.

CAUTION! This chartlet not intended for use in navigation.

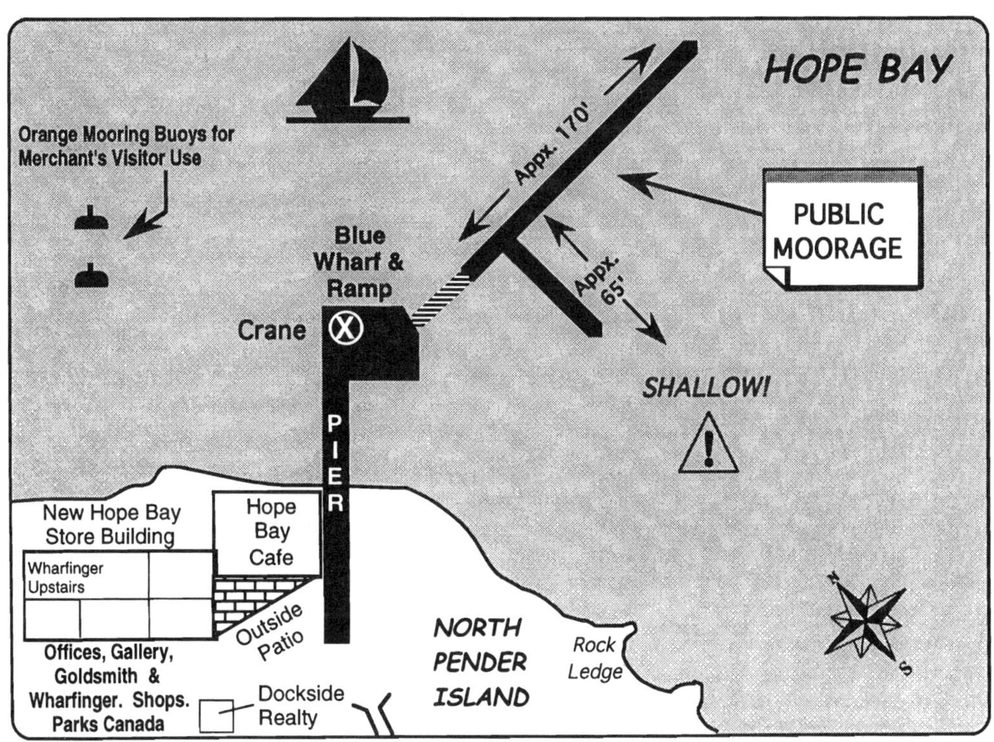

North Pender Is.

NAME OF MARINA: *OTTER BAY MARINA* RADIO: VHF 66A
TELEPHONE: 250-629-3579 MGR: Chuck Spence
E-MAIL: None FAX: 250-629-3589 (Fax)
ADDRESS: RR #1, 2311 MacKinnon Rd. Pender Island, B.C. Canada V0N 2M1

SHORT DESCRIPTION & LOCATION:

48°48.00' - 123°19.50' Located in Hyashi Cove in Otter Bay on N.W. side of North Pender Island. This popular marina offers many boater amenities. The well maintained slips & floats are protected with a rock breakwater covered by a beautiful deck with flags flying.

GUEST BOAT CAPACITY: ....Appx 40+ boats
DOCKSIDE DEPTH AT ZERO TIDE: ....10 ft.
SEASON: ....All year
RESERVATION POLICY: ....Accepts
AMT W/ELECTRICITY: ....All
FUEL DOCK: ....None
MARINE REPAIRS: ....Close by
TOILETS: ....Yes
HOT SHOWERS: ....Yes
RESTAURANT: Close by-shuttle service available
PICNIC AREA: ....Yes
BASIC STORE: ....Yes
BROADBAND/WI-FI: ....None
DAILY RATE: ....Moderate (75¢-$1.25/foot)

GUEST DOCK: ....300 ft. plus slips
GUEST SLIPS: 20
WATER: ....Yes
AMPS: ....15-30 A
PUMP OUT STATION ....None
HAUL OUT: ....None
BOAT RAMP: ....Yes
LAUNDRY: ....Yes
BAR: ....Close by
POOL: ....Yes
GOLF: Close by-shuttle available
PET FRIENDLY: ....Good
OTHER: Family & group oriented, function area gazebo, close to ferry, fishing supplies, kayak and bicycle rentals, espresso.

CAUTION! This chartlet not intended for use in navigation.

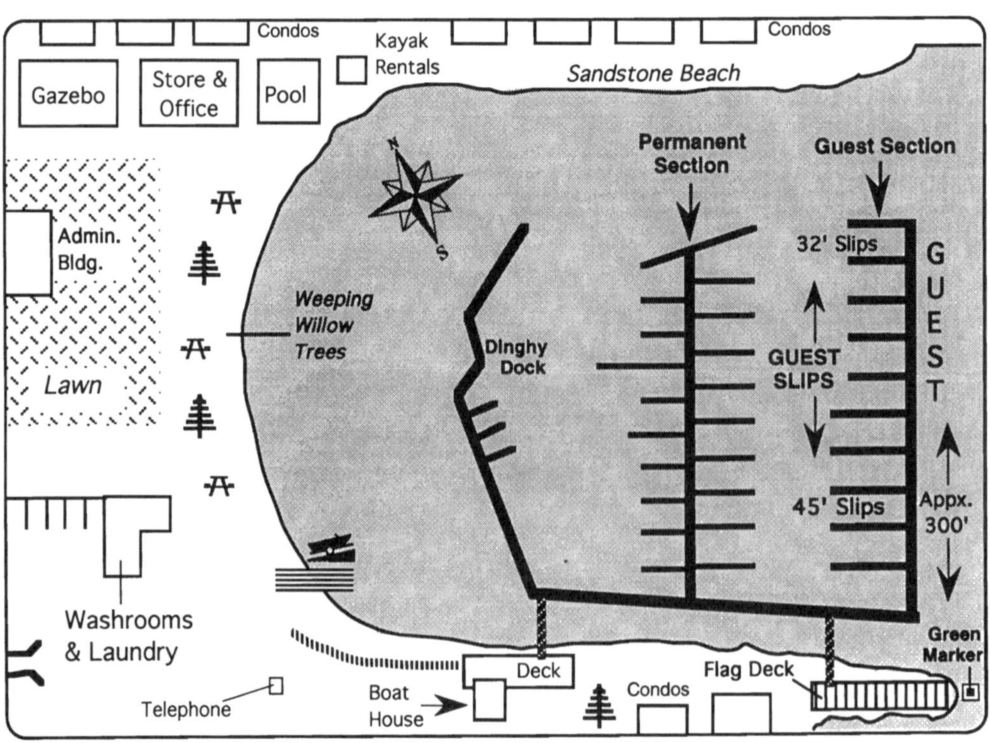

North Pender Is.

NAME OF MARINA: *PORT WASHINGTON DOCK* RADIO: None
TELEPHONE: 250-629-6111 MGR: Rod McLean, Wharfinger
E-MAIL: None FAX: None
ADDRESS: Mail: Site 21 C-56 Mayne Island, B.C. Canada V0N 2J0
SHORT DESCRIPTION & LOCATION: **www.crd.bc.ca/smallcraft**

48°48.6' - 123°19.2 Located in the quiet & peaceful small settlement of Port Washington in Grimmer Bay on N.W. side of North Pender Island. The Gov't float is good for short stopovers, but open to swells and could be uncomfortable for overnight stays.

GUEST BOAT CAPACITY: ....Appx 5-7 boats
DOCKSIDE DEPTH AT ZERO TIDE: ....4 ft. +
SEASON: ....All year
RESERVATION POLICY: ....None
AMT W/ELECTRICITY: ....None
FUEL DOCK: ....None
MARINE REPAIRS: ....None
TOILETS: ....None
HOT SHOWERS: ....None
RESTAURANT: ....20 Minute walk
PICNIC AREA: ....None
BASIC STORE: ....None
BROADBAND/WI-FI: ....None
DAILY RATE: ....Economical (Under 75¢/foot)

4 HOURS FREE

GUEST DOCK: ....100 ft.
GUEST SLIPS: ....Dock only
WATER: ....None
AMPS: ....None
PUMP OUT STATION ....None
HAUL OUT: ....None
BOAT RAMP: ....None
LAUNDRY: ....None
BAR: ....20 Minute walk
POOL: ....None
GOLF: ....20 Minute walk
PET FRIENDLY: ....Good
OTHER: Historic store planned to open in the future. Appx. 6 mi. from Driftwood Shopping Centre.
SeaAir: 1-800-44-SEAIR

CAUTION! This chartlet not intended for use in navigation.

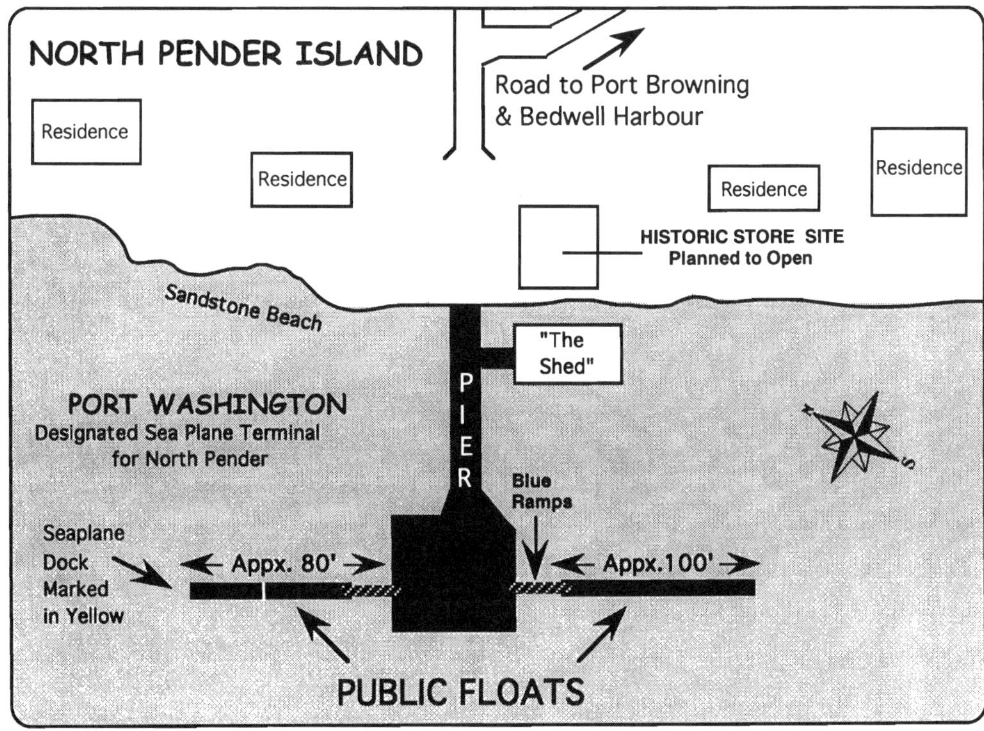

Saturna Island

NAME OF MARINA: ***SATURNA POINT LANDING*** **RADIO:** VHF Ch. 68

TELEPHONE: 250-539-5725 / 539-2229 **MGR:** Gloria Manzano

E-MAIL: gmanzano@telus.net **FAX:** 250-539-2229

ADDRESS: 101 East Point Rd. Saturna Island, B.C. Canada V0N 2Y0

SHORT DESCRIPTION & LOCATION:

48°47.90' - 123°11.00' The Lyall Hbr. Public Float is located at Saturna Pt. on the NW side of Saturna Is. at the S. entrance of Lyall Hbr. next to Saturna Ferry Landing. This good little overnight or short stay stopover is handy to groceries, pub, fuel & basic services.

GUEST BOAT CAPACITY: ....Appx. 8-10 boats
DOCKSIDE DEPTH AT ZERO TIDE: 20 ft. plus
SEASON: ....All year
RESERVATION POLICY: ....None
AMT W/ELECTRICITY: ....None
FUEL DOCK: ....Gas & Diesel
MARINE REPAIRS: ....Close by
TOILETS: ....Yes
HOT SHOWERS: ....None
RESTAURANT: ....Pub
PICNIC AREA: ....None
BASIC STORE: ....Yes
BROADBAND/WI-FI: ....None
DAILY RATE: ....Economical (Under 75¢/foot)

GUEST DOCK: ....Appx. 180 ft.
GUEST SLIPS: ....Dock only
WATER: ....None
AMPS: ....None
PUMP OUT STATION ....None
HAUL OUT: ....None
BOAT RAMP: ....None
LAUNDRY: ....None
BAR: ....Pub
POOL: ....None
GOLF: ....None
PET FRIENDLY: ....Good
OTHER: Liquor store appx. 1.25 mile walk, well stocked store & pub at head of dock. Ferry service off island. Rafting allowed.

CAUTION! This chartlet not intended for use in navigation.

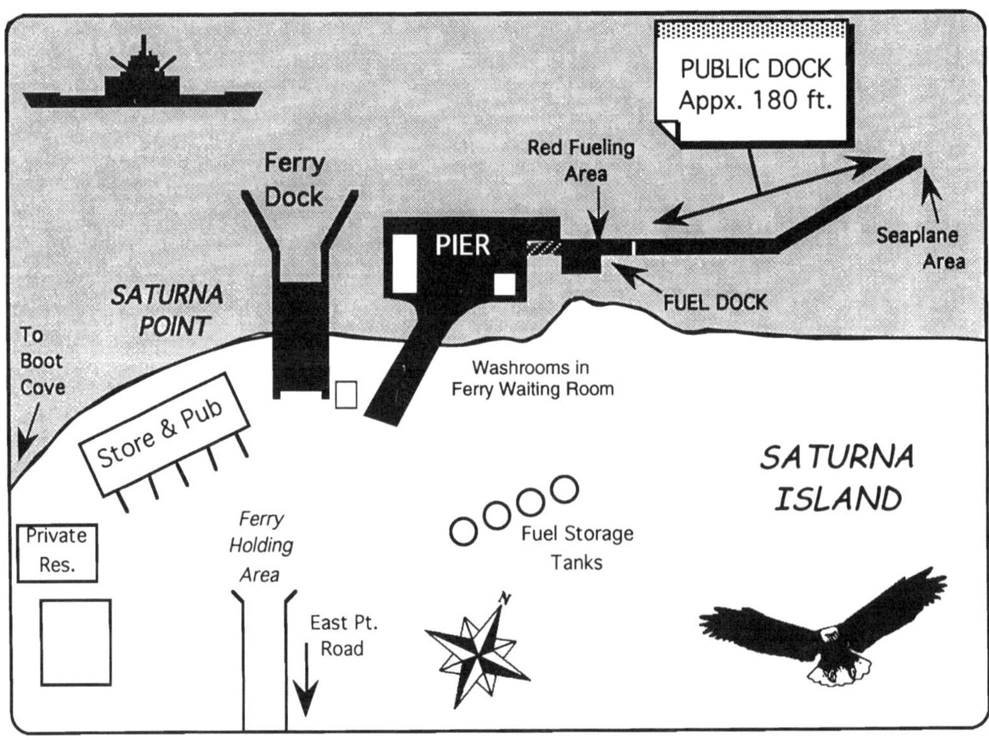

Mayne Island

NAME OF MARINA: ***HORTON BAY GOVERNMENT WHARF*** RADIO: None
TELEPHONE: 250-539-5425 MGR: Frank Fulford
E-MAIL: None FAX: None
ADDRESS: RR #1 Site 4 Comp 19 Mayne Island, B.C. Canada V0N 2J0

SHORT DESCRIPTION & LOCATION:

48°49.60' - 123°14.70' This sm. busy Gov't Float is situated on the SE side of Horton Bay on the SE side of Mayne Island. Floats are protected in the quiet & peaceful bay but the approach can be hazardous. Good charts & current tables are a must.

GUEST BOAT CAPACITY: ....Appx 5-10 boats
DOCKSIDE DEPTH AT ZERO TIDE: ....15 ft.
SEASON: ....All year
RESERVATION POLICY: ....None
AMT W/ELECTRICITY: ....None
FUEL DOCK: ....None
MARINE REPAIRS: ....None
TOILETS: ....None
HOT SHOWERS: ....None
RESTAURANT: ....None
PICNIC AREA: ....None
BASIC STORE: ....None
BROADBAND/WI-FI: ....None
DAILY RATE: ....Economical (Under 75¢/foot)

RAFTING MANDATORY

GUEST DOCK: ....Appx. 100 ft.
GUEST SLIPS: ....Dock only
WATER: ....None
AMPS: ....None
PUMP OUT STATION ....None
HAUL OUT: ....None
BOAT RAMP: ....None
LAUNDRY: ....None
BAR: ....None
POOL: ....None
GOLF: ....None
PET FRIENDLY: ....Good
OTHER: Customs Port of Entry for Canpass holders, good anchorage in bay if docks full, taxi service.

CAUTION! This chartlet not intended for use in navigation.

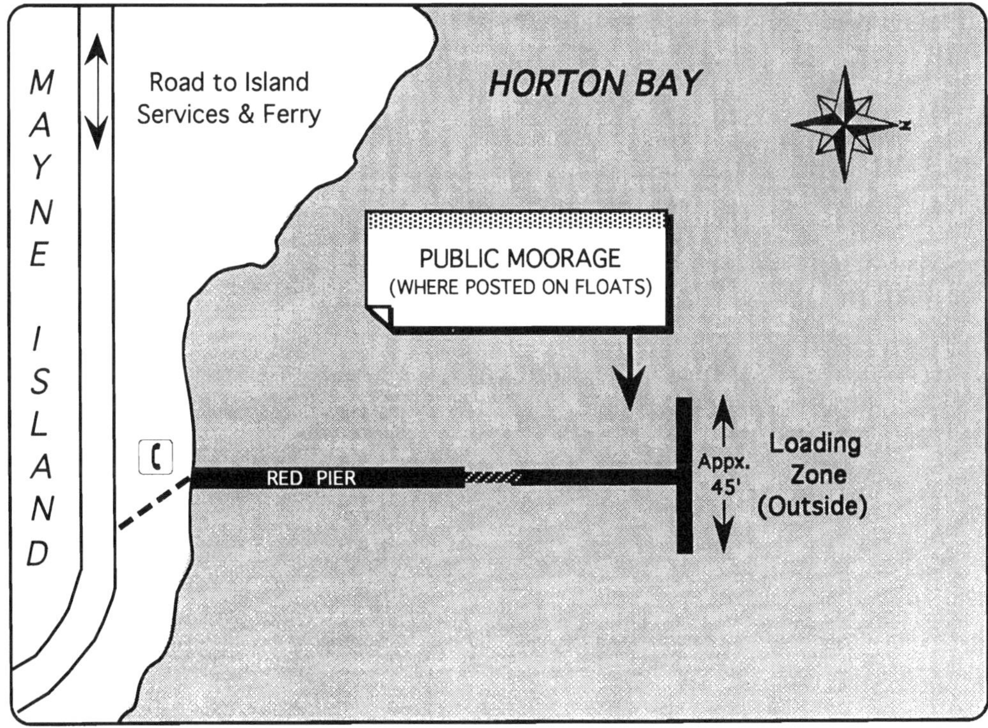

Montague Harbour

NAME OF MARINA: ***MONTAGUE HARBOUR MARINA*** **RADIO:** VHF 66A
TELEPHONE: **250-539-5733** MGR: Marilyn Breeze
E-MAIL: montaguemarina@gulfislands.com FAX: 250-539-3593 (Fax)
ADDRESS: S-17, C-57, RR #1 Galiano, B.C.. Canada VON 1PO
SHORT DESCRIPTION & LOCATION: **www.montagueharbour.com**

48°53.50' - 123°23.40' Friendly & well maintained marina located in lovely & protected Montague Hbr. on SE side of Galiano Island. The marina offers many boater amenities & is a short walk to beautiful Montague Marine Park. Ride the famous Pub Bus to Hummingbird Inn.

GUEST BOAT CAPACITY: ....Appx. 25 boats
DOCKSIDE DEPTH AT ZERO TIDE: ....20 ft.
SEASON: ....May thru Sept. - Closed Winter
RESERVATION POLICY: ....Encouraged
AMT W/ELECTRICITY: ....All
FUEL DOCK: ....Gas & Diesel
MARINE REPAIRS: ...."On Call"
TOILETS: ....Yes
HOT SHOWERS: ....None
RESTAURANT: ....The Harbour Grill
PICNIC AREA: ....None
BASIC STORE: ....Yes
BROADBAND/WI-FI: ....None
DAILY RATE: ....Moderate (75¢-$1.25/foot)

NO TRASH DROP

GUEST DOCK: 1000 ft. Total
GUEST SLIPS: 25
WATER:........................Yes - limited
AMPS: ....15-30 A
PUMP OUT STATION ....None
HAUL OUT: ....None
BOAT RAMP: ....Close by
LAUNDRY: ....None
BAR: ....Licensed Sundeck
POOL: ....None
GOLF: Close by/shuttle available
PET FRIENDLY: ....Good
OTHER: Well stocked grocery store, gift shop, books & charts, nice sportswear. Kayak, boat, scooter, moped & car rentals.

CAUTION! This chartlet not intended for use in navigation.

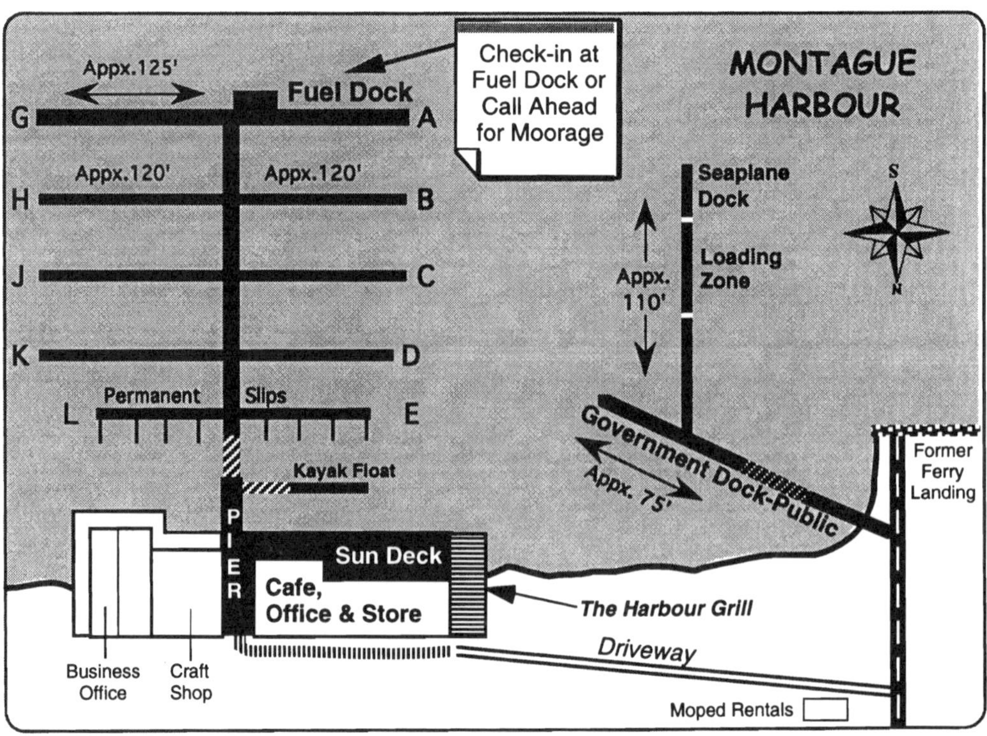

Salt Spring Island

NAME OF MARINA: ***MUSGRAVE LANDING*** *GOV'T WHARF* RADIO: None
TELEPHONE: 250-537-5711 MGR: Salt Spring Harbour Authority
E-MAIL: ssha@saltspring.com FAX: 250-537-5711 (Fax)
ADDRESS: 127 Fulford-Ganges Rd Salt Spring Is., B.C. Canada V8K 2T9
SHORT DESCRIPTION & LOCATION: **www.dfo-mpo.gc.ca/sch/HB_BC_e.asp**

48°44.90' - 123°33.95' This is a small public float on the S.W. side of Salt Spring Island adjacent to S. entrance of Sansum Narrows. The quiet and peaceful cove is a popular location during the summer months. The facility is park-like & very picturesque.

GUEST BOAT CAPACITY: ....Appx. 8-10 boats
DOCKSIDE DEPTH AT ZERO TIDE: 10 ft. plus
SEASON: ....All year
RESERVATION POLICY: ....None
AMT W/ELECTRICITY: ....None
FUEL DOCK: ....None
MARINE REPAIRS: ....None
TOILETS: ....None
HOT SHOWERS: ....None
RESTAURANT: ....None
PICNIC AREA: ....None
BASIC STORE: ....None
BROADBAND/WI-FI: ....None
DAILY RATE: ....Economical (Under 75¢/foot)

NOTE: BE PREPARED TO RAFT

GUEST DOCK: ....Appx. 175 ft.
GUEST SLIPS: ....Dock only
WATER: ....None
AMPS: ....None
PUMP OUT STATION ....None
HAUL OUT: ....None
BOAT RAMP: ....None
LAUNDRY: ....None
BAR: ....None
POOL: ....None
GOLF: ....None
PET FRIENDLY: Good-Use leash
OTHER: Rafting is customary, hiking trails on adjacent logging roads, deer & free range sheep. Fishing grounds close by.

CAUTION! This chartlet not intended for use in navigation.

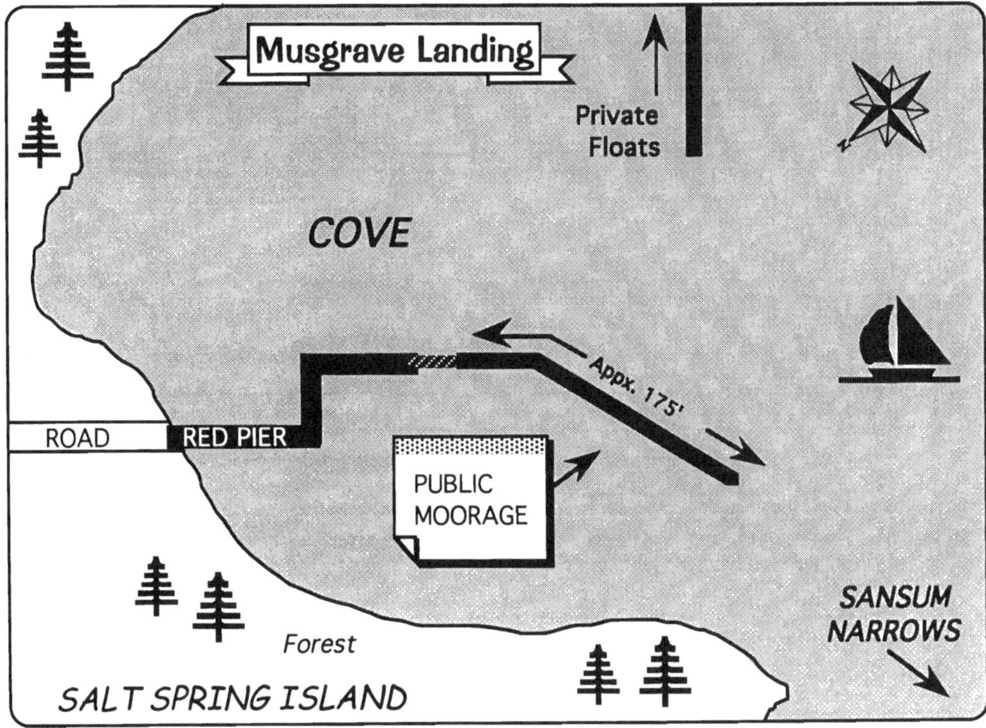

Salt Spring Is.

NAME OF MARINA: ***FULFORD MARINA LTD.*** **RADIO:** VHF 66A
TELEPHONE: **250-653-4467** **MGR:** Bill & Gay Perry
E-MAIL: fulfordmarina@saltspring.com **FAX:** 250-653-4457 (Fax)
ADDRESS: 5-2810 Fulford-Ganges Rd. Saltspring Is., B.C. Canada V8K 1Z2
SHORT DESCRIPTION & LOCATION: **www.saltspring.com/fulfordmarina**

48°46.20' - 123°27.00' Located near head of Fulford Harbour on S. end of Salt Spring Island, this modern marina is situated on beautifully landscaped grounds. Just a short walk away is quaint Fulford Village w/ shops, local artists, grocery, videos, espresso, pub & pizza.

GUEST BOAT CAPACITY: ....Appx 30 boats
DOCKSIDE DEPTH AT ZERO TIDE: ....8 ft.
SEASON: ....Guest moorage April - Sept. only
RESERVATION POLICY: ....Recommended
AMT W/ELECTRICITY: ....All
FUEL DOCK: ....Gas & Diesel next door
MARINE REPAIRS: ....Mobile service
TOILETS: ....Yes
HOT SHOWERS: ....Yes
RESTAURANT: ....2 Close by
PICNIC AREA: ....Yes
BASIC STORE: ....Close by
BROADBAND/WI-FI: ....None
DAILY RATE: ....Moderate
(75¢-$1.25/foot)
OTHER: Full Liquor Store close by

GUEST DOCK: ....500' plus slips
GUEST SLIPS: ....Appx. 30
WATER: ....Yes
AMPS: ....20-30 A
PUMP OUT STATION ....None
HAUL OUT: ....None
BOAT RAMP: ....None
LAUNDRY: ....Fairly Close by
BAR: ....Close by
POOL: ....None
GOLF: ....10 K away
PET FRIENDLY: ....Excellent
OTHER: Gazebos, tennis, 10 min. walk to swimming lake. Close to coffee house, B.C. Ferry to Sidney. Brewery & Wine Tours 3 mi.

CAUTION! This chartlet not intended for use in navigation.

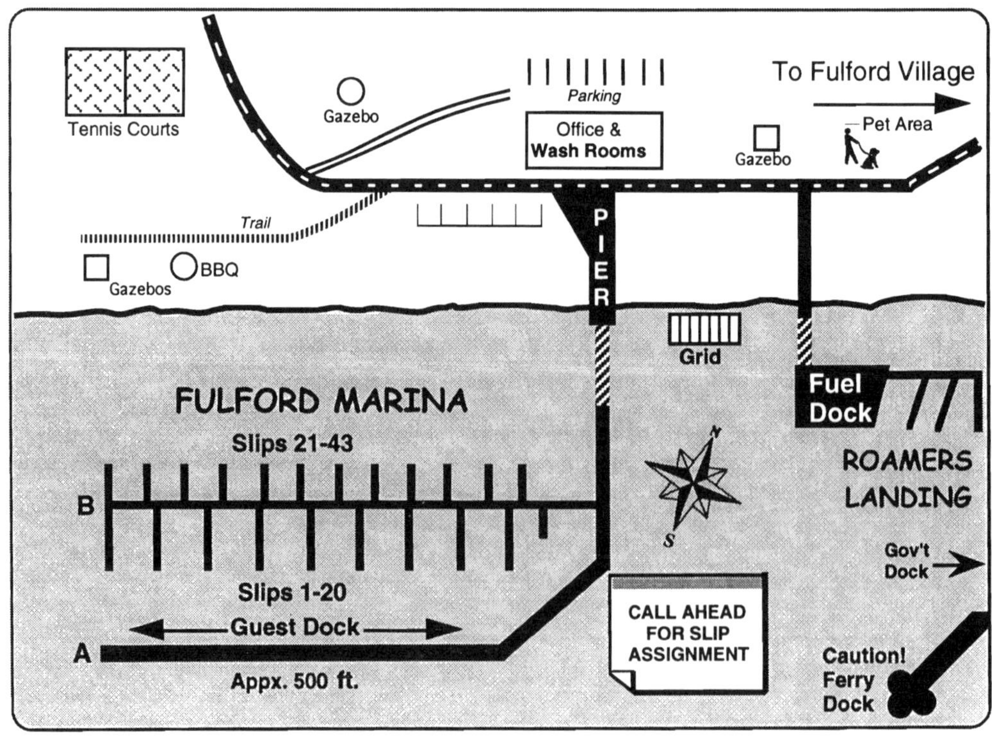

Ganges

NAME OF MARINA: ***GANGES CENTENNIAL WHARF*** RADIO: VHF Ch.09
TELEPHONE: 250-537-5711 MGR: Bart Terwiel
E-MAIL: ssha@saltspring.com FAX: 250-537-5711 (Fax)
ADDRESS: 127 Fulford-Ganges Rd Saltspring Island, B.C. Canada V8K 2T9
SHORT DESCRIPTION & LOCATION: **www.dfo-mpo.gc.ca/sch/HB_BC_e.asp**

48°51.10' - 123°29.70' Salt Spring Is. Harbour Authority moorage located behind the breakwater just S. of Grace Peninsula adjacent to town of Ganges. This often full harbour serves both commercial & recreational vessels and has designated areas for each.

GUEST BOAT CAPACITY: ....Varies
DOCKSIDE DEPTH AT ZERO TIDE: ......8-10 ft.
SEASON: ....All year
RESERVATION POLICY: ....Call First
AMT W/ELECTRICITY: ....All
FUEL DOCK: ....Close by
MARINE REPAIRS: ....Close by
TOILETS: ....Yes
HOT SHOWERS: ....Yes
RESTAURANT: ....Close by
PICNIC AREA: ....Yes
BASIC STORE: ....Close by
BROADBAND/WI-FI: ....None
DAILY RATE: ....Economical (Under 75¢/foot)

MOSTLY PERMANENT & COMMERCIAL MOORAGE.

GUEST DOCK: 3 docks appx. 200'
GUEST SLIPS: ....Varies
WATER: ....Yes
AMPS: ....20 A
PUMP OUT STATION Close by
HAUL OUT: ....None
BOAT RAMP: ....Yes
LAUNDRY: ....Close by
BAR: ....Close by
POOL: ....None
GOLF: ....Close by
PET FRIENDLY: ....Fair
OTHER: Busy but convenient location. Moorage only with permission of wharfinger. Rafting is mandatory as available.

CAUTION! This chartlet not intended for use in navigation.

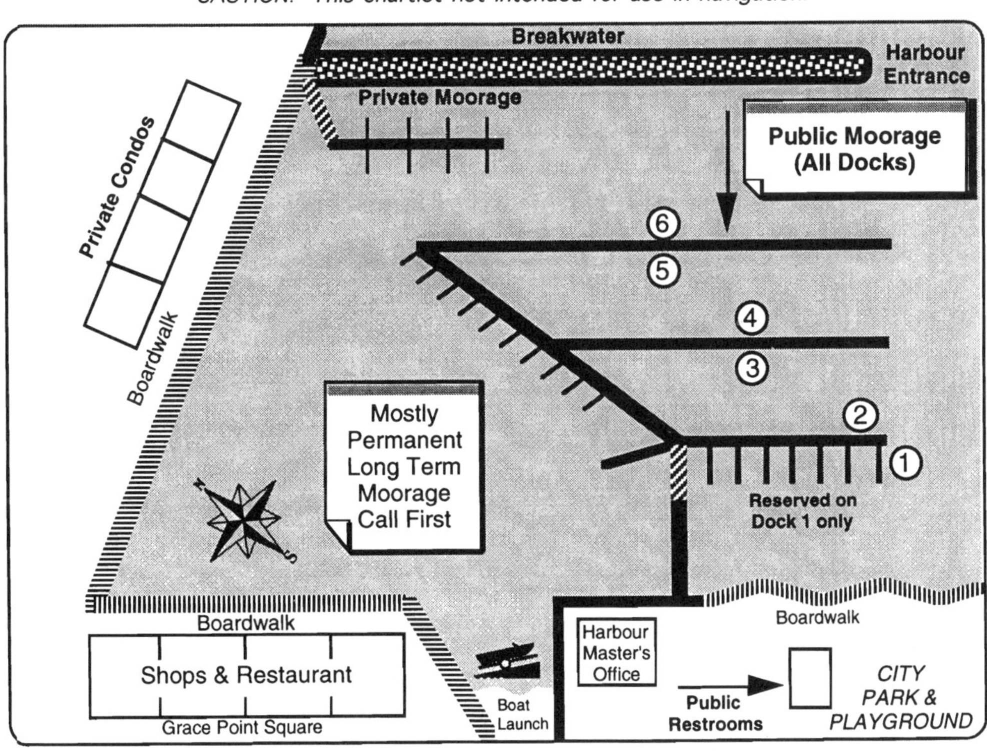

Ganges

NAME OF MARINA: ***GANGES KANAKA VISITOR WHARF*** **RADIO:** VHF Ch.09
TELEPHONE: **250-537-5711** **MGR:** Bart Terwiel
E-MAIL: ssha@saltspring.com **FAX:** 250-537-5711 (Fax)
ADDRESS: 127 Fulford-Ganges Rd Saltspring Island, B.C. Canada V8K 2T9
SHORT DESCRIPTION & LOCATION: **www.dfo-mpo.gc.ca/sch/HB_BC_e.asp**

48°51.20' - 123°29.80' Moorage is operated by the **Salt Spring Island Harbour Authority**. These government floats are located on north side of Grace Peninsula directly in front of picturesque town of Ganges located at head of Ganges Harbour. Canpass dock.

GUEST BOAT CAPACITY: ....Appx. 25 boats
DOCKSIDE DEPTH AT ZERO TIDE: 6-12 ft.
SEASON: ....All year
RESERVATION POLICY: ....None
AMT W/ELECTRICITY: ....Most
FUEL DOCK: ....Close by
MARINE REPAIRS: ....Close by
TOILETS: ....None
HOT SHOWERS: ....Close by
RESTAURANT: ....Close by
PICNIC AREA: ....Yes
BASIC STORE: ....Close by
BROADBAND/WI-FI: ....BroadbandXpress
DAILY RATE: ....Moderate (70¢-$1.25/foot)

NOTE: DO NOT ANCHOR CLOSE TO FLOAT PLANE DOCK.

GUEST DOCK: Appx. 1000' total
GUEST SLIPS: ....Docks only
WATER: ....Yes
AMPS: ....20 A
PUMP OUT STATION By Appt.
HAUL OUT: ....Close by
BOAT RAMP: ....Close by
LAUNDRY: ....Close by
BAR: ....Close by
POOL: ....None
GOLF: ....Close by
PET FRIENDLY: ....Fair
OTHER: Convenient location for short or overnight stay. Close to many nice shops and restaurants. Best provisioning in Gulf Islands.

CAUTION! This chartlet not intended for use in navigation.

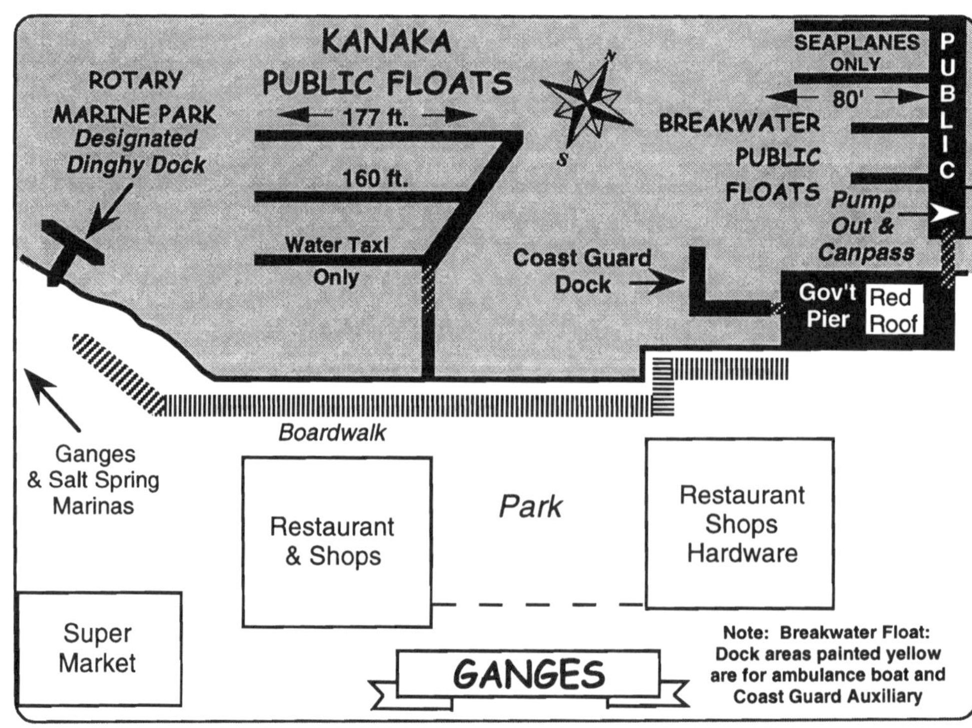

Ganges

NAME OF MARINA: *GANGES MARINA* RADIO: VHF 66A
TELEPHONE: 250-537-5242 MGR: Rick Barbieri/Suzanne Guinness
E-MAIL: info@ganges-marina.com FAX: 250-538-1719 (Fax)
ADDRESS: 161 Lower Ganges Rd. Ganges, B.C. Canada V8K 2T2
SHORT DESCRIPTION & LOCATION: **www.ganges-marina.com**

48°51.30' - 123°29.90' Friendly marina situated about 3 blocks north of heart of picturesque town of Ganges at head of Ganges Harbour. Walking distance to many tourist oriented activities, shops, and galleries. Complimentary coffee & muffins each morning!

GUEST BOAT CAPACITY: ....Appx. 125 boats
DOCKSIDE DEPTH AT ZERO TIDE: ....15 ft.
SEASON: ....All year
RESERVATION POLICY: ....Accepts
AMT W/ELECTRICITY: ....All
FUEL DOCK: ....Gas, Dsl, & LP
MARINE REPAIRS: ....Close by
TOILETS: ....Yes
HOT SHOWERS: ....Yes
RESTAURANT: ....Close by
PICNIC AREA: ....Yes (On dock)
BASIC STORE: ....Yes
BROADBAND/WI-FI: ....BroadbandXpress
DAILY RATE: ....Premium (Over $1.25/foot)

GUEST DOCK: 500 ft. plus slips
GUEST SLIPS: ....Appx. 120
WATER: ....Yes
AMPS: ....15-30 & 50 A
PUMP OUT STATION Close by
HAUL OUT: ....None
BOAT RAMP: ....None
LAUNDRY: ....Yes
BAR: ....Close by
POOL: ....Yes
GOLF: ....Close by
PET FRIENDLY: ....Challenging
OTHER: Boat rentals, ice. Internet service. Party docks. Gift shop & conv. store, Friendly helpful staff.

CAUTION! This chartlet not intended for use in navigation.

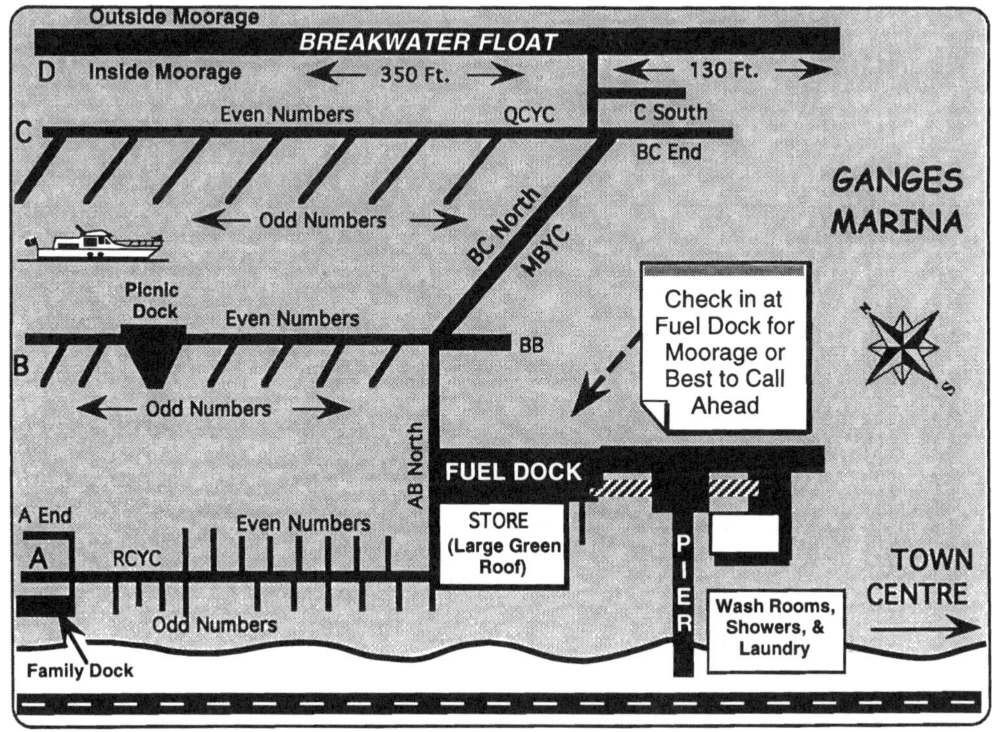

Ganges

NAME OF MARINA: ***SALT SPRING MARINA*** **RADIO:** VHF 66A
TELEPHONE: 250-537-5810 **MGR:** Lesley Cheeseman
E-MAIL: info@saltspringmarina.com **FAX:** 250-537-5809
ADDRESS: 124 Upper Ganges Rd. Saltspring Island, B.C.. Canada V8K 2S2
SHORT DESCRIPTION & LOCATION: **www.saltspringmarina.com**

48°51.50' - 123°29.90' Located at head of Ganges Harbour appx. 1/4 mile N. of downtown Ganges. Walking distance to business center w/complete boater amenities & provisions. New ownership assures continued good service for boaters with improvements.

GUEST BOAT CAPACITY: ....Appx. 50 boats
DOCKSIDE DEPTH AT ZERO TIDE: ....20 ft.
SEASON: ....All year
RESERVATION POLICY: ....Accepts
AMT W/ELECTRICITY: ....All
FUEL DOCK: ....Close by
MARINE REPAIRS: ....On premises
TOILETS: ....Yes
HOT SHOWERS: ....Yes
RESTAURANT: ....Pub on premises
PICNIC AREA: ....Yes
BASIC STORE: ....Close by
BROADBAND/WI-FI: ....BroadbandXpress
DAILY RATE:............Moderate (75¢-$1.25/foot)
Winter: $15/boat - Any Size

Toll Free Tel. #1-800-334-6629

GUEST DOCK: ....380 ft. plus slips
GUEST SLIPS: .....Varies-Appx. 50
WATER: ....Yes
AMPS: ....15-30 A
PUMP OUT STATION Close by
HAUL OUT: ....Up to 35' boats
BOAT RAMP: ....Yes
LAUNDRY: ....Yes
BAR: ....Pub
POOL: ....None
GOLF: ....Close by
PET FRIENDLY: ....Good
OTHER: Marine sales and service, kayak, scooter & car rentals, scuba air refills, fishing charters, marine chandlery & supplies.

CAUTION! This chartlet not intended for use in navigation.

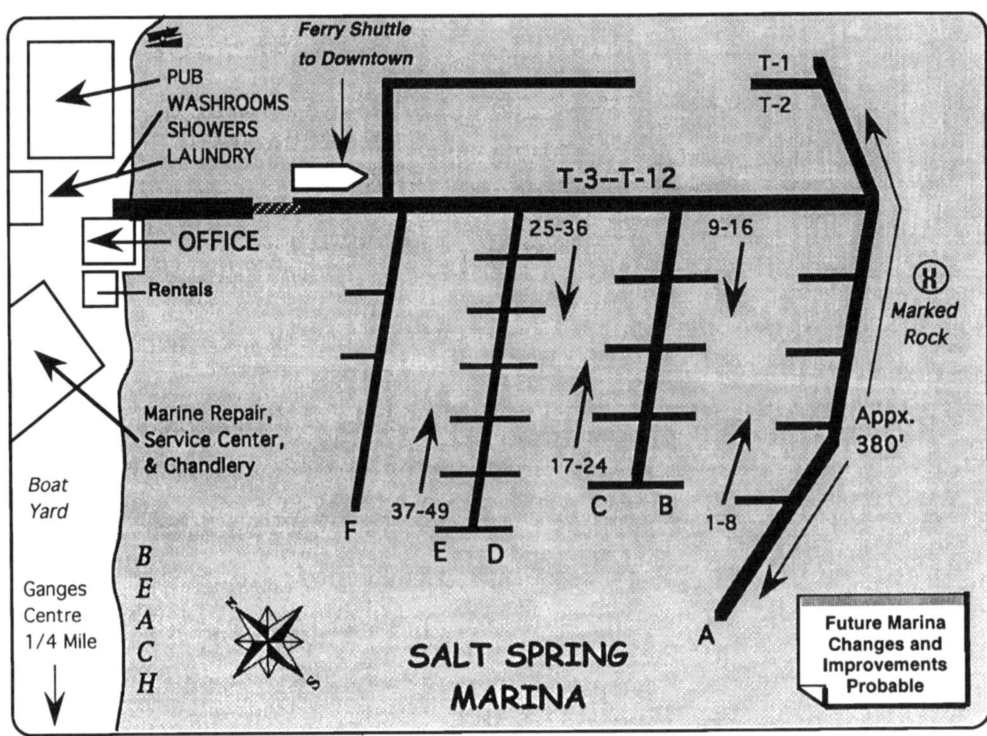

Wallace Island

NAME OF PARK: ***WALLACE ISLAND PROVINCIAL MARINE PARK - CONOVER COVE***
ADDRESS: c/o K-2 Parks, S3 - C9, Galiano Island, B.C. Canada V0N 1P0
TELEPHONE: 250-539-2115 **MGR:** K2 Park Services
SHORT DESCRIPTION & LOCATION: **http://www.env.gov.bc.ca/bcparks**
48°56.10' - 123°32.60' Historic & pristine low-lying island park in Trincomali Channel between Galiano Is. & the northern tip of Saltspring Is. Dock is in Conover Cove on W. side of Island. Conover Cove & Princess Bay offer great stern tie anchorages as well.

GUEST BOAT CAPACITY: ....At dock 4-6 boats
DOCKSIDE DEPTH AT ZERO TIDE: Appx. 5 ft.
SEASON: ....Closed winter
AMT W/ELECTRICITY: ....None
TOILETS: ....Pit toilets
HOT SHOWERS: ....None
PICNIC AREA: ....Yes
PLAY AREA: ....None
BASIC STORE: ....None
DAILY RATE: Appx. 65¢/ft.- $2.00/metre(CDN)
NOTE: Beware of tricky entrance to Cove. Consult chart & tide table. NW point has shallow underwater ledge extending into Cove.

GUEST DOCK: Appx. 200 ft. TTL
MOORING BUOYS: ....None
WATER: Hand Pump - Boil 5 min.
PAY PHONES: ....None
BOAT RAMP: ....None
PICNIC SHELTER: ....Yes
BBQ: ....None
PUMP OUT STATION: ......None
PET FRIENDLY: Good - On leash
OTHER: NO FIRES YEAR AROUND, Great hiking trails, orchard with picnic area, camping. NO sewage discharge.

CAUTION! This chartlet not intended for use in navigation.

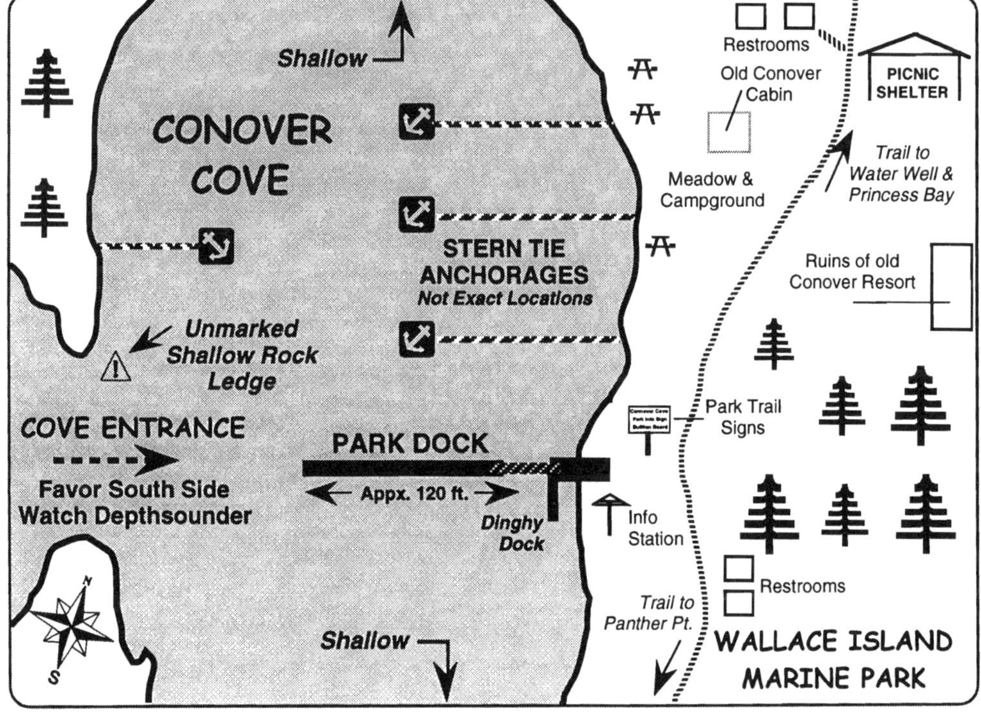

Thetis Island

NAME OF MARINA: *TELEGRAPH HARBOUR MARINA* **RADIO:** VHF 66A
TELEPHONE: 800-246-6011/250-246-9511 **MGR:** Ron & Barbara Williamson
E-MAIL: boating@telegraphharbour.com **FAX:** 250-246-2668 (Fax)
ADDRESS: Box 7-10, Thetis Island, B.C. Canada VOR 2YO
SHORT DESCRIPTION & LOCATION: **www.telegraphharbour.com**

48°58.90' - 123°40.20' A friendly family owned marina situated at back of Telegraph Harbour on S. end of Thetis Island. Ample & well maintained docks located in a park like setting offering many amenities for groups & kids. Cafe serves light lunches, pizza & espresso.

GUEST BOAT CAPACITY: ....Appx. 60 boats
DOCKSIDE DEPTH AT ZERO TIDE: ....8 ft.
SEASON: ....Open April - October
RESERVATION POLICY: ....Accepts
AMT W/ELECTRICITY: ....All
FUEL DOCK: ....Gas & Diesel
MARINE REPAIRS: ....Close by
TOILETS: ....Yes
HOT SHOWERS: ....For moorage guests only
RESTAURANT: ....Cafe & Soda Fountain
PICNIC AREA: ....Yes
BASIC STORE: ....Yes
BROADBAND/WI-FI: ....BroadbandXpress
DAILY RATE:............................Call for current prices

NOTE: SUMMER FARMERS MARKET ON SUNDAYS

GUEST DOCK: ....2700 ft. total
GUEST SLIPS: ....Docks only
WATER: For moorage guests only
AMPS: ....15-30 A
PUMP OUT STATION ....None
HAUL OUT: ....None
BOAT RAMP: ....Close by
LAUNDRY: For moorage guests
BAR: ....None
POOL: ....None
GOLF: ....In Chemainus
PET FRIENDLY: ....Excellent
OTHER: Covered pavilion, ice cream, home baked pies, gift shop w/books, charts, & crafts. Game area. 3/4 mile to Van. Is. ferry.

CAUTION! This chartlet not intended for use in navigation.

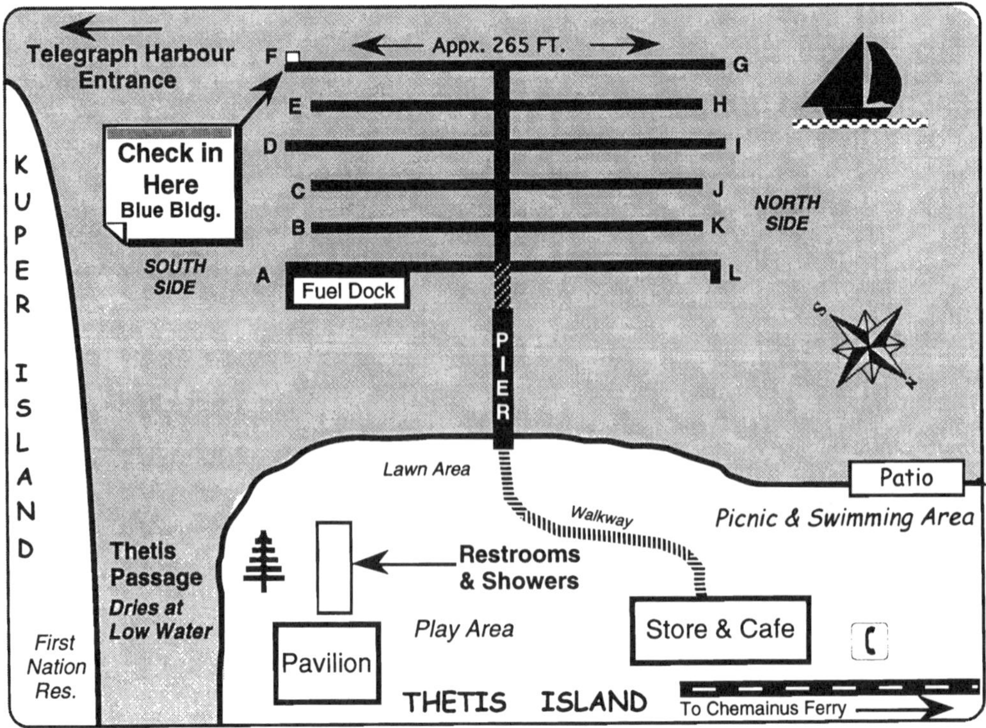

Thetis Island

NAME OF MARINA: ***THETIS ISLAND MARINA & PUB*** RADIO: VHF 66A
TELEPHONE: **250-246-3464** MGR: Paul Deacon, Owner
E-MAIL: marina@thetisisland.com FAX: 250-246-1433 (Fax)
ADDRESS: General Delivery Thetis Island, B.C. Canada VOR 2YO
SHORT DESCRIPTION & LOCATION: **www.thetisisland.com**

48°58.70' - 123°40.10' Located half-way down the entrance of Telegraph Harbour on port side, this family oriented marina offers many services for vacationing boaters. Noted for great food & warm, friendly Pub atmosphere. Marina is a short walk to Chemainus ferry.

GUEST BOAT CAPACITY: ....Appx. 60 boats
DOCKSIDE DEPTH AT ZERO TIDE: ....6 ft.
SEASON: ....All year
RESERVATION POLICY: ....Accepts
AMT W/ELECTRICITY: ....All
FUEL DOCK: ....Gas, Dsl, & LP
MARINE REPAIRS: ....None
TOILETS: ....Yes
HOT SHOWERS: ....Yes
RESTAURANT: ....Yes
PICNIC AREA: ....Yes
BASIC STORE: ....Yes
BROADBAND/WI-FI: ....BroadbandXpress
DAILY RATE: ....Moderate (75¢-$1.25/foot)

GUEST DOCK: (S)............3000 ft.
GUEST SLIPS: ....3000 ft.
WATER: Ltd. - For moored guests
AMPS: ....15-30 A
PUMP OUT STATION ....None
HAUL OUT: ....None
BOAT RAMP: ....Close by
LAUNDRY: ....Yes
BAR: ....Yes
POOL: ....None
GOLF: ....Ferry to Chemainus
PET FRIENDLY: ....Good
OTHER: Post Office, Pub with satellite TV, clamming & oysters close by. Pig roasts a specialty. LIQUOR STORE.

CAUTION! This chartlet not intended for use in navigation.

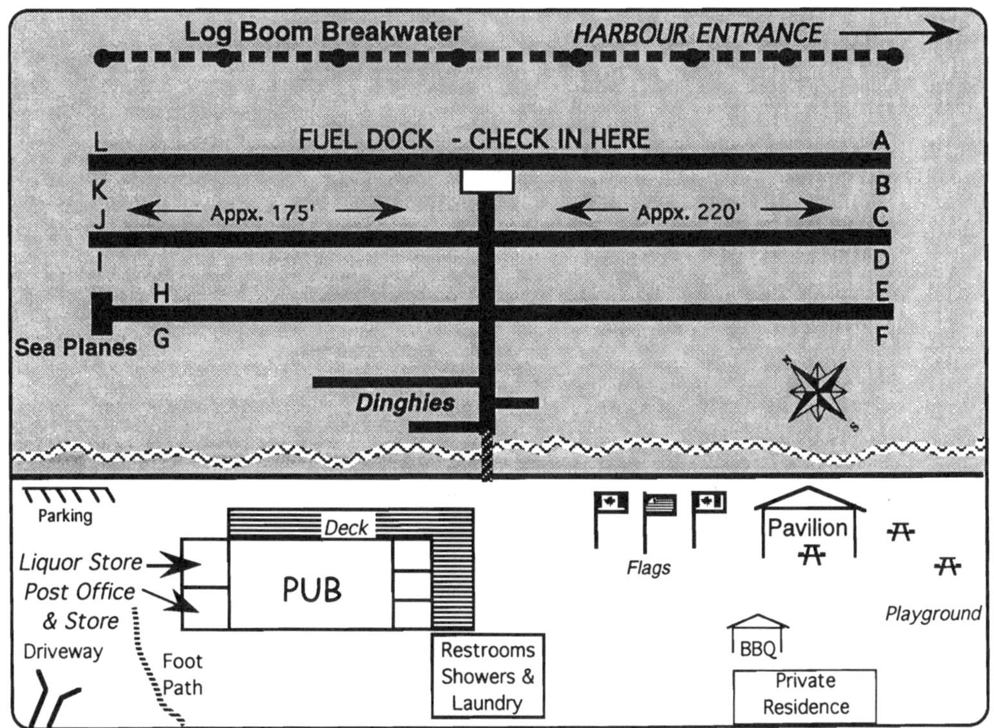

Gabriola Island

NAME OF MARINA: ***DEGNEN BAY HARBOUR AUTHORITY*** RADIO: None
TELEPHONE: 250-247-8380 MGR: Peter Sakich
E-MAIL: None FAX: None
ADDRESS: P.O. Box 51 Gabriola Island, B.C. Canada VOR 1X0
SHORT DESCRIPTION & LOCATION: **http://www.haa.bc.ca/harbours/South_Island**

49°08.40' - 123°42.65' The public floats are located at the head of Degnen Bay on the N. side of Gabriola Passage on S. side of Gabriola Island. The floats are about 1-1/2 miles from both Silva Bay & Drumbeg Park by road. Bay offers nice anchorage - Dock is often full.

GUEST BOAT CAPACITY: ....Limited
DOCKSIDE DEPTH AT ZERO TIDE: Unknown
SEASON: ....All year
RESERVATION POLICY: ....None
AMT W/ELECTRICITY: ....All
FUEL DOCK: ....None
MARINE REPAIRS: ....None
TOILETS: ....None
HOT SHOWERS: ....None
RESTAURANT: ....None
PICNIC AREA: ....None
BASIC STORE: ....None
BROADBAND/WI-FI: ....None
DAILY RATE: ....Economical (Under 75¢/foot)

NOTE: DOCKS OFTEN FULL W/LOCAL BOATS. FISHING BOATS HAVE PRIORITY.

GUEST DOCK: ....Appx. 200 ft.
GUEST SLIPS: ....None
WATER: ....None
AMPS: ....15-20 A
PUMP OUT STATION ....None
HAUL OUT: ....Grid
BOAT RAMP: ....None
LAUNDRY: ....None
BAR: ....None
POOL: ....None
GOLF: ....Yes
PET FRIENDLY: ....Good
OTHER: Tidal grid, crane, taxi service & phone. A killer whale petroglyph is near head of the bay.

CAUTION! This chartlet not intended for use in navigation.

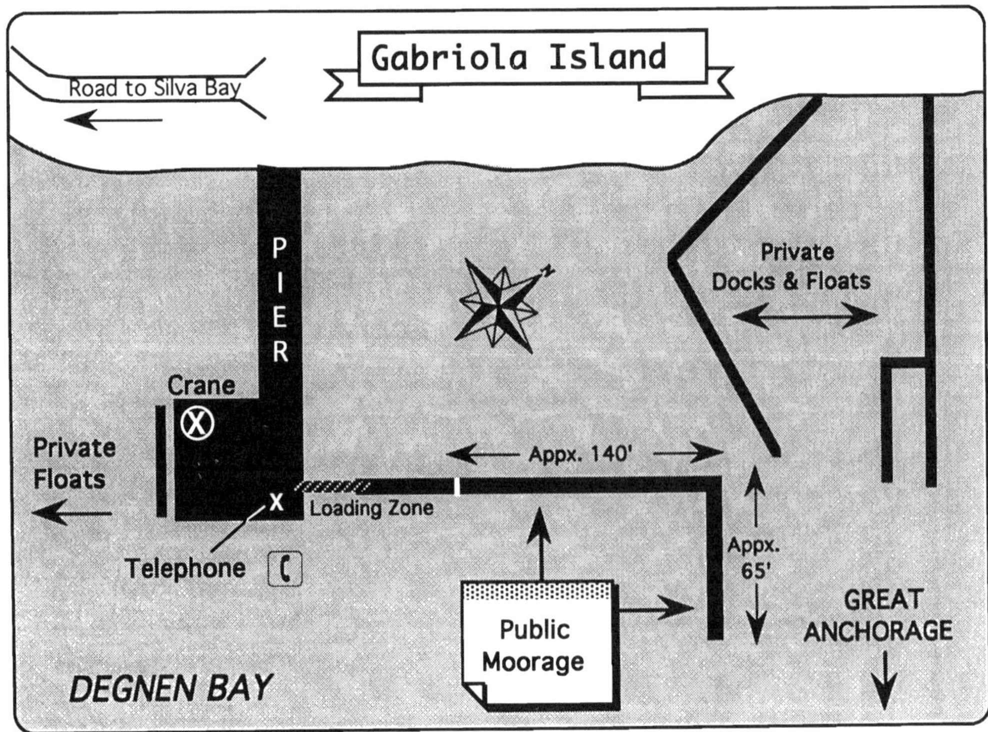

Silva Bay

NAME OF MARINA: ***PAGE'S MARINA (at Silva Bay)*** RADIO: None
TELEPHONE: **250-247-8931** MGR: Ted & Phyllis Reeve
E-MAIL: mail@pagesresort.com FAX: 250-247-8997 (Fax)
ADDRESS: 3350 Coast Rd. Gabriola Island, B.C. Canada V0R 1X7
SHORT DESCRIPTION & LOCATION: **www.pagesresort.com**

49°09.00' - 123°42.20' Silva Bay is a fairly well protected harbour on the SE end of Gabriola Is. inside of the Flattop Isl. group. Page's is the southerly marina of the 3 in the harbour. Open slips are rented to guest boaters. Local large scale charts are needed.

GUEST BOAT CAPACITY: ....Appx. 10 boats
DOCKSIDE DEPTH AT ZERO TIDE: ....12-17 ft.
SEASON: ....All year
RESERVATION POLICY: ....Accepts
AMT W/ELECTRICITY: ....All
FUEL DOCK: ....Gas & Diesel
MARINE REPAIRS: ....Close by
TOILETS: ....Yes
HOT SHOWERS: ....Yes
RESTAURANT: ....Close by
PICNIC AREA: ....Yes
BASIC STORE: ....Close by
BROADBAND/WI-FI: ....None
DAILY RATE: ....Moderate (75¢-$1.25/foot)

GUEST DOCK: ....Varies
GUEST SLIPS: ....Varies
WATER: ....Limited
AMPS: ....15 A
PUMP OUT STATION ....None
HAUL OUT: ....None
BOAT RAMP: ....None
LAUNDRY: ....Yes
BAR: ....Close by
POOL: ....None
GOLF: ....Close by
PET FRIENDLY: ....Excellent
OTHER: Cottage rentals, campground, marine charts & books, taxi service, good diving. Close to Drumbeg Park.

CAUTION! This chartlet not intended for use in navigation.

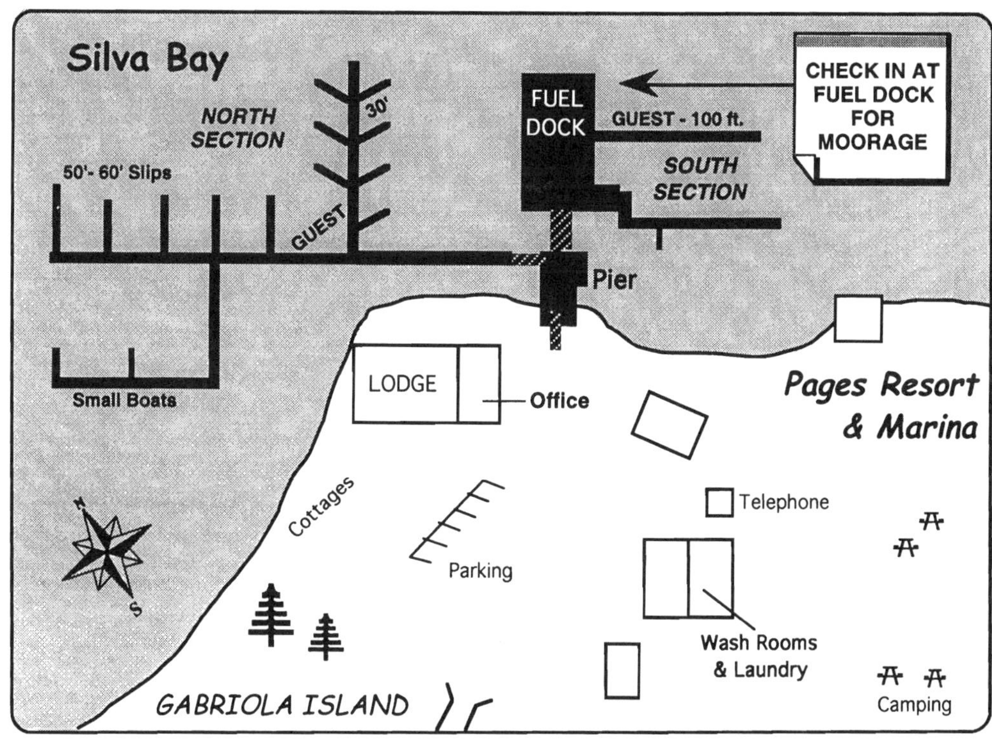

Silva Bay

NAME OF YACHT CLUB: *SILVA BAY YACHT CLUB*

ADDRESS: P.O. Box 154 Gabriola Island, B.C. Canada V0R 1X0

TELEPHONE: Marina - 250-247-8931 **PERSON IN CHARGE:** Commodore

SHORT DESCRIPTION & LOCATION: www.silvabayyachtclub.com

49°09.90' - 123°41.81' SBYC provides reciprocal moorage at Pages Marina in Silva Bay. Page's is the southerly marina of the 3 in the harbour. Reciprocal dock is 46 ft. & clearly marked with SBYC signage. Extra moorage may be available if member slips are vacant. Please register at marina office ASAP and show your current membership card.

RECIPROCAL BOAT CAPACITY: ....1-3 Boats

DOCKSIDE DEPTH AT ZERO TIDE: 12-17 ft.

SEASON: ....All year

RESERVATION POLICY: Call Marina before 1700

TOILETS: ....Yes

HOT SHOWERS: ....Yes

RESTAURANT: ....Close by

BROADBAND/WI-FI: ....Silva Bay Anchorage

DAILY RATE: Electricity $3.50/Day. 2 Days per month Free, may be consecutive except for Fri. & Sat.

RECIPROCAL DOCK: ....46 ft.

RECIPROCAL SLIPS: ....Varies

WATER: ....Limited

AMT W/ELECTRICITY: ....All

AMPS: ....15 A

BAR: ....Close by

PET FRIENDLY: ....Excellent

OTHER: Cottage rentals, campground, marine charts & books, taxi service. Good diving.

***NOTE:* THIS IS PRIVATE MOORAGE AND ONLY AVAILABLE TO MEMBERS OF RECIPROCAL YACHT CLUBS. YOUR CLUB MUST HAVE RECIPROCAL PRIVILEGES AND YOU MUST FLY YOUR BURGEE.**

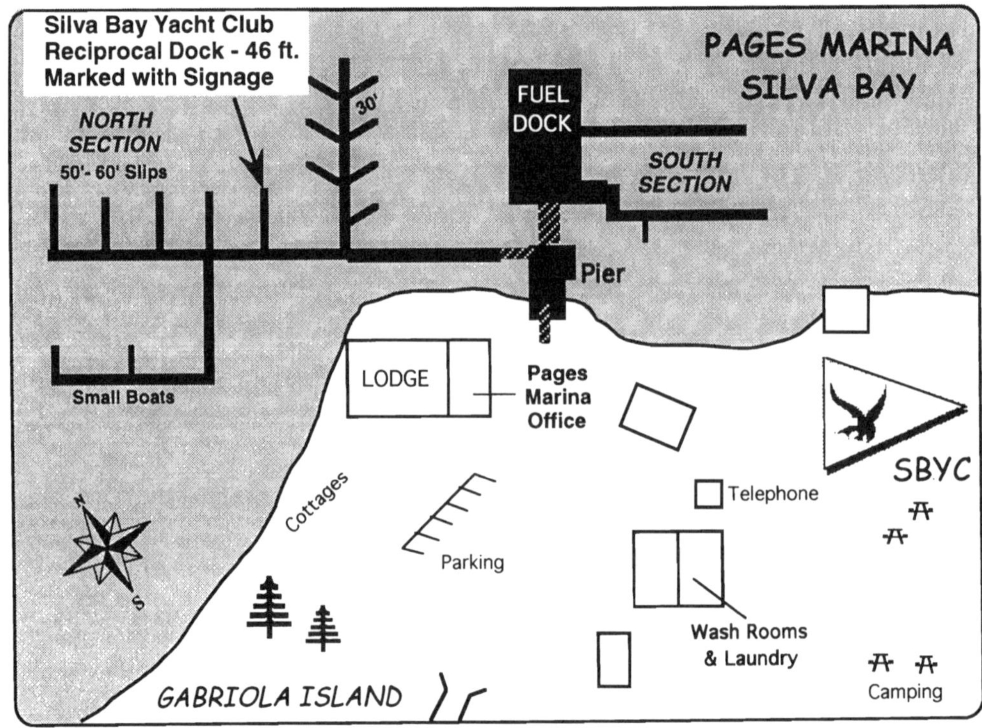

Silva Bay

NAME OF MARINA: *SILVA BAY RESORT & MARINA* RADIO: VHF 66A
TELEPHONE: 250-247-8662 MGR: Janice Fuller
E-MAIL: silvabay@canada.com FAX: 250-247-8663 (Fax)
ADDRESS: 3383 South Road Gabriola, B.C. Canada V0R 1X7
SHORT DESCRIPTION & LOCATION: **www.silvabay.com**

49°09.00' - 123°42.20' Silva Bay is a harbour located on the S.E. end of Gabriola Is. inside of the Flattop Island group. The marina is the largest of the 3 marinas in the harbor & situated in the middle. Friendly pub & family dining. Large scale charts are needed.

GUEST BOAT CAPACITY: ....Appx. 40 boats
DOCKSIDE DEPTH AT ZERO TIDE: ....7-8 ft.
SEASON: ....All year
RESERVATION POLICY: ....Accepts
AMT W/ELECTRICITY: ....All
FUEL DOCK: ....None
MARINE REPAIRS: ....On premises
TOILETS: ....Yes
HOT SHOWERS: ....Yes
RESTAURANT: ....Cafe and Silva Bay Bar & Grill
PICNIC AREA: ....Yes
BASIC STORE: ....Close by
BROADBAND/WI-FI: ....BroadbandXpress
DAILY RATE: ....Moderate (75¢-$1.25/foot)

GUEST DOCK: 110 ft. plus slips
GUEST SLIPS: ....Appx. 40
WATER: ....Yes
AMPS: ....30 A
PUMP OUT STATION ....None
HAUL OUT: ....Rail & Travel-Lift
BOAT RAMP: ....None
LAUNDRY: ....Yes
BAR: ....Yes
POOL: ....None
GOLF: ....Close by
PET FRIENDLY: ....Good
OTHER: Wooden Boat School, shipyard, gift shop, fishing charters, tennis courts, Sunday Farmers & Artisans Market, Family Dining.

CAUTION! This chartlet not intended for use in navigation.

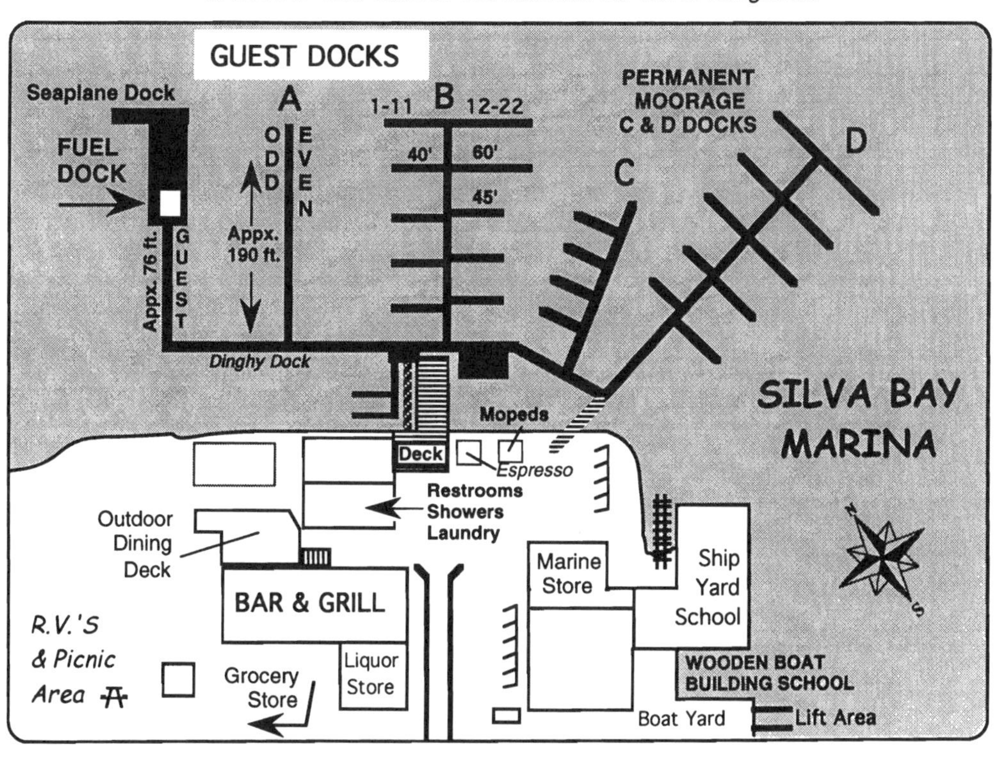

Silva Bay

NAME OF MARINA: ***SILVA BAY INN (formerly Boatel)*** **RADIO:** None
TELEPHONE: 250-247-9351 **MGR:** Paula Maddison
E-MAIL: None **FAX:** None
ADDRESS: 3415 South Road Gabriola, B.C. Canada VOR 1X7
SHORT DESCRIPTION & LOCATION: SMALL BOAT MOORAGE ONLY

49°09.00' - 123°42.20' ***Under new ownership.*** Refurbished motel & quality food market with limited moorage. This is smallest marina of the 3 in the harbour & situated at the far N.W. side. Floats shallow at low tide - **4.0 ft. maximum draft allowed at dock**.

GUEST BOAT CAPACITY: Appx. 10 small boats
DOCKSIDE DEPTH AT ZERO TIDE: ....4 ft.
SEASON: ....All year
RESERVATION POLICY: ....Accepts
AMT W/ELECTRICITY: ....All
FUEL DOCK: ....None
MARINE REPAIRS: ....Close by
TOILETS: ....None
HOT SHOWERS: ....None
RESTAURANT: ....Close by
PICNIC AREA: ....Yes
BASIC STORE: ....Yes
BROADBAND/WI-FI: ....None
DAILY RATE: ....Moderate (75¢-$1.25/foot)

GUEST DOCK: ....Varies
GUEST SLIPS: ....Dock only
WATER: ....Limited
AMPS: ....15 A
PUMP OUT STATION ....None
HAUL OUT: ....None
BOAT RAMP: ....None
LAUNDRY: ....Close by
BAR: ....Close by
POOL: ....None
GOLF: ....Close by
PET FRIENDLY: ....Excellent
OTHER: Motel accommodations, taxi service, groceries, retail shops, coffee bar.

RECOMMENDED FOR SMALL BOATS ONLY

CAUTION! This chartlet not intended for use in navigation.

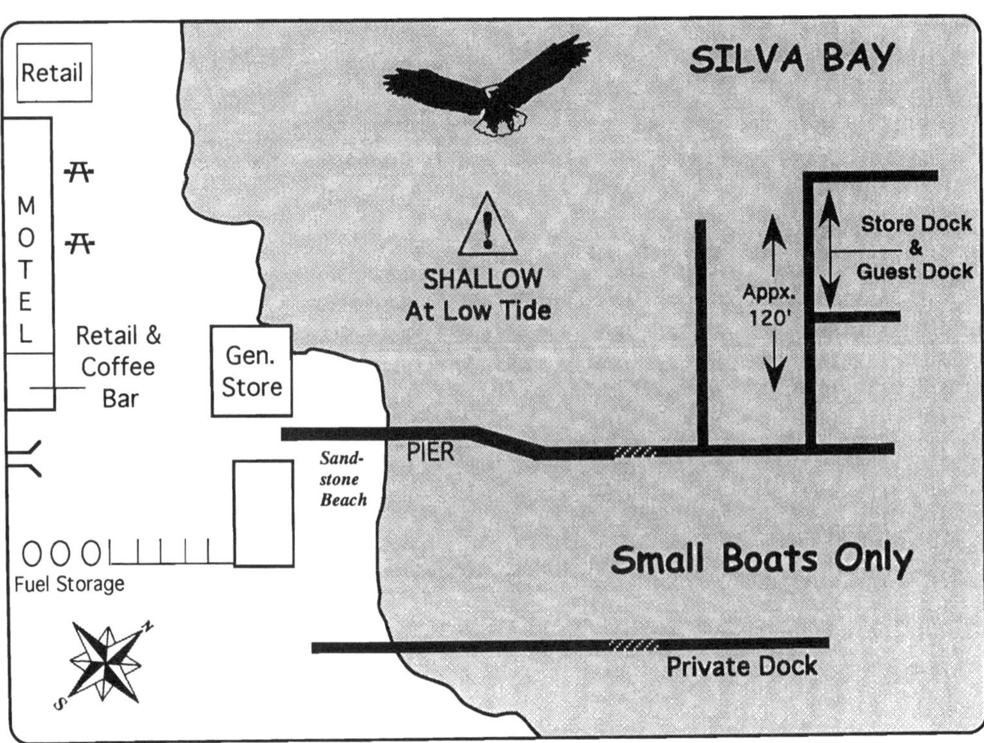

CHAPTER 7

VANCOUVER ISLAND
SOUTHERN PART

Chapter Map - Page 3

Victoria

NAME OF YACHT CLUB: ***CANADIAN FORCES SAILING ASSOCIATION***

ADDRESS: 1001 Maple Bank Road Victoria, B.C. Canada V9A 4M2

TELEPHONE: **250-385-2646 or 7833** PERSON IN CHARGE: Foreshore Chair.

SHORT DESCRIPTION & LOCATION: **www.cfsa.shawbiz.ca**

48°25.60' - 123°25.90' Moorage facilities located in Esquimalt Hbr. (E side) in Constance Cove (NW side) just N of Naval Dockyards. Best to call ahead to make arrangements. Also, upon entering Esq. Hbr, call Port Security on VHF 10 to advise your intentions. **If Visitor Dock full, tie up on Work Dock & register at club bar.**

RECIPROCAL BOAT CAPACITY: ....Appx. 2-4

DOCKSIDE DEPTH AT ZERO TIDE: ....30 ft.

SEASON: ....All year

RESERVATION POLICY: ....Call in advance

TOILETS: ....Yes

HOT SHOWERS: ....Yes

RESTAURANT: ....None

BROADBAND/WI-FI: ....None

DAILY RATE: First 2 days free, further moorage 50¢/ft. Max 7 Days. Gate code available at check-in at Bar.

RECIPROCAL DOCK: ....70 ft.

RECIPROCAL SLIPS: ....Varies

WATER: ....Yes

AMT W/ELECTRICITY: ....All

AMPS: ....15 A

BAR: Bar open Tue. thru Sun.

PET FRIENDLY: ....Good

OTHER: Restaurants, groceries, laundry & liquor store within 10-15 minute walk.

***NOTE:* THIS IS PRIVATE MOORAGE AND ONLY AVAILABLE TO MEMBERS OF RECIPROCAL YACHT CLUBS. YOUR CLUB *MUST* HAVE RECIPROCAL PRIVILEGES AND YOU MUST FLY YOUR BURGEE.**

CAUTION! This chartlet not intended for use in navigation.

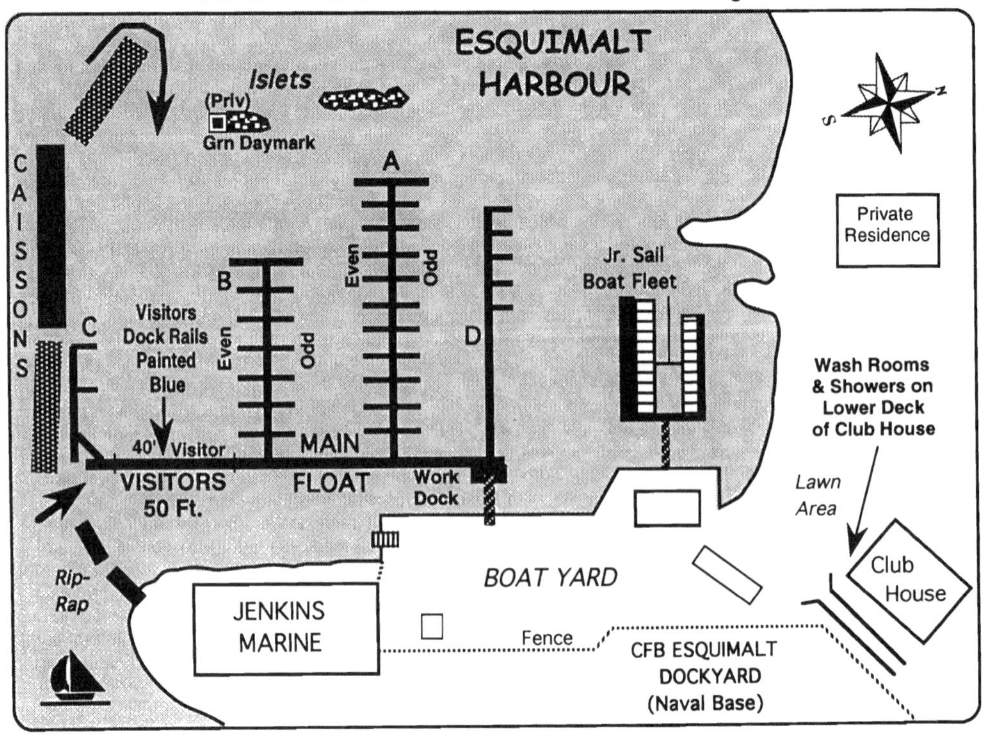

Victoria

NAME OF MARINA: ***COAST HARBOURSIDE MARINA*** **RADIO:** VHF 66A
TELEPHONE: **250-360-1211** **MGR:** Gail Windle
E-MAIL: victoriamarina@coasthotels.com **FAX:** 250-360-1418 (Fax)
ADDRESS: 146 Kingston St. Victoria, B.C. Canada V8V 1V4
SHORT DESCRIPTION & LOCATION: www.nwboat.com/coastharbourside/coast.htm

48°25.40' - 123°22.80' Located about 1km W. of Victoria Empress Floats on S. side of harbour entrance just before Laurel Pt. Upscale hotel & marina is within walking distance to all downtown Victoria attractions in quiet waterfront neighborhood with boardwalks.

GUEST BOAT CAPACITY: ....40-50 boats
DOCKSIDE DEPTH AT ZERO TIDE: ....4 ft. +
SEASON: ....All year
RESERVATION POLICY: ....Recommended
AMT W/ELECTRICITY: ....All
FUEL DOCK: ....Close by
MARINE REPAIRS: ....Close by
TOILETS: ....Yes
HOT SHOWERS: ....Yes
RESTAURANT: ....Yes
PICNIC AREA: ....Close by
BASIC STORE: ....Close by
BROADBAND/WI-FI: ....BroadbandXpress
DAILY RATE:.................Premium (Over $1.25/foot)
Note: Register at Front Desk of Hotel

GUEST DOCK: ...Slips 30' to 120'
GUEST SLIPS: 40-50
WATER: ....Yes
AMPS: ....30-50 A
PUMP OUT STATION ....Yes
HAUL OUT: ....None
BOAT RAMP: ....None
LAUNDRY: ....Close by
BAR: ....Yes
POOL: ....Yes
GOLF: ....Close by
PET FRIENDLY: ....Good
OTHER: **Water taxi** stop to downtown Vic. Jacuzzi, sauna, & fitness room avail. for marina patrons.

CAUTION! This chartlet not intended for use in navigation.

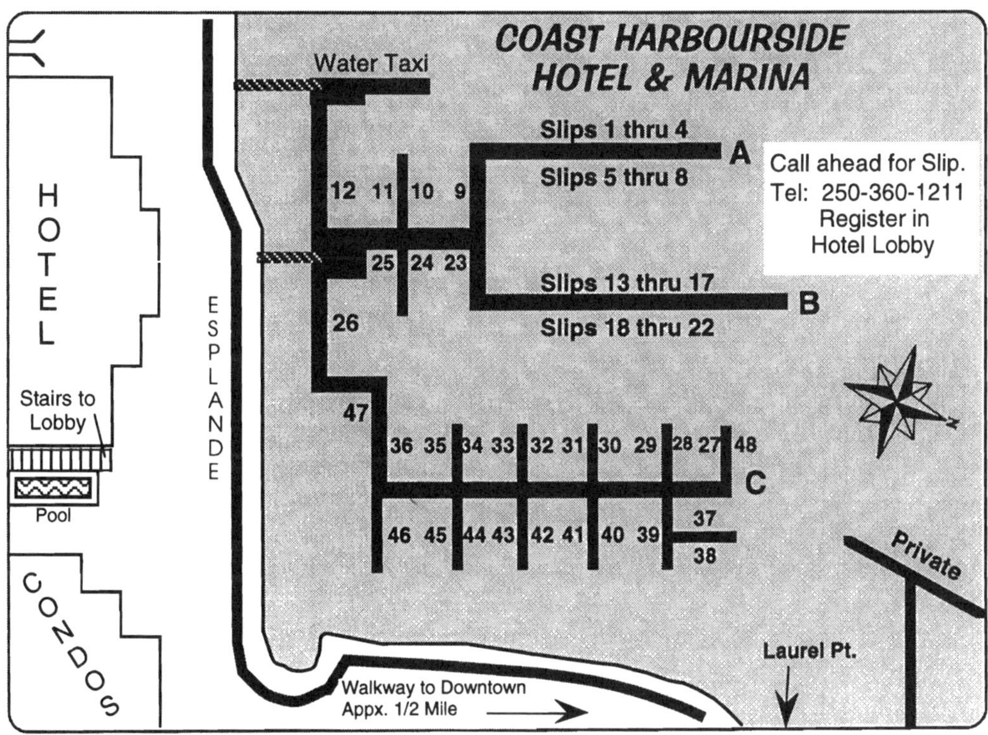

Victoria

NAME OF MARINA: ***CAUSEWAY FLOATS - Harbour Authority*** RADIO: VHF 66A
TELEPHONE: 250-383-8326 / 877-783-8300 MGR: Sheila Neapole
E-MAIL: moorage@victoriaharbour.org FAX: 250-383-8306 (Fax)
ADDRESS: #202-468 Bellevue St. Victoria, B.C. Canada V8V 1W9
SHORT DESCRIPTION & LOCATION: **www.victoriaharbour.org**
48°25.80' - 123°22.30' Located in the heart of the Inner Harbour directly in front of the historic and beautiful Empress Hotel and adjacent to the B.C. Parliament Buildings. This location offers the best vantage point for viewing the night lights of Victoria.

GUEST BOAT CAPACITY: ....Appx. 50 boats
DOCKSIDE DEPTH AT ZERO TIDE: ......5-15 ft.
SEASON: ....All year
RESERVATION POLICY: Groups & vessels 65'+
AMT W/ELECTRICITY: ....All
FUEL DOCK: ....Close by
MARINE REPAIRS: ....Close by
TOILETS: ....Yes
HOT SHOWERS: ....At Wharf St. Floats
RESTAURANT: ....Close by
PICNIC AREA: ....Close by
BASIC STORE: ....Close by
BROADBAND/WI-FI: ....BroadbandXpress
DAILY RATE: ....Premium
(Over $1.25/foot)
NOTE: RAFTING IS MANDATORY UPON REQUEST.

GUEST DOCK: ......2000 ft. Total
GUEST SLIPS: ....7 large slips
WATER: ....Yes
AMPS: ....30 A
PUMP OUT STATION Close by
HAUL OUT: ....Close by
BOAT RAMP: ....Close by
LAUNDRY: .....At Wharf St. Floats
LOUNGE: ....Close by
POOL: ....Close by
GOLF: ....Close by
PET FRIENDLY: ....Fair
OTHER: Customs Port-of-Entry, fine museums, several sightseeing tours, shopping & night life, planes & ferries to mainland.

NAME OF MARINA: ***WHARF ST. FLOATS - Harbour Authority*** RADIO: VHF 66A
ADDRESS: TOLL FREE: 877-783-8300 Victoria, B.C. Canada
TELEPHONE: 250-383-8326 MGR: Sheila Neapole
SHORT DESCRIPTION & LOCATION:
48°26.40' - 123°22.30' Located about 2 blocks north of the Empress Floats next door to the Customs Dock. This location is also very convenient to the many amenities of beautiful Victoria.. Security gate. **RAFTING IS MANDATORY UPON REQUEST.**

GUEST BOAT CAPACITY: ....Appx. 100 boats
DOCKSIDE DEPTH AT ZERO TIDE: ....10-25 ft.
SEASON: ....All year
RESERVATION POLICY: Groups & vessels 65'+
AMT W/ELECTRICITY: ....All
FUEL DOCK: ....Close by
MARINE REPAIRS: ....Close by
TOILETS: ....Yes
HOT SHOWERS: ....Yes
RESTAURANT: ....Close by
PICNIC AREA: ....Close by
BASIC STORE: ....Close by
BROADBAND/WI-FI: ....BroadbandXpress
DAILY RATE: ....Premium
(Over $1.25/foot)
NOTE: RAFTING IS MANDATORY UPON REQUEST.

GUEST DOCK: .....1780 ft. Total
GUEST SLIPS: ....Appx. 100
WATER: ....Yes
AMPS: ....20-50 Amp
PUMP OUT STATION Close by
HAUL OUT: ....Close by
BOAT RAMP: ....Close by
LAUNDRY: ....Yes
LOUNGE: ....Close by
POOL: ....Close by
GOLF: ....Close by
PET FRIENDLY: ....Fair
OTHER: Customs Port-of-Entry, fine museums, several sightseeing tours, shopping & night life, planes & ferries to mainland.

Victoria

CAUTION! This chartlet not intended for use in navigation.

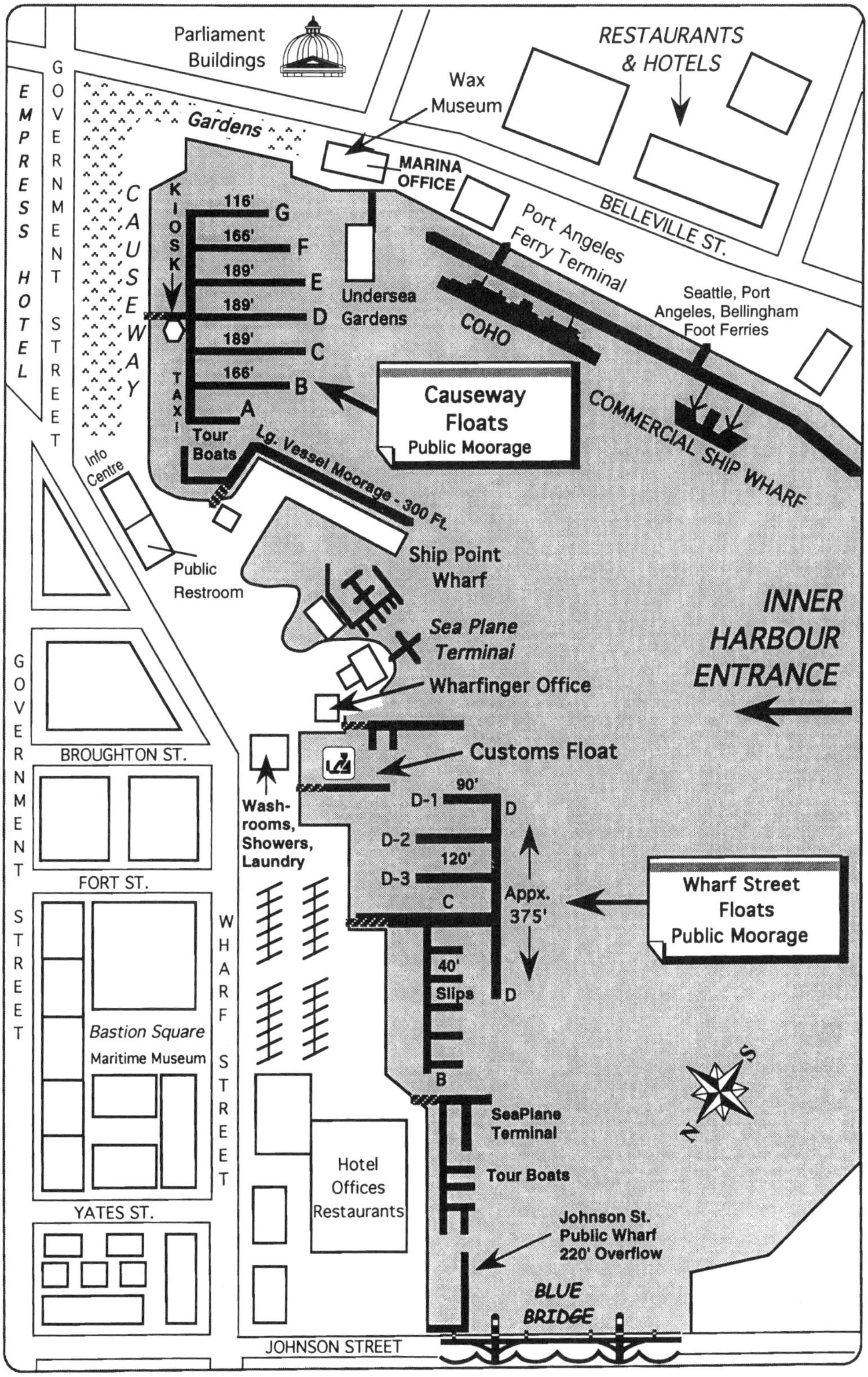

Parliament Buildings

RESTAURANTS & HOTELS

Wax Museum

Gardens

MARINA OFFICE

BELLEVILLE ST.

EMPRESS HOTEL

GOVERNMENT STREET

CAUSEWAY

KIOSK

TAXI

116' G

166' F

189' E

189' D

189' C

166' B

A

Undersea Gardens

Port Angeles Ferry Terminal

COHO

Seattle, Port Angeles, Bellingham Foot Ferries

Causeway Floats
Public Moorage

COMMERCIAL SHIP WHARF

Tour Boats

Lg. Vessel Moorage - 300 Ft.

Info Centre

Public Restroom

Ship Point Wharf

Sea Plane Terminal

INNER HARBOUR ENTRANCE

Wharfinger Office

Customs Float

BROUGHTON ST.

Wash-rooms, Showers, Laundry

90' D-1 D

D-2

120'

D-3

C

Appx. 375'

Wharf Street Floats
Public Moorage

FORT ST.

WHARF STREET

40' Slips

D

Bastion Square

Maritime Museum

B

SeaPlane Terminal

Tour Boats

Hotel Offices Restaurants

YATES ST.

Johnson St. Public Wharf 220' Overflow

BLUE BRIDGE

JOHNSON STREET

N S

Victoria

NAME OF MARINA: ***OAK BAY MARINA*** **RADIO:** VHF 66 A
TELEPHONE: **250-598-3369** MGR: Marina Manager
E-MAIL: obm@obmg.com FAX: 250-598-1361 (Fax)
ADDRESS: 1327 Beach Drive Victoria, B.C. Canada V8S 2N4
SHORT DESCRIPTION & LOCATION: **www.oakbaymarina.com**

48°26.60' - 123°18.00' Located between Gonzales & Cattle Pts. in Oak Bay on SE tip of Van. Is., The upscale marina is appx 3 road mi. from downtown Victoria and has resident seals which can be fed & enjoyed by adults & kids. Rents vacant slips to guest boaters.

GUEST BOAT CAPACITY: ....Varies
DOCKSIDE DEPTH AT ZERO TIDE: 20 ft. +
SEASON: ....All year
RESERVATION POLICY: ....None
AMT W/ELECTRICITY: ....All
FUEL DOCK: ....Gas & Diesel
MARINE REPAIRS: ....On premises
TOILETS: ....Yes
HOT SHOWERS: ....Yes
RESTAURANT: ....Marina Restaurant
PICNIC AREA: ....Yes
BASIC STORE: ....Yes
BROADBAND/WI-FI: ....Yes
DAILY RATE: ....Moderate (75¢-$1.25/foot)
Key Deposit Required
<u>CUSTOMS PORT-OF-ENTRY</u>

GUEST DOCK: ....Varies
GUEST SLIPS: ....Varies
WATER: ....Yes
AMPS: ....15-30 A
PUMP OUT STATION ....None
HAUL OUT: ....Rail
BOAT RAMP: ....Yes
LAUNDRY: ....Yes
BAR: ....Yes
POOL: ....Close by
GOLF: ....Close by
PET FRIENDLY: ....Good
OTHER: Seal watching, arts & gift store, marine chandlery, cafe, shopping centre nearby, bus service to city center.

CAUTION! This chartlet not intended for use in navigation.

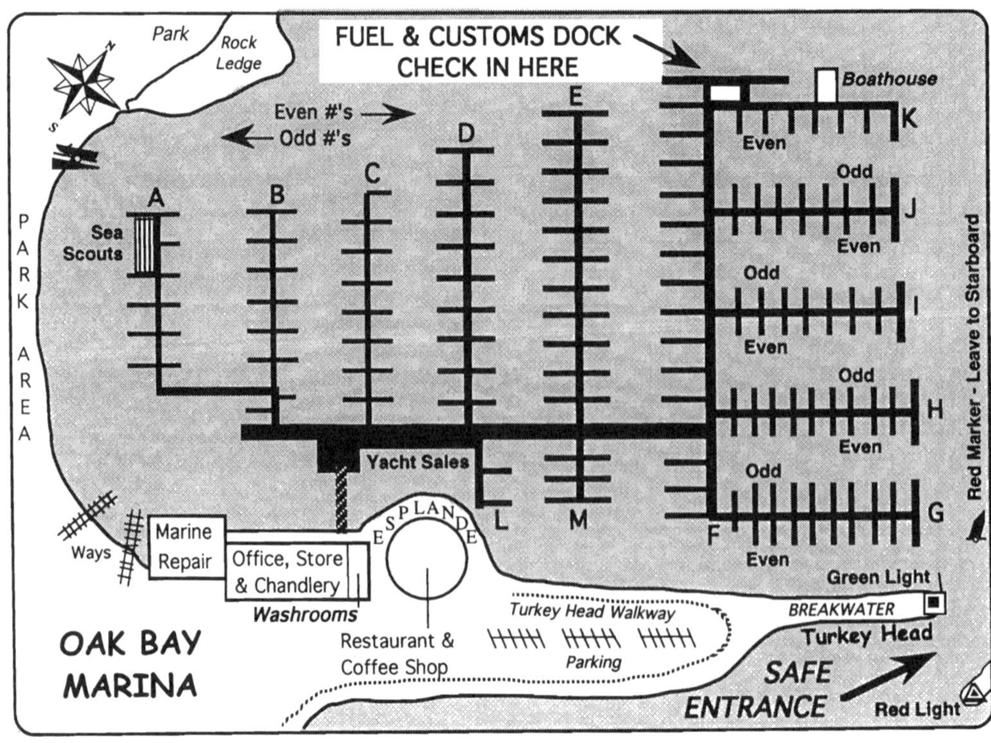

Victoria

NAME OF YACHT CLUB: *ROYAL VICTORIA YACHT CLUB*

ADDRESS: 3475 Ripon Rd. Victoria, B.C., Canada V8R 6HI

TELEPHONE: 250-592-3433/2441(office) PERSON IN CHARGE: Staff Captain

SHORT DESCRIPTION & LOCATION: **www.rvyc.bc.ca**

48°27.09' - 123°17.36' RVYC is in Cadboro Bay in the N.W. corner of Oak Bay appx. 10 water miles (4 land miles) from downtown Victoria. Quiet location. Upon arrival, moor on Visitor Dock & register at Foreshore Office (250-592-3433), Club Office or Bar. Visitors may be assigned vacant berths. Reciprocal priveleges vary depending on your club status.

RECIPROCAL BOAT CAPACITY:.....Appx 6 boats

DOCKSIDE DEPTH AT ZERO TIDE: ....22 ft.

SEASON: ....April to October

RESERVATION POLICY: ....None

TOILETS: ....Yes

HOT SHOWERS: & LAUNDRY:..........................Yes

RESTAURANT: ....Yes - Wed. thru Sun.

BROADBAND/WI-FI: ....None

DAILY RATE: RVYC offers 2 levels of reciprocity, full & basic priveleges. Check with your home club for your status.

RECIPROCAL DOCK: Appx. 60 ft.

RECIPROCAL SLIPS: ....Varies

WATER: ....Yes

AMT W/ELECTRICITY: Limited

AMPS: ....15 Amp if available

BAR: ....Nice Clubhouse & Bar

PET FRIENDLY: ....Good

OTHER: Customs port-of-entry, Key deposit required for washrooms and gate. DINING ROOM SERVICE

***NOTE:* THIS IS PRIVATE MOORAGE AND ONLY AVAILABLE TO MEMBERS OF RECIPROCAL YACHT CLUBS. YOUR CLUB MUST HAVE RECIPROCAL PRIVILEGES AND YOU MUST FLY YOUR BURGEE.**

CAUTION! This chartlet not intended for use in navigation.

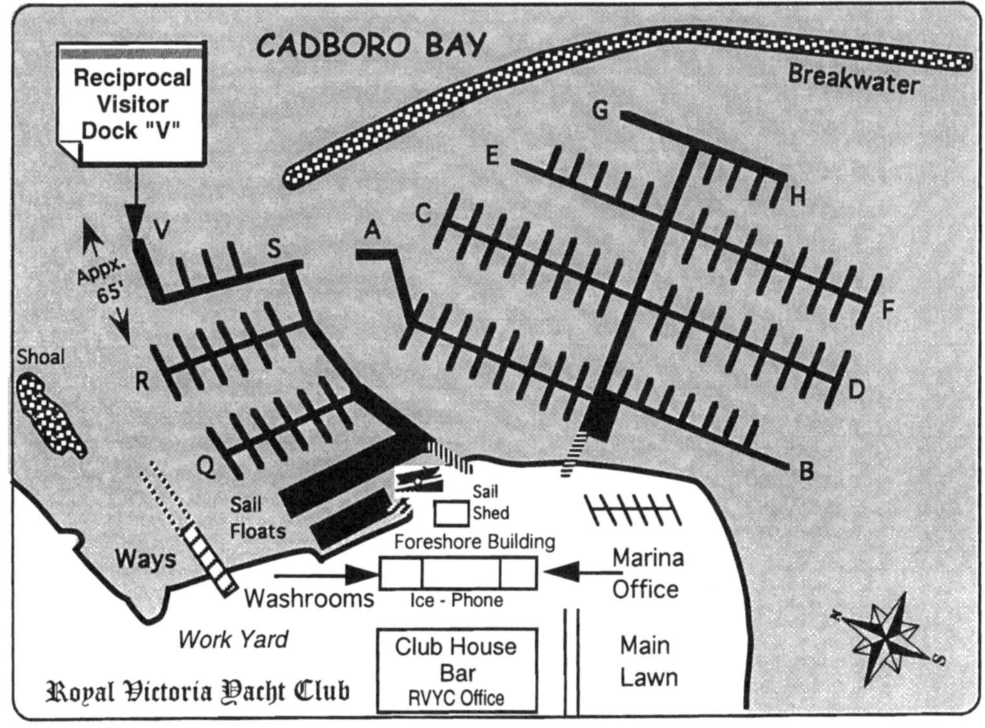

Victoria

NAME OF MARINA: **GOLDSTREAM BOATHOUSE** RADIO: VHF 66A
TELEPHONE: **250-478-4407** MGR: Lida Seymonsbergen
E-MAIL: info@goldstreamboathousemarina.com FAX: 250-478-6882 (Fax)
ADDRESS: 3540 Trans-Can Hwy Victoria, B.C. Canada V9B 6H6
SHORT DESCRIPTION & LOCATION: **www.goldstreamboathousemarina.com**

48°30.40' - 123°33.00' Quiet marina in fjord like setting located at head (S. end) of Saanich Inlet on W side in Finlayson Arm. Consult chart and provide caution for shoal waters adjacent to marina. Docks have been recently rebuilt to provide good deep water moorage.

GUEST BOAT CAPACITY: ....Appx. 10 boats
DOCKSIDE DEPTH AT ZERO TIDE: Appx. 10'
SEASON: ....All year
RESERVATION POLICY: ....Accepts
AMT W/ELECTRICITY: ....All
FUEL DOCK: ....Gas & Diesel
MARINE REPAIRS: ....On premises
TOILETS: ....Yes
HOT SHOWERS: ....None
RESTAURANT: ....Close by
PICNIC AREA: ....Yes
BASIC STORE: ....Yes
BROADBAND/WI-FI: ....None
DAILY RATE: ....Moderate (75¢-$1.25/foot)

GUEST DOCK: ....Varies
GUEST SLIPS: ....Varies
WATER: ....Yes
AMPS: ....15-30 A
PUMP OUT STATION ....None
HAUL OUT: ....Yes
BOAT RAMP: ....Yes
LAUNDRY: ....None
BAR: ....None
POOL: ....None
GOLF: ....Close by
PET FRIENDLY: ....Good
OTHER: Marine & fishing supplies, Provincial park closeby, bus & taxi to downtown Victoria. Excellent prawning & crab fishing.

CAUTION! This chartlet not intended for use in navigation.

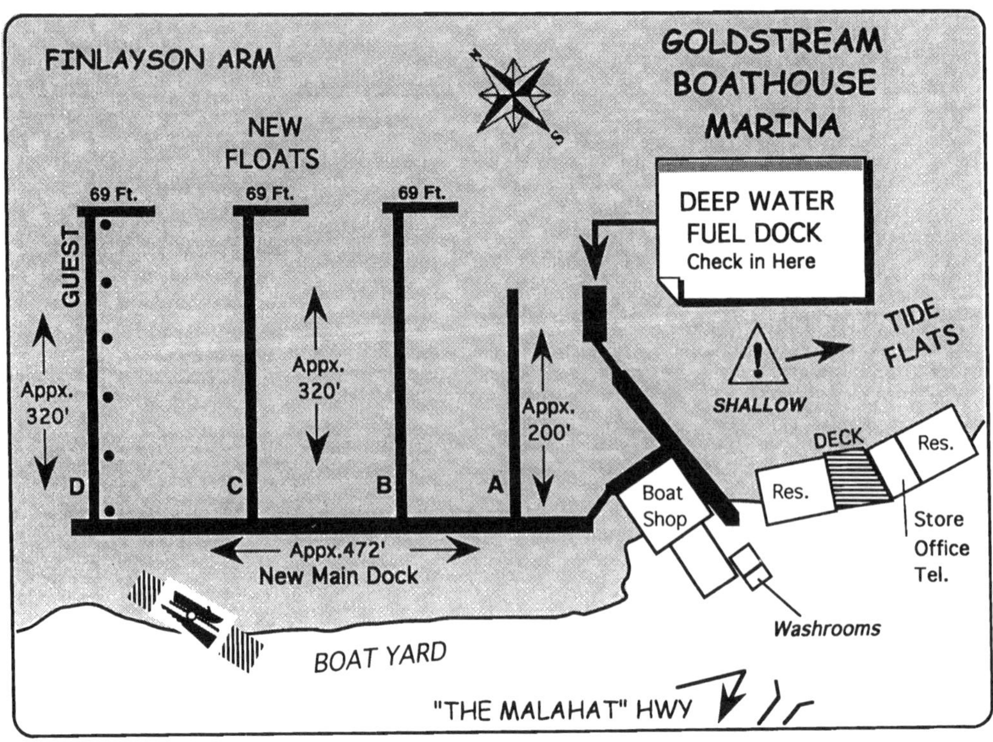

Sidney

NAME OF PARK: *SIDNEY SPIT - GULF ISLANDS NATIONAL PARK RESERVE*
ADDRESS: 2220 Harbour Rd., Sidney, B.C Canada V8L 2P6
TELEPHONE: 250-654-4000 **MGR:** Gulf Islands National Park Resv.
SHORT DESCRIPTION & LOCATION: **http://www.parkscanada.gc.ca**

48°38.50' - 123°19.90' Located on the NW end of Sidney Island about 2 miles east of Sidney. This 1000 acre marine park offers beautiful sandy beaches and wooded trails with an abundance of birds and wildlife. Now part of Canada's new Gulf Islands National Park Res.

GUEST BOAT CAPACITY: ....Appx. 10-15 boats
DOCKSIDE DEPTH AT ZERO TIDE: Appx. 5 ft.
SEASON: ....May 15 - Sep. 30
AMT W/ELECTRICITY: ....None
TOILETS: ....Pit Toilets
HOT SHOWERS: ....None
PICNIC AREA: ....Yes
PLAY AREA: ....Yes
BASIC STORE: ....None
DAILY RATE: ....Appx. 70¢/ft.- $1.95/metre
.......Buoys - $9.95 per night
.......Campsites - $14 per night
PARK FACILITY OPERATOR WILL COME TO BOAT TO COLLECT FEES

GUEST DOCK: ....Appx. 275 ft.
MOORING BUOYS: 21
WATER: ....Yes
PAY PHONES: ....None
BOAT RAMP: ....None
PICNIC SHELTER: ....Yes
BBQ: ....None
PUMP OUT STATION: ......None
PET FRIENDLY: Good - On leash
OTHER: Great swimming beach, good crabbing, campsites, hiking trails, bird watching, Amphitheater.

CAUTION! This chartlet not intended for use in navigation.

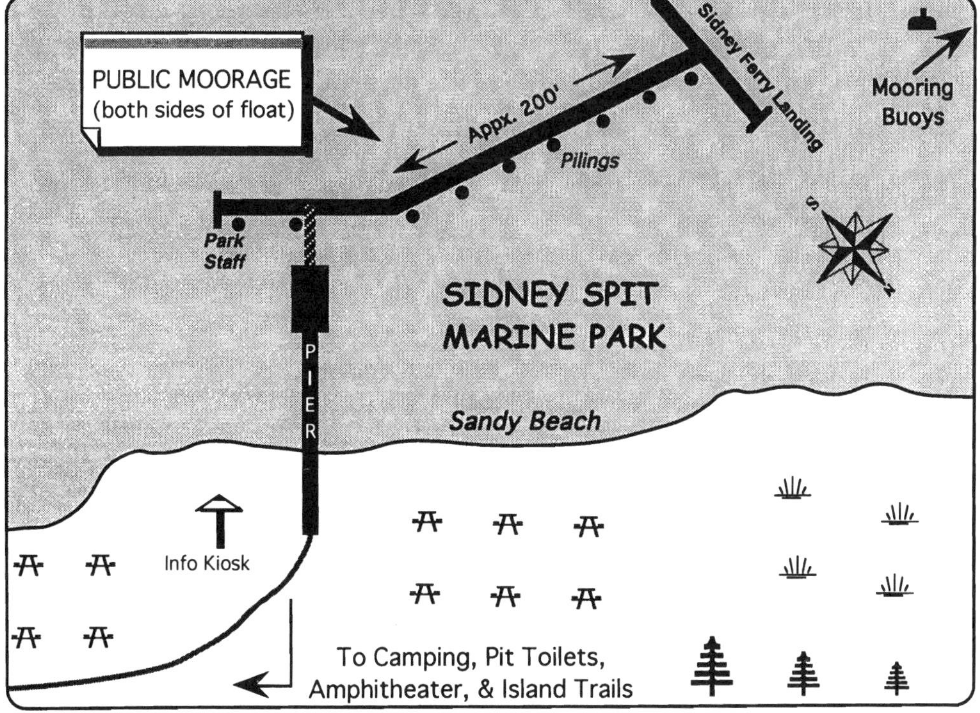

Sidney

NAME OF MARINA: ***PORT SIDNEY MARINA*** **RADIO:** VHF 66A
TELEPHONE: 250-655-3711 **MGR:** Lyndell Curry
E-MAIL: admin@portsidney.com **FAX:** 250-655-3771 (Fax)
ADDRESS: 9835 Seaport Place Sidney, B.C.. Canada V8L 4X3
SHORT DESCRIPTION & LOCATION: **www.portsidney.com**

48°39.10' - 123°23.55' Located 1 block N. of Sidney city center. Spacious and modern full service marina. Walking distance to shops, restaurants, and city amenities. Public transportation avail. to Victoria. Convenient to Victoria Int'l Airport & ferries to mainland.

GUEST BOAT CAPACITY: ....Appx. 300 boats
DOCKSIDE DEPTH AT ZERO TIDE: ....9 ft.
SEASON: ....All year
RESERVATION POLICY: ....Accepts
AMT W/ELECTRICITY: ....All
FUEL DOCK: ....Close by
MARINE REPAIRS: ....Close by
TOILETS: ....Yes
HOT SHOWERS: ....Yes
RESTAURANT: ....Yes
PICNIC AREA: ....Close by
BASIC STORE: ....Close by
BROADBAND/WI-FI: ....BroadbandXpress
DAILY RATE: ....Premium (Over $1.25/foot)

GUEST DOCK: ..... 100' plus slips
GUEST SLIPS: ....Appx. 200
WATER: ....Yes
AMPS: ....30-50 A
PUMP OUT STATION ....Yes
HAUL OUT: ....Close by
BOAT RAMP: ....Close by
LAUNDRY: ....Yes
BAR: ....Close by
POOL: ....None
GOLF: ....Close by
PET FRIENDLY: ....Fair
OTHER: Customs port-of entry, gift shop & boutique, market, cable T.V, public transportation to Vancouver Island sights.

CAUTION! This chartlet not intended for use in navigation.

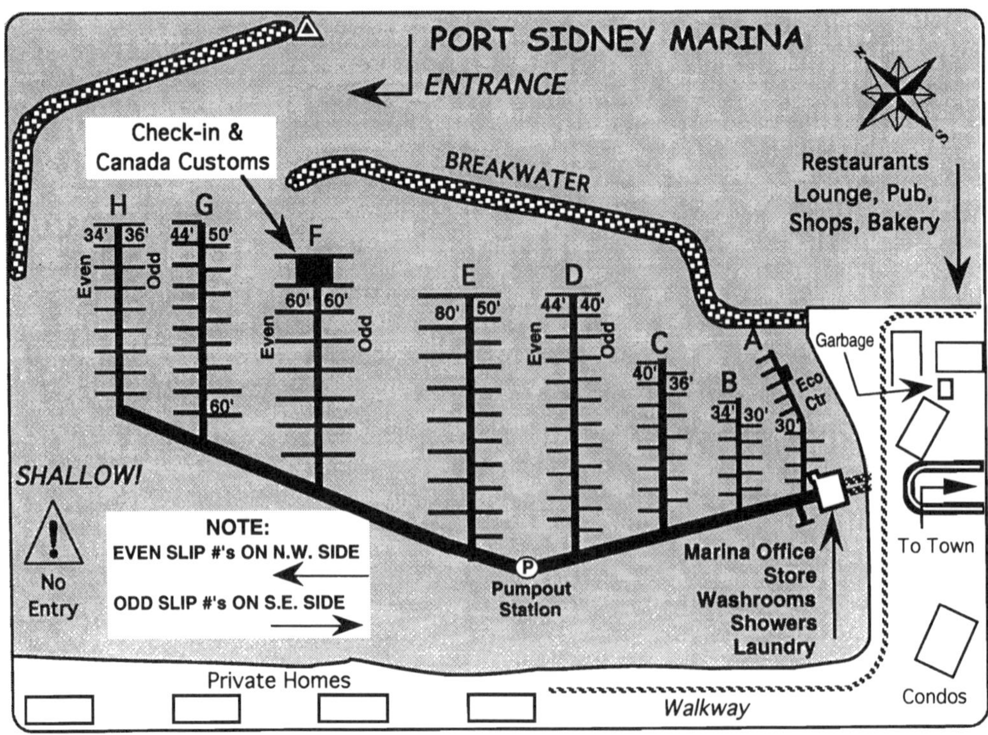

Sidney

NAME OF MARINA: ***VAN ISLE MARINA*** **RADIO:** VHF 66A
TELEPHONE: **250-656-1138** MGR: Mark Dickinson
E-MAIL: info@vanislemarina.com FAX: 250-656-0182 (Fax)
ADDRESS: 2320 Harbour Rd. Sidney, B.C. Canada V8L 2P6
SHORT DESCRIPTION & LOCATION: **www.vanislemarina.com**

48°40.18' - 123°24.30' Located in Tsehum Harbour appx 1 1/4 mi. N of Sidney. This is the first marina on the port side upon entering the harbour & offers complete repair & shipyard services. Guest moorage consists of open slips primarily on "D" docks.

GUEST BOAT CAPACITY: ....Appx. 75 boats
DOCKSIDE DEPTH AT ZERO TIDE: ....18 ft.
SEASON: ....All year
RESERVATION POLICY: ....Accepts
AMT W/ELECTRICITY: ....All
FUEL DOCK: ....Gas & Diesel
MARINE REPAIRS: ....Excellent service
TOILETS: ....Yes
HOT SHOWERS: ....Yes
RESTAURANT: ....Yes
PICNIC AREA: ....Close by
BASIC STORE: ....Close by
BROADBAND/WI-FI: ....BroadbandXpress
DAILY RATE: ....Premium (Over $1.25/foot)
Note: **Discounted winter rates available**

GUEST DOCK: ....Slips only
GUEST SLIPS: ....Varies
WATER: ....Yes
AMPS: 15-30-50-100A
PUMP OUT STATION ....Yes
HAUL OUT: ....Yes
BOAT RAMP: ....Yes
LAUNDRY: ....Yes
BAR: ....Yes
POOL: ....Close by
GOLF: ....Close by
PET FRIENDLY: ....Good
OTHER: Customs port-of-entry, fishing-marine supplies, close to shopping & services in Sidney, Internet broadband.

CAUTION! This chartlet not intended for use in navigation.

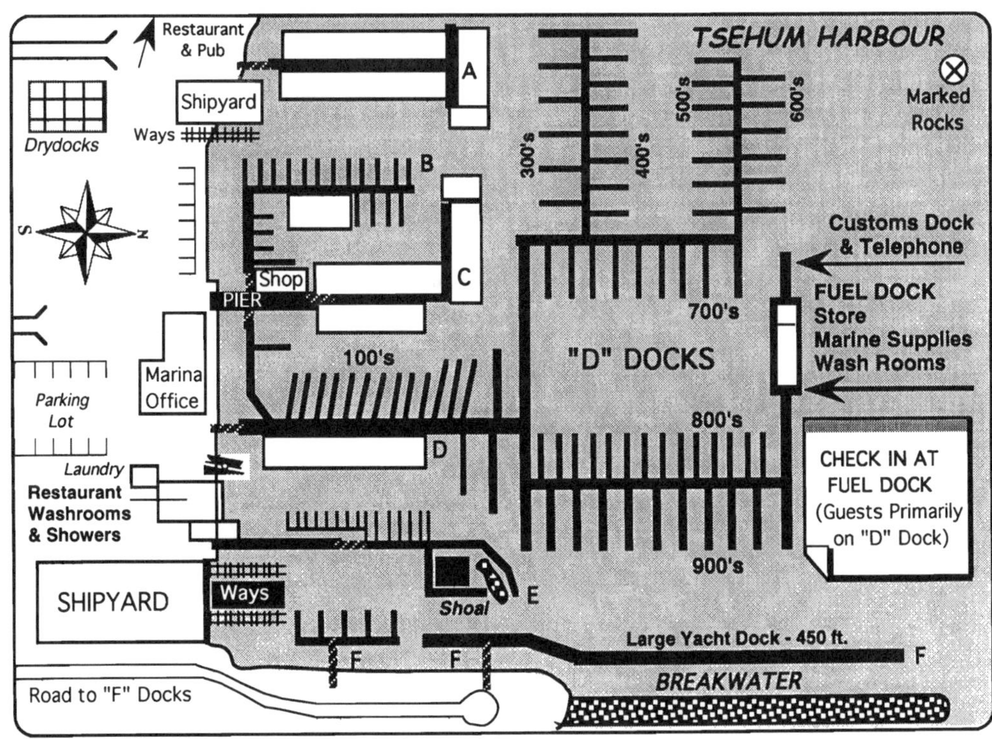

Sidney

NAME OF YACHT CLUB: *SIDNEY NORTH SAANICH YACHT CLUB*

ADDRESS: MAIL: P.O. Box 2521 Sidney, B.C. Canada V8L 4B9

TELEPHONE: 250-656-4600 PERSON IN CHARGE: Recip. Director

SHORT DESCRIPTION & LOCATION: **www.snsyc.ca**

48°40.40'-123°25.0' Located in Tsehum Harbour on the N side of Blue Heron Basin, SNSYC offers reciprocal moorage at N. Saanich Marina on the inside of a big 260 ft. dock. Register upon arrival as per instructions on visitors information posted on the Shed on the reciprocal dock. Vacant member slips in marina can also be assigned by foreshoreman.

RECIPROCAL BOAT CAPACITY: ....Appx. 10-15

DOCKSIDE DEPTH AT ZERO TIDE: ....8 ft.

SEASON: ....All year

RESERVATION POLICY: ....None

TOILETS: ....Yes

HOT SHOWERS: ....Yes

RESTAURANT: ......Yes-Open Wed. thru Sun. eves

BROADBAND/WI-FI: ....None

DAILY RATE: Free for 48 hrs. - 50¢ /ft. (CDN) after 48 hrs. Power charge $3/day ....(Key Deposit $25.00)

RECIPROCAL DOCK: Appx. 260'

RECIPROCAL SLIPS: ....Varies

WATER: ....Yes

AMT W/ELECTRICITY: ....All

AMPS: ....15 A

BAR: ....Yes

PET FRIENDLY: ....Excellent

OTHER: Bar open Wed-Sun, bus & taxi available to town. Rafting OK-2 boats deep.

***NOTE:* THIS IS PRIVATE MOORAGE AND ONLY AVAILABLE TO MEMBERS OF RECIPROCAL YACHT CLUBS. YOUR CLUB *MUST* HAVE RECIPROCAL PRIVILEGES AND YOU MUST FLY YOUR BURGEE.**

CAUTION! This chartlet not intended for use in navigation.

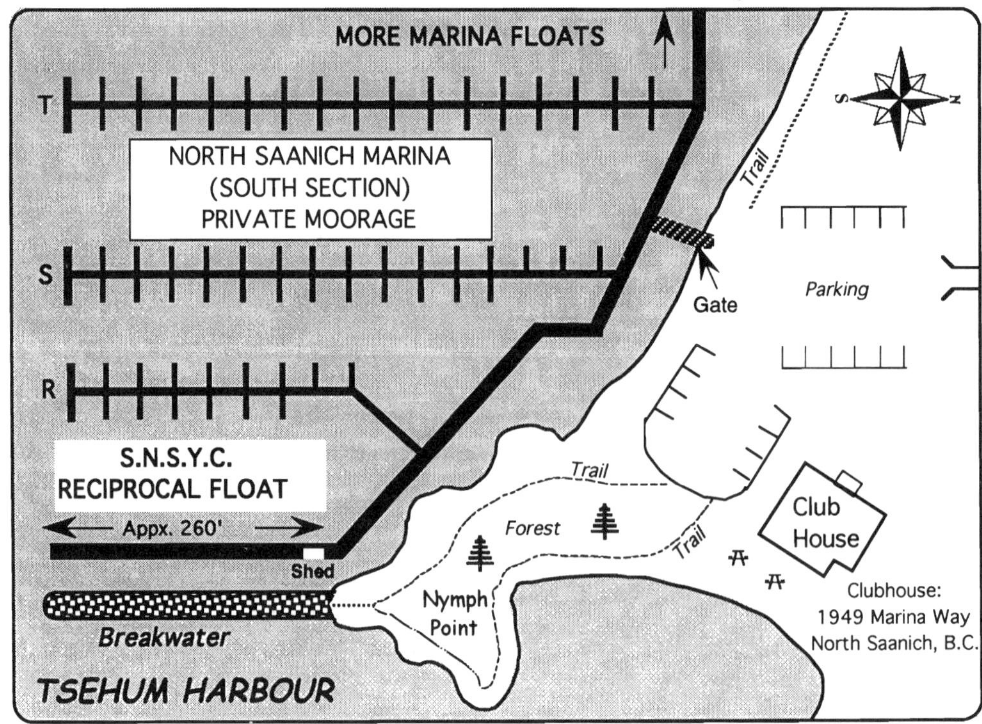

Sidney

NAME OF YACHT CLUB: ***CAPITAL CITY YACHT CLUB***

ADDRESS: 10630 Blue Heron Rd. RR#3 Sidney, B.C. Canada V8L 5S6

TELEPHONE: 250-656-4512 **PERSON IN CHARGE:** Director

SHORT DESCRIPTION & LOCATION:

48°40.30' - 123°25.10' Located in Tsehum Hbr, Blue Heron Basin. CCYC provides a 100 ft. Visitor's Dock plus open member slips for reciprocals. Check in at visitors dock (painted red), go to visitors registration booth (top of B Dock) to see list of available slips, or stay at visitors dock if late & no other slip is available. Ask any member for key.

RECIPROCAL BOAT CAPACITY:.....6-10 (Varies)

DOCKSIDE DEPTH AT ZERO TIDE: ....5 ft.

SEASON: ....All year

RESERVATION POLICY: ....None

TOILETS: ....Yes

HOT SHOWERS: ....Yes

RESTAURANT: ....Close by-Appx. 1 mile

BROADBAND/WI-FI: ....None

DAILY RATE: 2 Days Free twice/year. After 48 hrs - 50¢/ft. (Limit 7 days/year). Power - $3.00/day.

RECIPROCAL DOCK: ....100 ft.

RECIPROCAL SLIPS: ....Varies

WATER: ....Yes

AMT W/ELECTRICITY: ....All

AMPS: ....30 A

BAR: ....None

PET FRIENDLY: ....Good

OTHER: Water taxi to downtown Sidney. BBQ available. **NO BOATS OVER 50' LONG & 14.5' BEAM**

NOTE: THIS IS PRIVATE MOORAGE AND ONLY AVAILABLE TO MEMBERS OF RECIPROCAL YACHT CLUBS. YOUR CLUB MUST HAVE RECIPROCAL PRIVILEGES AND YOU MUST FLY YOUR BURGEE.

CAUTION! This chartlet not intended for use in navigation.

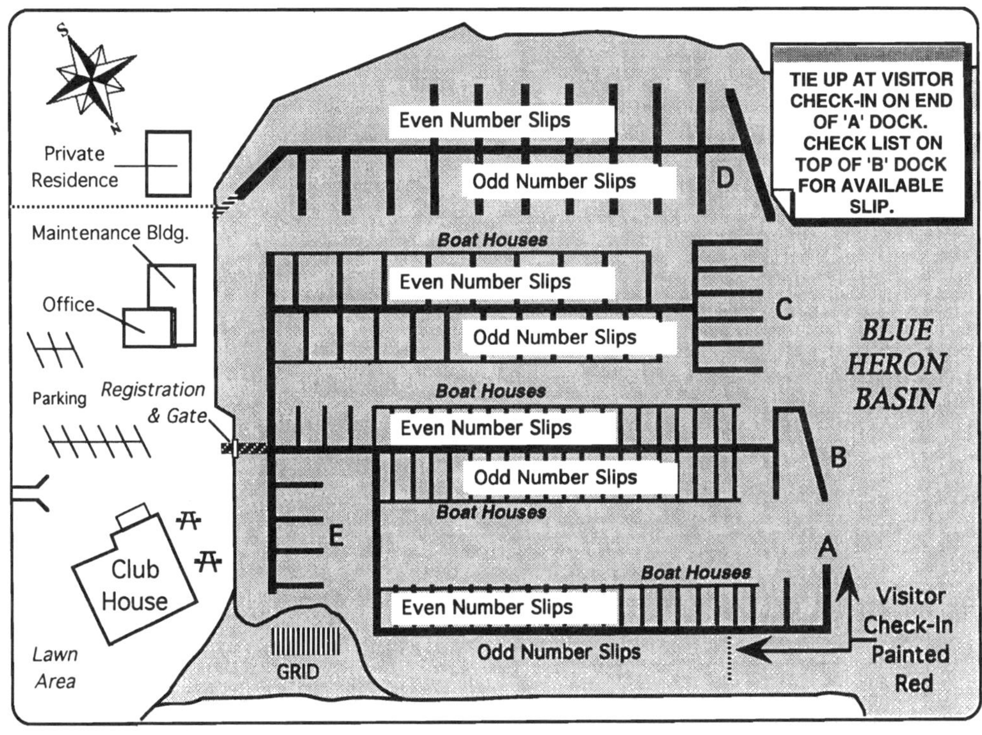

NOTES

Sidney

NAME OF MARINA: *CANOE COVE MARINA* **RADIO:** VHF 66A
TELEPHONE: **250-656-5566** **MGR:** Terry Crawford, Harbourmaster
E-MAIL: wharfage@canoecovemarina.com **FAX:** 250-655-7197 (Fax)
ADDRESS: 2300 Canoe Cove Rd. Sidney, B.C., Canada V8L 3X9
SHORT DESCRIPTION & LOCATION: **www.canoecovemarina.com**

48°41.00' - 123°24.30' Located in beautiful Canoe Bay appx. 2 nautical miles north of Sidney. One of the largest full service boat repair facilities in B.C. Walking distance to Swartz Bay ferry terminal. Guest moorage is available in vacated permanent tenant slips.

GUEST BOAT CAPACITY: Space available basis
DOCKSIDE DEPTH AT ZERO TIDE: 5-10 ft.**
SEASON: .... All year
RESERVATION POLICY: .... Recommended
AMT W/ELECTRICITY: .... All
FUEL DOCK: .... Gas, Dsl, & LP
MARINE REPAIRS: .... On premises
TOILETS: .... Yes
HOT SHOWERS: .... Yes
RESTAURANT: .... Cafe & English Style Pub
PICNIC AREA: .... None
BASIC STORE: .... None
BROADBAND/WI-FI: .... None
DAILY RATE: .... Moderate (75¢-$1.25/foot)
****DEPTH AT ZERO TIDE:** 5' AT FUEL DOCK, 10' AT CUSTOMS DOCK

GUEST DOCK: None-slips only
GUEST SLIPS: Space available
WATER: .... Yes
AMPS: .... 15-30 A
PUMP OUT STATION .... None
HAUL OUT: .... Travel-Lift
BOAT RAMP: .... None
LAUNDRY: .... Yes
BAR: .... Pub
POOL: .... None
GOLF: .... None
PET FRIENDLY: .... Good
OTHER: Marine chandlery & hardware store, complete marine repair yard & ship wright, harbour taxi. Customs port of entry.

CAUTION! This chartlet not intended for use in navigation.

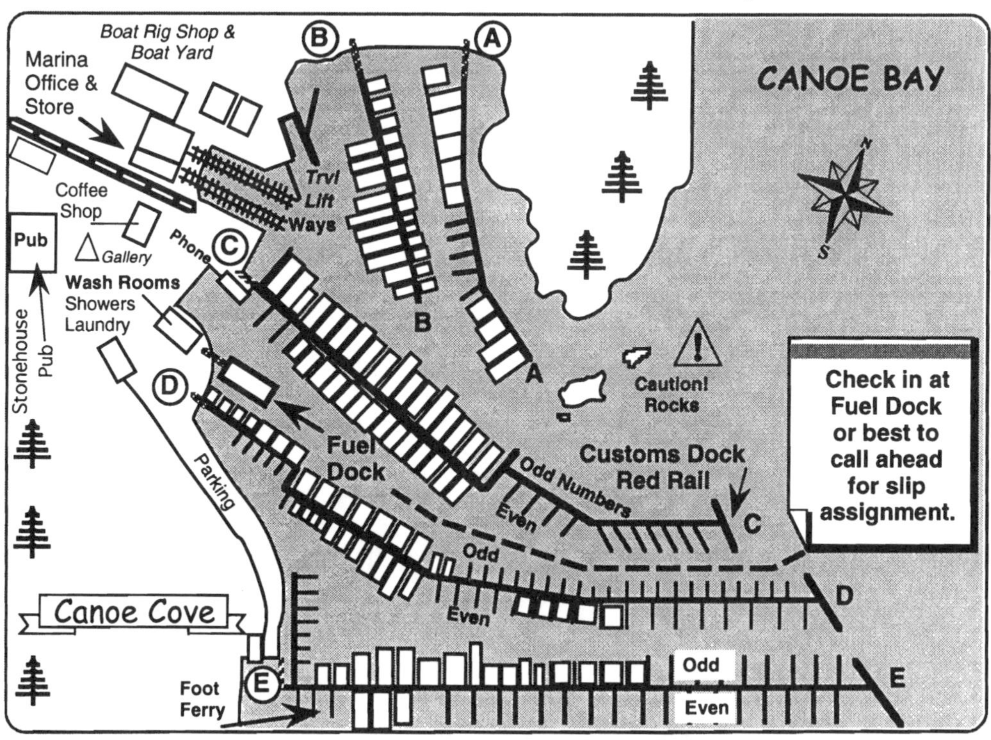

Brentwood Bay

NAME OF MARINA: ***BRENTWOOD BAY MARINA*** RADIO: VHF 66A
TELEPHONE: 250-652-3151 MGR: Matt Smiley
E-MAIL: marina@brentwoodbaylodge.com FAX: 250-652-3151
ADDRESS: 849 Verdier Brentwood Bay, B.C. Canada V8M 1C5
SHORT DESCRIPTION & LOCATION: **www.brentwoodbaylodge.com**

48°34.65' - 123°27.90' This is one of the nearest marinas to Butchart Gardens, located at head of Brentwood Bay on E side of Saanich Inlet on Vancouver Is. Upscale Restaurant, Pub, Spa, and Lodge. Uphill 1 mile walk or bus to supermarket, Post Office, & services.

GUEST BOAT CAPACITY: ....Appx. 15-30 boats
DOCKSIDE DEPTH AT ZERO TIDE: ....12 ft.
SEASON: ....All year
RESERVATION POLICY: ....Accepts
AMT W/ELECTRICITY: ....All
FUEL DOCK: ....None
MARINE REPAIRS: ....Close by
TOILETS: ....Yes
HOT SHOWERS: ....Yes
RESTAURANT: Fine Dining and Casual Cafe/Pub
PICNIC AREA: ....Yes
BASIC STORE: ....Yes
BROADBAND/WI-FI: ....BroadbandXpress
DAILY RATE: ....Premium (Over $1.25/foot)

***MOBILE PUMPOUT BOAT "PUMPTY DUMPTY" ON CALL

GUEST DOCK: 3 @ Appx. 100 ft.
GUEST SLIPS: ....10 plus
WATER: ....Yes
AMPS: ....30-50 A
PUMP OUT STATION Mobile***
HAUL OUT: ....Close by
BOAT RAMP: ....Close by
LAUNDRY: ....Yes
BAR: ....Pub
POOL: ....Adult Pool
GOLF: ....Close by
PET FRIENDLY: ....Good
OTHER: Lodge, spa, taxi & bus service. Scuba activities. Close by: Post Office, Liq. Store, shopping, & doctor. Water Taxi to Butchart.

CAUTION! This chartlet not intended for use in navigation.

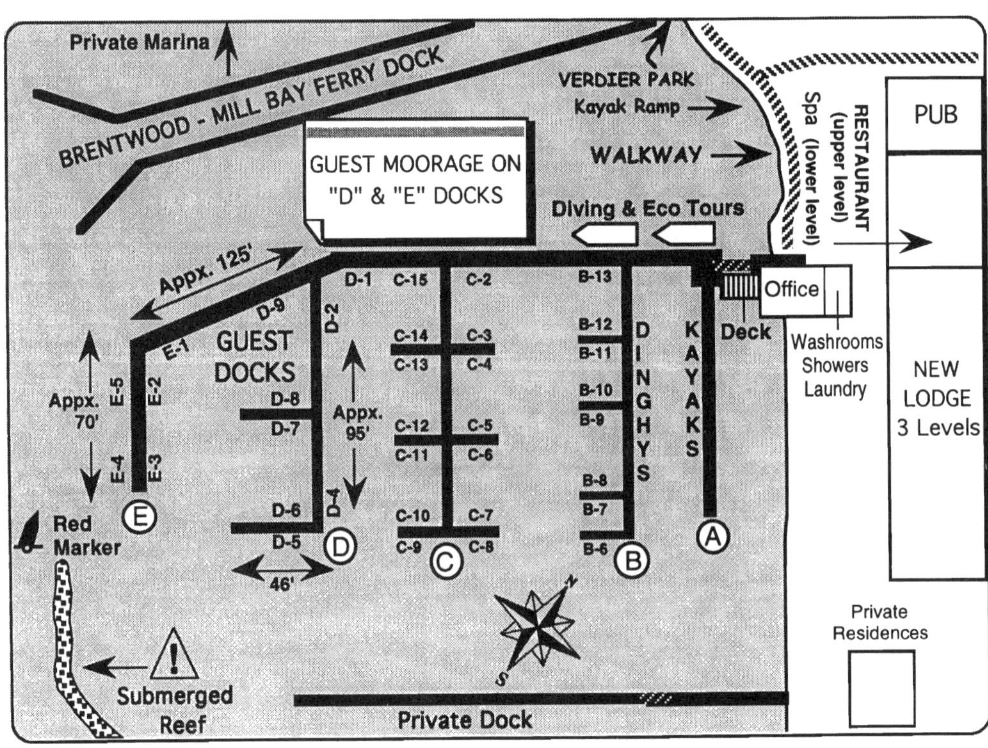

Mill Bay

NAME OF MARINA: *MILL BAY MARINA* RADIO: VHF 66 A
TELEPHONE: 250-743-4112 MGR: Jim Young
E-MAIL: FAX: 250-743-4122
ADDRESS: Box 231, 740 Handy Road, Mill Bay, B.C. Canada V0R 2P0

SHORT DESCRIPTION & LOCATION:

48°.39.10 - 123°.33.10 Located in Mill Bay on W side of Saanich Inlet on Vancouver Isl. 25 Miles N. of Victoria. Small friendly marina is within walking distance of modern & large shopping centre w/ supermarket, liquor store, bank, & shops. 3 Miles N of Mill Bay ferry dock.

GUEST BOAT CAPACITY: ....Appx. 20-25 boats
DOCKSIDE DEPTH AT ZERO TIDE: ......20 ft. +
SEASON: ....All year
RESERVATION POLICY: ....Accepts
AMT W/ELECTRICITY: ....All
FUEL DOCK: ....Gas & Diesel
MARINE REPAIRS: ....None
TOILETS: ....Yes
HOT SHOWERS: ....Yes
RESTAURANT: ....Close by
PICNIC AREA: ....Yes
BASIC STORE: ....Yes
BROADBAND/WI-FI: ....None
DAILY RATE: ....Moderate (75¢-$1.25/foot)

NOTE: MARINA & FORESHORE DEVELOPMENT IN PROGRESS. CHANGES ARE IMMINENT.

GUEST DOCK: ....250 ft.
GUEST SLIPS: ....Varies
WATER: ....Yes
AMPS: ....15 A
PUMP OUT STATION ....None
HAUL OUT: ....None
BOAT RAMP: ....Yes
LAUNDRY: ....Yes
BAR: ....Pub close by
POOL: ....None
GOLF: ....Close by
PET FRIENDLY: ....Good
OTHER: Chandlery, fishing tackle, prawning traps & gear, charts, & marine supplies. Shopping centre nearby.

CAUTION! This chartlet not intended for use in navigation.

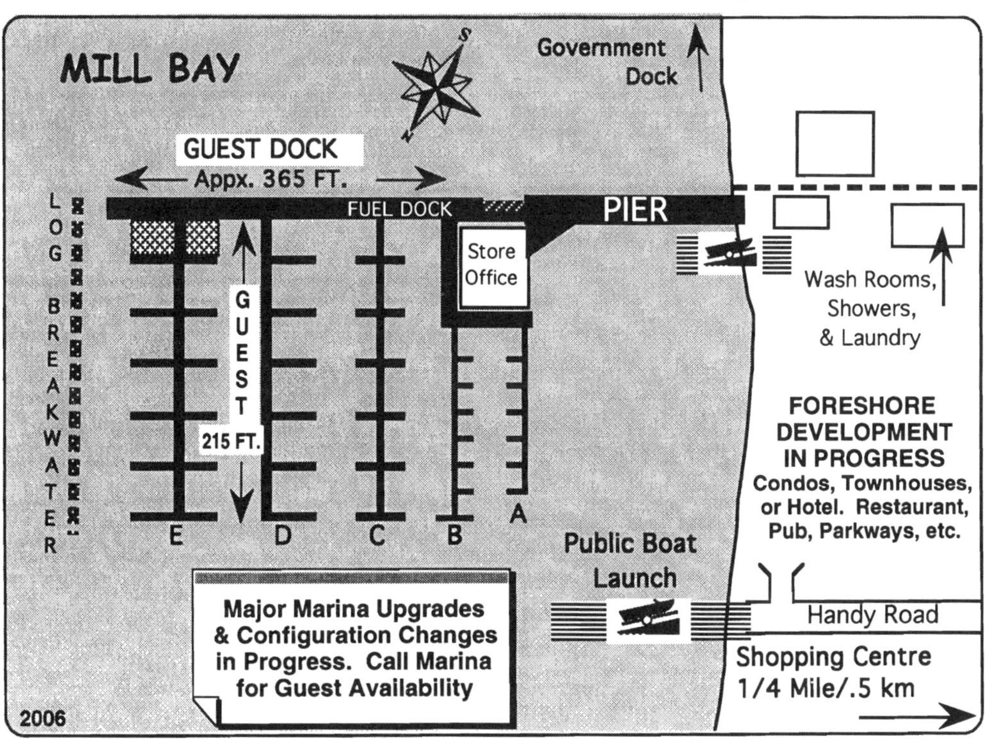

Cowichan Bay

NAME OF MARINA: *COWICHAN BAY FISHERMEN'S WHARF* RADIO: VHF 66A
TELEPHONE: 250-746-5911 MGR: Chuck Von-Haas Cell 250.755.6763
E-MAIL: cbfwa@shaw.ca FAX: 250-701-0729 (Fax)
ADDRESS: P.O. Box 52 Cowichan Bay, B.C. Canada V0R 1N0
SHORT DESCRIPTION & LOCATION: **www.haa.bc.ca**

48°44.29' - 123°37.10' Located at far E. end of town, main public moorage is behind a new floating breakwater & operated by the Wharf Ass'n. Cow. Bay is a picturesque town famous for Cowichan sweaters. Improvements planned for upgrading & expansion of docks.

GUEST BOAT CAPACITY: ....Appx. 35-40 boats
DOCKSIDE DEPTH AT ZERO TIDE: Appx. 15'
SEASON: ....All year
RESERVATION POLICY: ....None
AMT W/ELECTRICITY: ....All
FUEL DOCK: ....Close by
MARINE REPAIRS: ....Close by
TOILETS: ....Yes
HOT SHOWERS: ....Yes
RESTAURANT: ....Close by
PICNIC AREA: ....Close by
BASIC STORE: ....Close by
BROADBAND/WI-FI: Internet terminal in office
DAILY RATE:........Moderate (75¢-$1.25/foot)

FISHING BOATS HAVE PRIORITY

GUEST DOCK: Appx. 1000 ft.
GUEST SLIPS: ....Docks only
WATER: ....Yes
AMPS: ....20 A
PUMP OUT STATION ....Yes
HAUL OUT: ....Close by
BOAT RAMP: ....Close by
LAUNDRY: ....Yes
BAR: ....Close by
POOL: ....Close by
GOLF: ....Close by
PET FRIENDLY: ....Fair
OTHER: Museum, Wooden Boat Foundation, public eco-station, fish market, liquor store, several good restaurants and pubs.

CAUTION! This chartlet not intended for use in navigation.

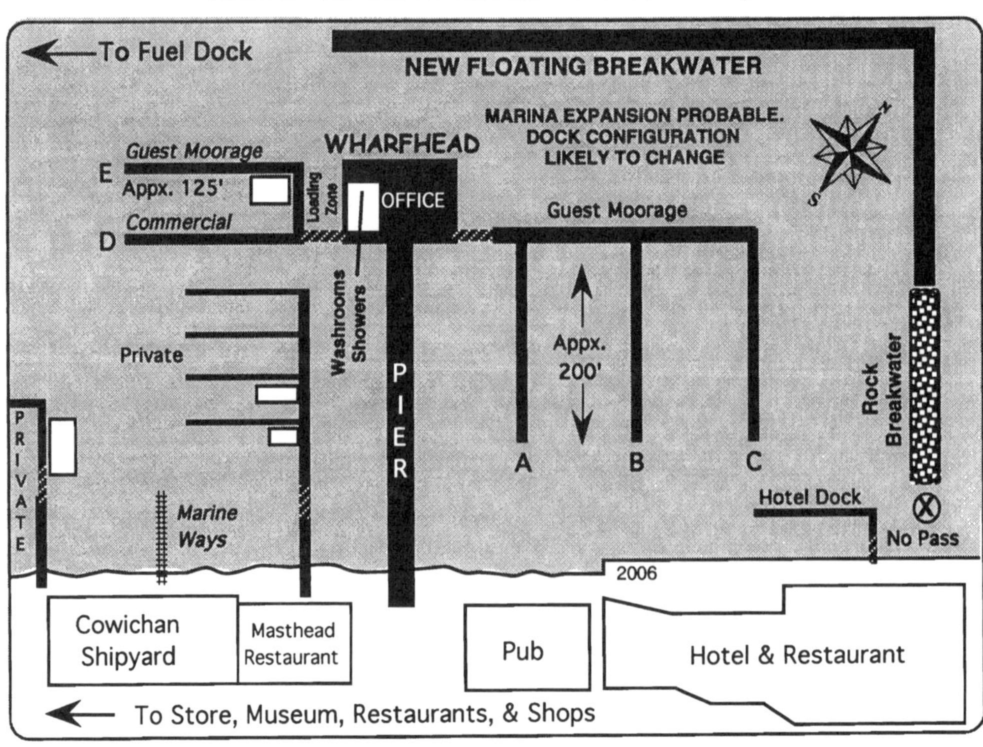

Cowichan Bay

NAME OF MARINA: ***DUNGENESS MARINA*** **RADIO:** VHF 66A
TELEPHONE: **250-748-6789** MGR: Bob Hokanson
E-MAIL: info@dungenessmarina.com FAX: 250-748-9869 (Fax)
ADDRESS: Box 51 Cowichan Bay, B.C. Canada V0R 1N0
SHORT DESCRIPTION & LOCATION: **www.dungenessmarina.com**

48°44.27' - 123°37.17' Newly updated first class small marina located in the middle of town in between the Fuel Dock & Maritime Centre. Cowichan Bay is a picturesque little town, rich with maritime history, with several shops and services for boaters.

GUEST BOAT CAPACITY: ....Appx. 4-6 boats
DOCKSIDE DEPTH AT ZERO TIDE: 8 Ft. plus
SEASON: ....All year
RESERVATION POLICY: ....Accepts
AMT W/ELECTRICITY: ....All
FUEL DOCK: ....Next door
MARINE REPAIRS: ....Close by
TOILETS: ....Yes
HOT SHOWERS: ....Yes
RESTAURANT: ....Yes
PICNIC AREA: ....None
BASIC STORE: ....Close by
BROADBAND/WI-FI: ....None
DAILY RATE: ....Moderate (75¢-$1.25/foot)

GUEST DOCK: .....92 Ft. plus slips
GUEST SLIPS: ....Varies
WATER: ....Yes
AMPS: ....30 A
PUMP OUT STATION ....Yes
HAUL OUT: ....Close by
BOAT RAMP: ....Close by
LAUNDRY: ....Close by
BAR: ....Pubs close by
POOL: ....Close by
GOLF: ....Close by
PET FRIENDLY: ....Fair
OTHER: Close to Museum, markets, liquor store, post office, several good pubs & restaurants. Telephone & cable TV on docks.

CAUTION! This chartlet not intended for use in navigation.

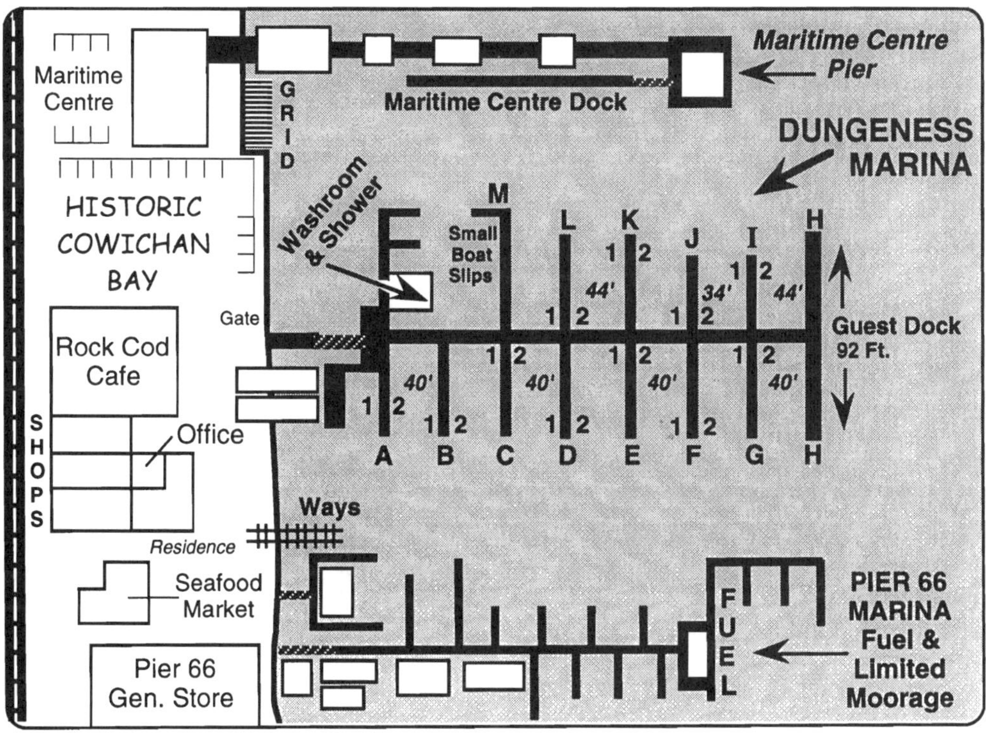

Genoa Bay

NAME OF MARINA: ***GENOA BAY MARINA*** **RADIO:** VHF 66 A
TELEPHONE: **800-572-6481** MGR: Will & Ben Kiedash
E-MAIL: info@genoabaymarina.com FAX: 250-746-7621 (Fax)
ADDRESS: 5100 Genoa Bay Rd. Unit 1 Duncan. B.C. Canada V9L 5Y8
SHORT DESCRIPTION & LOCATION: **www.genoabaymarina.com**

48°45.60' - 123°35.60' Marina is protected behind a little peninsula in SW Genoa Bay which opens to the N. end of Cowichan Bay on Vancouver Is. The family oriented facility has a well stocked store & offers quiet relaxation in the picturesque bay. Very friendly staff.

GUEST BOAT CAPACITY: ....Appx. 30-40 boats
DOCKSIDE DEPTH AT ZERO TIDE: ....20 ft.+
SEASON: ....All year
RESERVATION POLICY: ....Accepts
AMT W/ELECTRICITY: ....All
FUEL DOCK: ....Close by
MARINE REPAIRS: ....Close by
TOILETS: ....Yes
HOT SHOWERS: ....Yes
RESTAURANT: ....Yes
PICNIC AREA: ....Yes
BASIC STORE: ....Yes
BROADBAND/WI-FI: ....None
DAILY RATE: ....Moderate (75¢-$1.25/foot)

NOTE: SHUTTLE SERVICE TO GOLF & FARMERS MARKET

GUEST DOCK: ....1200 ft. Total
GUEST SLIPS: ....3 floats
WATER: ....Yes
AMPS: ....15-30 A
PUMP OUT STATION ....None
HAUL OUT: ....None
BOAT RAMP: ....Yes
LAUNDRY: ....Yes
BAR: ....Licenced Restaurant
POOL: ....None
GOLF: ....Close by
PET FRIENDLY: ....Good
OTHER: Crabbing, walking trails, fishing tackle, fresh baked goods. caters to clubs. Art Gallery. Genoa Bay Cafe known for fine dining.

CAUTION! This chartlet not intended for use in navigation.

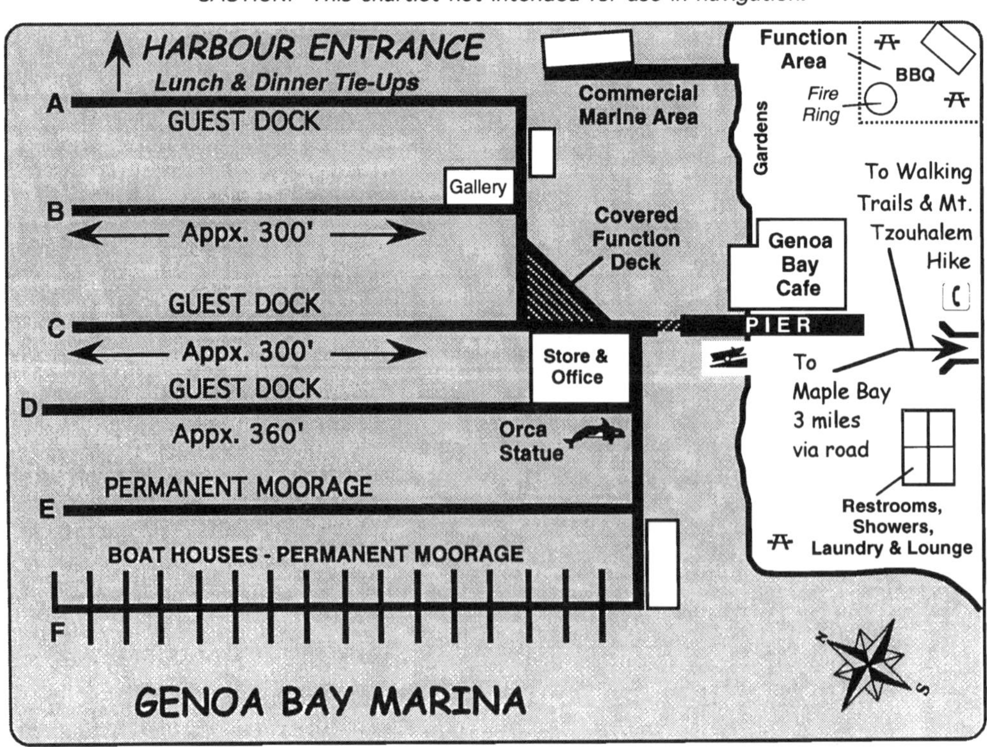

Maple Bay

NAME OF MARINA: ***MAPLE BAY PUBLIC WHARF*** **RADIO:** None
TELEPHONE: **250-715-8186** **MGR:** Harmen Bootsma
E-MAIL: hbootsma@shaw.ca **FAX:**
ADDRESS: P,O, Box 193 Chemainus, B.C. Canada V0R 1K0

SHORT DESCRIPTION & LOCATION:

48°49.20' - 123°36.15' Public Wharf and Village lie in the N.W. corner of Maple Bay. Public moorage is somewhat exposed, but can be excellent in settled weather conditions. Area popular with kayakers and swimmers in summer. Excellent Pub just to the north.

GUEST BOAT CAPACITY:	Appx. 4-6 boats	**GUEST DOCK:**	Appx. 250 ft TTL
DOCKSIDE DEPTH AT ZERO TIDE:	Appx. 6 ft.	**GUEST SLIPS:**	Dock only
SEASON:	All year	**WATER:**	Yes
RESERVATION POLICY:	None	**AMPS:**	None
AMT W/ELECTRICITY:	None	**PUMP OUT STATION**	None
FUEL DOCK:	Close by	**HAUL OUT:**	Close by
MARINE REPAIRS:	Close by	**BOAT RAMP:**	Yes
TOILETS:	None	**LAUNDRY:**	Close by
HOT SHOWERS:	None	**BAR:**	Pub close by
RESTAURANT:	Close by	**POOL:**	None
PICNIC AREA:	Yes	**GOLF:**	Close by
BASIC STORE:	Yes	**PET FRIENDLY:**	Good
BROADBAND/WI-FI:	None	**OTHER:**	Swimming beach, park, tennis courts close by. Excellent Pub nearby.
DAILY RATE:	Economical (Under 75¢/foot)		

CAUTION! This chartlet not intended for use in navigation.

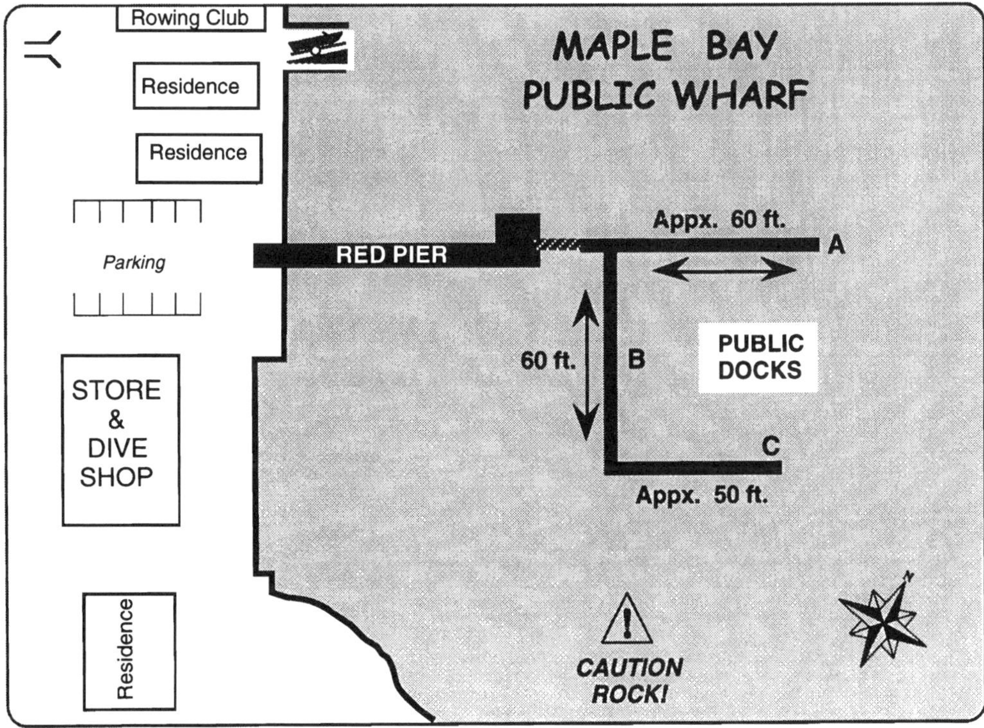

Maple Bay

NAME OF YACHT CLUB: *MAPLE BAY YACHT CLUB*

ADDRESS: 6337 Genoa Bay Road Duncan, B.C. Canada V9L 5Y4

TELEPHONE: 250-746-4521 **PERSON IN CHARGE:** Dockmaster

SHORT DESCRIPTION & LOCATION: **www.mbyc.bc.ca**

48°48.18' - 123°36.13' The Y.C. is located in Birds Eye Cove in Maple Bay about 1 mi N of the Maple Bay Marina. Tie up on Breakwater (BW) Float & sign in at Club House or w/Dockmaster. Moorage allowed on outside of Breakwater Float only, unless inside slip authorized by Dockmaster. No vessels over 60 ft. Note: 100' on N. end of BW reserved.

RECIPROCAL BOAT CAPACITY: ....Appx. 6-10

DOCKSIDE DEPTH AT ZERO TIDE: Appx. 50'

SEASON: ....All year except July 1 & Labour Day

RESERVATION POLICY: ....None

TOILETS: ....Yes

HOT SHOWERS: ....Yes

RESTAURANT: ....Close by

BROADBAND/WI-FI: ....In Clubhouse

DAILY RATE: 2 Days per year Free. Facilities charge $5/ea. day. Add'l nights 50¢/ft. Max. stay: 7 Days/year.

RECIPROCAL DOCK: ....400 ft.

RECIPROCAL SLIPS: Dock only

WATER: ....Yes

AMT W/ELECTRICITY: ....All

AMPS: ....15 A

BAR: ....Yes

PET FRIENDLY: ....Good

OTHER: Boat repairs & fuel close by. Clubhouse & bar open daily. Comfortable lounge, big screen TV, Ice & BBQ.

***NOTE:* THIS IS PRIVATE MOORAGE AND ONLY AVAILABLE TO MEMBERS OF RECIPROCAL YACHT CLUBS. YOUR CLUB *MUST* HAVE RECIPROCAL PRIVILEGES AND YOU MUST FLY YOUR BURGEE.**

CAUTION! This chartlet not intended for use in navigation.

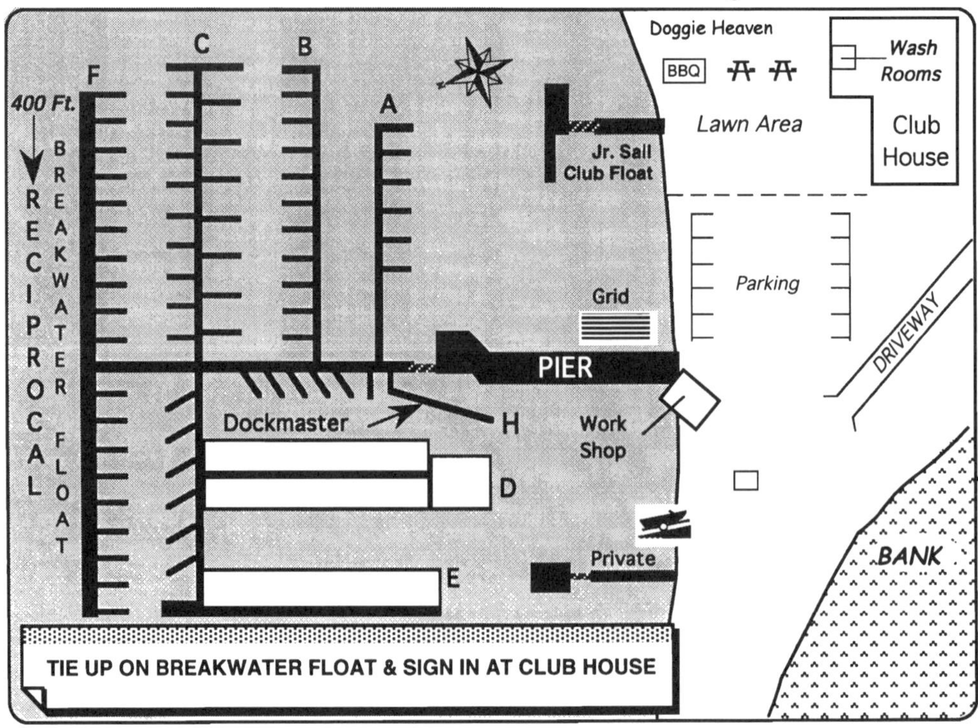

Maple Bay

NAME OF MARINA: **MAPLE BAY MARINA** RADIO: VHF 66 A
TELEPHONE: 250-746-8482 MGR: David & Carol Messier, Owners
E-MAIL: info@maplebaymarina.com FAX: 250-746-8490 (Fax)
ADDRESS: 6145 Genoa Bay Road, Duncan, B.C Canada V9L 5T7
SHORT DESCRIPTION & LOCATION: www.maplebaymarina.com

48°47.73' - 123°36.03' Full service marina situated in picturesque Birds Eye Cove located at head of Maple Bay. Marina plans many upgrades & reconstruction, so it's best to call ahead for slip assignment & directions. Excellent & very clean facilities, friendly staff.

GUEST BOAT CAPACITY: ....Appx. 60 boats
DOCKSIDE DEPTH AT ZERO TIDE: ....7 ft. +
SEASON: ....All year
RESERVATION POLICY: ....Recommended
AMT W/ELECTRICITY: ....All
FUEL DOCK: ....Gas, Dsl, CNG & LP
MARINE REPAIRS: ....On premises
TOILETS: ....Yes
HOT SHOWERS: ....Yes
RESTAURANT: ....Shipyard Pub & Restaurant
PICNIC AREA: ....Yes
BASIC STORE: ....Yes
BROADBAND/WI-FI: ....BroadbandXpress
DAILY RATE: ....Moderate (75¢-$1.25/foot)

GUEST DOCK: ....85' plus slips
GUEST SLIPS: 60
WATER: ....Yes
AMPS: ....15-30-50 A
PUMP OUT STATION ....None
HAUL OUT: ....Travel-Lift
BOAT RAMP: ....Close by
LAUNDRY: ....Yes
BAR: ....Pub
POOL: ....None
GOLF: ....Close by
PET FRIENDLY: ....Excellent
OTHER: Well stocked general store, charts, chandlery, gift shop, car rentals. Shipwright & boat yard. Internet service.

CAUTION! This chartlet not intended for use in navigation.

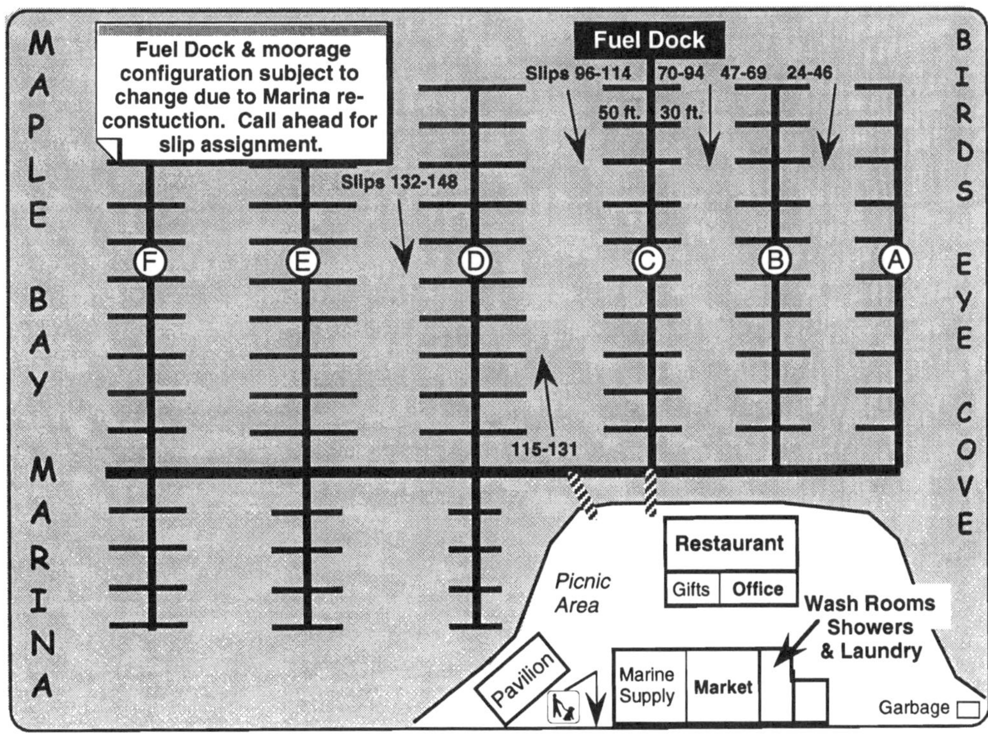

Crofton

NAME OF MARINA: ***CROFTON PUBLIC WHARF*** **RADIO:** Cell Phone
TELEPHONE: **250-246-4655** **MGR:** Harmen Bootsma 250-715-8186
E-MAIL: hbootsma@shaw.ca **FAX:** 250-246-1398 (Fax)
ADDRESS: P.O. Box 193 Chemainus, B.C. Canada VOR 1KO
SHORT DESCRIPTION & LOCATION: **www.dfo-mpo.gc.ca/sch/HB_BC_e.asp**

48°52.00' - 123°38.20' Crofton is located on Van. Is. in Osborne Bay just off Stuart Channel well marked by the huge Crofton Pulp and Paper Mill. The public dock is located behind the breakwater next to the ferry landing. **PRIMARILY A FISH BOAT FACILITY.**

GUEST BOAT CAPACITY: ....Appx. 20-25 boats
DOCKSIDE DEPTH AT ZERO TIDE: ....8 ft.
SEASON: ....All year
RESERVATION POLICY: ....None
AMT W/ELECTRICITY: ....All
FUEL DOCK: ....None
MARINE REPAIRS: ....None
TOILETS: ....Yes - Key at ferry pay booth
HOT SHOWERS: .....Yes - Key at ferry pay booth
RESTAURANT: ....Close by
PICNIC AREA: ....Yes
BASIC STORE: ....Close by
BROADBAND/WI-FI: ....None
DAILY RATE: ....Economical (Under 75¢/foot)

GUEST DOCK: Appx. 500 ft. TTL
GUEST SLIPS: ....Docks only
WATER: ....Yes
AMPS: ....20-30 A
PUMP OUT STATION ....None
HAUL OUT: ....None
BOAT RAMP: ....Yes
LAUNDRY: ....Close by
BAR: ....Close by
POOL: ....Close by
GOLF: ....Close by
PET FRIENDLY: ....Excellent
OTHER: Walking distance to Town offering groceries, restaurants, fishing supplies. Nearby playground, Rec Area & Trails.

CAUTION! This chartlet not intended for use in navigation.

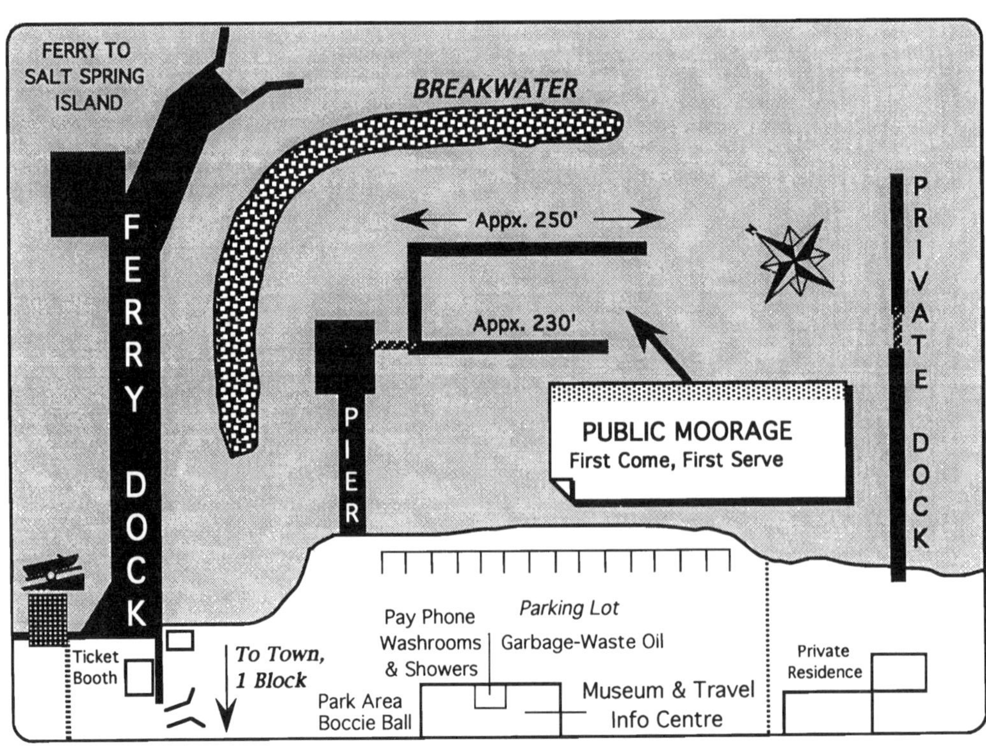

Chemainus

NAME OF MARINA: ***CHEMAINUS MUNICIPAL DOCK*** **RADIO:** VHF 66 A
TELEPHONE: **Cell Phone:** 250-715-8186 **MGR:** Harmen Bootsma
E-MAIL: **Office:** 250-246-4655 **FAX:** 250-246-1398 (Fax)
ADDRESS: P.O. Box 193 Chemainus, B.C. Canada V0R 1KO
SHORT DESCRIPTION & LOCATION: **www.chemainus.bc.ca**

48°55.50' - 123°42.70' Chemainus is located on the W side of Stuart Channel & is a tourist community with many attractions, especially the theatre and famous outdoor murals painted on buildings. The much improved public marina welcomes visiting overnight boaters.

GUEST BOAT CAPACITY: ....Appx. 20 boats
DOCKSIDE DEPTH AT ZERO TIDE: ......8-12 ft.
SEASON: ....All year
RESERVATION POLICY: ....Accepts
AMT W/ELECTRICITY: ....All
FUEL DOCK: ....Next door
MARINE REPAIRS: ....Close by
TOILETS: ....Yes
HOT SHOWERS: ....Yes
RESTAURANT: ....Close by
PICNIC AREA: ....Close by
BASIC STORE: ....Close by
BROADBAND/WI-FI: ....None
DAILY RATE: ....Moderate
(75¢-$1.25/foot)
$5-10 LANDING FEE
FOR DAY STOP

GUEST DOCK: ....700 ft. Total
GUEST SLIPS: ....12 slips
WATER: ....Yes
AMPS: ....30-50 A
PUMP OUT STATION ....None
HAUL OUT: ....Close by
BOAT RAMP: ....None
LAUNDRY: ....Close by
BAR: ....Close by
POOL: ....None
GOLF: ....Close by
PET FRIENDLY: ....Fair
OTHER: Guided tours of murals, walking & shopping tours, city parks, summer festivals, great theater, near hospital & groceries.

CAUTION! This chartlet not intended for use in navigation.

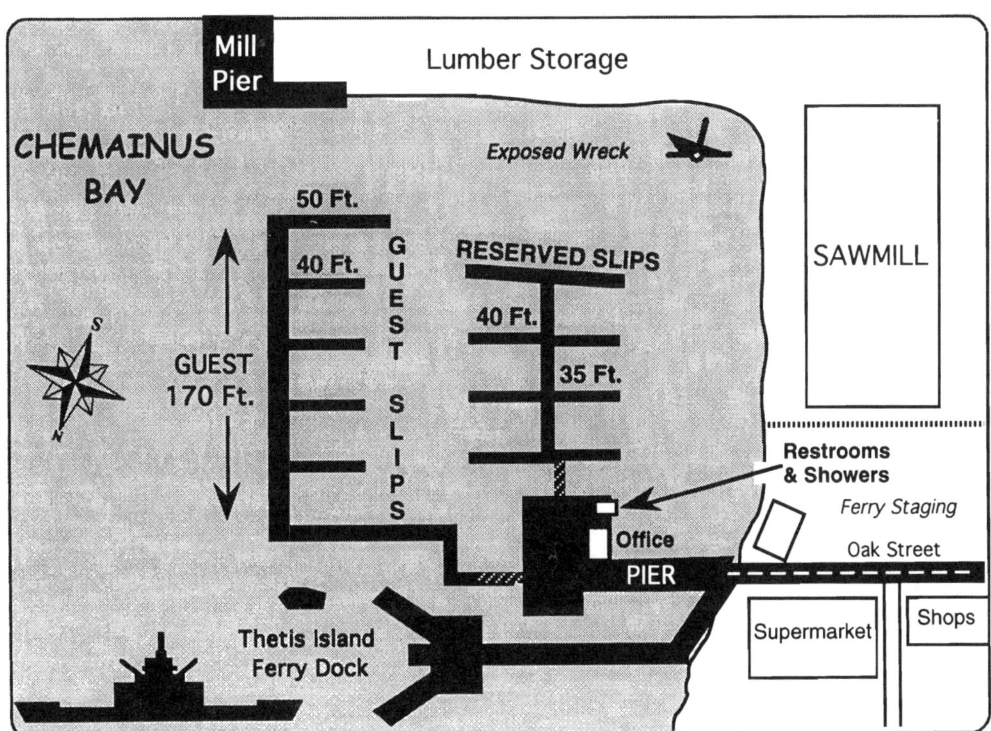

Ladysmith

NAME OF MARINA: ***LADYSMITH FISHERMAN'S WHARF*** RADIO: None
TELEPHONE: 250-245-7511 MGR: Ken Bryski
E-MAIL: lfwa@telus.net FAX: 250-245-7747 (Fax)
ADDRESS: P.O. Box 130 Ladysmith, B.C. Canada V9G 1A1
SHORT DESCRIPTION & LOCATION: **www.ladysmithfishermanswharf.com**

48°59.90' - 123°48.60' The Harbour Authority marina is located on SW side of Ladysmith Harbour appx. 6 miles N of Chemainus. This public wharf is the 2nd set of docks seen on W. side of harbour, behind rock breakwater. Fairly convenient location for shopping.

GUEST BOAT CAPACITY: ....Varies
DOCKSIDE DEPTH AT ZERO TIDE: ....12 ft.
SEASON: ....All year
RESERVATION POLICY: ....None
AMT W/ELECTRICITY: ....All
FUEL DOCK: ....Close by
MARINE REPAIRS: ....Close by
TOILETS: ....Yes
HOT SHOWERS: ....Yes
RESTAURANT: ....Close by
PICNIC AREA: ....Close by
BASIC STORE: Super Market has free shuttle
BROADBAND/WI-FI: ....Planned
DAILY RATE: ....Economical (Under 75¢/foot)

FISHING BOATS HAVE PRIORITY

GUEST DOCK: ....1100 ft. Total
GUEST SLIPS: ....6 large docks
WATER: ....Yes
AMPS: ....15-20 A
PUMP OUT STATION Planned
HAUL OUT: ....Tidal grid
BOAT RAMP: ....Yes
LAUNDRY: ....Yes
BAR: ....Close by
POOL: ....Close by
GOLF: ....Close by
PET FRIENDLY: ....Excellent
OTHER: Museum, shopping centre, restaurants & liquor store are all within walking distance or a short cab ride.

CAUTION! This chartlet not intended for use in navigation.

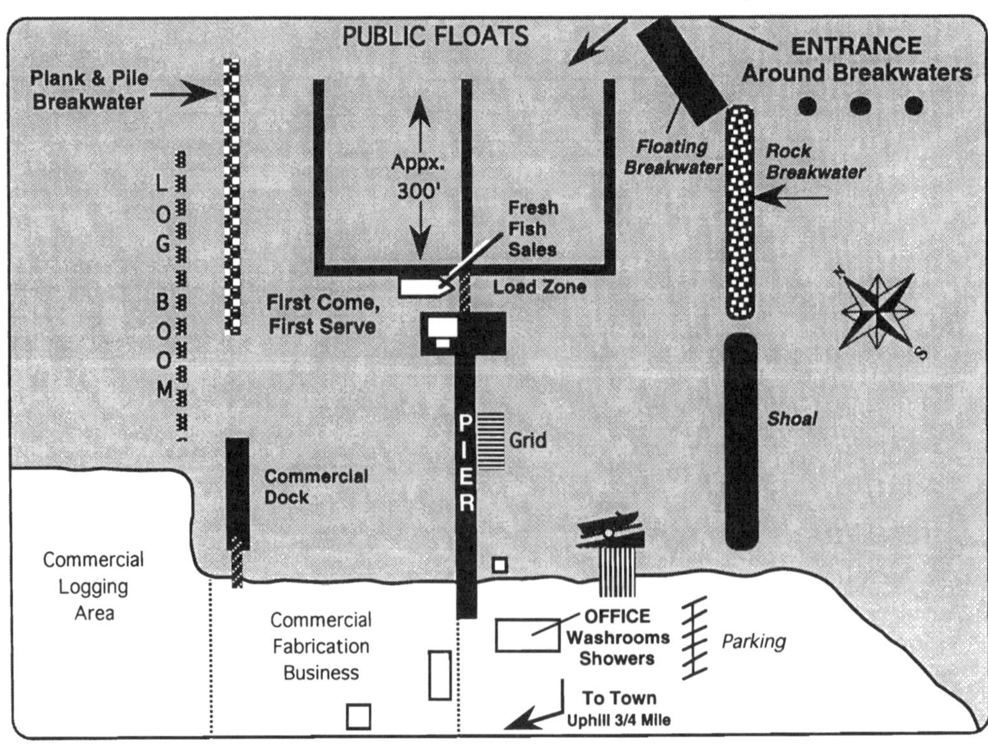

Ladysmith

NAME OF MARINA: *PAGE POINT INN & MARINA* *Formerly MAÑANA* **RADIO:** VHF 66 A
TELEPHONE: 250-245-2312 MGR: Lawrence & Lexie Lambert
E-MAIL: info@pagepointinn.com FAX: 250-245-7546 (Fax)
ADDRESS: 4760 Brenton-Page Rd. Ladysmith, B.C. Canada V9G 1L7
SHORT DESCRIPTION & LOCATION: **www.pagepointinn.com**

49°00.70' - 123°49.30' Located on NE side of Oyster (Ladysmith) Harbour across from town of Ladysmith at Page Pt. Recently upgraded docks. Many amenities offered for boaters. Friendly staff enhances the beautiful & secluded Vancouver Is. setting. Excellent restaurant.

GUEST BOAT CAPACITY: ....Appx. 20-30 boats
DOCKSIDE DEPTH AT ZERO TIDE: ....9 ft.
SEASON: ....All year
RESERVATION POLICY: ....Accepts
AMT W/ELECTRICITY: ....All
FUEL DOCK: ....Gas & Diesel
MARINE REPAIRS: ....Close by
TOILETS: ....Yes
HOT SHOWERS: ....Yes
RESTAURANT: ....Yes
PICNIC AREA: ....Yes
BASIC STORE: ....Close by
BROADBAND/WI-FI: ....None
DAILY RATE: ....Moderate (75¢-$1.25/foot)

GUEST DOCK: Appx. 1000 ft.
GUEST SLIPS: ....Varies
WATER: ....Limited
AMPS: ....15-30 A
PUMP OUT STATION ....None
HAUL OUT: ....Close by
BOAT RAMP: ....Close by
LAUNDRY: ....Yes
BAR: ....Yes
POOL: ....None
GOLF: ....Close by
PET FRIENDLY: ....Excellent
OTHER: B & B, cabins close by, fine dining, award winning chef. Nanaimo Airport close by.

CAUTION! This chartlet not intended for use in navigation.

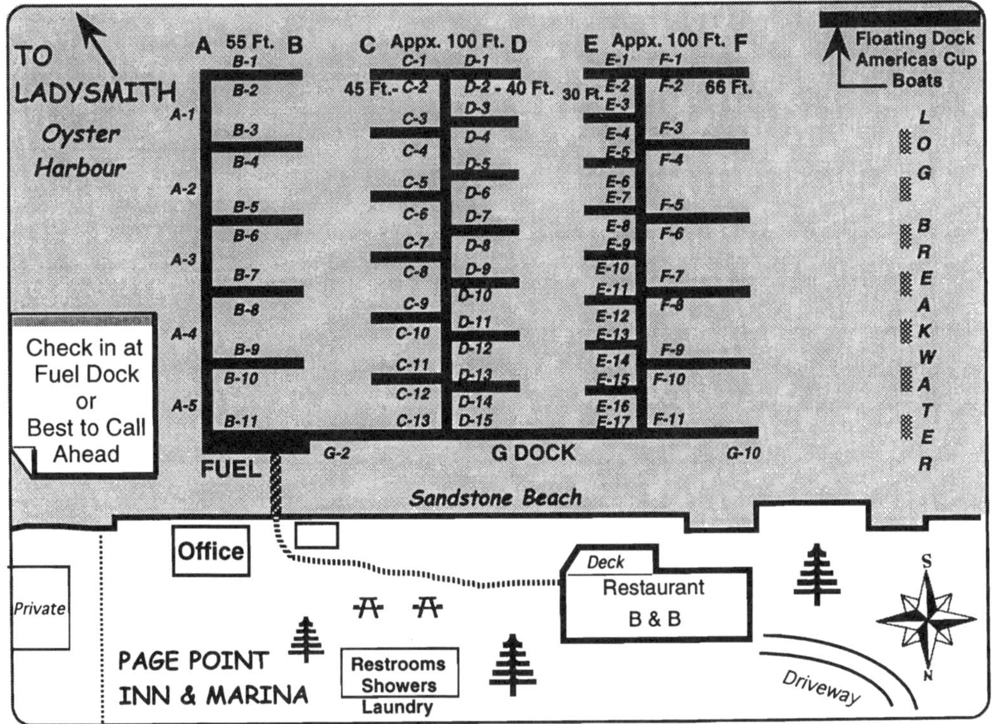

Nanaimo

NAME OF MARINA: ***NANAIMO PORT AUTHORITY*** **RADIO:** VHF Ch.67

TELEPHONE: 250-754-5053 **MGR:** David Mailloux

E-MAIL: marina@npa.ca **FAX:** 250-754-4186 (Fax)

ADDRESS: 10 Wharf St. Nanaimo, B.C. Canada V9R 2X3

SHORT DESCRIPTION & LOCATION: **www.npa.ca**

49°10.20' - 123°55.90' The protected boat harbour is located adjacent to downtown Nanaimo close to the base of the white round Bastion often used as a landmark. It is the first marina upon entering from the south and the area offers complete marine services.

GUEST BOAT CAPACITY: ....Appx. 250 boats
DOCKSIDE DEPTH AT ZERO TIDE: ....15 ft.
SEASON: ....All year
RESERVATION POLICY: ....Ltd.- Tel. 755-1216
AMT W/ELECTRICITY: ....All
FUEL DOCK: ....Gas & Diesel
MARINE REPAIRS: ....Close by
TOILETS: ....Yes
HOT SHOWERS: ....Yes with heated floors
RESTAURANT: ....Close by
PICNIC AREA: ....Yes
BASIC STORE: ....Close by
BROADBAND/WI-FI: ....BroadbandXpress
DAILY RATE: ....Moderate
(75¢-$1.25/foot)
3 Hrs. Complimentary - Call Ahead
CUSTOMS PORT-OF-ENTRY

GUEST DOCK: ....9000 ft. Total
GUEST SLIPS: ....15 lg. floats
WATER: ....Yes
AMPS: ....15-100 A
PUMP OUT STATION ....Yes
HAUL OUT: ....Close by
BOAT RAMP: ....Close by
LAUNDRY: ....Yes
BAR: ....Close by
POOL: ....Close by
GOLF: ....Close by
PET FRIENDLY: ....Fair
OTHER: Close to lg. shopping ctr, restaurants, pubs, lg. chart store, hardware & chandlery, ample transportation to mainland.

CAUTION! This chartlet not intended for use in navigation.

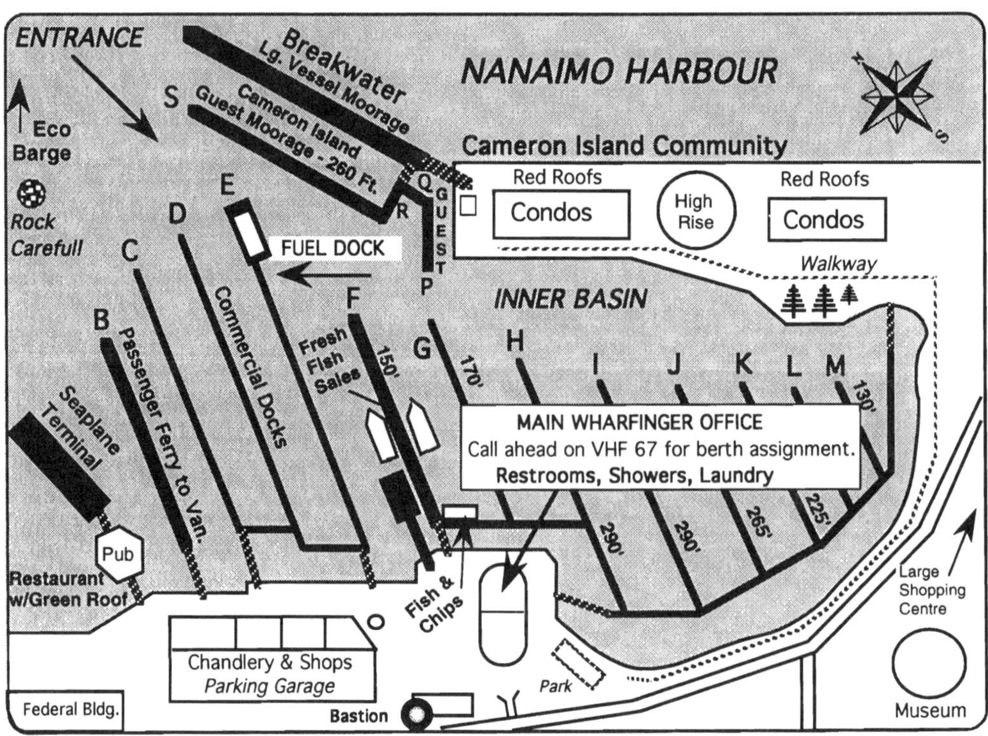

Nanaimo

NAME OF PARK: ***NEWCASTLE ISLAND PROVINCIAL MARINE PARK***
ADDRESS: Mail: P.O. Box 2174 Nanaimo, B.C. Canada V9R 6X6
TELEPHONE: 250-754-7893 **MGR:** Drew Chapman, B.C. Parks
SHORT DESCRIPTION & LOCATION: **http://www.newcastleisland.com**

49°10.70' - 123°55.70' This 720 acre island park is accessible only by water and is located on the N.E. side of Nanaimo Harbour about 1/2 mile across from downtown Nanaimo. The park has a colorful history explained at various park displays.

GUEST BOAT CAPACITY: ....Appx 30-35 boats
DOCKSIDE DEPTH AT ZERO TIDE:..........20 ft.
SEASON: ....All year
AMT W/ELECTRICITY: ....None
TOILETS: ....Yes
HOT SHOWERS: ....Yes
PICNIC AREA: ....Yes
PLAY AREA: ....Yes
BASIC STORE: ....None
DAILY RATE: ....Appx. 70¢/ft. - $2.00/metre
NOTE: For more park info contact B.C.Parks at 250-751-3206---http://www.env.gov.bc.ca/bcparks/

GUEST DOCK: .....100' long floats
MOORING BUOYS: ....None
WATER: ....Yes
PAY PHONES: ....Yes
BOAT RAMP: ....None
PICNIC SHELTER: ....Yes
BBQ: ....None
PUMP OUT STATION: ......None
PET FRIENDLY:.....Good-On leash
OTHER: Ample moorage, but busy in Summer. Ferry to Nanaimo, park interpreter, tent sites, trails, wildlife, snack bar, group pavilion, & swimming.

CAUTION! This chartlet not intended for use in navigation.

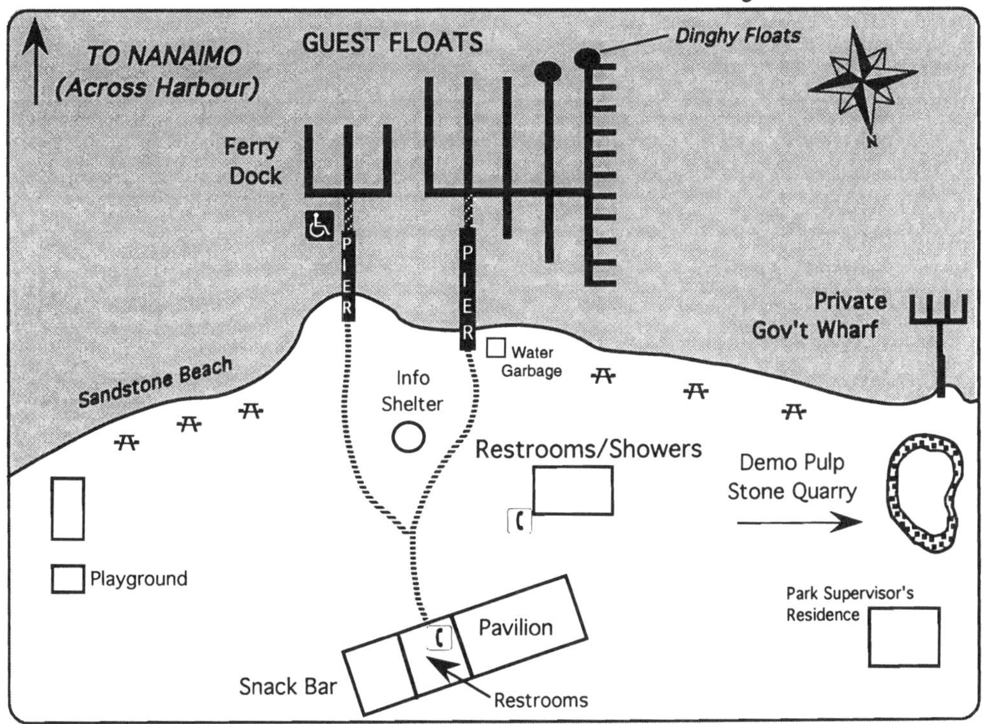

NOTES

Nanaimo

NAME OF YACHT CLUB: *NANAIMO YACHT CLUB*

ADDRESS: 400 Newcastle Ave. Nanaimo, B.C Canada V9S 4J1

TELEPHONE: 250-754-7011 PERSON IN CHARGE: Reciprocal Chair.

SHORT DESCRIPTION & LOCATION: **www.nanaimoyc.ca**

49°10.39' - 123°56.23' Located appx. 1 mi. N of city centre at south entrance of Newcastle Island Passage just across the channel from Bate Pt. The club offers 700 ft. of reciprocal & guest moorage on the outside of H, K. & L docks. Maximum boat size is 60 ft. Note: NYC offers visitor moorage (for a fee) to bona-fide members of non reciprocal clubs.

RECIPROCAL BOAT CAPACITY: ....Appx. 40

DOCKSIDE DEPTH AT ZERO TIDE: ....20 ft.

SEASON: ....All year

RESERVATION POLICY: ....None

TOILETS: ....Yes

HOT SHOWERS: ....Yes

RESTAURANT: ....Close by

BROADBAND/WI-FI: ....None

DAILY RATE: 1 Day free reciprocal only. After 1 day & visitor dock is 50¢/ft- Max 5 days/month. **Power $5/day.**

RECIPROCAL DOCK: Total 760'+

RECIPROCAL SLIPS: Docks only

WATER: ....Yes

AMT W/ELECTRICITY: ....All

AMPS: ....30 A

BAR: ....Close by

PET FRIENDLY: ....Good

OTHER: Marine services close by. No rafting is allowed. Security gate w/code. Thrifty's Foods will deliver.

NOTE: THIS IS PRIVATE MOORAGE AND ONLY AVAILABLE TO MEMBERS OF RECIPROCAL YACHT CLUBS. YOUR CLUB MUST HAVE RECIPROCAL PRIVILEGES AND YOU MUST FLY YOUR BURGEE.

CAUTION! This chartlet not intended for use in navigation.

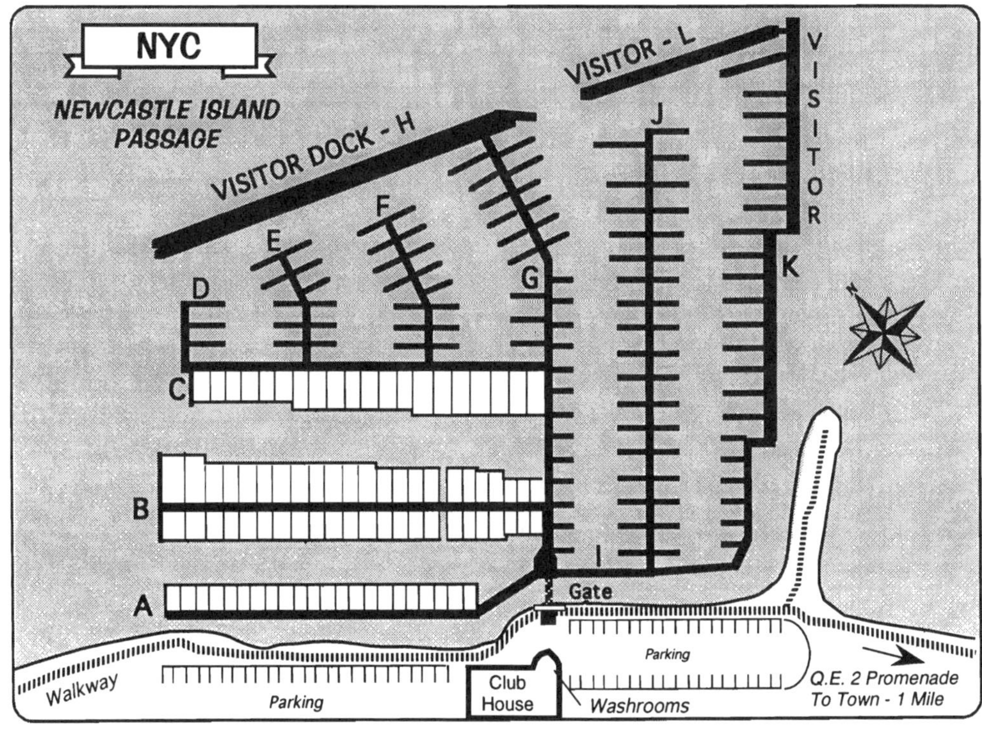

Schooner Cove

NAME OF MARINA: ***SCHOONER COVE MARINA*** **RADIO:** VHF 66 A
TELEPHONE: **1-250-468-5364** MGR: Wayne Newport
E-MAIL: marina@fairwinds.ca FAX: 250-468-5744
ADDRESS: 3521 Dolphin Drive Nanoose Bay, B.C. Canada V9P 9J7
SHORT DESCRIPTION & LOCATION: **www.fairwinds.ca**
49°17.28' - 124°07.90' Located appx. 12 mi. NW of Nanaimo & 2 mi. N of Nanoose Hbr. on Georgia Str. on E coast of Van. Is. on S. side of Nankivell Pt. This upscale & popular destination resort offers many amenities including highly ranked award winning golf course.

GUEST BOAT CAPACITY: ....50 plus
DOCKSIDE DEPTH AT ZERO TIDE: ....15 ft.
SEASON: ....All year
RESERVATION POLICY: ....Accepts
AMT W/ELECTRICITY: ....All
FUEL DOCK: ....Gas & Diesel
MARINE REPAIRS: ....Close by
TOILETS: ....Yes
HOT SHOWERS: ....Yes
RESTAURANT: & PUB: ....Yes
PICNIC AREA: ....Yes
BASIC STORE: ....Yes
BROADBAND/WI-FI: ....BroadbandXpress
DAILY RATE: ....Moderate (75¢-$1.25/foot)

GUEST DOCK: ....150 ft. + Slips
GUEST SLIPS: ....Appx. 20
WATER: ....Yes
AMPS: ....30-50 A
PUMP OUT STATION ....Yes
HAUL OUT: ....Tidal grid
BOAT RAMP: ....Yes
LAUNDRY: ....Yes
BAR: ....Pub
POOL: ....Yes
GOLF: ....Yes
PET FRIENDLY: ....Excellent
OTHER: Hotel, exercise room, hot tub, fishing & marine supplies, scuba diving, hiking, tennis, bike rentals hores riding, & kayaking,

NAME OF YACHT CLUB: ***SCHOONER COVE YACHT CLUB***
CLUB ADDRESS: P.O. Box 9, Nanoose Bay, B.C. V9P 9J9 Canada
CLUB TELEPHONE: PERSON IN CHARGE: Secretary
LOCATION & SPECIAL NOTES: **www.scyc.ca**
SCYC by agreement with Schooner Cove Marina offers reciprocal moorage & facilities to yacht clubs having a reciprocal moorage agreement with SCYC. Available dock space includes 150 ft. of reciprocal moorage on floats "B", "C", & "D" & any vacant member slips. Call Marina for slip assignment. Upon arrival register at marina office & sign moorage book.

RECIPROCAL BOAT CAPACITY: ....Varies
DOCKSIDE DEPTH AT ZERO TIDE: ....15 ft.
RECIPROCAL SEASON: ....All year
RESERVATION POLICY: ....Recommended
TOILETS: ....At Marina
HOT SHOWERS: ....At Marina
RESTAURANT: ....Pub & Coffee Shop at Marina
DAILY RATE: ....Free moorage for 2 non-consecutive nights per calendar year. Marina rates thereafter.

RECIPROCAL DOCK: ....Varies
RECIPROCAL SLIPS: ....Varies
WATER: ....Yes
AMT W/ELECTRICITY: ....All
AMPS: ....30-50 A
BAR: ....At Marina
OTHER: Maximum yacht length is 150 ft.

***NOTE:* THIS IS PRIVATE MOORAGE *ONLY* AVAILABLE TO MEMBERS OF RECIPROCAL YACHT CLUBS! YOUR CLUB *MUST* HAVE RECIPROCAL PRIVILEGES AND YOU MUST FLY YOUR BURGEE!**

Schooner Cove

CAUTION! This chartlet not intended for use in navigation.

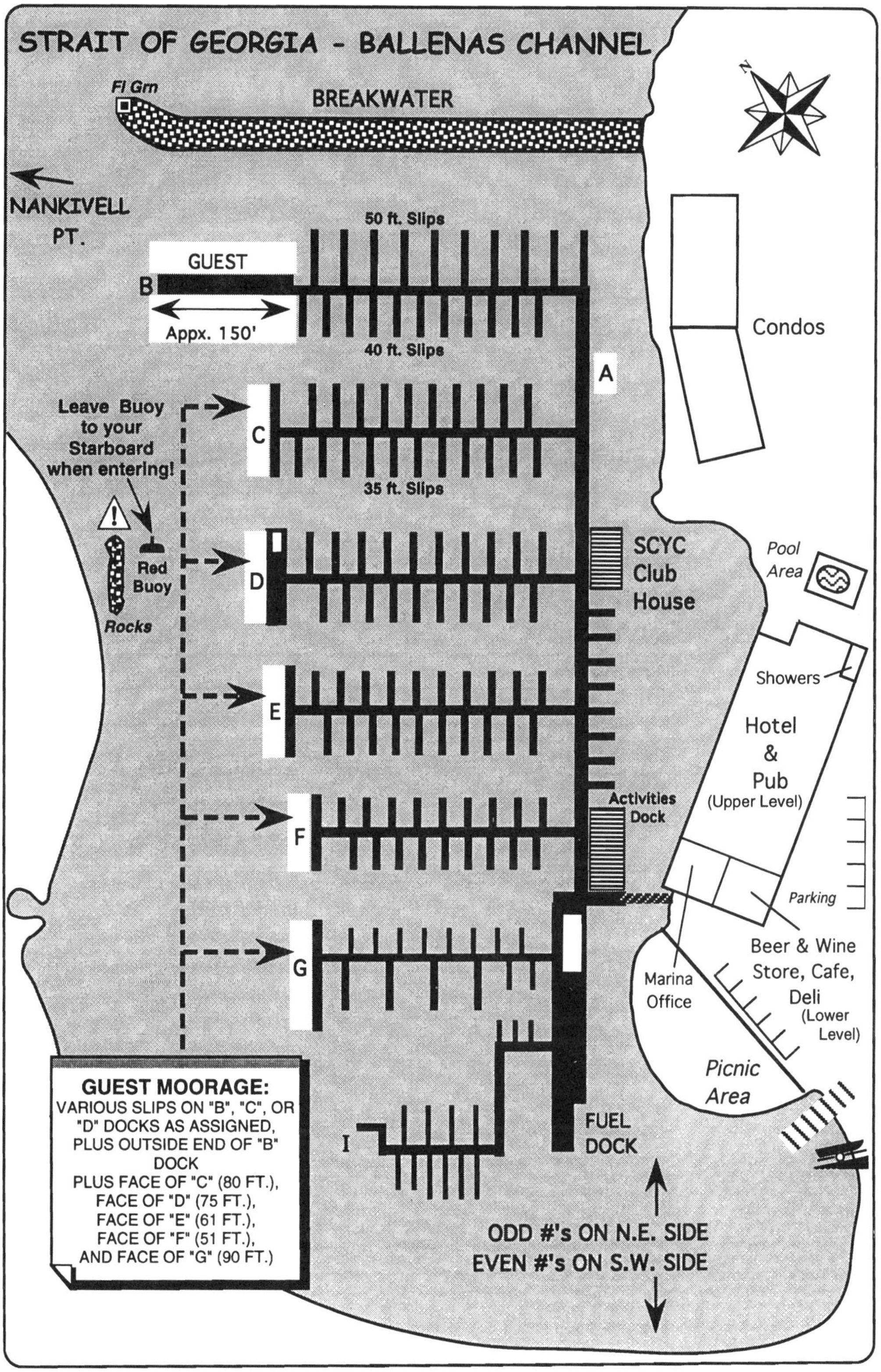

STRAIT OF GEORGIA - BALLENAS CHANNEL

Fl Grn

BREAKWATER

NANKIVELL PT.

50 ft. Slips

GUEST

B

Appx. 150'

40 ft. Slips

A

Condos

Leave Buoy to your Starboard when entering!

C

35 ft. Slips

D

Red Buoy

Rocks

SCYC Club House

Pool Area

Showers

E

Hotel & Pub (Upper Level)

F

Activities Dock

Parking

G

Beer & Wine Store, Cafe, Deli (Lower Level)

Marina Office

Picnic Area

I

FUEL DOCK

GUEST MOORAGE:
VARIOUS SLIPS ON "B", "C", OR "D" DOCKS AS ASSIGNED, PLUS OUTSIDE END OF "B" DOCK
PLUS FACE OF "C" (80 FT.),
FACE OF "D" (75 FT.),
FACE OF "E" (61 FT.),
FACE OF "F" (51 FT.),
AND FACE OF "G" (90 FT.)

ODD #'s ON N.E. SIDE
EVEN #'s ON S.W. SIDE

Comox

NAME OF MARINA: ***COMOX VALLEY HBR. AUTHORITY*** **RADIO:** VHF 66A
TELEPHONE: 250-339-6041 **MGR:** Elizabeth McLeod/Mo Nordstrom
E-MAIL: info@comoxfishermanswharf.com **FAX:** 250-339-6057 (Fax)
ADDRESS: 121 Port Augusta St. Comox, B.C. Canada V9N 3M8
SHORT DESCRIPTION & LOCATION: **www.comoxfishermanswharf.com**
49°40.10' - 124°55.80' Comox Hbr., entered between Gartley Pt. & Goose Spit, is a friendly working Port that welcomes pleasure boaters w/many boater amenities & local attractions. Main guest moorage is on E. side of causeway - go around E. end of breakwater.

GUEST BOAT CAPACITY:	Appx. 20-30 boats	GUEST DOCK:	1800 ft. TTL
DOCKSIDE DEPTH AT ZERO TIDE:	12 ft. +	GUEST SLIPS:	Docks only
SEASON:	Visitors - Mainly April - October	WATER:	Yes
RESERVATION POLICY:	Accepts	AMPS:	15-30 A
AMT W/ELECTRICITY:	All	PUMP OUT STATION	Yes
FUEL DOCK:	Gas & Diesel	HAUL OUT:	None
MARINE REPAIRS:	On premises	BOAT RAMP:	Yes
TOILETS:	Yes	LAUNDRY:	Yes
HOT SHOWERS:	Yes	LOUNGE:	3 Pubs close by
RESTAURANT:	Close by	POOL:	via Bus
PICNIC AREA:	Close by	GOLF:	Close by
BASIC STORE:	Close by	PET FRIENDLY:	Excellent
BROADBAND/WI-FI:	None	OTHER:	Call on VHF for docking assistance, fresh seafood in Harbour. Internet terminal. Nearby shopping, banks, hospital, chandlery, & parks.
DAILY RATE:	Moderate (75¢-$1.25/foot)		

ACTUAL OVERALL LENGTH PLEASE

Note: Good charts needed to approach Harbour & cross Comox Bar to the east.

NAME OF MARINA: ***GAS 'N GO MARINA & FUEL DOCK*** **RADIO:** VHF 66A
ADDRESS: P.O. Box 1296 Comox, B.C. Canada V9M 7Z8
TELEPHONE: 250-339-4664 **MGR:** Joan Benda
SHORT DESCRIPTION & LOCATION:
49°40.10' - 124°55.80' The small marina located within Comox Harbour operates the main Fuel Dock. Enter central breakwater opening & go to the Fuel Dock. Limited guest moorage by Fuel Dock. Usually has a few slips available when permanent tenants are away.

GUEST BOAT CAPACITY:	Appx. 3-4	GUEST DOCK:	Varies
DOCKSIDE DEPTH AT ZERO TIDE:	9 ft.	GUEST SLIPS:	Varies
SEASON:	Visitors - Mainly April - October	WATER:	Yes
RESERVATION POLICY:	Accepts	AMPS:	15 amp
AMT W/ELECTRICITY:	All	PUMP OUT STATION	Yes
FUEL DOCK:	Gas & Diesel	HAUL OUT:	None
MARINE REPAIRS:	On premises	BOAT RAMP:	Yes
TOILETS:	Yes	LAUNDRY:	Yes
HOT SHOWERS:	At The Edge Pub	LOUNGE:	3 Pubs close by
RESTAURANT:	Close by	POOL:	via Bus
PICNIC AREA:	Close by	GOLF:	Close by
BASIC STORE:	Yes	PET FRIENDLY:	Excellent
BROADBAND/WI-FI:	None	OTHER:	Marina & fuel dock also have small store with sandwiches, basic supplies & fishing tackle.
DAILY RATE:	Moderate (75¢-$1.25/foot)		

Note: Good charts needed to approach Harbour & cross Comox Bar to the east.

Comox

CAUTION! This chartlet not intended for use in navigation.

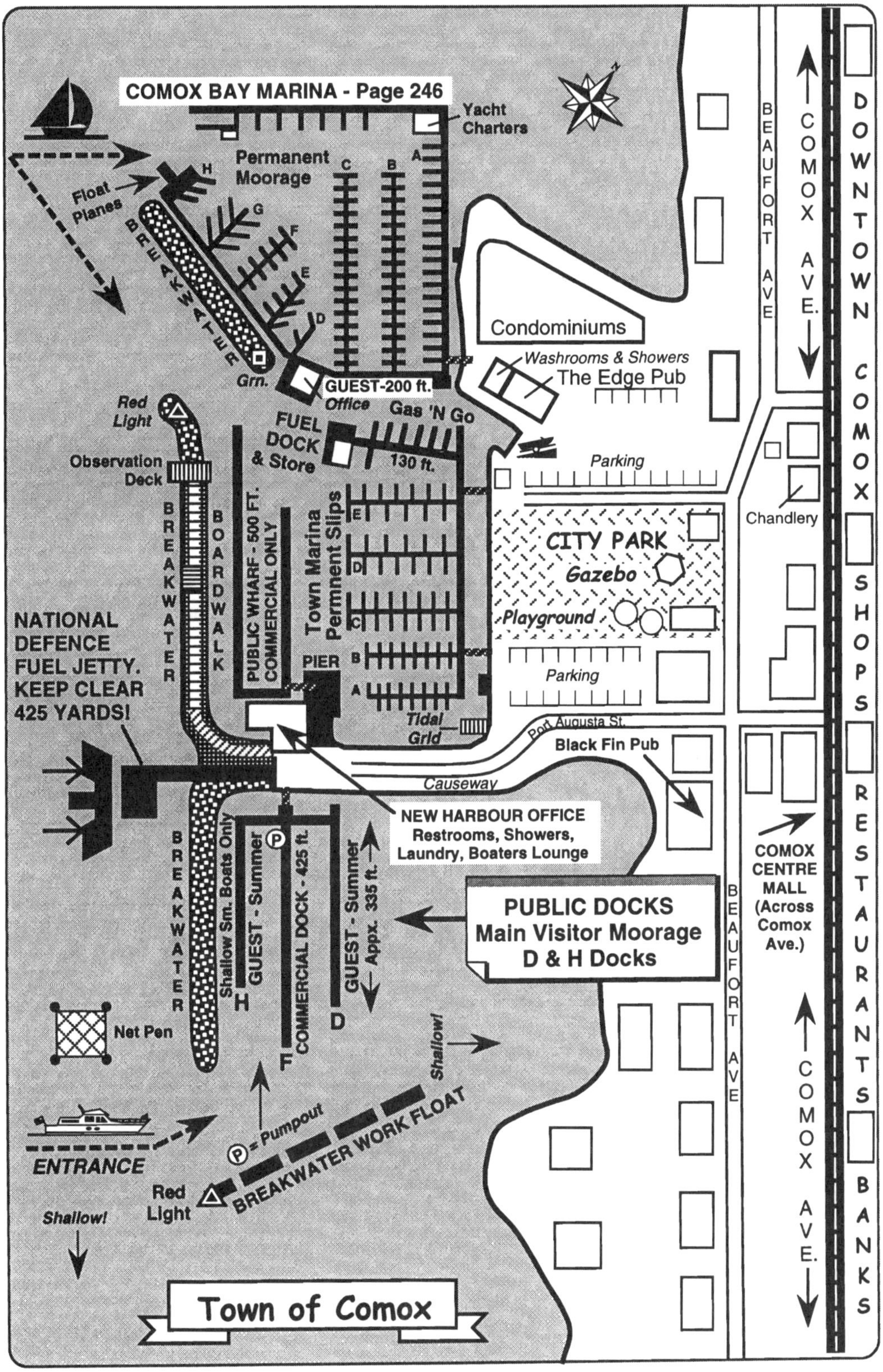

COMOX BAY MARINA - Page 246

Yacht Charters

Permanent Moorage

A B C D E F G H

Float Planes

BREAKWATER

Grn.

GUEST-200 ft.

Office

Gas 'N Go

FUEL DOCK & Store

130 ft.

Red Light

Observation Deck

BREAKWATER

BOARDWALK

PUBLIC WHARF - 500 FT. COMMERCIAL ONLY

Town Marina Permnent Slips

PIER

Tidal Grid

Condominiums

Washrooms & Showers

The Edge Pub

Parking

CITY PARK

Gazebo

Playground

Parking

Chandlery

Port Augusta St.

Black Fin Pub

Causeway

NATIONAL DEFENCE FUEL JETTY. KEEP CLEAR 425 YARDS!

NEW HARBOUR OFFICE
Restrooms, Showers,
Laundry, Boaters Lounge

BREAKWATER

Shallow Sm. Boats Only

GUEST - Summer H

COMMERCIAL DOCK - 425 ft. F

GUEST - Summer D

Appx. 335 ft.

PUBLIC DOCKS
Main Visitor Moorage
D & H Docks

COMOX CENTRE MALL (Across Comox Ave.)

Net Pen

Shallow!

P = Pumpout

BREAKWATER WORK FLOAT

ENTRANCE

Red Light

Shallow!

Town of Comox

BEAUFORT AVE

COMOX AVE.

DOWNTOWN COMOX

SHOPS

RESTAURANTS

BANKS

Comox

NAME OF MARINA: ***COMOX BAY MARINA*** **RADIO:** VHF 66A
TELEPHONE: **250-339-2930** **MGR:** Brad Jenkins, Cell 250-218-2827
E-MAIL: manager@comoxbaymarina.com **FAX:** 250-339-2930 (Fax)
ADDRESS: 1805 Beaufort Ave. Comox, B.C. Canada V9M 1R9

SHORT DESCRIPTION & LOCATION:

49°40.10' - 124°55.80' Located at W. end of Comox Harbour, entered between Gartley Pt. & Goose Spit. Marina consists of mostly permanent moorage, but has small guest dock next to marina office-across from Fuel Dock. Enter central breakwater opening.

GUEST BOAT CAPACITY: ....Appx. 4-6 boats
DOCKSIDE DEPTH AT ZERO TIDE: ....10 ft.
SEASON: ....All year
RESERVATION POLICY: ....Recommended
AMT W/ELECTRICITY: ....All
FUEL DOCK: ....Next door
MARINE REPAIRS: ....On call
TOILETS: ....Yes
HOT SHOWERS: ....Yes
RESTAURANT: ....Several close by
PICNIC AREA: ....Yes
BASIC STORE: ....Close by
BROADBAND/WI-FI: ....None
DAILY RATE: ....Moderate (75¢-$1.25/foot)

Note: Good charts needed to approach Harbour & cross Comox Bar to the east.

GUEST DOCK: ....Appx. 200 ft.
GUEST SLIPS: ....Varies
WATER: ....Yes
AMPS: ....15-30 A
PUMP OUT STATION Close by
HAUL OUT: Hydro-hoist-50' max
BOAT RAMP: ....Close by
LAUNDRY: ....Yes
BAR: ....3 Pubs close by
POOL: ....via Bus
GOLF: ....Yes
PET FRIENDLY: ....Excellent
OTHER: Call on VHF for Docking assistance. Fresh seafood in harbour. Nearby: Shopping, banks, hospital, chandlery, & parks.

NOTES

Marina Chartlet Page 245

CHAPTER

8

GREATER VANCOUVER & HOWE SOUND

Chapter Map - Page 2

Steveston

NAME OF MARINA: *STEVESTON HARBOUR AUTHORITY* **RADIO:** None
TELEPHONE: **604-272-5539** **MGR:** Bob Baziuk/Jim Jones
E-MAIL: info@stevestonharbour.com **FAX:** 604-271-6142 (Fax)
ADDRESS: 12740 Trites Road Richmond, B.C. Canada V7E 3R8
SHORT DESCRIPTION & LOCATION: **www.stevestonharbour.com**

49°07.40' - 123°11.30' Located appx. 5 mi. inside entrance of S. Arm & on N. side of Fraser River & Cannery Channel, inside of Steveston Island. Public wharf primarily for fishing boats but pleasure craft welcomed in this quaint village with shopping & nice restaurants.

GUEST BOAT CAPACITY: ....Varies
DOCKSIDE DEPTH AT ZERO TIDE: ....9 ft.
SEASON: ....All year
RESERVATION POLICY: ....Best to Call Ahead
AMT W/ELECTRICITY: ....All
FUEL DOCK: 2 Fuel Docks nearby w/Gas & Diesel
MARINE REPAIRS: ....Close by
TOILETS: ....Yes
HOT SHOWERS: ....Yes
RESTAURANT: ....Close by
PICNIC AREA: ....None
BASIC STORE: ....Close by
BROADBAND/WI-FI: ....None
DAILY RATE: ....Moderate (75¢-$1.25/foot)

GUEST DOCK: ....Varies
GUEST SLIPS: ....Varies
WATER: ....Yes
AMPS: ....20-30 A
PUMP OUT STATION ....None
HAUL OUT: ....Travel-Lift
BOAT RAMP: ....Close by
LAUNDRY: ....Close by
BAR: ....Close by
POOL: ....None
GOLF: ....Close by
PET FRIENDLY: Good-On leash
OTHER: Chandleries & shopping close by. Gulf of Georgia Cannery - National Historic Site with public exhibits is adjacent to marina.

CAUTION! This chartlet not intended for use in navigation.

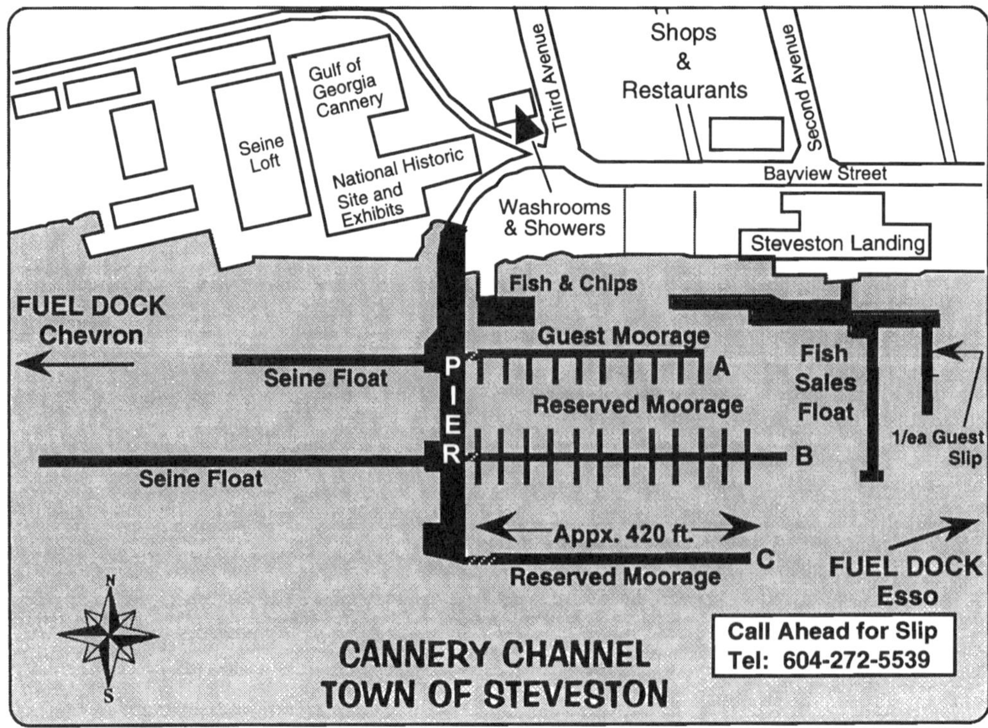

Ladner

NAME OF YACHT CLUB: *LADNER YACHT CLUB*

ADDRESS: 5011 River Road West Delta, B.C. Canada V4K 1S7

TELEPHONE: 604-946-4056 **PERSON IN CHARGE:** Caretaker

SHORT DESCRIPTION & LOCATION:
49°05.70'-123°05.0' Enter the South Arm of Fraser River (near high tide) at Sand Heads. Proceed upriver 10 miles using Sea Reach to access Ladner Slough. Use the reciprocal directions sent to your yacht club & local chart 3490 - both a must. Call ahead for caretaker to assign vacant slip. Walking distance to historic village of Ladner.

RECIPROCAL BOAT CAPACITY: ....Varies
DOCKSIDE DEPTH AT ZERO TIDE: ....5 feet
SEASON: ....All year
RESERVATION POLICY: ....Recommended
TOILETS: ....Yes
HOT SHOWERS: ....None
RESTAURANT: ....Several close by
BROADBAND/WI-FI: ....None
DAILY RATE: No Charge for 48 hours, thereafter 20¢/ft for max of 1 week.
MAX BOAT SIZE 50 FEET.

RECIPROCAL DOCK: Vacant slips
RECIPROCAL SLIPS: ....Varies
WATER: ....Yes
AMT W/ELECTRICITY: ....All
AMPS: ....15 A
BAR: ....Close by
PET FRIENDLY: ....Good
OTHER: Museum, chandlery, dining & shopping close by. Scenic kayak/dinghy loop at high tide.

***NOTE:* THIS IS PRIVATE MOORAGE AND ONLY AVAILABLE TO MEMBERS OF RECIPROCAL YACHT CLUBS. YOUR CLUB *MUST* HAVE RECIPROCAL PRIVILEGES AND YOU MUST FLY YOUR BURGEE.**

CAUTION! This chartlet not intended for use in navigation.

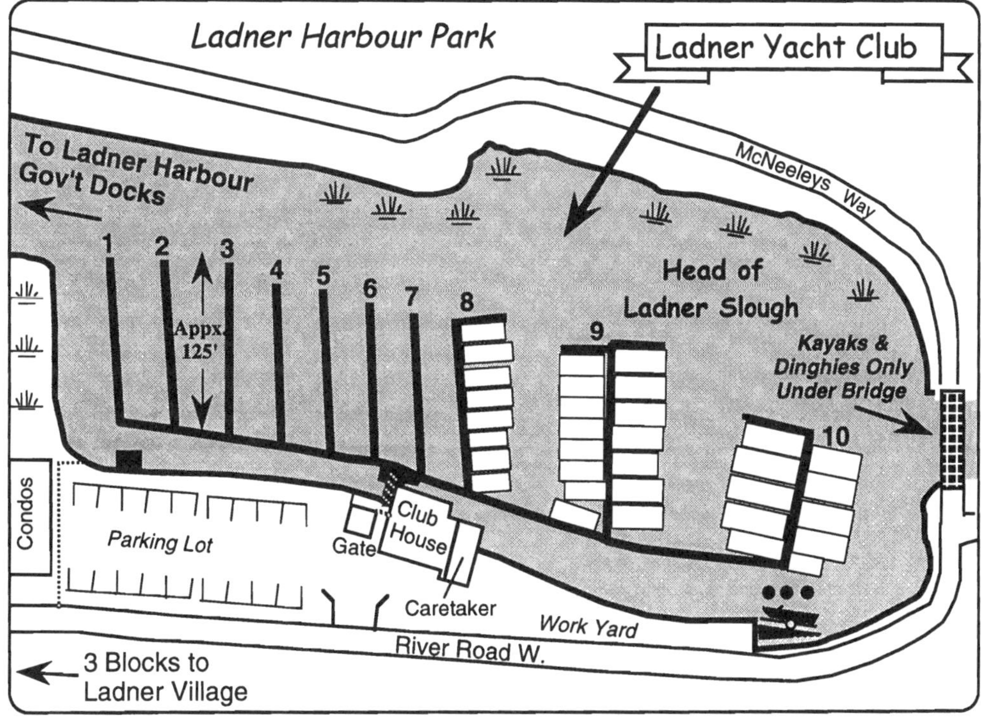

Ladner

NAME OF MARINA: ***CAPTAIN'S COVE MARINA*** **RADIO:** None
TELEPHONE: **604-946-1244** **MGR:** Elizabeth Model
E-MAIL: info@captainscovemarina.ca **FAX:** 604-946-1246 (Fax)
ADDRESS: 6100 Ferry Road Delta, B.C. Canada V4K 3M9
SHORT DESCRIPTION & LOCATION: **www.captainscovemarina.ca**

49°09.80' - 122°59.30' Located in community of Ladner on S. Arm of Fraser River, appx. 10 miles up river on south side of Deas Slough. Marina offers well protected fresh water facility with boatyard and Pub on site. Directly across from Deas Island Provincial Park.

GUEST BOAT CAPACITY: ....Appx. 10 boats
DOCKSIDE DEPTH AT ZERO TIDE: Appx. 10'
SEASON: ....All year
RESERVATION POLICY: ....None
AMT W/ELECTRICITY: ....All
FUEL DOCK: ....Gas & Diesel
MARINE REPAIRS: ....On call
TOILETS: ....Yes
HOT SHOWERS: ....Yes
RESTAURANT: ....Yes
PICNIC AREA: ....None
BASIC STORE: ....Close by
BROADBAND/WI-FI: ....None
DAILY RATE: ....Moderate (75¢-$1.25/foot)

GUEST DOCK: Appx. 500 ft. TTL
GUEST SLIPS: ....10 Slips
WATER: ....Yes
AMPS: ....30 A
PUMP OUT STATION ....None
HAUL OUT: ....Travel-Lift
BOAT RAMP: ....Yes
LAUNDRY: ....Yes
BAR: ....Rusty Anchor Pub
POOL: ....None
GOLF: ....Yes
PET FRIENDLY: ....Excellent
OTHER: Deas Is. Park has trails, picnicing, wildlife. Close by (via taxi /bus): Chandlery, shopping, medical services, post office, & banks.

CAUTION! This chartlet not intended for use in navigation.

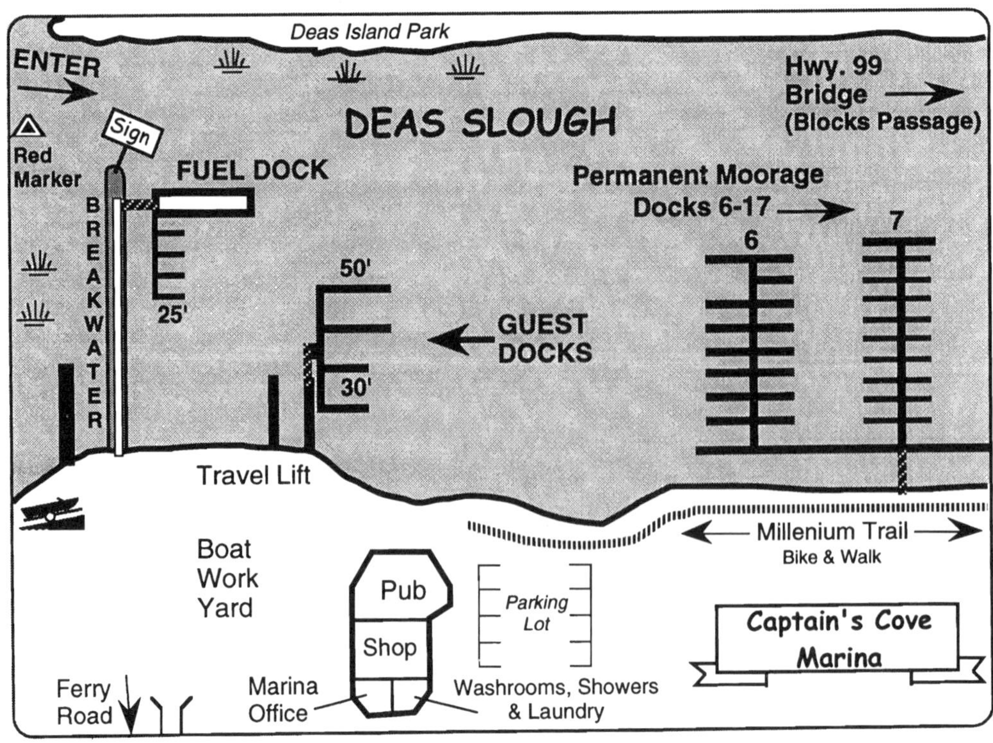

Richmond

NAME OF YACHT CLUB: *RICHMOND YACHT CLUB*

ADDRESS: 7471 River Rd. Richmond, B.C. Canada V6X 2W4

TELEPHONE: 604-278-1014 PERSON IN CHARGE: Reciprocal Chair

SHORT DESCRIPTION & LOCATION: **www.richmondyachtclub.ca**

49°10.90'- 123°08.50' RYC is on Fraser River Middle Arm. Enter N. Arm at Pt. Grey. Go 4 miles to stb just past Chevron fuel dock, entering Middle Arm. Pass under bridge & go 1 mi. 50' dock is on outside of boat shed. No rafting. Register with Don York aboard Summer Wind at foot of ramp. Vacant member slips sometimes available too.

RECIPROCAL BOAT CAPACITY: Appx. 2 boats

DOCKSIDE DEPTH AT ZERO TIDE: ....25 ft.

SEASON: ....All year

RESERVATION POLICY: First come first served

TOILETS: ....Yes

HOT SHOWERS: ....Yes

RESTAURANT: ....Close by

BROADBAND/WI-FI: ....None

DAILY RATE: ....3 DAYS FREE MOORAGE
................Power $2.00/night
Swing Bridge VHF Ch. 74

RECIPROCAL DOCK: ....Appx. 50'

RECIPROCAL SLIPS: ....Varies

WATER: ....Yes

AMT W/ELECTRICITY: ....All

AMPS: ....15 A

BAR: ....Close by

PET FRIENDLY: ....Excellent

OTHER: Taxi telephone number is 278-8444. Pet walk and garbage on the dyke.

***NOTE:* THIS IS PRIVATE MOORAGE AND ONLY AVAILABLE TO MEMBERS OF RECIPROCAL YACHT CLUBS. YOUR CLUB *MUST* HAVE RECIPROCAL PRIVILEGES AND YOU MUST FLY YOUR BURGEE.**

CAUTION! This chartlet not intended for use in navigation.

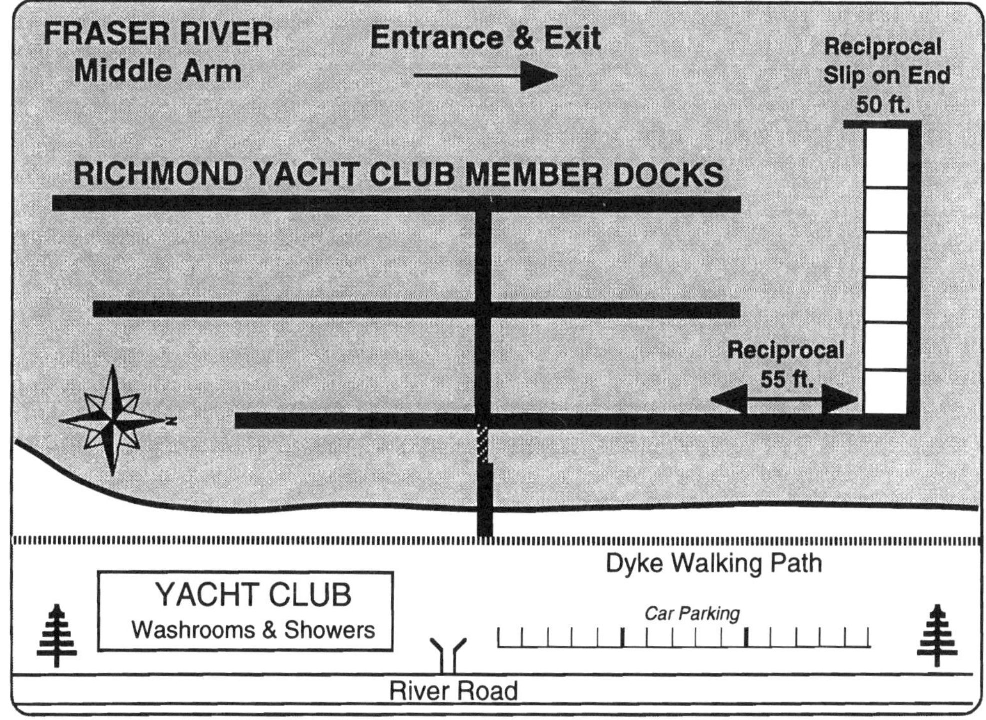

Richmond

NAME OF MARINA: ***SHELTER ISLAND MARINA*** **RADIO:** VHF 66A
TELEPHONE: **604-270-6272** MGR: Rich Hutton
E-MAIL: infodesk@shelterislandmarina.com FAX: 604-273-6282 (Fax)
ADDRESS: 6911 Graybar Rd. Richmond, B.C. Canada V6W 1H3
SHORT DESCRIPTION & LOCATION: **www.shelterislandmarina.com**

49°09.80' - 122°59.30' Located in Richmond on S. Arm of Fraser River, appx. 14 mi. up river on northern side. Marina offers a modern well protected fresh water facility with repairs, Travel Lift to 150 tons, huge boatyard, covered storage & well stocked chandlery.

GUEST BOAT CAPACITY: ....Appx 6-8 boats
DOCKSIDE DEPTH AT ZERO TIDE: 8 ft. plus
SEASON: ....All year
RESERVATION POLICY: ....Recommended
AMT W/ELECTRICITY: ....All
FUEL DOCK: ....None
MARINE REPAIRS: ....On premises
TOILETS: ....Yes
HOT SHOWERS: ....Yes
RESTAURANT: ....Yes
PICNIC AREA: ....Yes
BASIC STORE: ....Yes
BROADBAND/WI-FI: ....None
DAILY RATE: ....Economical (Under 75¢/foot)

GUEST DOCK: ....Varies
GUEST SLIPS: ....Varies
WATER: ....Yes
AMPS: ....20-50 A
PUMP OUT STATION ....Yes
HAUL OUT: ....Travel-Lift
BOAT RAMP: ....None
LAUNDRY: ....Yes
BAR: ....Pub
POOL: ....None
GOLF: ....Close by
PET FRIENDLY: ....Excellent
OTHER: Beer and Wine Store, chandlery, boat storage, skilled marine trades people on site.

NOTES

Richmond

CAUTION! This chartlet not intended for use in navigation.

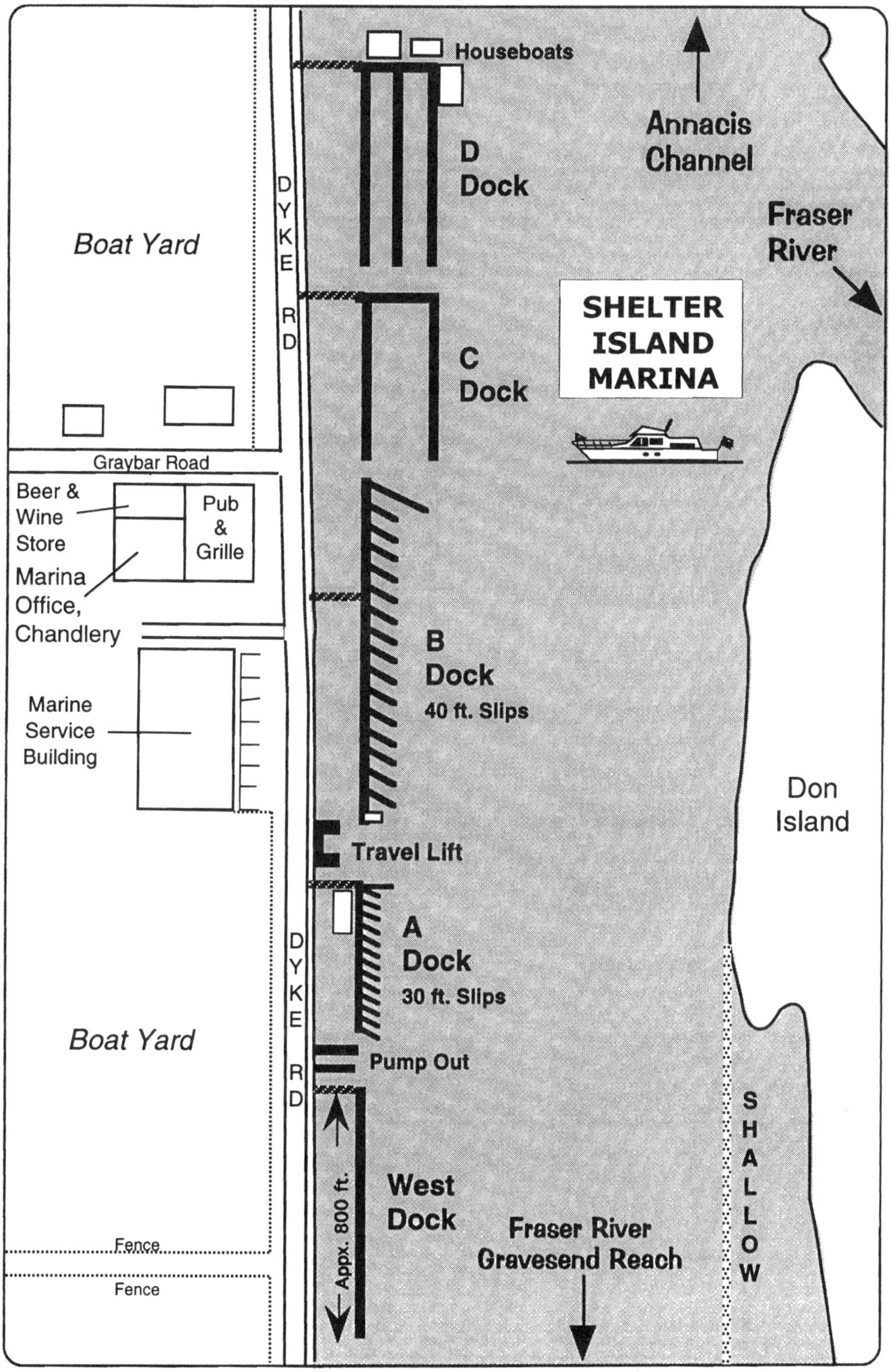

Houseboats

Annacis Channel

D Dock

Fraser River

Boat Yard

DYKE RD

SHELTER ISLAND MARINA

C Dock

Graybar Road

Beer & Wine Store

Pub & Grille

Marina Office, Chandlery

B Dock

40 ft. Slips

Marine Service Building

Don Island

Travel Lift

A Dock

30 ft. Slips

DYKE RD

Boat Yard

Pump Out

SHALLOW

Appx. 800 ft.

West Dock

Fraser River Gravesend Reach

Fence

Fence

Richmond

NAME OF MARINA: ***RIVER ROCK MARINA** (formerly Great Canadian)* RADIO: None
TELEPHONE: **604-273-8560** MGR: Gary Cross, Marina Coordinator
E-MAIL: info@riverrock.com FAX: 604-276-8000 (Fax)
ADDRESS: 8831 River Rd, P.O. Box 102 Richmond, B.C. Canada V6X 1Y6
SHORT DESCRIPTION & LOCATION: **www.riverrock.com**

49°11.50' - 123°07.70' Fairly modern freshwater marina located between Sea Isle & Oak Street bridge on S. side of Fraser River at junction of North Arm & Middle Arm. The Marina is operated by the River Rock Casino Resort with full resort amenities.

GUEST BOAT CAPACITY: ....Appx. 10 boats
DOCKSIDE DEPTH AT ZERO TIDE: ....6 Ft.
SEASON: ....All year
RESERVATION POLICY: ....Recommended
AMT W/ELECTRICITY: ....All
FUEL DOCK: ....Close by
MARINE REPAIRS: ....Close by
TOILETS: ....Yes
HOT SHOWERS: ....Yes
RESTAURANT: ....Yes
PICNIC AREA: ....None
BASIC STORE: ....Close by
BROADBAND/WI-FI: ....Planned
DAILY RATE: ....Moderate
(75¢-$1.25/foot)

GUEST DOCK: Appx. 500 ft total
GUEST SLIPS: ....Dock only
WATER: ....Yes
AMPS: ....15-30 A
PUMP OUT STATION ....None
HAUL OUT: ....None
BOAT RAMP: ....None
LAUNDRY: ....Yes
BAR: ....Yes
POOL: ....Yes
GOLF: ....None
PET FRIENDLY: ....Good
OTHER: Moorage includes use of Resort & Casino facilities, Swimming Pool, Hot Tub, Fitness facility.

CAUTION! This chartlet not intended for use in navigation.

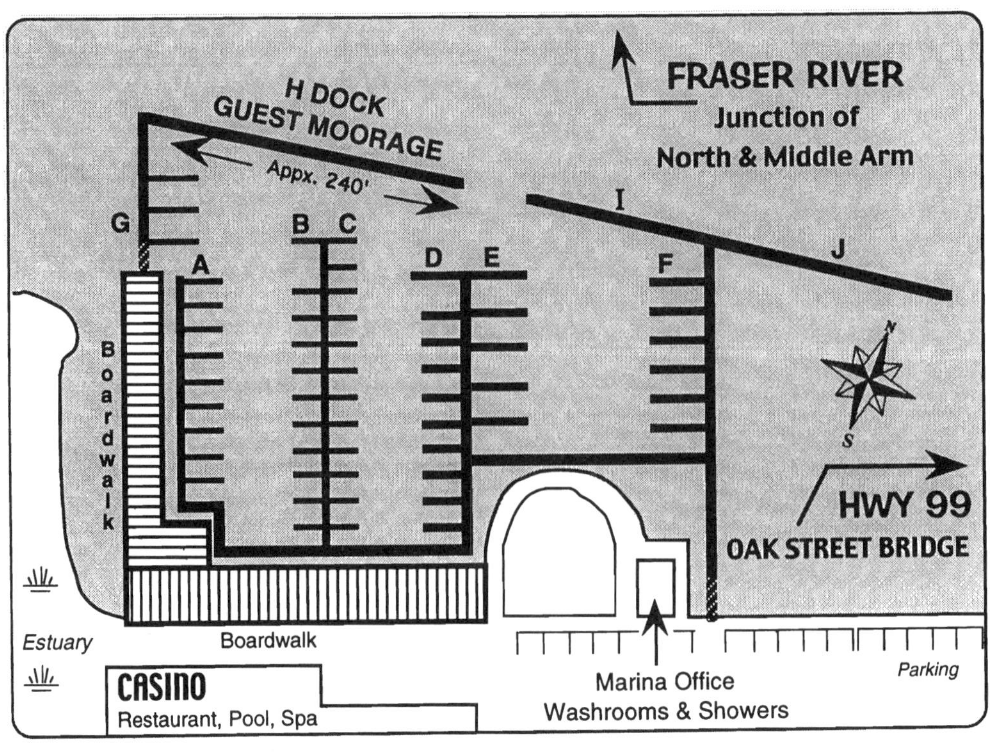

Richmond

NAME OF MARINA: ***DELTA MARINA*** RADIO: See note**
TELEPHONE: 604-273-4211 MGR: Rick Cockburn
E-MAIL: deltacharters@telus.net FAX: 604-273-7251 (Fax)
ADDRESS: 3500 Cessna Drive Richmond, B.C. Canada V7B 1C7
SHORT DESCRIPTION & LOCATION: **www.deltacharters.com**

49°11.40' - 123°08.34' Located on Sea Island on Fraser R. Middle Arm. Enter North Arm at Pt. Grey. Go about 6 mi. & turn (right) down Middle Arm. Pass under Middle Arm swing bridge & see marina immediately on right. Marina & large hotel are near Vancouver Airport.

GUEST BOAT CAPACITY: ....Varies
DOCKSIDE DEPTH AT ZERO TIDE: Appx. 9 Ft.
SEASON: ....All year
RESERVATION POLICY: ....Recommended
AMT W/ELECTRICITY: ....All
FUEL DOCK: ....Close by
MARINE REPAIRS: ....On premises
TOILETS: ....Yes
HOT SHOWERS: ....None
RESTAURANT: ....Nice Pub and Restaurants
PICNIC AREA: ....None
BASIC STORE: ....None
BROADBAND/WI-FI: ....Across River
DAILY RATE: ....Moderate (75¢-$1.25/foot)

**NOTE: Middle Arm Swing Bridge closes 7-9AM & 4-6PM. Call on VHF 74 for opening.

GUEST DOCK: Rents Open Slips
GUEST SLIPS: ....Varies
WATER: ....Yes
AMPS: ....20-30 A
PUMP OUT STATION ....None
HAUL OUT: ....50 Ton Lift
BOAT RAMP: ....None
LAUNDRY: ....None
BAR: ....Yes
POOL: ....Yes
GOLF: ....Close by
PET FRIENDLY: ....Good
OTHER: Elephant & Castle Pub. Hotel, restaurants, bar, pool. Liquor store closeby. Shuttle to airport & shop center. Customs Clearance.

CAUTION! This chartlet not intended for use in navigation.

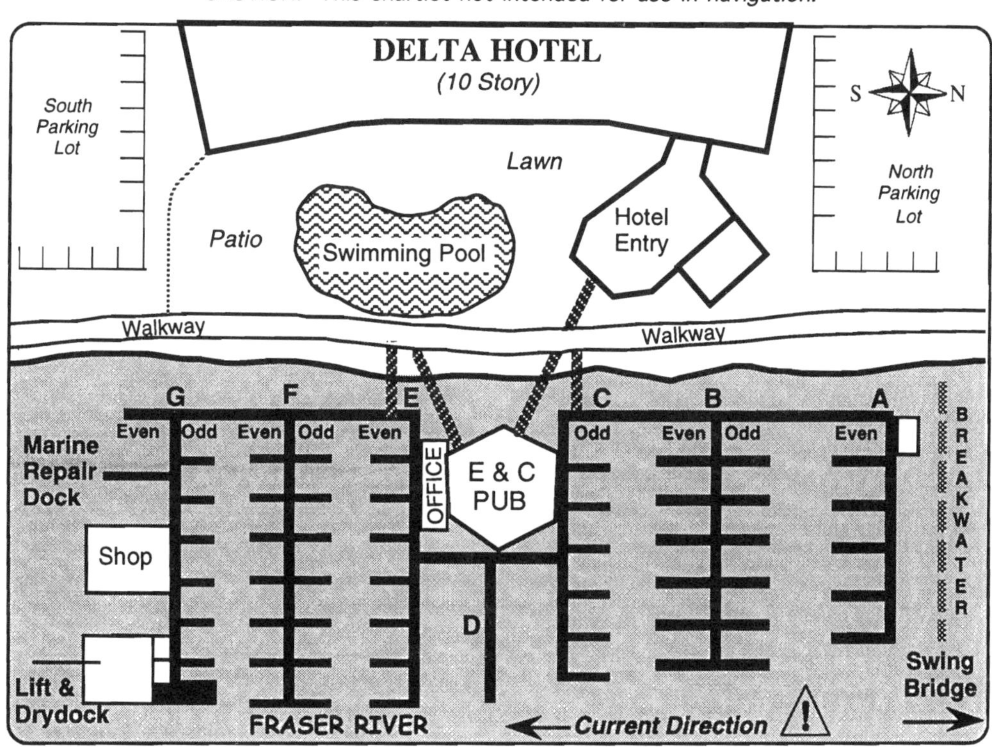

Vancouver

NAME OF YACHT CLUB: *FALSE CREEK YACHT CLUB* RADIO: VHF 66A

ADDRESS: 1661 Granville Street Vancouver, B.C. V6Z 1N3

TELEPHONE: 604-648-2628 (Welcome Ctr) PERSON IN CHARGE: Dockmaster

SHORT DESCRIPTION & LOCATION: **www.fcyc.com**

49°16.37' -123°07.96' OPEN TO PUBLIC. Located on N. side False Creek directly under Granville Bridge. False Creek Yacht Club offers reciprocal club members one night free moorage per season. Additional nights are available at going rate. Marina gate key obtained with key deposit-charge for lost key. Moorage assigned by Boating Welcome Centre. Busy.

RECIPROCAL BOAT CAPACITY..........5 boats +

DOCKSIDE DEPTH AT ZERO TIDE: ....5 Ft. +

SEASON: ....All year

RESERVATION POLICY: ....Accepts

TOILETS: ....Yes

HOT SHOWERS: ....Yes

RESTAURANT: ....Yes

BROADBAND/WI-FI: ....Yes

DAILY RATE: 1 Day free for reciprocal guests.
Public Moorage:
..............Premium (Over $1.25/foot)

RECIPROCAL DOCK: ....Slips

RECIPROCAL SLIPS: ....Varies

WATER: ....Yes

AMT W/ELECTRICITY: ....All

AMPS: ....15-30 A

BAR: ....Yes

PET FRIENDLY: ....Fair

OTHER: Pump-out available. Excellent restaurant at Club. Aqua Bus to Granville Island amenities.

***NOTE:* THIS IS PUBLIC MOORAGE MANAGED BY THE VANCOURVER BOATING WELCOME CENTRE. RECIPROCAL MOORAGE IS ALSO AVAILABLE FOR RECIPROCAL CLUBS IN GOOD STANDING.**

NOTES

Vancouver-False Creek Y.C.

CAUTION! This chartlet not intended for use in navigation.

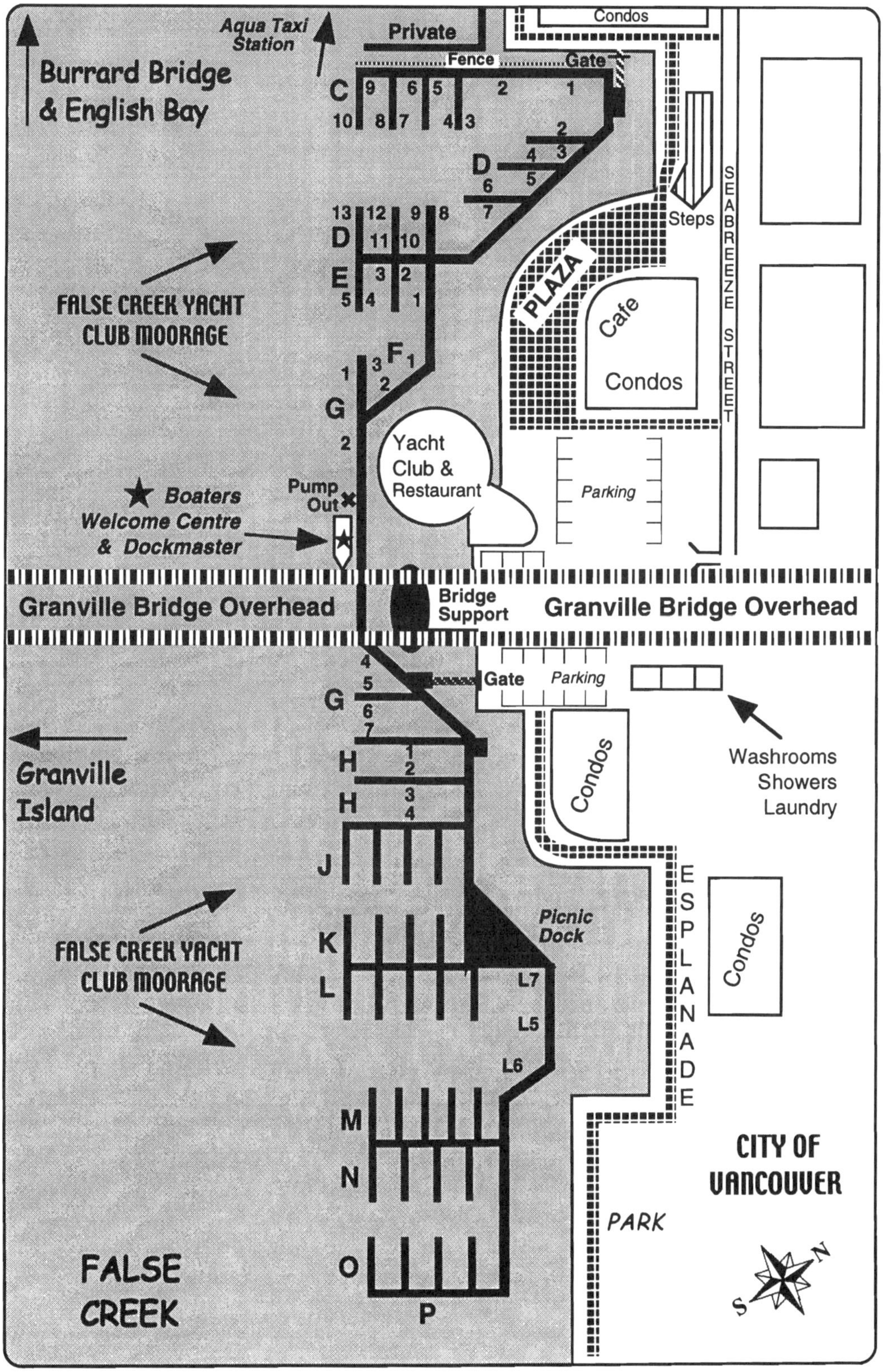

Burrard Bridge & English Bay

Aqua Taxi Station

Private

Fence

Gate

Condos

C 9 6 5 2 1

10 8 7 4 3

D 2 3 4 5 6 7

Steps

SEABREEZE STREET

13 12 9 8

D 11 10

E 3 2

5 4 1

PLAZA

Cafe

Condos

FALSE CREEK YACHT CLUB MOORAGE

F 1 3 2

1

G

2

Yacht Club & Restaurant

Parking

★ Boaters Welcome Centre & Dockmaster

Pump Out

Granville Bridge Overhead

Bridge Support

Granville Bridge Overhead

4 5 G 6 7

Gate

Parking

H 1 2

H 3 4

Granville Island

Washrooms Showers Laundry

Condos

J

FALSE CREEK YACHT CLUB MOORAGE

K

L

Picnic Dock

L7

L5

L6

ESPLANADE

Condos

M

N

O

P

FALSE CREEK

CITY OF VANCOUVER

PARK

N S

Vancouver-Granville Is.

NAME OF MARINA: ***MARITIME MARKET & MARINA*** RADIO: None
TELEPHONE: 604-408-0100 MGR: L. Jacinto, Property Manager
E-MAIL: maritimemarket@telus.net FAX: 604-408-0112 (Fax)
ADDRESS: #100-1676 Duranleau St. Vancouver, B.C. Canada V6H 3S4
SHORT DESCRIPTION & LOCATION: **www.granvilleisland.com**
49°16.35' - 123°08.20' Located in False Creek on W. side of **Granville Is.** next to Bridges Restaurant. There are 3 moorage facilities on west side of Island with access to all Granville amenities. **Maritime Market Marina** rents open slips available on their docks.

GUEST BOAT CAPACITY: ....Varies
DOCKSIDE DEPTH AT ZERO TIDE: 7 ft. plus
SEASON: ....All year
RESERVATION POLICY: ....Recommended
AMT W/ELECTRICITY: ....All
FUEL DOCK: ....Close by
MARINE REPAIRS: ....On premises
TOILETS: ....Yes
HOT SHOWERS: ....Yes
RESTAURANT: ....Several close by
PICNIC AREA: ....Yes
BASIC STORE: ....Several close by
BROADBAND/WI-FI: ....None
DAILY RATE: ....Premium (Over $1.25/foot)

GUEST DOCK: ....Varies
GUEST SLIPS: ....Varies
WATER: ....Yes
AMPS: ....30 A
PUMP OUT STATION Close by
HAUL OUT: ....Travel-Lift
BOAT RAMP: ....Close by
LAUNDRY: ....Close by
BAR: ....Several close by
POOL: ....None
GOLF: ....None
PET FRIENDLY: ....Fair
OTHER: Shopping, books, crafts, theaters, art, lg. market, fresh produce, seafood, meat, baking, beer & wine sales, chandleries.

NOTES

Within Maritime Market Marina on Granville Island there are two additional marine facilities that often rent open slips to visiting boaters - phone ahead to check availability:

1. **Blue Pacific Yacht Charters**
   Tel: 604-682-2161 (summer only)
   www.bluepacificcharters.com

2. **Cooper Boating Center**
   Tel: 604-687-4110
   www.cooperboating.com

Vancouver-Granville Island

CAUTION! This chartlet not intended for use in navigation.

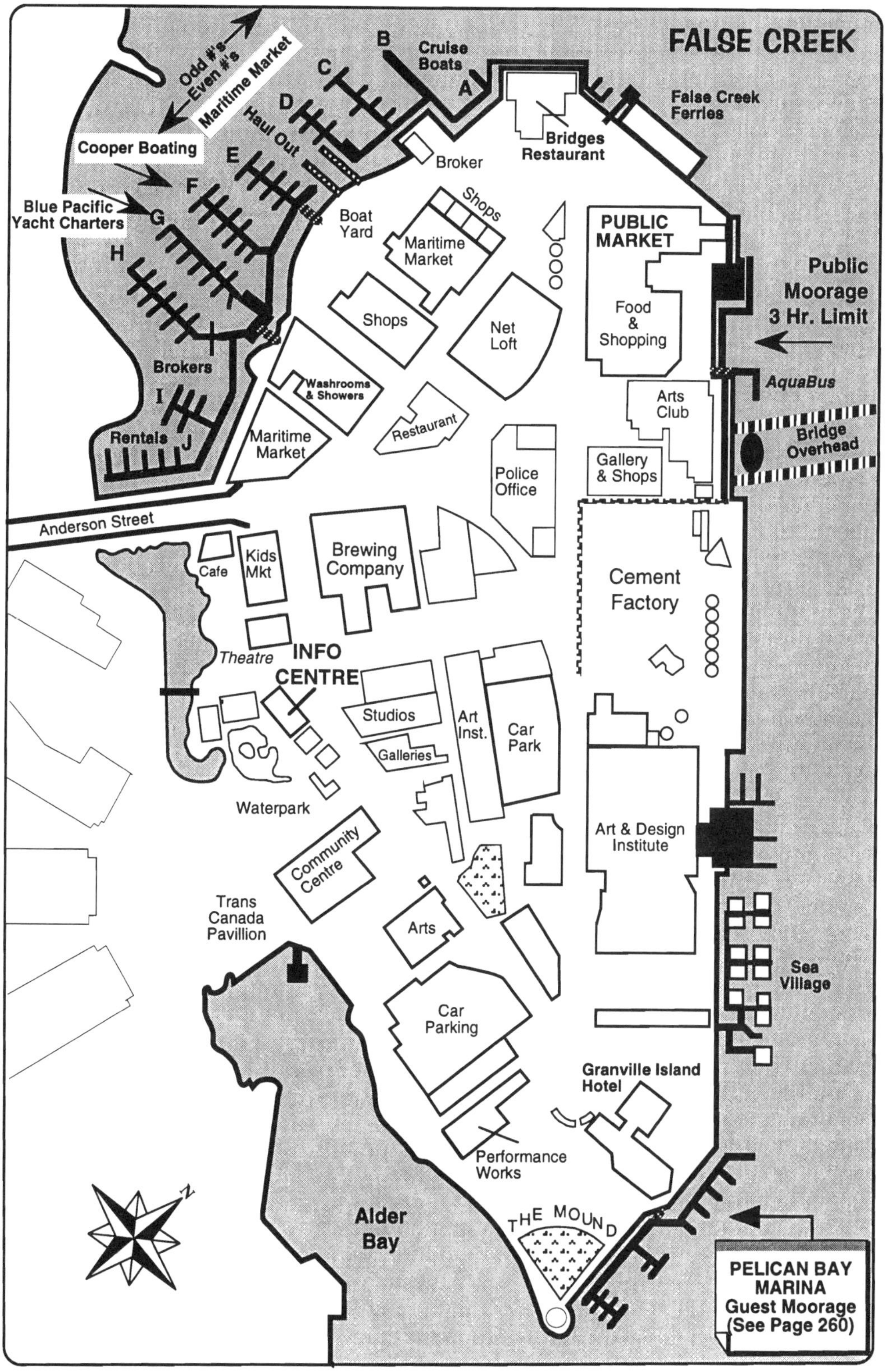

FALSE CREEK
Odd #'s
Even #'s
Maritime Market
B
Cruise Boats
A
C
D
Haul Out
E
Cooper Boating
F
Blue Pacific Yacht Charters
G
H
Brokers
I
J
Rentals
Broker
Bridges Restaurant
False Creek Ferries
Boat Yard
Shops
Maritime Market
PUBLIC MARKET
Food & Shopping
Public Moorage 3 Hr. Limit
AquaBus
Bridge Overhead
Shops
Net Loft
Washrooms & Showers
Restaurant
Maritime Market
Arts Club
Gallery & Shops
Police Office
Anderson Street
Cafe
Kids Mkt
Brewing Company
Cement Factory
Theatre
INFO CENTRE
Studios
Art Inst.
Car Park
Galleries
Waterpark
Art & Design Institute
Community Centre
Trans Canada Pavillion
Arts
Sea Village
Car Parking
Granville Island Hotel
Performance Works
Alder Bay
THE MOUND
N
PELICAN BAY MARINA
Guest Moorage
(See Page 260)

Vancouver-Granville Is.

NAME OF MARINA: ***PELICAN BAY MARINA LTD.*** **RADIO:** None
TELEPHONE: **604-729-1442, 682-7420** **MGR:** Mike & Elaine Jensen, Owners
E-MAIL: mejens@shaw.ca **FAX:** 604-682-7433 (Fax)
ADDRESS: 1708 West 6th Vancouver, B.C. Canada V6J 5E8
SHORT DESCRIPTION & LOCATION: **www.globalairphotos.com/pelican_bay_marina/**

49°16.25' - 123°07.80' Located in False Creek at the Granville Island Hotel on E. end of Granville Is., locally called the "quiet end of the Island". Modern, protected & visitor friendly marina rents open slips during the summer months - phone for reservation.

GUEST BOAT CAPACITY: ....Varies
DOCKSIDE DEPTH AT ZERO TIDE: 4 ft. plus
SEASON: ....All year
RESERVATION POLICY: ....Recommended
AMT W/ELECTRICITY: ....All
FUEL DOCK: ....Close by
MARINE REPAIRS: ....Close by
TOILETS: ....Yes
HOT SHOWERS: ....None
RESTAURANT: ....Close by
PICNIC AREA: ....Close by
BASIC STORE: ....Close by
BROADBAND/WI-FI: ....None
DAILY RATE: ....Premium (Over $1.25/foot)

GUEST DOCK: ....Varies
GUEST SLIPS: ....Varies
WATER: ....Yes
AMPS: ....30-50 A
PUMP OUT STATION Close by
HAUL OUT: ....Close by
BOAT RAMP: ....Close by
LAUNDRY: ....None
BAR: ....Close by
POOL: ....None
GOLF: ....None
PET FRIENDLY: ....Good
OTHER: Cable TV, electronic security, access to all Granville Island amenities & Maritime Museum. False Creek Ferry to downtown.

CAUTION! This chartlet not intended for use in navigation.

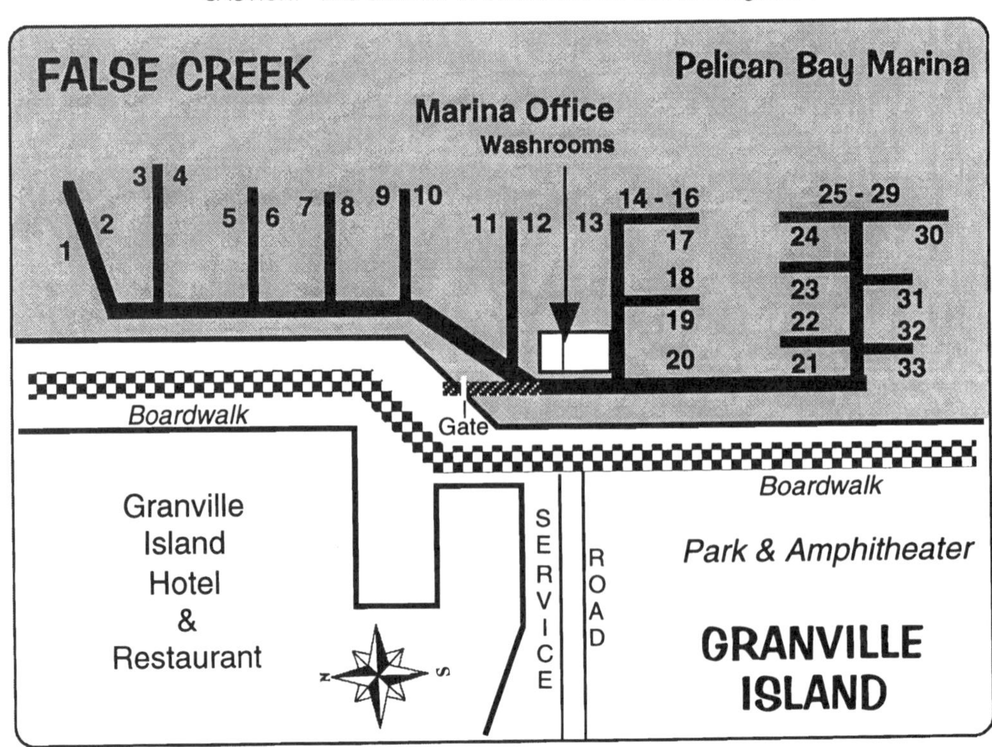

Vancouver

NAME OF MARINA: ***QUAYSIDE MARINA*** **RADIO:** VHF 66A
TELEPHONE: **604-681-9115** **MGR:** Dockmaster
E-MAIL: qsmarina@ranchogroup.com **FAX:** 604-681-1932
ADDRESS: 1088 Marinaside Crescent, Vancouver, B.C. Canada V6Z 3C4
SHORT DESCRIPTION & LOCATION: **www.quaysidemarina.ca**

49°16.30' - 123°07.10 Located on N. side False Creek, just E. of Cambie Bridge. Marina is quiet (mostly condo slips) situated in the upscale highrise Yaletown district. Convenient to good restaurants & downtown Vancouver shopping & services. (Pronounced "Keyside")

GUEST BOAT CAPACITY: ....Appx. 5-10 boats
DOCKSIDE DEPTH AT ZERO TIDE: 10 ft. +
SEASON: ....All year
RESERVATION POLICY: ....Recommended
AMT W/ELECTRICITY: ....All
FUEL DOCK: ....Close by
MARINE REPAIRS: ....Close by
TOILETS: ....Yes
HOT SHOWERS: ....Yes
RESTAURANT: ....Close by
PICNIC AREA: ....Close by
BASIC STORE: ....Close by
BROADBAND/WI-FI: ....None
DAILY RATE: ....Premium (Starting at $1.85/foot)

GUEST DOCK: Appx. 300-400 ft.
GUEST SLIPS: ....None
WATER: ....Yes
AMPS: ....30 A
PUMP OUT STATION ....Yes
HAUL OUT: ....None
BOAT RAMP: ....None
LAUNDRY: ....Yes
BAR: ....Close by
POOL: ....None
GOLF: ....Close by
PET FRIENDLY: ....Good
OTHER: Handy Aqua Bus serves Granville Is., NO customs clearance. Diver, yacht maintenance & marine mechanic available.

CAUTION! This chartlet not intended for use in navigation.

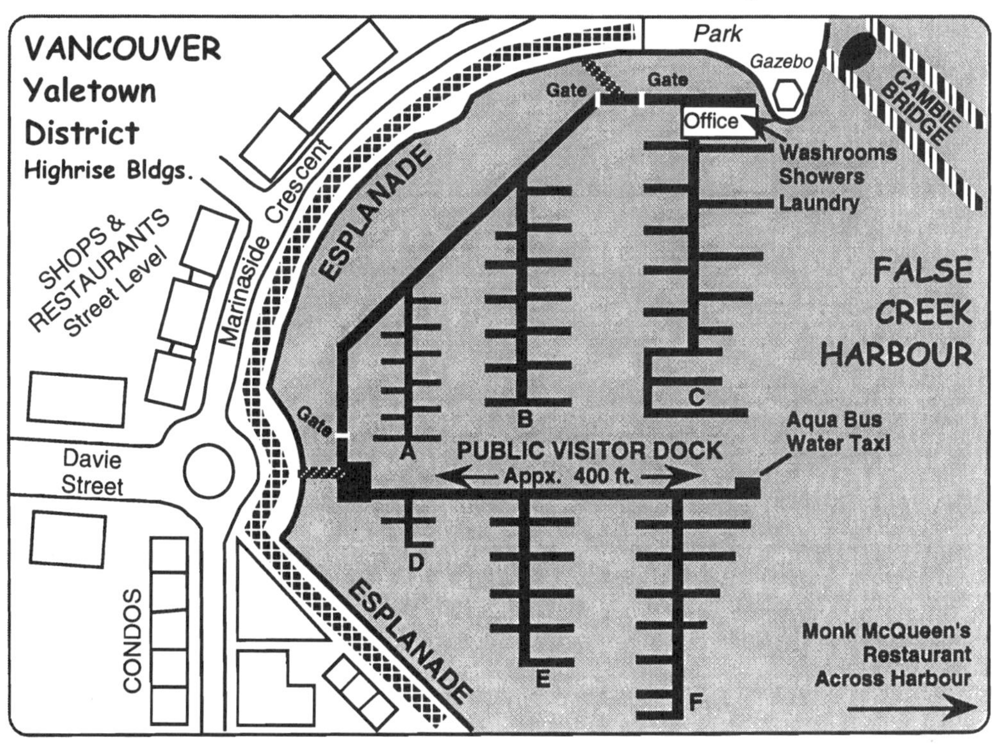

Vancouver

NAME OF MARINA: ***COAL HARBOUR MARINA*** **RADIO:** VHF 66A
TELEPHONE: **604-681-2628** **MGR:** Marina Manager
E-MAIL: info@coalharbourmarina.com **FAX:** 604-681-4666 Fax
ADDRESS: 1525 Coal Harbour Quay Vancouver, B.C. Canada V6G 3E7
SHORT DESCRIPTION & LOCATION: **www.coalharbourmarina.com**

49°17.50' - 123°07.60' Strategically located in the heart of Vancouver's downtown waterfront. Marina is a full service facility offering boaters a large range of yachting amenities. Shopping, gift shops, theaters, fine dining & night clubs are walking distance.

GUEST BOAT CAPACITY: ....Appx. 20 boats
DOCKSIDE DEPTH AT ZERO TIDE: 20 ft. plus
SEASON: ....All year
RESERVATION POLICY: ....Required
AMT W/ELECTRICITY: ....All
FUEL DOCK: ....Close by
MARINE REPAIRS: ....Close by
TOILETS: ....Yes
HOT SHOWERS: ....Yes
RESTAURANT: ....Yes
PICNIC AREA: ....Close by in Stanley Park
BASIC STORE: ....Close by
BROADBAND/WI-FI: ....BroadbandXpress
DAILY RATE: ....Premium (Over $1.25/foot)

GUEST DOCK: ....Up to 330 ft.
GUEST SLIPS: ....Varies
WATER: ....Yes
AMPS: ....30-100 A
PUMP OUT STATION ....Yes
HAUL OUT: ....None
BOAT RAMP: ....None
LAUNDRY: ....Yes
BAR: ....Pub
POOL: ....None
GOLF: ....None
PET FRIENDLY: ....Good
OTHER: Cable TV, telephone hookups, chandlery, boat rentals, & security. Close to Stanley Park's trails, gardens, & Aquarium.

CAUTION! This chartlet not intended for use in navigation.

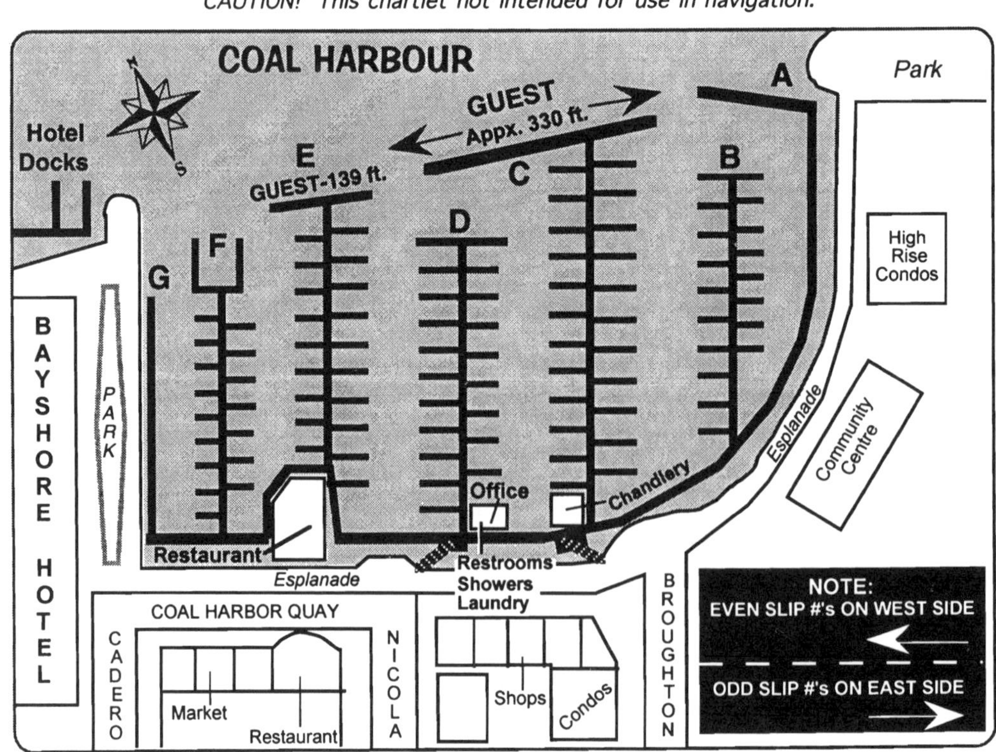

Vancouver

NAME OF MARINA: ***HARBOUR CRUISES MARINA*** **RADIO:** None
TELEPHONE: **604-682-9558** **MGR:** Fred Hercules
E-MAIL: james@harbourcruises.com **FAX:** 604-683-0684 (Fax)
ADDRESS: #1 N. Foot of Denman Vancouver, B.C. Canada V6G 2W9
SHORT DESCRIPTION & LOCATION: **www.boatcruises.com**

49°17.70' - 123°07.90' Located in Coal Harbour next to Stanley Park. Marina has no designated guest moorage but rents open slips to visiting boats. Close to downtown Vancouver, a short walk or cab ride to shops, galleries, nightclubs, and restaurants.

GUEST BOAT CAPACITY: ....Varies
DOCKSIDE DEPTH AT ZERO TIDE: 7 ft. plus
SEASON: ....All year
RESERVATION POLICY: ....Accepts
AMT W/ELECTRICITY: ....All
FUEL DOCK: ....Close by
MARINE REPAIRS: ....Close by
TOILETS: ....None
HOT SHOWERS: ....None
RESTAURANT: ....Close by
PICNIC AREA: ....Close by
BASIC STORE: ....Close by
BROADBAND/WI-FI: ....None
DAILY RATE: ....Premium (Over $1.25/foot)

GUEST DOCK: ....Slips only
GUEST SLIPS: ....Varies
WATER: ....Yes
AMPS: ....15-30 A
PUMP OUT STATION ....None
HAUL OUT: ....None
BOAT RAMP: ....None
LAUNDRY: ....Close by
BAR: ....Close by
POOL: ....None
GOLF: ....None
PET FRIENDLY: ....Excellent
OTHER: Close to Stanley Park's trails, gardens, & the Vancouver Aquarium.

CAUTION! This chartlet not intended for use in navigation.

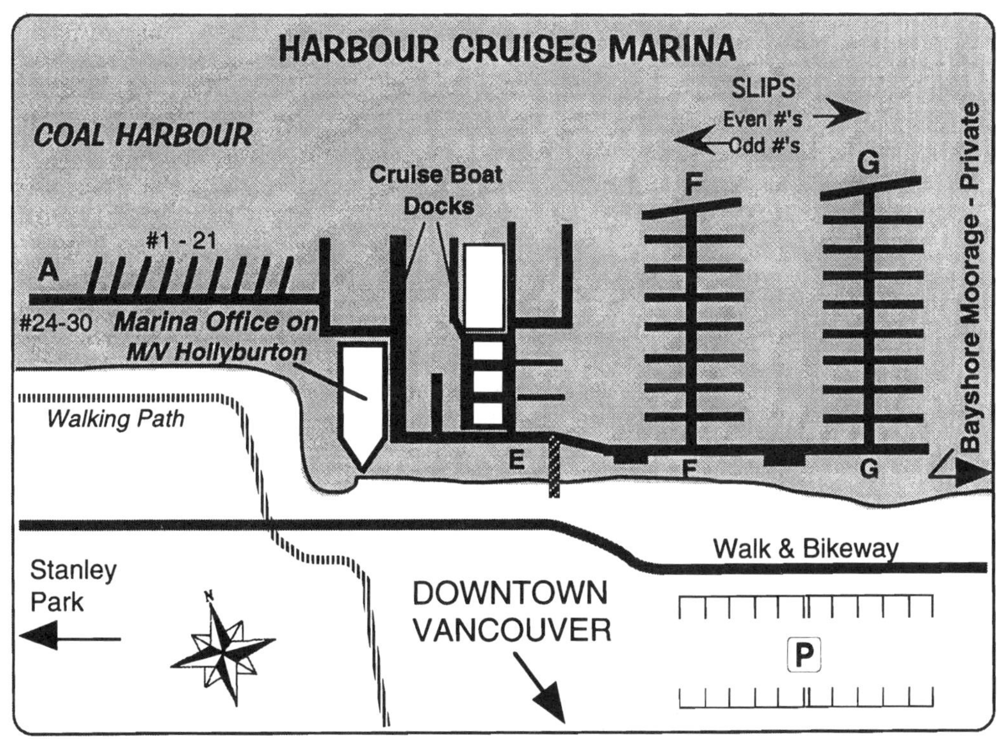

Vancouver - North Van

NAME OF YACHT CLUB: *BURRARD YACHT CLUB*

ADDRESS: 10 Gostick Place North Vancouver, B.C. V7M 3G3

TELEPHONE: 604-988-0817 PERSON IN CHARGE: Staff Captain

SHORT DESCRIPTION & LOCATION: **www.burrardyachtclub.com**

49°18.65' -123°05.57' Located in North Van. on N. shore of Burrard Inlet by entering First Narrows under Lions Gate Bridge. Club moorage protected by breakwaters & former Navy ship. Absent member slips are used for visitors. Tie up at visitor check-in (inside breakwater) & register at Office for slip (if available). Otherwise use 120' on breakwater.

RECIPROCAL BOAT CAPACITY: ....Varies

DOCKSIDE DEPTH AT ZERO TIDE: ....9 ft. +

SEASON: ....All year

RESERVATION POLICY: ....Call ahead

TOILETS: ....Yes

HOT SHOWERS: ....Yes

RESTAURANT: ....Close by

BROADBAND/WI-FI: ....None

DAILY RATE: First 2 days free, 25¢/ft. thereafter. 14 Days maximum per year. Electricity $4/each day.

RECIPROCAL DOCK: ....120 Ft.

RECIPROCAL SLIPS: ....Varies

WATER: ....Yes

AMT W/ELECTRICITY: ....All

AMPS: ....15-30 A

BAR: ....Pubs close by

PET FRIENDLY: ....Good

OTHER: Short cab or dinghy ride to restaurants, shopping, local bus service, & Seabus to downtown Vancouver. Laundry & ice.

***NOTE:* THIS IS PRIVATE MOORAGE AND ONLY AVAILABLE TO MEMBERS OF RECIPROCAL YACHT CLUBS. YOUR CLUB *MUST* HAVE RECIPROCAL PRIVILEGES AND YOU MUST FLY YOUR BURGEE.**

CAUTION! This chartlet not intended for use in navigation.

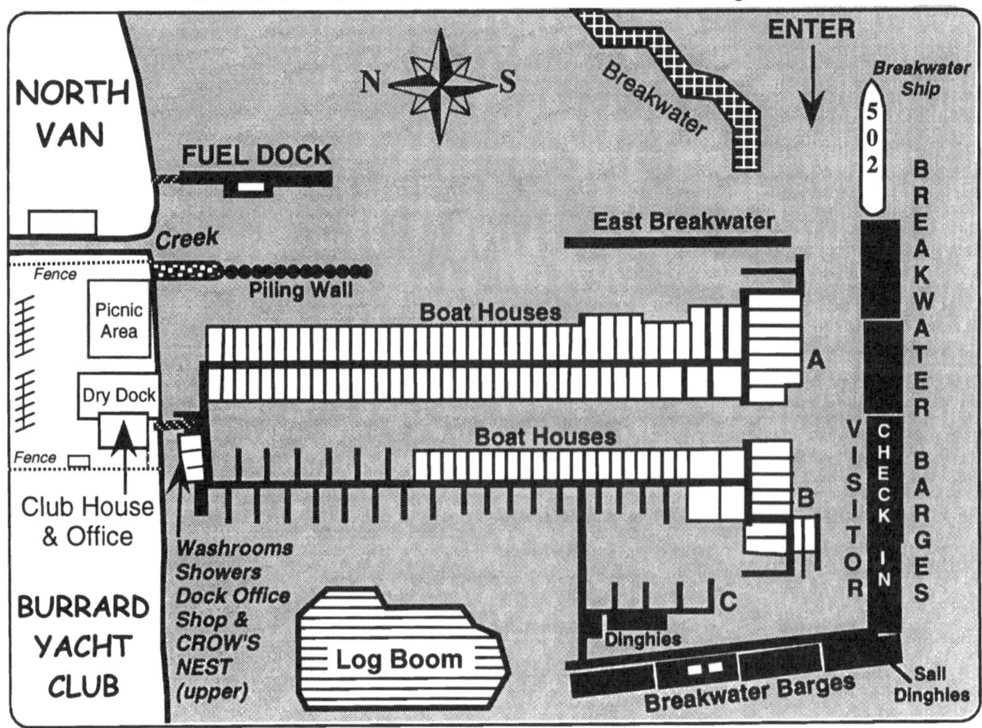

Deep Cove

NAME OF YACHT CLUB: *DEEP COVE YACHT & SPORT CLUB*
ADDRESS: 4420 Gallant Avenue North Vancouver, Canada V7G 1L2
TELEPHONE: 604-929-1009/999-9960 PERSON IN CHARGE: Secretary
SHORT DESCRIPTION & LOCATION: **www.deepcoveyc.com**
49°19.68' -122°56.77' Deep Cove is a village at the head of the Cove with the same name just through the entrance of scenic Indian Arm.. Yacht club & floats are directly in front of town next to small govt. float, surrounded by a bearutiful park & plaza. Reciprocal moorage consists of the Breakwater Float. Check in & get gate code from caretaker.

RECIPROCAL BOAT CAPACITY: Appx. 8 boats
DOCKSIDE DEPTH AT ZERO TIDE: 40 ft. plus
SEASON: ....All year
RESERVATION POLICY: .....FCFS - Can call ahead
TOILETS: ....Yes
HOT SHOWERS: ....Yes
RESTAURANT: ....Yes - Closed Mon. & Tue
BROADBAND/WI-FI: ....None
DAILY RATE: Free for 48 hrs. - 50¢ ft. after 48 hours. Power charge $3.00/day

RECIPROCAL DOCK: ....250 ft.
RECIPROCAL SLIPS: ....Varies
WATER: ....Yes
AMT W/ELECTRICITY: ....Limited
AMPS: ....15 A
BAR: ....Yes
PET FRIENDLY: ....Good
OTHER: Upscale village is rich with arts, theatres, shops & restaurants.

***NOTE:* THIS IS PRIVATE MOORAGE AND ONLY AVAILABLE TO MEMBERS OF RECIPROCAL YACHT CLUBS. YOUR CLUB *MUST* HAVE RECIPROCAL PRIVILEGES AND YOU MUST FLY YOUR BURGEE.**

CAUTION! This chartlet not intended for use in navigation.

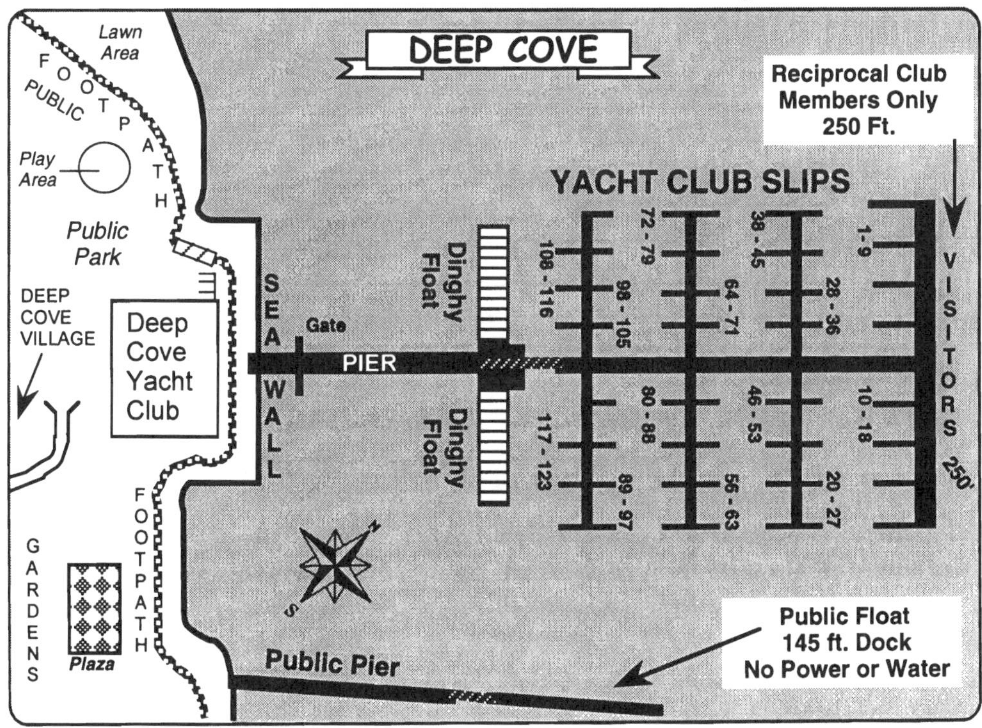

Deep Cove

NAME OF MARINA: *DEEP COVE NORTH SHORE MARINA* RADIO: None
TELEPHONE: 604-929-1251 MGR: Rick Barbieri
E-MAIL: None FAX: 604-929-7862 (Fax)
ADDRESS: 2890 Panorama Drive North Vancouver, B.C. Canada V7G 1V6
SHORT DESCRIPTION & LOCATION: **www.deepcovemarina.com**

49°19.90' - 122°56.34' Deep Cove is a village at head of Cove with same name just through the entrance of scenic Indian Arm. Marina located 1/2 mi N of village but within walking distance. Marina has no designated guest moorage, but rents out any open slips.

GUEST BOAT CAPACITY: ....Appx . 6-7 boats
DOCKSIDE DEPTH AT ZERO TIDE: 100 ft.
SEASON: ....All year
RESERVATION POLICY: ....Recommended
AMT W/ELECTRICITY: ....All
FUEL DOCK: ....Gas & Diesel
MARINE REPAIRS: ....On premises
TOILETS: ....Yes
HOT SHOWERS: ....Yes
RESTAURANT: ....Close by
PICNIC AREA: ....Close by
BASIC STORE: ....Yes
BROADBAND/WI-FI: ....Yes
DAILY RATE: ....Premium (Over $1.25/foot)

GUEST DOCK: ......Uses open slips
GUEST SLIPS: ....Varies
WATER: ....Yes
AMPS: ....30 A
PUMP OUT STATION ....Yes
HAUL OUT: ....Small boats
BOAT RAMP: ....Yes
LAUNDRY: ....Yes
BAR: ....Close by
POOL: ....Close by
GOLF: ....Close by
PET FRIENDLY: ....Fair
OTHER: Walking distance to upscale Deep Cove Village, rich with arts, theaters, shops & restaurants.

CAUTION! This chartlet not intended for use in navigation.

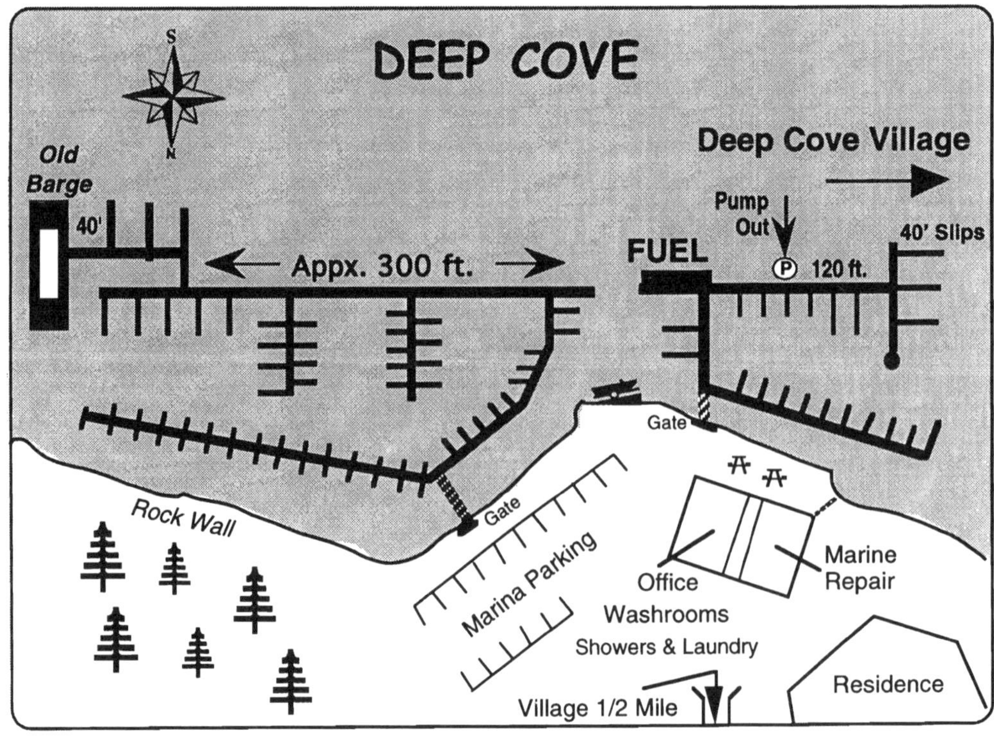

Horseshoe Bay

NAME OF MARINA: ***SEWELL'S MARINA*** **RADIO:** VHF 81A
TELEPHONE: **604-921-3474** **MGR:** Megan Sewell
E-MAIL: info@sewellsmarina.com **FAX:** 604-921-7027 (Fax)
ADDRESS: 6409 Bay Street West Vancouver, B.C. Canada V7W 3H5
SHORT DESCRIPTION & LOCATION: **www.sewellsmarina.com**

49°22.65' - 123°16.40' Located just inside Howe Sound in Horseshoe Bay inside of Tyee Pt. on E side of Queen Charlotte Ch. Marina is breakwater protected located next to big B.C. Ferry Dock that serves the Gulf Islands, Vancouver Island, and Sunshine Coast.

GUEST BOAT CAPACITY: ....Appx. 6-10 boats
DOCKSIDE DEPTH AT ZERO TIDE: 40 ft. plus
SEASON: ....All year
RESERVATION POLICY: ....None
AMT W/ELECTRICITY: ....All
FUEL DOCK: ....Gas & Diesel
MARINE REPAIRS: ....Close by
TOILETS: ....Yes
HOT SHOWERS: ....None
RESTAURANT: ....Close by
PICNIC AREA: ....Close by
BASIC STORE: ....Close by
BROADBAND/WI-FI: ....None
DAILY RATE: ....Moderate (75¢-$1.25/foot)

GUEST DOCK: Appx. 100 ft TTL
GUEST SLIPS: ....Varies
WATER: ....Yes
AMPS: ....15-30 A
PUMP OUT STATION ....None
HAUL OUT: ....None
BOAT RAMP: ....Close by
LAUNDRY: ....None
BAR: ....Close by
POOL: ....None
GOLF: ....Close by
PET FRIENDLY: ....Good
OTHER: Chandlery, fishing charters, licenses, bait, tackle, boat rentals. Adjacent restaurants, shops, & ferries.

CAUTION! This chartlet not intended for use in navigation.

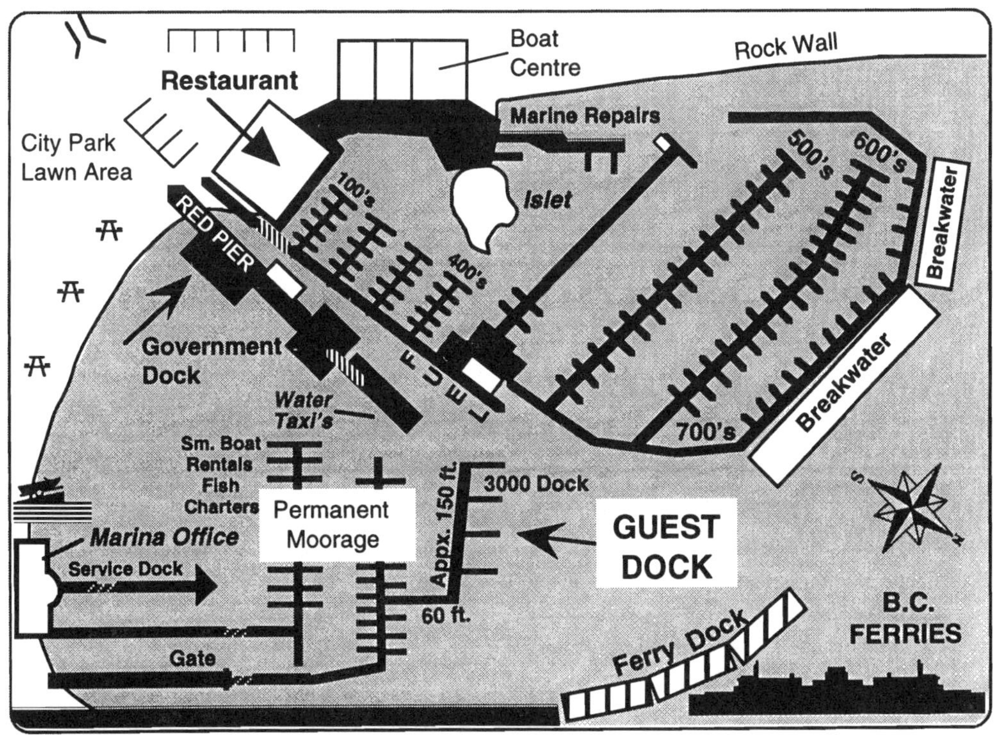

Bowen Island

NAME OF MARINA: ***BOWEN ISLAND MARINA*** **RADIO:** VHF 66A
TELEPHONE: **604-947-9710** **MGR:** Norma Dallas
E-MAIL: norma@bowen-island.com **FAX:** 604-947-9710 (Ph/Fax)
ADDRESS: 375 Cardena Dr.,RR 1 A 1 Bowen Island, B.C. Canada V0N 1G0
SHORT DESCRIPTION & LOCATION: **www.bowen-island.com**

49°22.80' - 123°19.85' Located in Snug Cove on Bowen Island in Howe Sound. Cozy small marina just to the RIGHT of the ferry dock. Check in at Ice Cream Shop or brown house w/white railings above marina. Many services for boaters.

GUEST BOAT CAPACITY: ....Limited
DOCKSIDE DEPTH AT ZERO TIDE: 9 ft. +
SEASON: ....All year
RESERVATION POLICY: ....Recommended
AMT W/ELECTRICITY: ....All
FUEL DOCK: ....None
MARINE REPAIRS: ....Yes
TOILETS: ....Close by
HOT SHOWERS: ....Close by
RESTAURANT: ....Close by
PICNIC AREA: ....Yes
BASIC STORE: ....Close by
BROADBAND/WI-FI: ....Yes
DAILY RATE:....Moderate (75¢-$1.25/foot)

Check in at Ice Cream Shop or brown house w/white railings above marina.

GUEST DOCK: ....60 ft. Total
GUEST SLIPS: ....Varies
WATER: ....Yes
AMPS: ....15-30 A
PUMP OUT STATION ....None
HAUL OUT: ....None
BOAT RAMP: ....Close by
LAUNDRY: ....Close by
BAR: ....Pub-Close by
POOL: ....None
GOLF: ....Close by
PET FRIENDLY: ....Good
OTHER: Fishing licenses, tackle, bait, ice, kayak rentals. Ice Cream Shoppe, Tacos, Gifts, Historical Display, Summertime live music.

CAUTION! This chartlet not intended for use in navigation.

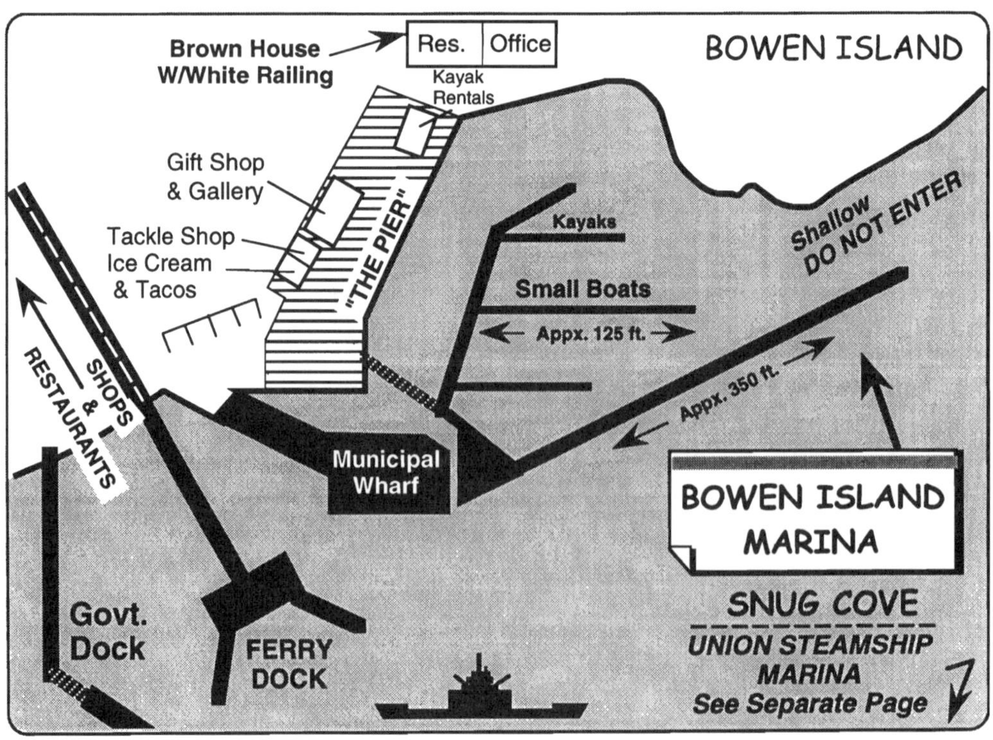

Bowen Island

NAME OF MARINA: *UNION STEAMSHIP MARINA* RADIO: VHF 66A
TELEPHONE: 604-947-0707 MGR: Rondy Dike
E-MAIL: ussc@shaw.ca FAX: 604-947-0708 (Fax)
ADDRESS: P.O. Box 250 Bowen Island, B.C. Canada V0N 1G0
SHORT DESCRIPTION & LOCATION: **www.steamship-marina.bc.ca**

49°22.90' - 123°19.80' Located on W side of Bowen Island in Howe Sound at head of Snug Cove, the southerly and smaller part of a double headed bay. First Class marina next to ferry dock, with plenty of space for visiting boats from 20 to 200 ft. long.

GUEST BOAT CAPACITY: ....Appx. 60 boats
DOCKSIDE DEPTH AT ZERO TIDE: ....8 Ft. +
SEASON: ....All year
RESERVATION POLICY: ....Accepts
AMT W/ELECTRICITY: ....All
FUEL DOCK: ....None
MARINE REPAIRS: ....Close by
TOILETS: ....Yes
HOT SHOWERS: ....Yes
RESTAURANT: ....Yes
PICNIC AREA: ....Yes
BASIC STORE: ....Yes
BROADBAND/WI-FI: ....Yes
DAILY RATE: ....Premium
(Over $1.25/foot)
WINTER RATES AVAILABLE

GUEST DOCK: ....1500 ft. Total
GUEST SLIPS: ....Varies - 10 +
WATER: ....Yes
AMPS: ....30-50 A
PUMP OUT STATION ....Portable
HAUL OUT: ....None
BOAT RAMP: ....Yes
LAUNDRY: ....Yes
BAR: ....Pub
POOL: ....None
GOLF: ....Close by
PET FRIENDLY: ....Excellent
OTHER: Shops, cottages, boardwalks, chandlery, 600 acre park nearby. Ferry & bus service to Vancouver.

CAUTION! This chartlet not intended for use in navigation.

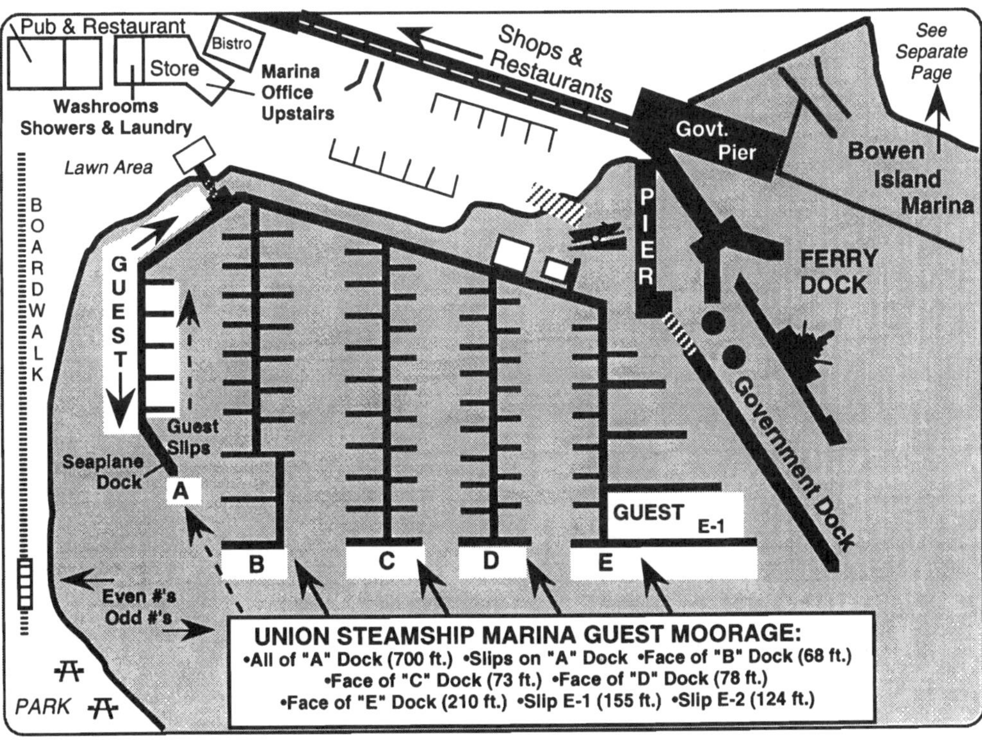

Lions Bay

NAME OF MARINA: *LIONS BAY MARINA LTD.* **RADIO:** None
TELEPHONE: **604-921-7510** **MGR:** Ken Wolder
E-MAIL: lionsbaymarina@telus.net **FAX:** 604-921-0782 (Fax)
ADDRESS: Box 262, 60 Lions Bay Ave., Lions Bay, B.C. Canada V0N 2E0
SHORT DESCRIPTION & LOCATION: **www.lionsbaymarina.com**

49°27.30' - 123°14.30' Located midway up Howe Sound on mainland side appx. 5 miles N of Horseshoe Bay. Marina provides adequate summer overnight moorage, however mainly a dry storage small boat facility. Summer floats somewhat open to boat wakes.

GUEST BOAT CAPACITY: ....Appx. 6 boats
DOCKSIDE DEPTH AT ZERO TIDE: ...6 ft. plus
SEASON: ....Closed Dec. 15 to Jan. 15
RESERVATION POLICY: ....Accepts
AMT W/ELECTRICITY: ....None
FUEL DOCK: ....Gas & LP
MARINE REPAIRS: ....On premises
TOILETS: ....Yes
HOT SHOWERS: ....None
RESTAURANT: ....Close by
PICNIC AREA: ....Yes
BASIC STORE: ....Yes
BROADBAND/WI-FI: ....None
DAILY RATE: ....Economical (Under 75¢/foot)

GUEST DOCK: ....400 ft. total
GUEST SLIPS: ....Dock only
WATER: ....Yes
AMPS: ....None
PUMP OUT STATION ....None
HAUL OUT: ....To 32 ft.
BOAT RAMP: ....Yes
LAUNDRY: ....None
BAR: ....Close by
POOL: ....None
GOLF: ....None
PET FRIENDLY: ....Good
OTHER: Bait, Fishing Tackle, Full size grocery store and Post Office close by.

CAUTION! This chartlet not intended for use in navigation.

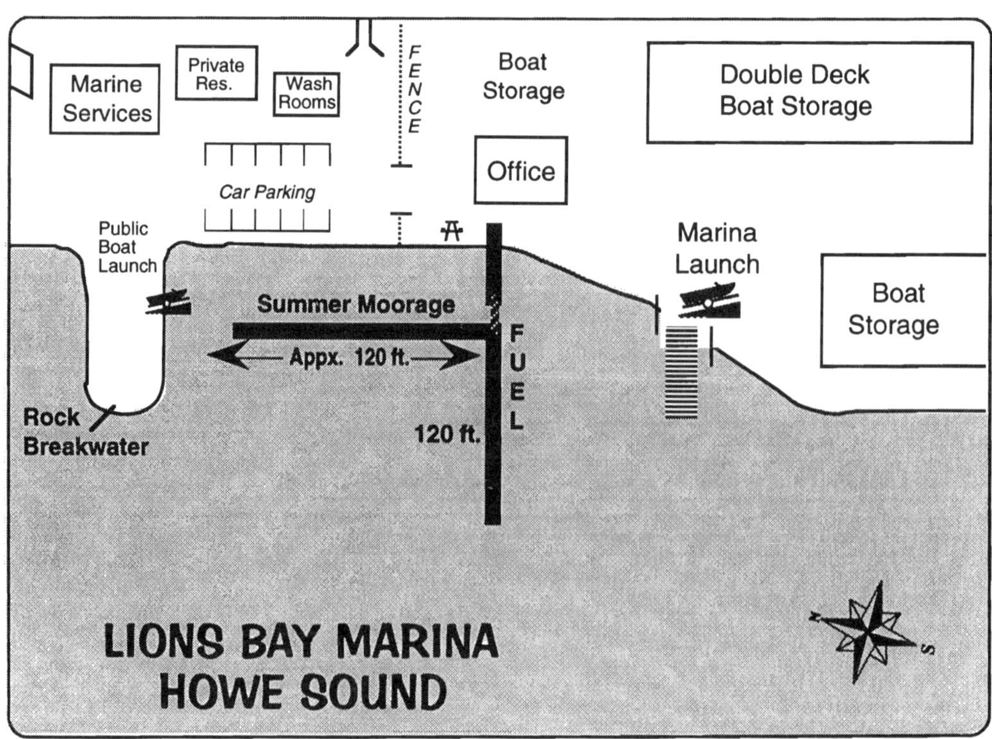

Squamish

NAME OF MARINA: ***SQUAMISH YACHT CLUB (Y.C.)*** **RADIO:** None
TELEPHONE: Y.C: 604-892-4101 ***& HARBOUR AUTHORITY (H.A.)***
TELEPHONE H.A: 604-892-3275 **MGR:** Bill McEnery **FAX:** 604-892-3942 (Y.C. Fax)
ADDRESS: Y.C.-P.O. Box 1681 H.A.-P.O. Box 97 Squamish, B.C. Canada V0N 3G0
SHORT DESCRIPTION & LOCATION: **Y.C. WEBSITE: www.squamishyachtclub.com**

49°41.67' - 123°09.25' Best to preplan visit by checking info on website. Situated at N. end of Howe Sound. Enter thru easternmost entrance of E. Arm of the Squamish River between Chemical Wharf & range lights. Public wharf & Y.C. are 1 mi. NNE of entrance light.

GUEST BOAT CAPACITY: ....Appx. 4-6 boats
DOCKSIDE DEPTH AT ZERO TIDE: ....3-7 ft.
SEASON: ....All year
RESERVATION POLICY:............Best to call ahead
AMT W/ELECTRICITY: ....Limited
FUEL DOCK: ....None
MARINE REPAIRS: ....Limite
TOILETS: ....Yes
HOT SHOWERS: ....Yes
RESTAURANT: ....Many close by
PICNIC AREA: ....Close by
BASIC STORE: ....Close by
BROADBAND/WI-FI: ....None
DAILY RATE:.............Economical (Under 75¢/foot)
If public float full, or if you're reciprocal, tie up in an empty outside slip & find a member to help you.

GUEST DOCK:H.A.: Appx. 200 ft.
GUEST SLIPS:.........Y.C. slips vary
WATER: ....Yes
AMPS: ....15 A
PUMP OUT STATION ....Yes
HAUL OUT: ....Grid
BOAT RAMP:..........Yes
LAUNDRY: ....Close by
BAR: ....Close by
POOL:..........Close by
GOLF: ....Close by
PET FRIENDLY: Good-On leash
OTHER: Hospitable yacht club usually allows overnight moorage for visitors. Retail stores, restaurants, pubs, hotels, & hospital nearby.

CAUTION! This chartlet not intended for use in navigation.

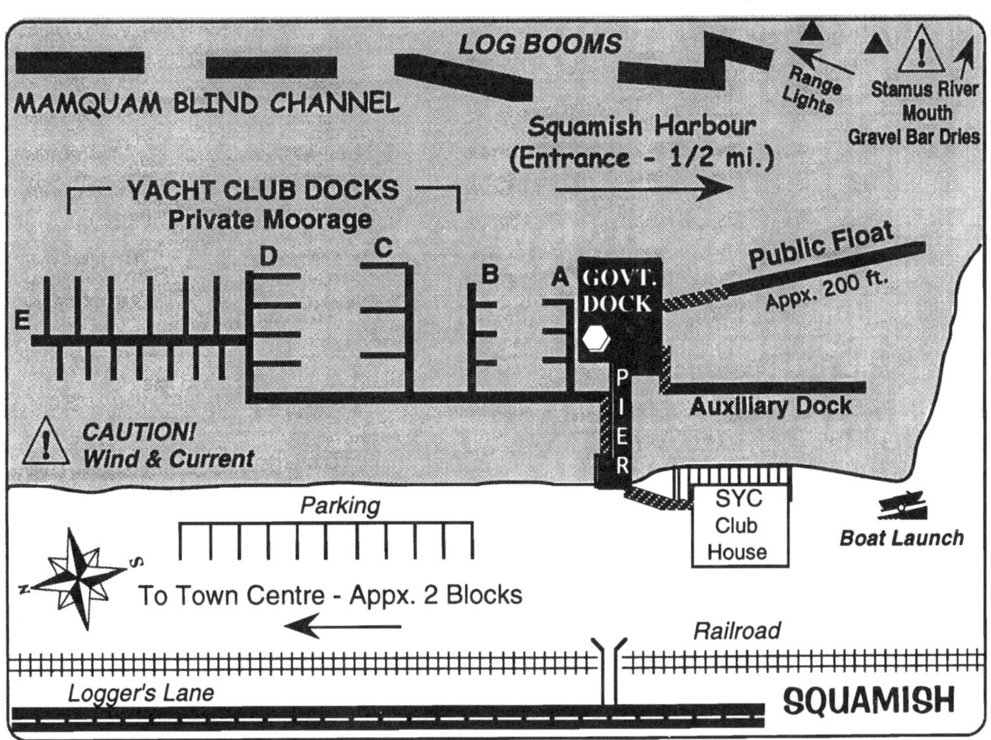

NOTES

CHAPTER 9

THE SUNSHINE COAST

Chapter Map - Page 2

Gibsons

NAME OF MARINA: ***GIBSONS MARINA*** RADIO: VHF 66A
TELEPHONE: **604-886-8686** MGR: Arthur McGinnis
E-MAIL: None FAX: 604-886-8686 (Fax)
ADDRESS: P.O. Box 1520, Gibsons, B.C. Canada V0N 1V0

SHORT DESCRIPTION & LOCATION:

49°23.98' - 123°30.30' Town on NW side of Shoal Channel, N of Steep Bluff. Marina located on left (S) side upon clearing entrance between 2 rock breakwaters. Downtown Gibsons offers shops, restaurants, & a pub. Marina staff friendly & town is charming.

GUEST BOAT CAPACITY: ....Appx. 20-30 boats
DOCKSIDE DEPTH AT ZERO TIDE: 7 ft. plus
SEASON: ....All year
RESERVATION POLICY:........Accepts
AMT W/ELECTRICITY: ....All
FUEL DOCK: ....Close by
MARINE REPAIRS: ....Close by
TOILETS: ....Yes
HOT SHOWERS: ....Yes
RESTAURANT: ....Close by
PICNIC AREA: ....Close by
BASIC STORE: ....Yes
BROADBAND/WI-FI: ....BroadbandXpress
DAILY RATE:........Moderate (75¢-$1.25/foot)

GUEST DOCK: Appx. 130' + slips
GUEST SLIPS: ....Appx. 20-25
WATER: ....Yes
AMPS: ....15 A
PUMP OUT STATION Portable
HAUL OUT: ....Close by
BOAT RAMP: ....Yes
LAUNDRY: ....Yes
BAR: ....Close by
POOL: ....None
GOLF: ....Close by
PET FRIENDLY: ....Good
OTHER: Chandlery, security on premises. Restaurants, stores, tennis courts & picnic areas within easy walking distance.

NOTES

Gibsons Landing

CAUTION! This chartlet not intended for use in navigation.

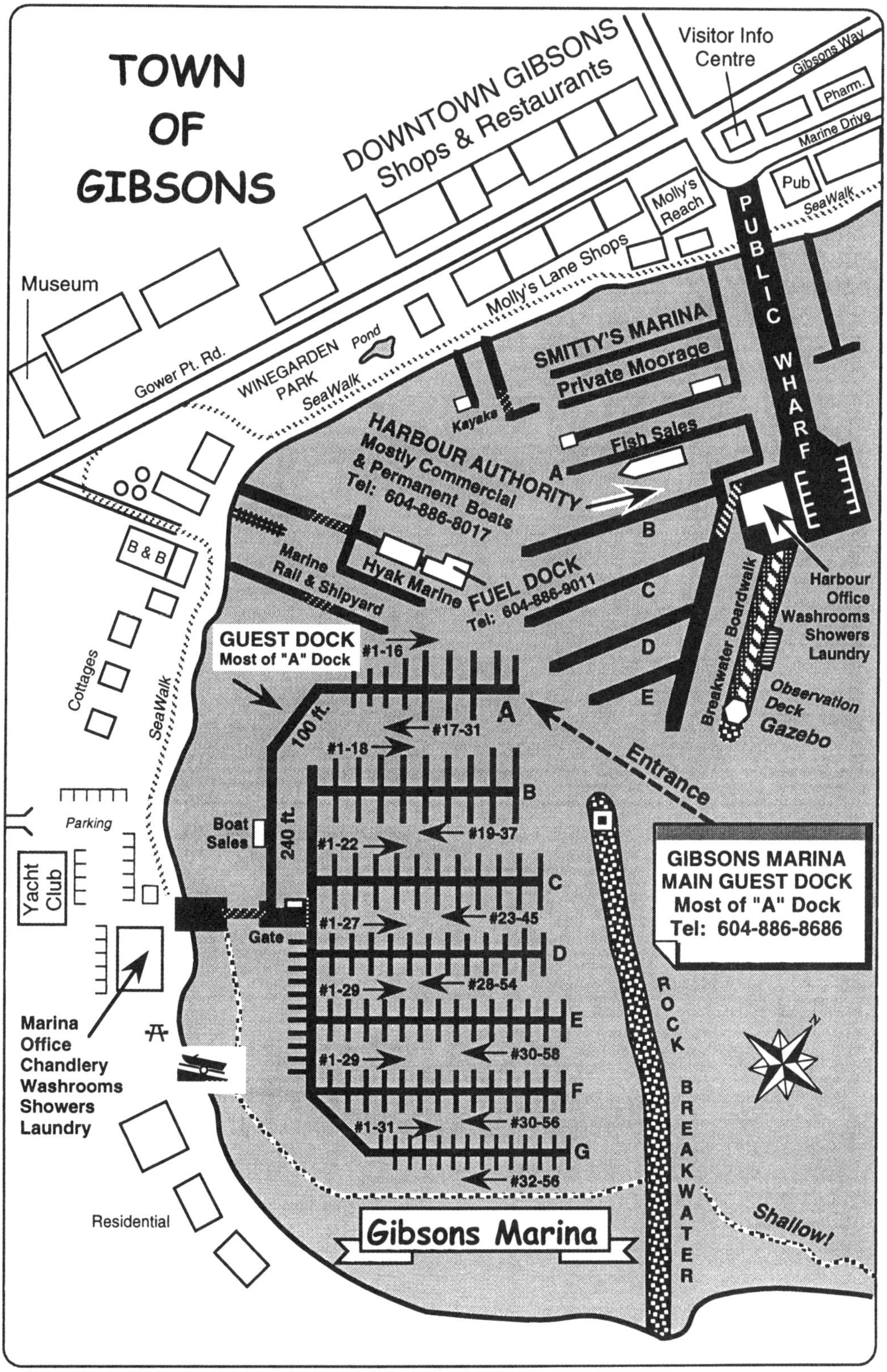

TOWN OF GIBSONS

DOWNTOWN GIBSONS
Shops & Restaurants

Visitor Info Centre
Gibsons Way
Pharm.
Marine Drive
Pub
SeaWalk
Molly's Reach
Molly's Lane Shops
PUBLIC WHARF
Museum
Gower Pt. Rd.
WINEGARDEN PARK
Pond
SeaWalk
SMITTY'S MARINA
Private Moorage
Kayaks
Fish Sales
HARBOUR AUTHORITY
Mostly Commercial & Permanent Boats
Tel: 604-886-8017
A
B
C
D
E
Marine Rail & Shipyard
Hyak Marine
FUEL DOCK
Tel: 604-886-9011
Breakwater Boardwalk
Harbour Office
Washrooms
Showers
Laundry
Observation Deck
Gazebo
B & B
Cottages
SeaWalk

GUEST DOCK
Most of "A" Dock

#1-16
100 ft.
A
#17-31
#1-18
B
Entrance
Boat Sales
240 ft.
#1-22
#19-37
C
#1-27
#23-45
Gate
D
#1-29
#28-54
E
#1-29
#30-58
F
#1-31
#30-56
G
#32-56

Parking
Yacht Club

GIBSONS MARINA
MAIN GUEST DOCK
Most of "A" Dock
Tel: 604-886-8686

ROCK BREAKWATER
N
Shallow!

Marina Office
Chandlery
Washrooms
Showers
Laundry

Residential

Gibsons Marina

Secret Cove

NAME OF MARINA: **SECRET COVE MARINA** RADIO: VHF 66A
TELEPHONE: **604-885-3533** MGR: Scott Rowland
E-MAIL: info@secretcovemarina.com FAX: 604-885-6037 (Fax)
ADDRESS: Box 1118, Sechelt, B.C. Canada V0N 1V0
SHORT DESCRIPTION & LOCATION: **www.secretcovemarina.com**

49°32.10' - 123°57.90' Located on Sechelt Peninsula. Entered between S. end of Turnagain Is. & Jack Tolmie Is. at S. end of Malaspina Strait near Welcome Pass. Full service marina in N. arm of Cove - floating store & restaurant characterized by a Quonset hut roof.

GUEST BOAT CAPACITY: ....Appx. 40 boats
DOCKSIDE DEPTH AT ZERO TIDE: 20 ft. plus
SEASON: ....**Open April 1 - Sept. 30**
RESERVATION POLICY: ....Accepts
AMT W/ELECTRICITY: ....All
FUEL DOCK: ....Gas & Diesel
MARINE REPAIRS: ....Close by
TOILETS: ....Yes
HOT SHOWERS: ....Yes
RESTAURANT: ....Upper Deck Cafe - Licensed
PICNIC AREA: ....None
BASIC STORE: ....Yes
BROADBAND/WI-FI: ....Planned
DAILY RATE: ....Moderate (75¢-$1.25/foot)

GUEST DOCK: ...125 ft. plus slips
GUEST SLIPS: 40
WATER: ....Yes
AMPS: ....15-30 A
PUMP OUT STATION ....None
HAUL OUT: ....None
BOAT RAMP: ....None
LAUNDRY: ....None
BAR: ....Licensed Cafe
POOL: ....None
GOLF: ....Close by - Shuttle
PET FRIENDLY: ....Good
OTHER: Well stocked store, govt. liquor agency, fishing tackle & licenses, clothing, charts, golf shuttles, moped & bike rentals.

CAUTION! This chartlet not intended for use in navigation.

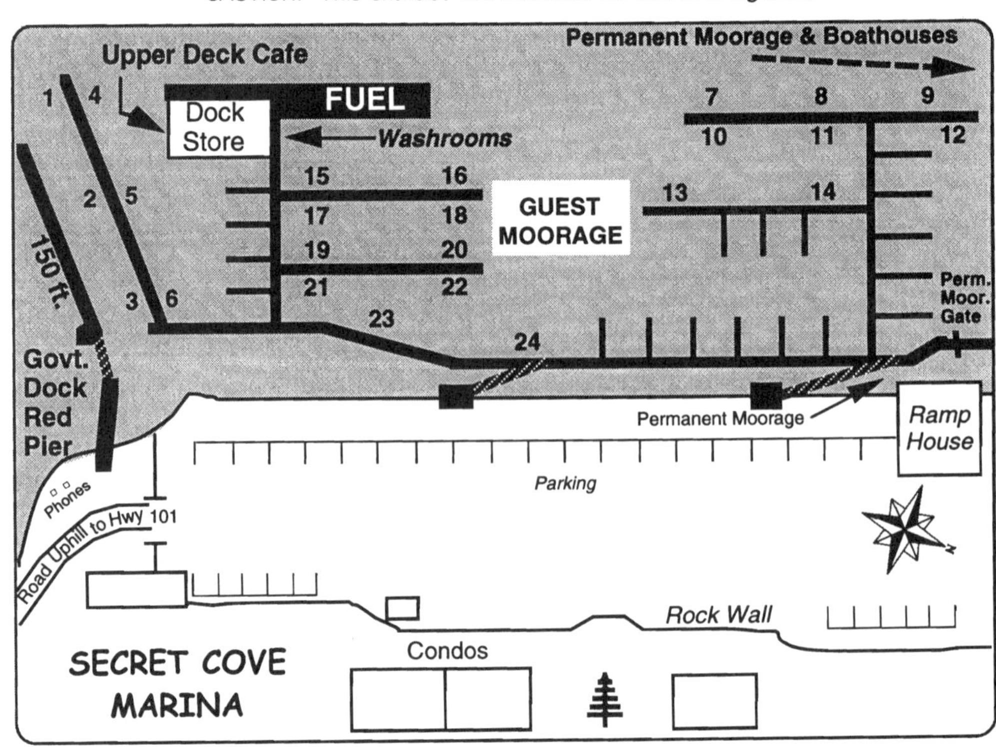

Pender Harbour

NAME OF MARINA: ***PENDER HARBOUR RESORT*** **RADIO:** None
TELEPHONE: **604-883-2424** **MGR:** Office Manager
E-MAIL: info@penderharbourresort.com **FAX:** 604-883-2414 (Fax)
ADDRESS: 4686 Sinclair Bay Rd. RR#1 Garden Bay, B.C. Canada V0N 1S0
SHORT DESCRIPTION & LOCATION: **www.penderharbourresort.com**

49°38.00' - 124°02.50' Located in Duncan Cove on N. side of Pender Harbor, just past Irvines Landing on chart. Small picturesque marina resort in wooded park-like setting. Great destination for sm. rendezvous's combining boats, R/V's, & motel-cottage accommodations.

GUEST BOAT CAPACITY: Appx. 20 boats
DOCKSIDE DEPTH AT ZERO TIDE: 60 ft. +
SEASON: All year
RESERVATION POLICY: Accepts
AMT W/ELECTRICITY: All
FUEL DOCK: Close by
MARINE REPAIRS: Close by
TOILETS: Yes
HOT SHOWERS: Yes
RESTAURANT: Close by
PICNIC AREA: Yes
BASIC STORE: Yes
BROADBAND/WI-FI: None
DAILY RATE: Moderate (75¢-$1.25/foot)
NOTE: FORMERLY NAMED DUNCAN COVE RESORT

GUEST DOCK: 2400 ft. TTL
GUEST SLIPS: 10-15 slips + dock
WATER: Yes
AMPS: 15-30 A
PUMP OUT STATION PortaPot
HAUL OUT: None
BOAT RAMP: Yes
LAUNDRY: Yes
BAR: Pub close by
POOL: Yes
GOLF: Close by - Ask for Shuttle
PET FRIENDLY: Good
OTHER: Waterfront cottages, R/V & tenting, motel, boat rentals tackle, bait, meeting/common room, swimming lakes nearby.

CAUTION! This chartlet not intended for use in navigation.

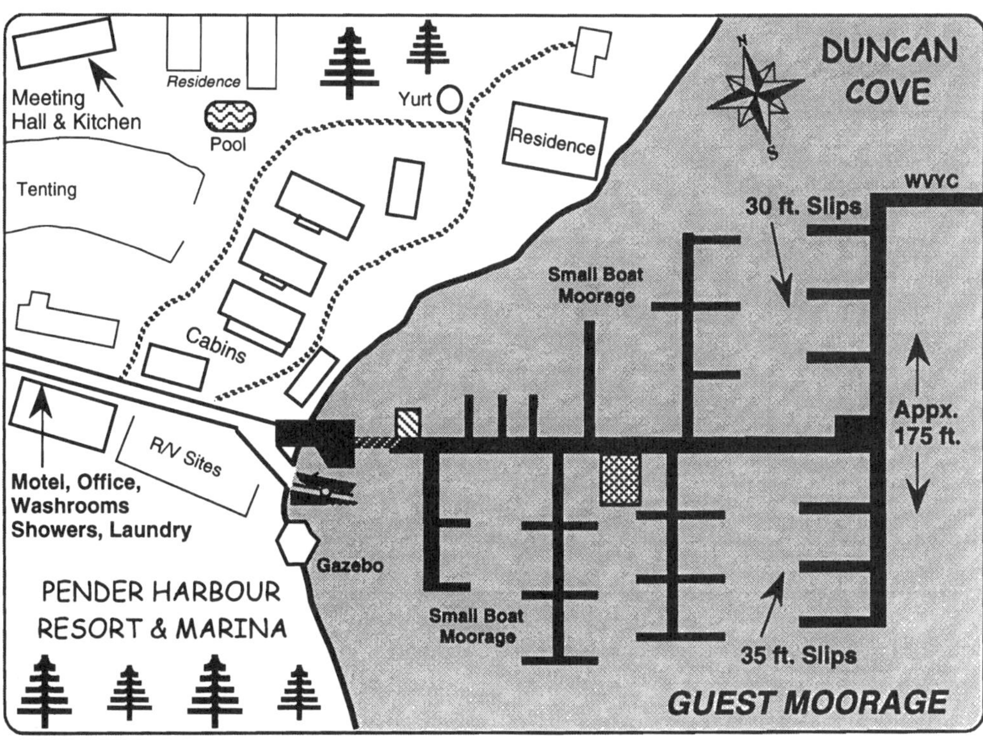

Pender Harbour

NAME OF MARINA: ***HOSPITAL BAY PUBLIC WHARF*** RADIO: VHF 66A
TELEPHONE: **604-883-2234** MGR: Wharfinger
E-MAIL: camas@telus.net FAX: 604-883-2234
ADDRESS: Harbor Auth. P.O. Box 118 Madeira Park, B.C. Canada V0N 2H0
SHORT DESCRIPTION & LOCATION: **http://www.dfo-mpo.gc.ca/sch/HB_BC_e.asp**

49°38.07' - 124°02.25' Located in Hospital Bay on N side of Pender Harbour. Well kept public wharf with familiar red gangway to docks. John Henry's fuel dock and store next door has shopping open 7 days per week with Post Office and Liquor Store on premises.

GUEST BOAT CAPACITY: ....Appx. 10 boats
DOCKSIDE DEPTH AT ZERO TIDE: ......20 ft. +
SEASON: ....All year
RESERVATION POLICY: ....None
AMT W/ELECTRICITY: ....Most
FUEL DOCK: ....Gas, Dsl, & LP - Next Door
MARINE REPAIRS: ....Close by
TOILETS: ....Close by
HOT SHOWERS: ....Close by
RESTAURANT: ....Close by
PICNIC AREA: ....Next door
BASIC STORE: ....Next door
BROADBAND/WI-FI: ....BroadbandXpress
DAILY RATE: ....Economical (Under 75¢/foot)
MOORAGE BOX ON TOP OF DOCK.
-----WHARF MANAGED BY HARBOUR AUTHORITY OF PENDER HARBOUR.

GUEST DOCK: ....200 ft. total
GUEST SLIPS: ....Dock only
WATER: ....Yes
AMPS: ....20 A
PUMP OUT STATION ...Close by
HAUL OUT: ....Close by
BOAT RAMP: ....Close by
LAUNDRY: ....Close by
BAR: ....Close by
POOL: ....None
GOLF: ....Close by
PET FRIENDLY: ....Good
OTHER: John Henry Store next door w/produce, meats & groceries, fishing supplies, fax, office services, Harbour Electronics.

CAUTION! This chartlet not intended for use in navigation.

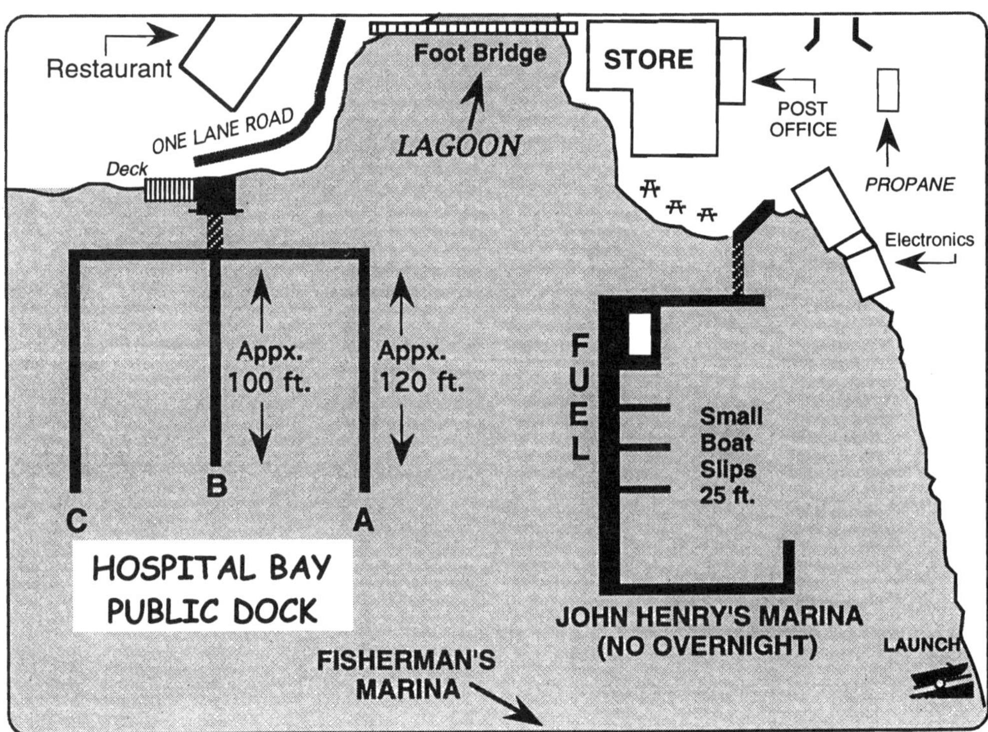

Pender Harbour

NAME OF MARINA: ***FISHERMAN'S RESORT & MARINA*** **RADIO:** VHF 66A
TELEPHONE: **604-883-2336** MGR: David Pritchard
E-MAIL: fishermans@dccnet.com FAX: 604-883-2336
ADDRESS: P.O. Box 68 Garden Bay, B.C. Canada V0N 1S0
SHORT DESCRIPTION & LOCATION: **www.fishermansresortmarina.com**

49°37.89' - 124°02.01' Destination marina resort in Pender Harbour's Hospital Bay. Well maintained docks & beautifully landscaped grounds adjacent to John Henry's fuel dock & General Store. Walking distance to liquor store, good restaurants, pubs, & post office.

GUEST BOAT CAPACITY: ....Appx. 30-40 boats
DOCKSIDE DEPTH AT ZERO TIDE: ......40 ft. +
SEASON: ....April to September
RESERVATION POLICY: ....Accepts
AMT W/ELECTRICITY: ....All
FUEL DOCK: ....Close by
MARINE REPAIRS: ....Close by
TOILETS: ....Yes
HOT SHOWERS: ....Yes
RESTAURANT: ....Close by
PICNIC AREA: ....Yes
BASIC STORE: ....Close by
BROADBAND/WI-FI: ....Yes
DAILY RATE: ....Moderate
(75¢-$1.25/foot)
"WEST COAST 1960's AMBIENCE"

GUEST DOCK: ....2300 ft. Total
GUEST SLIPS: Appx. 35 + docks
WATER: ....Yes
AMPS: ....20-30 A
PUMP OUT STATION ....None
HAUL OUT: ....None
BOAT RAMP: ....Yes
LAUNDRY: ....Yes
BAR: ....Pub close by
POOL: ....None
GOLF: ....Close by
PET FRIENDLY: ....Excellent
OTHER: Ice, bait, licences, tackle, charts, cottages, RV sites, float plane service, nearby hiking trails and fresh water lakes, boat sitting.

CAUTION! This chartlet not intended for use in navigation.

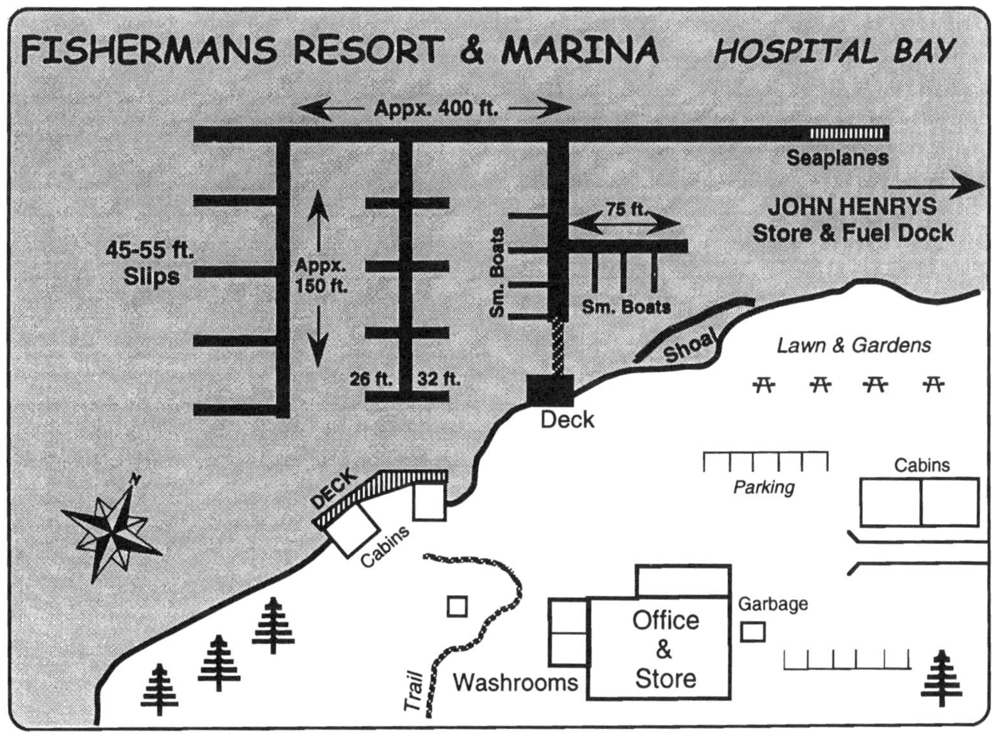

Pender Harbour

NAME OF MARINA: ***GARDEN BAY HOTEL & MARINA*** RADIO: VHF 66A
TELEPHONE: **604-883-2674** MGR: Ron Johnston, Owner
E-MAIL: gbhm@dccnet.com FAX: 604-883-2674
ADDRESS: P.O. Box 90 Garden Bay, B.C. Canada VON 1S0
SHORT DESCRIPTION & LOCATION: **www.gardenbaypub.com**

49°37.80' - 124°03.98' Nice marina located in Garden Bay, NE section of Pender Harbour, next door to large RVYC outstation. Marina offers fine dining in their waterfront restaurant and adjoining pub. Marine services & shopping available within walking distance.

GUEST BOAT CAPACITY: 30-40 boats
DOCKSIDE DEPTH AT ZERO TIDE: 20 ft. +
SEASON: All year
RESERVATION POLICY: Accepts
AMT W/ELECTRICITY: All
FUEL DOCK: Close by
MARINE REPAIRS: Close by
TOILETS: Yes
HOT SHOWERS: None
RESTAURANT: Yes
PICNIC AREA: None
BASIC STORE: Close by
BROADBAND/WI-FI: BroadbandXpress
DAILY RATE: Moderate (75¢-$1.25/foot)
REGISTER AT PUB

GUEST DOCK: 1200 ft total
GUEST SLIPS: Docks only
WATER: Yes
AMPS: 15-30 A
PUMP OUT STATION Close by
HAUL OUT: None
BOAT RAMP: Close by
LAUNDRY: Marina next door
BAR: Pub
POOL: None
GOLF: Close by
PET FRIENDLY: Good
OTHER: Liquor Store on premise. Gift shop, art gallery, churches, & P.O. within walking. Public transp. to Vancouver & Powell River.

CAUTION! This chartlet not intended for use in navigation.

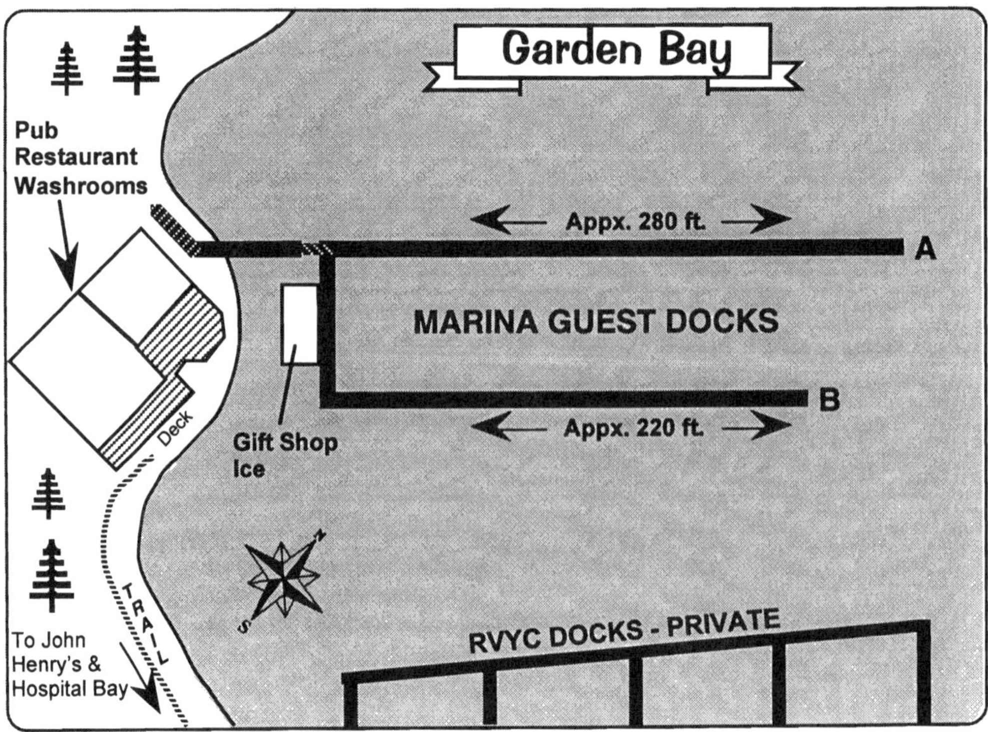

Pender Harbour

NAME OF MARINA: ***SPORTSMAN'S MARINA & RESORT*** **RADIO:** None
TELEPHONE: **604-883-2479** **MGR:** Owners
E-MAIL: **FAX:** None
ADDRESS: P.O. Box 6 Garden Bay, B.C. Canada V0N 1S0

SHORT DESCRIPTION & LOCATION:

49°37.88' - 124°1.55' Small older marina located in Garden Bay in Pender Harbour in between the Garden Bay Hotel & Marina. and the Garden Bay Marine Park. Caters to yacht club outstations, but has some guest moorage. Friendly hosts.

GUEST BOAT CAPACITY: ....Varies
DOCKSIDE DEPTH AT ZERO TIDE: ....40ft. +
SEASON: ....All year
RESERVATION POLICY: ....Accepts
AMT W/ELECTRICITY: ....All
FUEL DOCK: ....Close by
MARINE REPAIRS: ....On call
TOILETS: ....Yes
HOT SHOWERS: ....Yes
RESTAURANT: ....Close by
PICNIC AREA: ....Close by
BASIC STORE: ....Close by
BROADBAND/WI-FI: ....BroadbandXpress
DAILY RATE: ....Moderate (75¢-$1.25/foot)

GUEST DOCK: Appx. 200 ft. TTL
GUEST SLIPS: ....Varies
WATER: ....Yes
AMPS: ....15-30 A
PUMP OUT STATION ....None
HAUL OUT: ....None
BOAT RAMP: ....None
LAUNDRY: ....Yes
BAR: ....Pub close by
POOL: ....None
GOLF: ....Close by
PET FRIENDLY: ....Good
OTHER: Next door to marine park and trails, rental cabins.

CAUTION! This chartlet not intended for use in navigation.

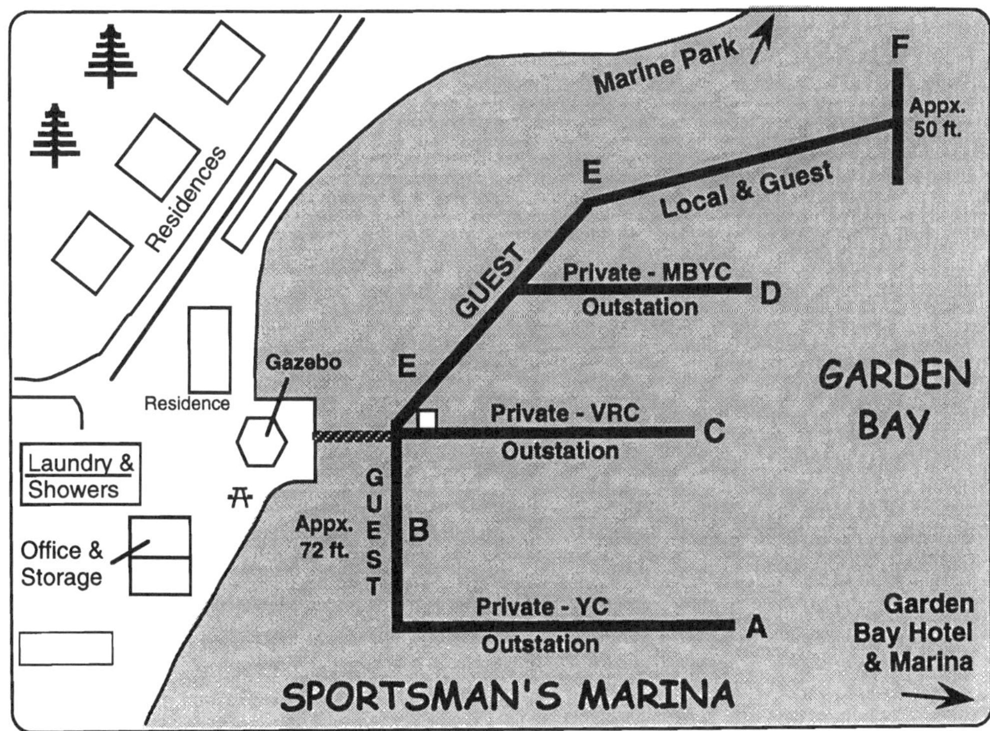

Pender Harbour

NAME OF MARINA: ***MADEIRA PARK PUBLIC WHARF*** **RADIO:** VHF 66A
TELEPHONE: 604-883-2234/9878 **MGR:** Wharfinger
E-MAIL: camas@telus.net **FAX:** 604-883-2234 (Fax)
ADDRESS: Harbor Auth. P.O. Box 118, Madeira Park, B.C. Canada V0N 2H0
SHORT DESCRIPTION & LOCATION: **http://www.dfo-mpo.gc.ca/sch/HB_BC_e.asp**

49°37.41' - 124°01.53' Located in Wellborne Cove on S side of Pender Hbr. Busy public wharf w/designated & upgraded pleasure craft moorage. This is one of the nicest government docks on the coast. Good reprovisioning location, close to lg. shopping center.

GUEST BOAT CAPACITY: ....Appx. 10-15 boats
DOCKSIDE DEPTH AT ZERO TIDE: ......15 ft. +
SEASON: ....All year
RESERVATION POLICY: ....None
AMT W/ELECTRICITY: ....Most
FUEL DOCK: ....Close by
MARINE REPAIRS: ....Close by
TOILETS: ....Yes
HOT SHOWERS: ....Yes
RESTAURANT: ....Close by
PICNIC AREA: ....Yes
BASIC STORE: ....Close by
BROADBAND/WI-FI: ....Marginal
DAILY RATE:.................Moderate (75¢-$1.25/foot)

GUEST DOCK: ....400 ft. total
GUEST SLIPS: ....Dock only
WATER: ....Yes
AMPS: ....15-30-50 A
PUMP OUT STATION ....Yes
HAUL OUT: ....Close by
BOAT RAMP: ....Yes
LAUNDRY: ....None
BAR: ....Close by
POOL: ....None
GOLF: ....Close by
PET FRIENDLY: ....Fair
OTHER: Tidal Grid (fee). Walking distance to restaurant, supermarket, drugstore, liquor store, post office & medical clinic.

CAUTION! This chartlet not intended for use in navigation.

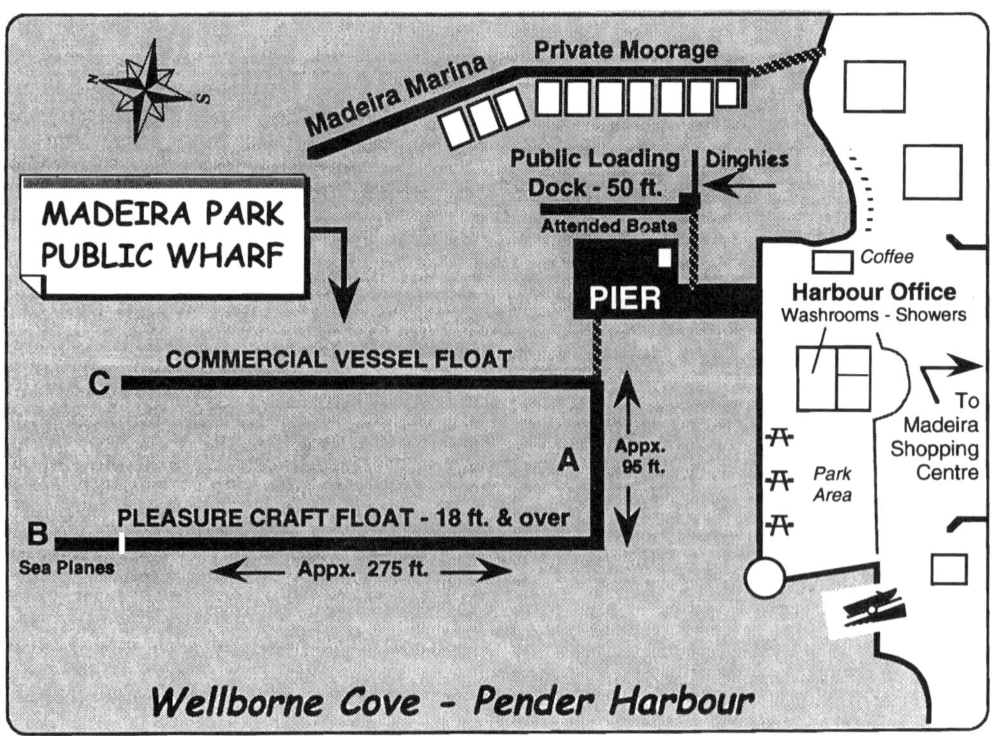

Egmont

NAME OF MARINA: ***EGMONT MARINA RESORT*** **RADIO:** VHF 66A
TELEPHONE: **800-626-0599** **MGR:** John & Margaret Mills
E-MAIL: info@egmont-marina.com **FAX:** 800-626-0599 (Fax)
ADDRESS: 16660 Backeddy Rd. Egmont, B.C. Canada V0N 1N0
SHORT DESCRIPTION & LOCATION: **www.egmont-marina.com**

49°45.10' - 123°55.70' Located on Sechelt Peninsula on Skookumchuck Narrows, south shore across from Sutton Islets, and near the entrance to Sechelt Rapids. Full service marina hosts guest boats up to 90 ft. Last provisioning stop before Princess Louisa Inlet.

GUEST BOAT CAPACITY: ....Appx. 20 boats
DOCKSIDE DEPTH AT ZERO TIDE: 8 ft. plus
SEASON: ....All year
RESERVATION POLICY: ....Accepts
AMT W/ELECTRICITY: ....All
FUEL DOCK: ....Gas & Diesel
MARINE REPAIRS: ....On premises
TOILETS: ....Yes
HOT SHOWERS: ....Yes
RESTAURANT: ....Yes
PICNIC AREA: ....Yes
BASIC STORE: ....Yes
BROADBAND/WI-FI: ....None
DAILY RATE: ....Moderate
(75¢-$1.25/foot)
BOAT/US MARINA

GUEST DOCK: Appx. 600 ft. TTL
GUEST SLIPS: ....Varies
WATER: ....Yes
AMPS: ....15-30 A
PUMP OUT STATION ....None
HAUL OUT: ....None
BOAT RAMP: ....Yes
LAUNDRY: ....Yes
BAR: ....Backeddy Pub
POOL: ....None
GOLF: ....Close by
PET FRIENDLY: ....Excellent
OTHER: Cabins & motel, RV Park, campground, fishing tackle, licenses, dive air, water taxi.

CAUTION! This chartlet not intended for use in navigation.

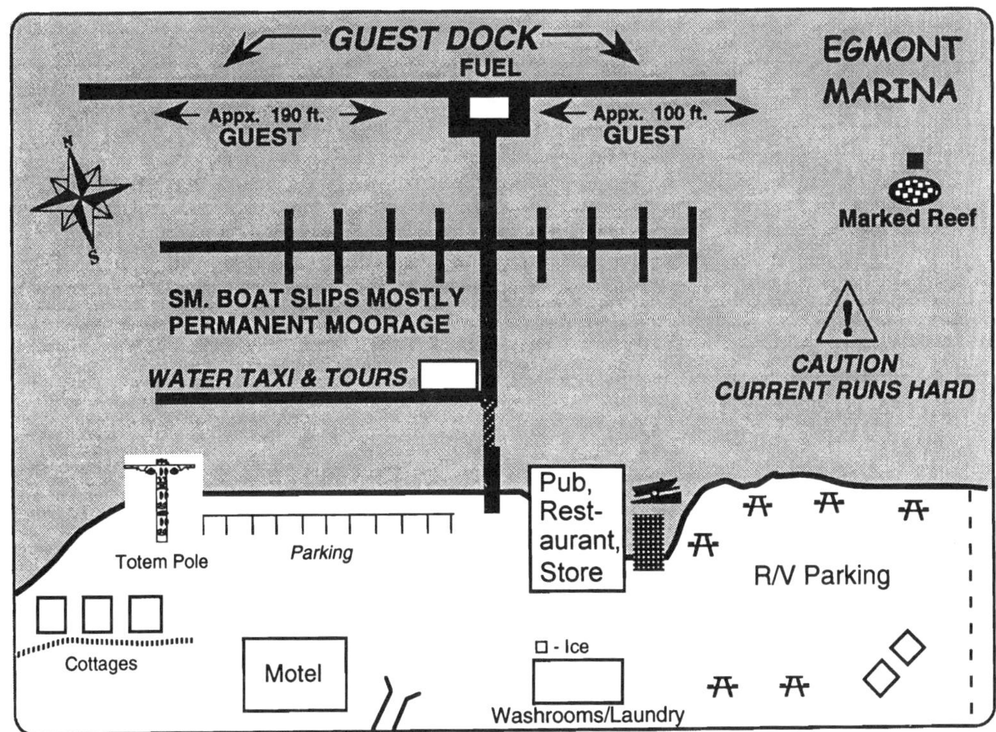

Egmont

NAME OF MARINA: ***BATHGATE GEN. STORE & MARINA*** **RADIO:** VHF 66A
TELEPHONE: **604-883-2222** **MGR:** Doug & Vicki Martin
E-MAIL: info@bathgate.com **FAX:** 604-883-2750 (Fax)
ADDRESS: 6781 Bathgate Rd. Egmont, B.C. Canada V0N 1N0
SHORT DESCRIPTION & LOCATION: **www.bathgate.com**

49°45.00' - 123°55.70' Located in Secret Bay - Sechelt Penn. near entrance of Skookumchuck Narrows & Sechelt Rapids. Rents open moorage to overnighters. Well stocked Gen. Store, w/produce & liquor. Caution- Reef at entrance, leave red daymarker to starboard.

GUEST BOAT CAPACITY: ....Varies
DOCKSIDE DEPTH AT ZERO TIDE: ....6 ft. +
SEASON: ....All year
RESERVATION POLICY: ....Recommended
AMT W/ELECTRICITY: ....All
FUEL DOCK: ....Gas, Dsl, & LP
MARINE REPAIRS: ....On premises
TOILETS: ....For moorage guests only
HOT SHOWERS: ....Yes
RESTAURANT: ....Close by
PICNIC AREA: ....Yes
BASIC STORE: ....Yes
BROADBAND/WI-FI: ....Yes
DAILY RATE: ....Moderate (75¢-$1.25/foot)

GUEST DOCK: ....Varies
GUEST SLIPS: ....Varies
WATER: ....Yes
AMPS: ....15-20-30 A
PUMP OUT STATION ....None
HAUL OUT: Ways - Up to 40 ft.
BOAT RAMP: ....Close by
LAUNDRY: ....Yes
BAR: ....Pub close by
POOL: ....None
GOLF: ....Close by
PET FRIENDLY: ....Excellent
OTHER: Liquor store, motel, camping, airplane service, boat rentals, & highway connection. Postage & fax service.

CAUTION! This chartlet not intended for use in navigation.

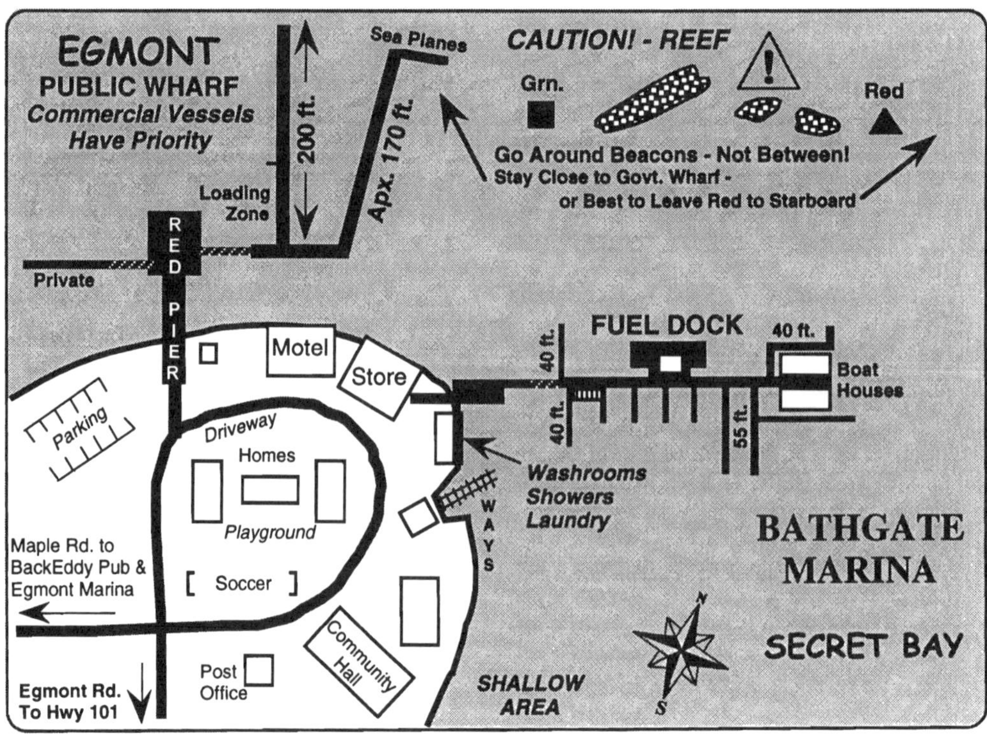

Princess Louisa

NAME OF PARK: *PRINCESS LOUISA MARINE PARK - (B.C. PARKS)*
ADDRESS: http://wlapwww.gov.bc.ca /bcparks/explore/ parkpgs /princesl.
TELEPHONE: 250-952-0932 MGR: Park Ranger
SHORT DESCRIPTION & LOCATION: **www.princesslouisa.bc.ca**

50°12.29' - 123°46.16' Entered thru Malibu Rapids, the Park dock is 4 miles up & at the head of beautiful & dramatic Princess Louisa Inlet, adjacent to spectaclular Chatterbox Falls. Princess Louisa Society supports the park. See website for membership & donations.

GUEST BOAT CAPACITY: ......Appx. 15-20 boats
DOCKSIDE DEPTH AT ZERO TIDE:.......40 Ft. +
SEASON: ....All year
AMT W/ELECTRICITY: ....None
TOILETS: ....Yes
HOT SHOWERS: ....None
PICNIC AREA: ....Yes
PLAY AREA: ....None
BASIC STORE: ....None
DAILY RATE: ....No Charge-DONATIONS APPRECIATED TO SUPPORT THE PARK. Donate via www.princesslouisa.bc.ca. NOTE: 55 Ft. maximum boat size at dock.

GUEST DOCK: ....760 Ft. Total
MOORING BUOYS: ....None
WATER: ....None
PAY PHONES: ....None
BOAT RAMP: ....None
PICNIC SHELTER: ....Yes
BBQ: ....None
PUMP OUT STATION: ......None
PET FRIENDLY: ....Excellent
OTHER: Walking & hiking trails, waterfalls, campsites, firepit, pavilion, historic & interesting tours available at Mailibu Camp. NO GARBAGE - NO DISCHARGE.

CAUTION! This chartlet not intended for use in navigation.

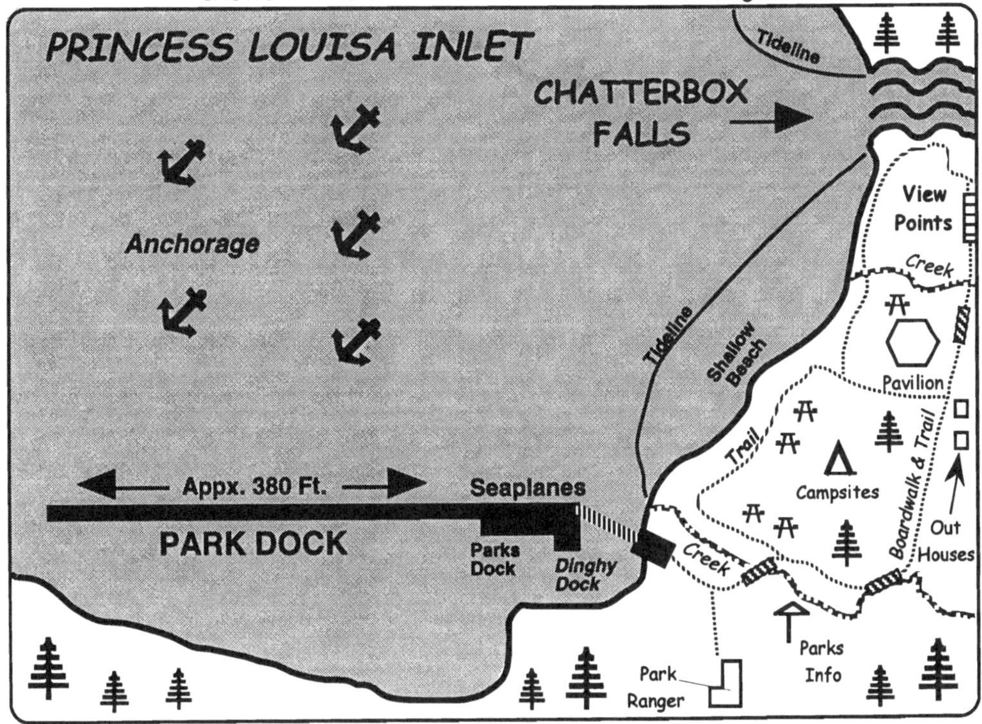

Saltery Bay

NAME OF MARINA: ***SALTERY BAY PUBLIC WHARF*** RADIO: None
TELEPHONE: 604-487-0663 MGR: Lynn Murray, Harbour Manager
E-MAIL: kpotvin@armourtech.com FAX:
ADDRESS: DFO, RR #3, Saltery Bay Powell River, B.C. Canada V8A 5C1
SHORT DESCRIPTION & LOCATION:

49°46.88' - 123°10.45' Located about 5 mi. inside Jervis Inlet on N side. The public wharf is next to ferry dock connecting the highway to & from Powell River. Remote - but provides quiet & secure moorage stopover on northern passage to or from Princess Louisa.

GUEST BOAT CAPACITY: ....Appx. 30-35 boats
DOCKSIDE DEPTH AT ZERO TIDE: ....Ample
SEASON: ....All year
RESERVATION POLICY: ....None
AMT W/ELECTRICITY: ....None
FUEL DOCK: ....None
MARINE REPAIRS: ....None
TOILETS: ....None
HOT SHOWERS: ....None
RESTAURANT: ....None
PICNIC AREA: ....Yes
BASIC STORE: ....None
BROADBAND/WI-FI: ....None
DAILY RATE: ....Economical (Under 75¢/foot)
COMMERCIAL VESSELS HAVE PRIORITY

GUEST DOCK: Appx. 600 ft TTL
GUEST SLIPS: ....Dock only
WATER: ....None
AMPS: ....None
PUMP OUT STATION ....None
HAUL OUT: ....None
BOAT RAMP: ....None
LAUNDRY: ....None
BAR: ....None
POOL: ....None
GOLF: ....None
PET FRIENDLY: ....Excellent
OTHER: Garbage service with fee,
NOTE: BE CAUTIOUS OF FERRY BACKING IN TO FERRY DOCK.

CAUTION! This chartlet not intended for use in navigation.

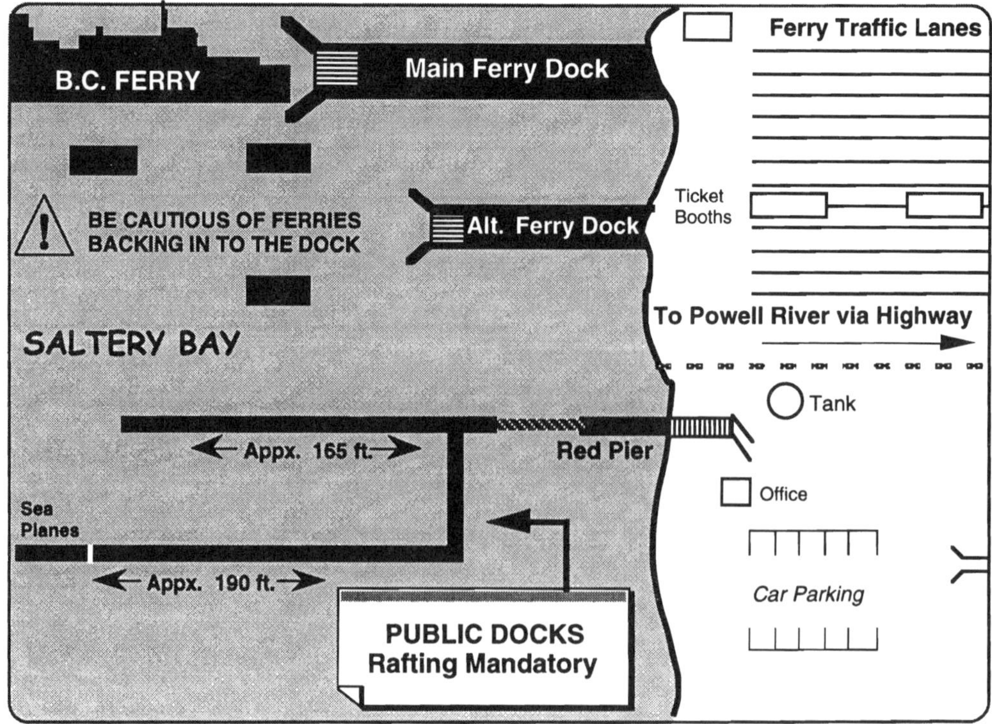

Texada Island

NAME OF MARINA: *TEXADA BOAT CLUB* **RADIO:** None
TELEPHONE: 604-486-7574 **MGR:** Wharfinger
E-MAIL: **FAX:**
ADDRESS: P.O. Box 196, Van Anda, B.C. Canada VON 3KO
SHORT DESCRIPTION & LOCATION: **www.texada.org**

49°45.60' - 124°33.80 Located in Sturt Bay, on Malaspina Strait near N end of Texada Island. Well protected, Texada Boat Club encourages visitors to use its ample moorage. If visitor dock full, check with wharfinger. Pay at head of dock or with wharfinger.

GUEST BOAT CAPACITY: ....Appx. 8-12 boats
DOCKSIDE DEPTH AT ZERO TIDE: ......80 ft. +
SEASON: ....All year
RESERVATION POLICY: ....None
AMT W/ELECTRICITY: ....Limited
FUEL DOCK: ....None
MARINE REPAIRS: ....Close by
TOILETS: ....None
HOT SHOWERS: ....At B & B and Hotel
RESTAURANT: ....15 Minute walk
PICNIC AREA: ....None
BASIC STORE: ....15 Minute walk
BROADBAND/WI-FI: Internet at Credit Union
DAILY RATE: ....Economical (Under 75¢/foot)

PUBLIC IS WELCOME

GUEST DOCK: ....370 ft. total
GUEST SLIPS: ....Dock only
WATER: ....Yes
AMPS: ....15 A
PUMP OUT STATION ....None
HAUL OUT: ....None
BOAT RAMP: ....Yes
LAUNDRY: ....Close by
BAR: ....Pub 15 Minute walk
POOL: ....None
GOLF: ....Appx. 5 miles
PET FRIENDLY: ....Excellent
OTHER: Walking distance to P.O., Liq. & Store, Hotel, Pub, & Restaurant. Convenient & quiet stopover enroute to Desolation - friendly staff.

CAUTION! This chartlet not intended for use in navigation.

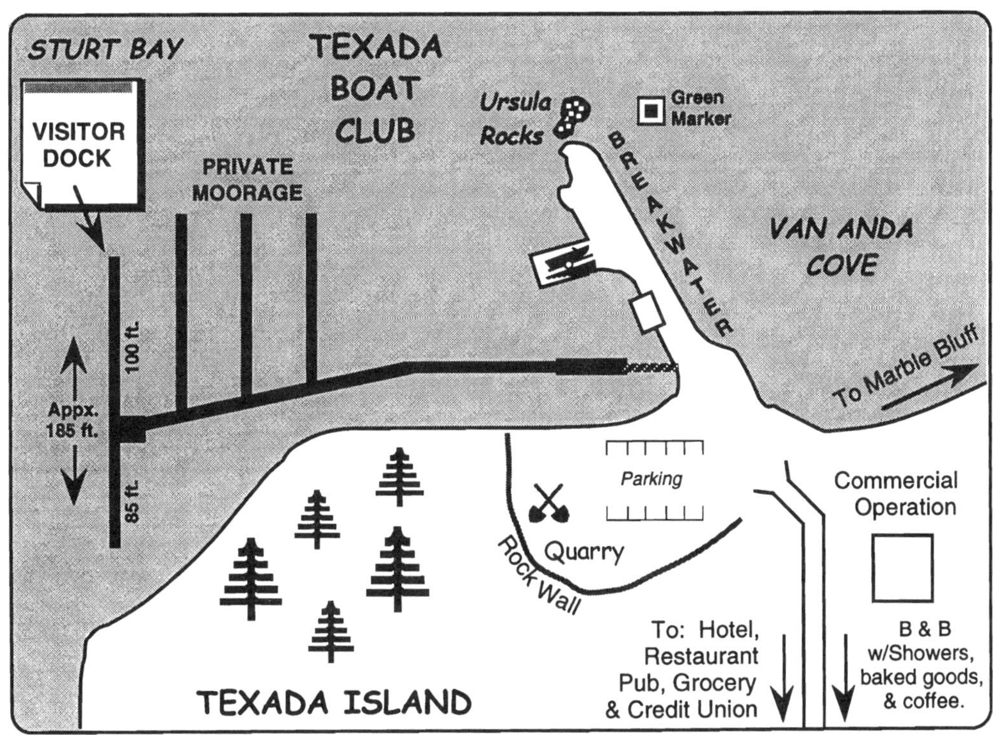

NOTES

Powell River

NAME OF MARINA: *BEACH GARDENS MARINA* RADIO: VHF 66A
TELEPHONE: 604-485-7734 MGR: Joan Barszczewski
E-MAIL: bgarden@beachgardens.com FAX: 604-485-2343 (Fax)
ADDRESS: 7074 Westminster Ave. Powell River, B.C. Canada V8A 1C5
SHORT DESCRIPTION & LOCATION: **www.beachgardens.com**

49°48.00' - 124°31.20' Small breakwater protected marina located S. of Powell River & just S. of Grief Pt. on E. side of Malaspina Strait. Resort has been recently upgraded. This is a popular & quiet stopover if heading to or from Desolation, offering good docks & amenities.

GUEST BOAT CAPACITY: ....Appx.10-20 boats
DOCKSIDE DEPTH AT ZERO TIDE: ....12 ft.
SEASON: ....All year
RESERVATION POLICY: ....Accepts
AMT W/ELECTRICITY: ....All
FUEL DOCK: ....Gas & Diesel-Summer Only
MARINE REPAIRS: ....Close by
TOILETS: ....Yes
HOT SHOWERS: ....Yes
RESTAURANT: ....Planned
PICNIC AREA: ....Yes
BASIC STORE: ....Close by
BROADBAND/WI-FI: ....None
DAILY RATE: ....Moderate (75¢-$1.25/foot)
OTHER: Summer courtesy van to shopping & restaurants.

GUEST DOCK: ....350 ft. + slips
GUEST SLIPS: ....Varies
WATER: ....Yes
AMPS: ....15-30 A
PUMP OUT STATION ....None
HAUL OUT: ....None
BOAT RAMP: ....None
LAUNDRY: ....Yes
BAR: ....None
POOL: ....None
GOLF: ....Close by
PET FRIENDLY: ....Good
OTHER: Fitness centre, liquor store, biking & hiking trails, kayaking centre, and lounge.

CAUTION! This chartlet not intended for use in navigation.

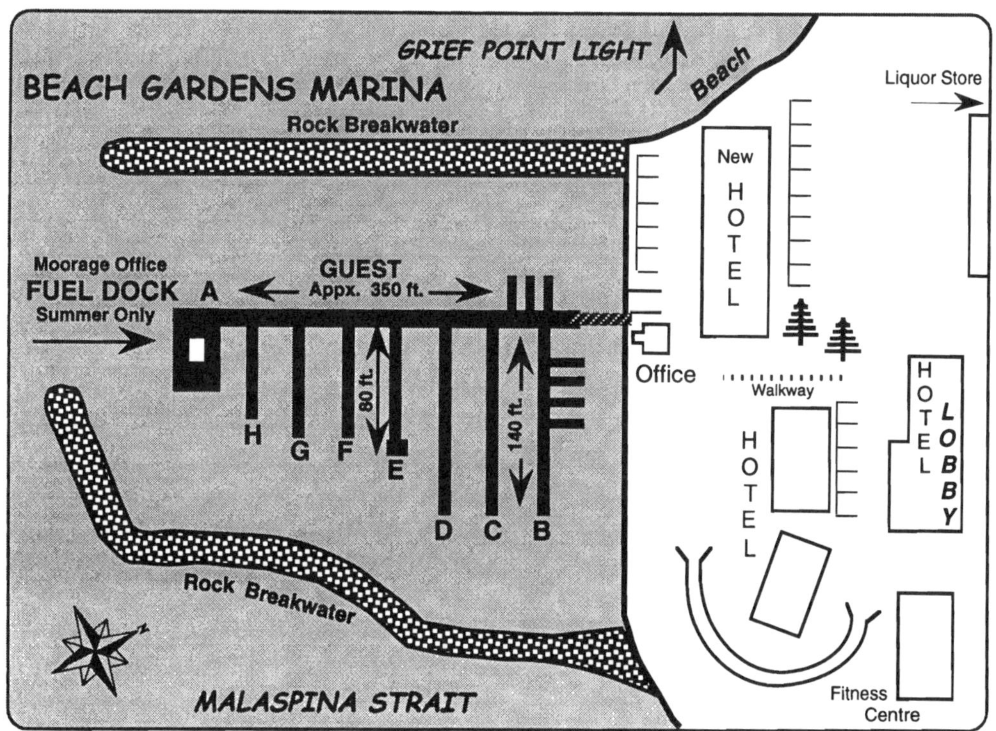

Powell River

NAME OF MARINA: ***WESTVIEW HARBOUR*** **RADIO:** **VHF 68 or 66A**
TELEPHONE: **604-485-5244** **MGR:** Jim Parsons Wharfinger
E-MAIL: jparsons@cdbr.bc.ca **FAX:** 604-485-5286 (fax)
ADDRESS: City Hall, 6910 Duncan St. Powell River, B.C. Canada V8A 1V4
SHORT DESCRIPTION & LOCATION: **http://www.haa.bc.ca**

49°50.10' - 124°31.85' Marina located 2 mi. S. of Paper Mill & protected by 3 breakwaters divided into 2 harbours w/ferry dock in between. Guest moorage in South Harbour. Shopping Ctr. a is a few blocks uphill but adjacent Marine Drive has shops and restaurants.

GUEST BOAT CAPACITY: ....Appx. 20-40 boats
DOCKSIDE DEPTH AT ZERO TIDE: 8 ft. plus
SEASON: ....All year
RESERVATION POLICY: First Come, First Serve
AMT W/ELECTRICITY: ....All
FUEL DOCK: ....Gas & Diesel
MARINE REPAIRS: ....On premises
TOILETS: ....Yes
HOT SHOWERS: ....Yes
RESTAURANT: ....Close by
PICNIC AREA: ....Close by
BASIC STORE: ....Close by
BROADBAND/WI-FI: ....Planned
DAILY RATE: ....Economical (Under 75¢/foot)
EXPECT RAFTING & MED TIES.
MARINA & FORESHORE REDEVELOPMENT IN PROGRESS.

GUEST DOCK: Appx.500 ft. TTL
GUEST SLIPS: Docks & Med Ties
WATER: ....Yes
AMPS: ....15-30 A
PUMP OUT STATION ....None
HAUL OUT: ....Ways to 45 ft.
BOAT RAMP: ....Yes
LAUNDRY: ....Yes
BAR: ....Pubs close by
POOL: ....2 Miles
GOLF: ....Close by - Free shuttle
PET FRIENDLY: ....Fair
OTHER: Marine & fishing supplies, shopping & restaurants within moderate walking. Summer shuttle bus. Historical museum nearby.

NOTES

Powell River

CAUTION! This chartlet not intended for use in navigation.

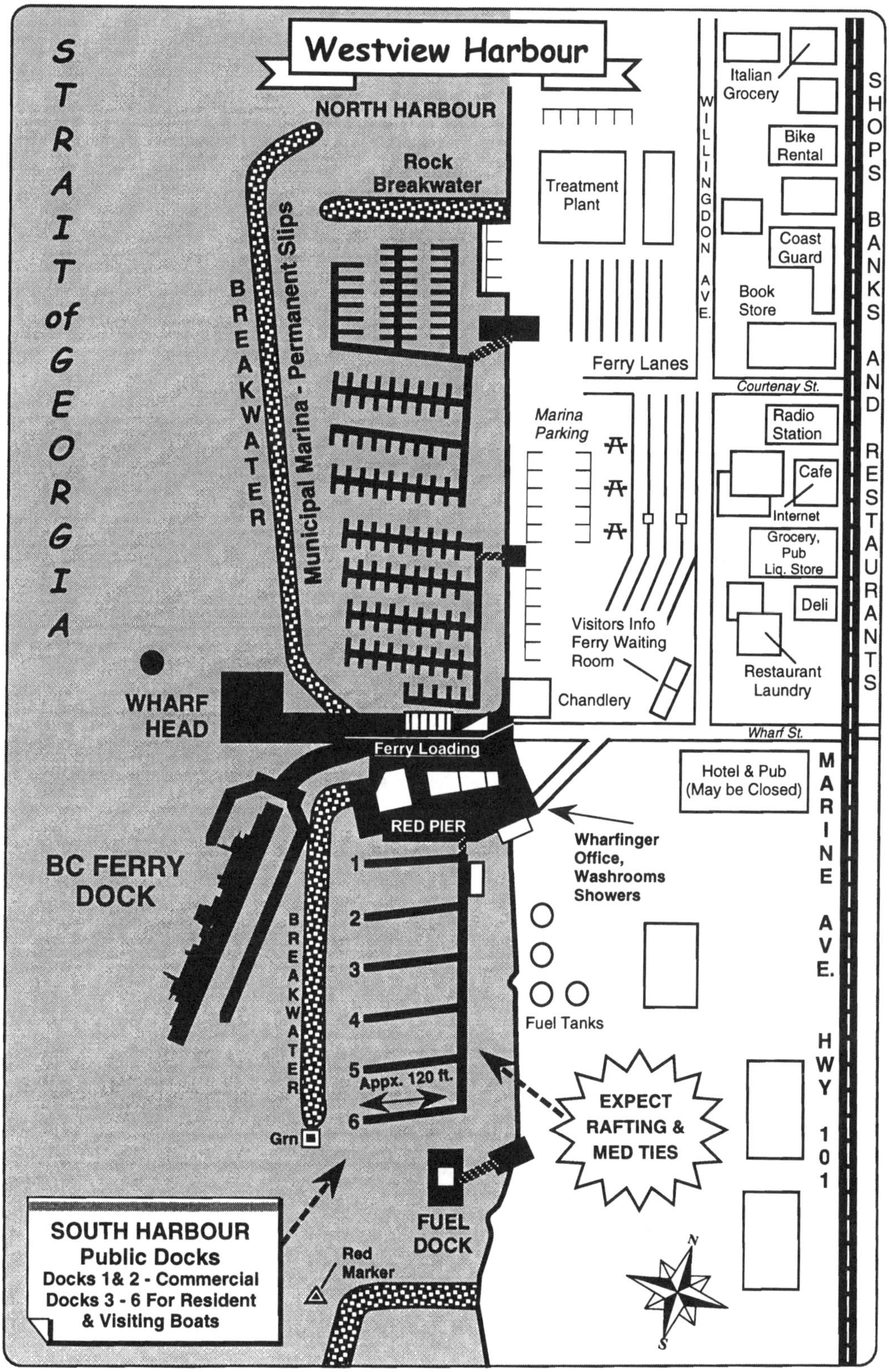

Westview Harbour

STRAIT of GEORGIA

NORTH HARBOUR

Rock Breakwater

BREAKWATER

Municipal Marina - Permanent Slips

Treatment Plant

WILLINGDON AVE.

Italian Grocery

Bike Rental

Coast Guard

Book Store

SHOPS BANKS AND RESTAURANTS

Ferry Lanes

Courtenay St.

Marina Parking

Radio Station

Cafe

Internet

Grocery, Pub Liq. Store

Deli

Restaurant Laundry

Visitors Info Ferry Waiting Room

Chandlery

WHARF HEAD

Ferry Loading

Wharf St.

Hotel & Pub (May be Closed)

MARINE AVE. HWY 101

RED PIER

Wharfinger Office, Washrooms Showers

BC FERRY DOCK

BREAKWATER

1

2

3

4

5

6

Appx. 120 ft.

Grn

Fuel Tanks

EXPECT RAFTING & MED TIES

FUEL DOCK

Red Marker

N

S

SOUTH HARBOUR Public Docks

Docks 1& 2 - Commercial

Docks 3 - 6 For Resident & Visiting Boats

Lund

NAME OF MARINA: *LUND HARBOUR AUTHORITY* **RADIO:** VHF 66A
TELEPHONE: 604-483-4711 **MGR:** Rosie O'Neill
E-MAIL: None **FAX:** No Fax
ADDRESS: General Delivery Lund, B.C. Canada V0N 2G0
SHORT DESCRIPTION & LOCATION: **www.dfo-mpo.gc.ca/sch/ha-ap-bc_e.asp**

49°58.80' - 124°45.80' Town of Lund located 1.3 mi. NW of Hurtedo Pt. on Malaspina Peninsula just S. of Copeland Islands. This is the end of the road & last of marine services & repairs before Desolation. Friendly public docks usually can accommodate most boats.

GUEST BOAT CAPACITY: ....Appx. 40-60 boats
DOCKSIDE DEPTH AT ZERO TIDE: 20 ft. plus
SEASON: ....All year
RESERVATION POLICY: ....None
AMT W/ELECTRICITY: ....All
FUEL DOCK: ....Gas & Diesel
MARINE REPAIRS: ....Yes - Haul out
TOILETS: ....Yes
HOT SHOWERS: ....Yes
RESTAURANT: ....Close by
PICNIC AREA: ....Yes
BASIC STORE: ....Yes
BROADBAND/WI-FI: Internet terminal at Hotel
DAILY RATE:..................Moderate (75¢-$1.25/foot)
.................RAFTING REQUIRED
Note: 375 ft. available on inside of breakwater floats.

GUEST DOCK: Appx. 600 ft TTL
GUEST SLIPS: Docks + Breakwter
WATER: ....Yes
AMPS: ....20-30 A
PUMP OUT STATION ....Yes
HAUL OUT: ....Yes
BOAT RAMP: ....Yes
LAUNDRY: ....Close by
BAR: ....Hotel Pub close by
POOL: ....None
GOLF: ....None
PET FRIENDLY: ....Fair
OTHER: Garbage drop at Fuel Dock (Fee), World Renowned Bakery, Historic Lund Hotel & Restaurant, Large well stocked store.

CAUTION! This chartlet not intended for use in navigation.

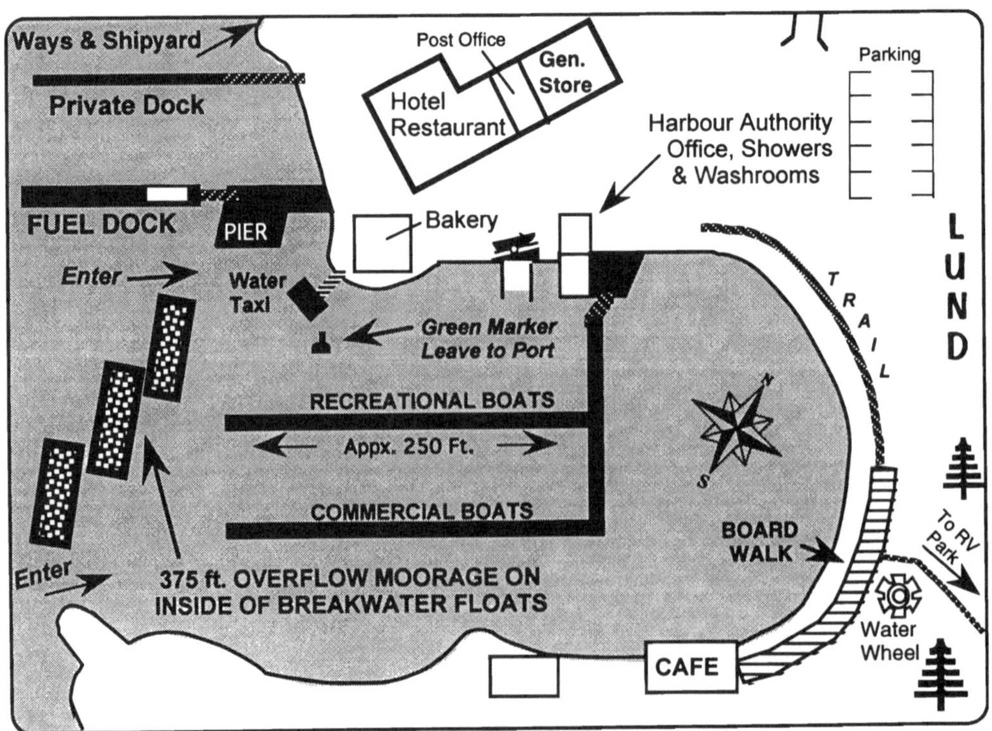

CHAPTER

10

DESOLATION SOUND & DISCOVERY PASSAGE

Chapter Map - Page 2

Campbell River

NAME OF MARINA: ***DISCOVERY HARBOUR MARINA*** **RADIO:** VHF 66A
TELEPHONE: **250-287-2614** MGR: Dean Drake
E-MAIL: mail@discvoveryharbourmarina.com FAX: 250-287-8939 (Fax)
ADDRESS: #392-1434 Island Hwy. Campbell River, B.C. Canada V9W 8C9
SHORT DESCRIPTION & LOCATION: **www.discoveryharbourmarina.com**
50°02.00' - 125°14.50' This is the third marina from south to north in city of Campbell River protected by a large breakwater. Adjacent to vast Discovery Harbour Centre Mall with several restaurants & large stores. Wide concrete docks & helpful staff.

GUEST BOAT CAPACITY: ....Appx. 60-100 boats
DOCKSIDE DEPTH AT ZERO TIDE: ......9 ft. +
SEASON: ....All year
RESERVATION POLICY: ....Accepts
AMT W/ELECTRICITY: ....All
FUEL DOCK: ....Gas & Diesel
MARINE REPAIRS: ....On premises
TOILETS: ....Yes
HOT SHOWERS: ....Yes
RESTAURANT: ....Yes
PICNIC AREA: ....Close by
BASIC STORE: ....Yes
BROADBAND/WI-FI: ....BroadbandXpress
DAILY RATE: ....Moderate (75¢-$1.25/foot)
Boats over 100 ft: $1.50/foot
OFF SEASON RATES AVAILABLE

GUEST DOCK: Appx. 5000 ft TTL
GUEST SLIPS: ....Varies -Ample
WATER: ....Yes
AMPS: ....20-50 A
PUMP OUT STATION Close by
HAUL OUT: ....None
BOAT RAMP: ....Close by
LAUNDRY: ....Yes
BAR: ....Pub
POOL: ....Close by
GOLF: ....Close by
PET FRIENDLY: ....Good
OTHER: Large shopping mall adjacent, groceries, liquor store, espresso, banks, fishing & diving charters, ice, full marine services.

NOTES

Campbell River

CAUTION! This chartlet not intended for use in navigation.

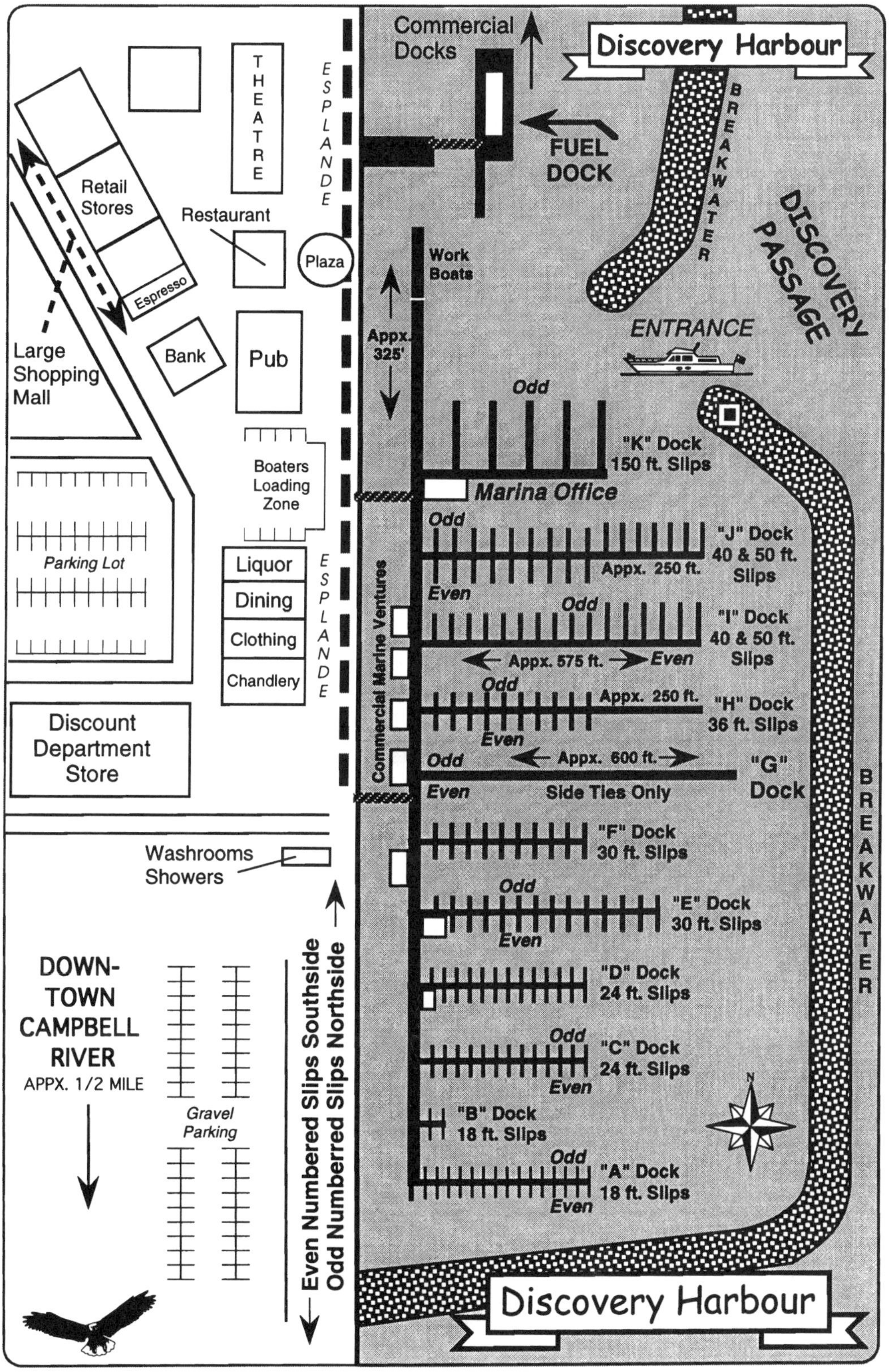

Commercial Docks

Discovery Harbour

FUEL DOCK

BREAKWATER

DISCOVERY PASSAGE

ENTRANCE

THEATRE

ESPLANADE

Retail Stores

Restaurant

Plaza

Espresso

Large Shopping Mall

Bank

Pub

Work Boats

Appx. 325'

Odd

"K" Dock 150 ft. Slips

Boaters Loading Zone

Marina Office

Parking Lot

Odd

"J" Dock 40 & 50 ft. Slips

Appx. 250 ft.

Even

Odd

"I" Dock 40 & 50 ft. Slips

Appx. 575 ft. Even

Liquor

Dining

Clothing

Chandlery

ESPLANADE

Commercial Marine Ventures

Odd

Appx. 250 ft.

"H" Dock 36 ft. Slips

Even

Discount Department Store

Odd

Appx. 600 ft.

"G" Dock

Even

Side Ties Only

"F" Dock 30 ft. Slips

Washrooms Showers

Odd

"E" Dock 30 ft. Slips

Even

BREAKWATER

DOWN-TOWN CAMPBELL RIVER

APPX. 1/2 MILE

"D" Dock 24 ft. Slips

Odd

"C" Dock 24 ft. Slips

Even

Gravel Parking

"B" Dock 18 ft. Slips

N

Odd

"A" Dock 18 ft. Slips

Even

Even Numbered Slips Southside

Odd Numberred Slips Northside

Discovery Harbour

Campbell River

NAME OF MARINA: ***COAST HOTEL & MARINA*** **RADIO:** VHF 66A
TELEPHONE: **250-287-7455** **MGR:** General Manager
E-MAIL: coastmarina@coasthotels.com **FAX:** 250-287-2213 (Hotel Fax)
ADDRESS: 975 Shoppers Row Campbell River, B.C. Canada V9W 2C4
SHORT DESCRIPTION & LOCATION: **www.coasthotels.com/home/sites/campbell river**

50°01,70' - 125°14.45' This is the middle marina breakwater from south to north in heart of downtown Campbell River. Newly rebuilt with wide concrete docks. Guests have hotel privileges and marina staff is friendly and helpful. Can handle boats up to 150 ft.

GUEST BOAT CAPACITY: ....Appx. 70 boats
DOCKSIDE DEPTH AT ZERO TIDE: ....8 ft. +
SEASON: ....All year
RESERVATION POLICY: ....Accepts
AMT W/ELECTRICITY: ....All
FUEL DOCK: ....Close by
MARINE REPAIRS: ....On premises
TOILETS: ....Yes
HOT SHOWERS: ....Yes
RESTAURANT: ....Yes
PICNIC AREA: ....Yes
BASIC STORE: ....Yes
BROADBAND/WI-FI: ....BroadbandXpress
DAILY RATE:..........Premium (Over $1.25/foot)
WINTER RATES AVAILABLE

GUEST DOCK: ....Appx. 1800 Ft.
GUEST SLIPS: ....Appx. 70
WATER: ....Yes
AMPS: 30-50-100A
PUMP OUT STATION ....None
HAUL OUT: ....Close by
BOAT RAMP: ....Close by
LAUNDRY: ....Yes
BAR: ....Yes
POOL: ....None
GOLF: ....None
PET FRIENDLY: ....Excellent
OTHER: Fitness facilities, hot tub, cable TV, Foreshore Park, shopping, & attractions. Golf packages & guided fishing charters.

CAUTION! This chartlet not intended for use in navigation.

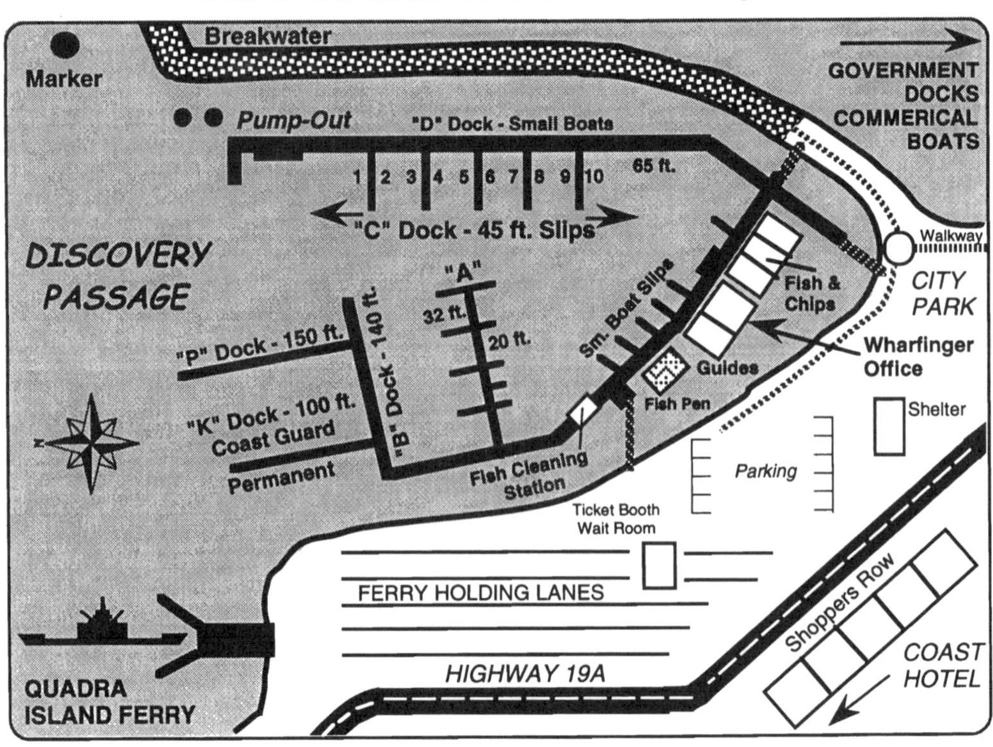

Campbell River

NAME OF MARINA: *APRIL POINT RESORT & MARINA* RADIO: VHF 66A
TELEPHONE: 250-285-2222 MGR: Lindsay Humber, Marina Manager
E-MAIL: info@obmg.com FAX: 250-285-2016 (Fax)
ADDRESS: 900 April Pt. Road Quathiaski Cove, B.C. Canada V0P 1N0
SHORT DESCRIPTION & LOCATION: **www.aprilpoint.com**

50°03.75' - 125°13.65' Located on Quadra Is. across Discovery Psg. from Campbell Riv. in the cove S of Gowland Harbour, near the resort. Marina caters to large yachts with full amenities. On entering **keep close** to the red marker & leave it on your starboard side!

GUEST BOAT CAPACITY: ....Ample
DOCKSIDE DEPTH AT ZERO TIDE: ....20 ft.+
SEASON: ....May - October
RESERVATION POLICY: ....Recommended
AMT W/ELECTRICITY: ....All
FUEL DOCK: ....Close by
MARINE REPAIRS: ....On Call
TOILETS: ....Yes
HOT SHOWERS: ....Yes
RESTAURANT: ....Close by
PICNIC AREA: ....Yes
BASIC STORE: ....Close by
BROADBAND/WI-FI: ....BroadbandXpress
DAILY RATE: ....Moderate (75¢-$1.25/foot)

NOTE: Marina guests have full use of amenities & facilities at both April Point Resort and Painters Lodge.

GUEST DOCK: ....4500 ft. total
GUEST SLIPS: ....Dock only
WATER: ....Yes
AMPS: ....15-30-50 A
PUMP OUT STATION ....None
HAUL OUT: Nearby - 50' or less
BOAT RAMP: ....Yes
LAUNDRY: ....Yes
BAR: ....Pub close by
POOL: ....Close by
GOLF: ....Campbell River
PET FRIENDLY: ....Good
OTHER: Cable TV, phone hook-ups, walking trails, fishing guides, hotel & bungalows, air service, scooter rentals, taxi service.

CAUTION! This chartlet not intended for use in navigation.

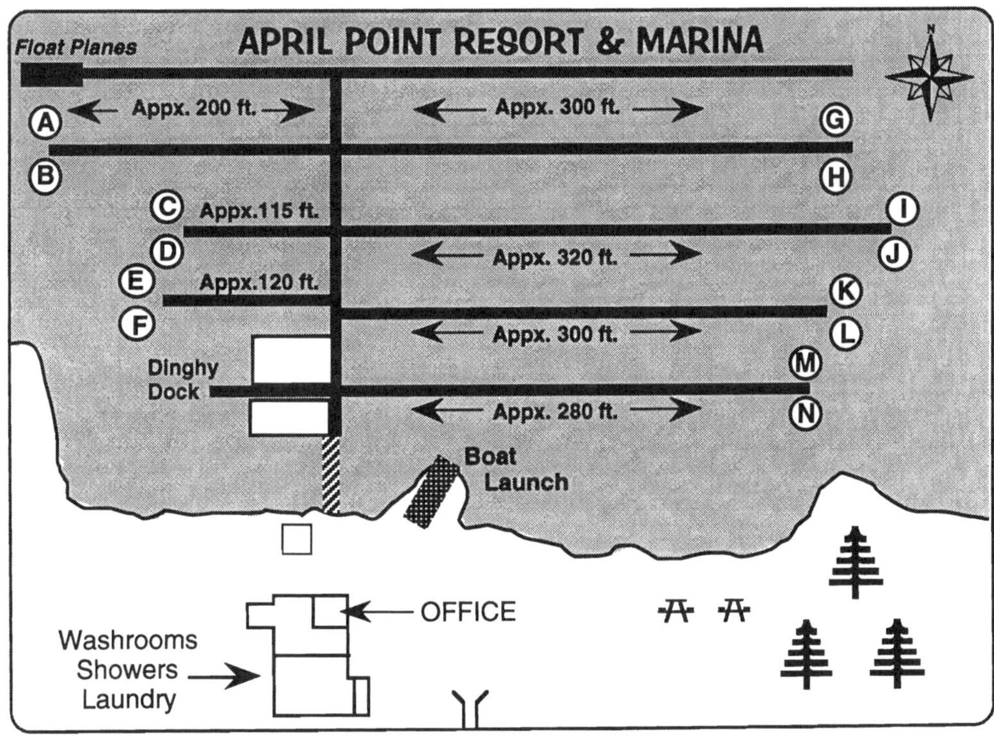

Campbell River

NAME OF MARINA: ***BROWNS BAY MARINA*** RADIO: VHF 66A
TELEPHONE: 250-286-3135 MGR: Jon Dawson
E-MAIL: FAX: 250-286-0951 (Fax)
ADDRESS: 2705 N. Island Hwy Campbell River, B.C. Canada V9W 2H4
SHORT DESCRIPTION & LOCATION: **www.brownsbayresort.com**

50°09.80' - 125°22.20' Located on Van. Is. at N. end of Seymour Narrows, 7.5 mi. N. of Campbell River. Floating breakwater of former tank cars protects this sport fishing oriented marina. Be cautious of strong tidal currents. Comfortable moorage for boats up to 90 ft.

GUEST BOAT CAPACITY: ....Varies
DOCKSIDE DEPTH AT ZERO TIDE: ....50 ft. +
SEASON: ....All year
RESERVATION POLICY: ....Accepts
AMT W/ELECTRICITY: ....All
FUEL DOCK: ....Gas & Diesel
MARINE REPAIRS: ....Close by
TOILETS: ....Yes
HOT SHOWERS: ....Yes
RESTAURANT: ....Summer only
PICNIC AREA: ....Yes
BASIC STORE: ....Yes
BROADBAND/WI-FI: ....None
DAILY RATE: ....Moderate (75¢-$1.25/foot)

GUEST DOCK: Appx. 2000 ft TTL
GUEST SLIPS: 50
WATER: ....Yes
AMPS: ....15-30 A
PUMP OUT STATION ....None
HAUL OUT: ....None
BOAT RAMP: ....Yes
LAUNDRY: ....Yes
BAR: ....Licensed restaurant
POOL: ....None
GOLF: ....Close by
PET FRIENDLY: ....Good
OTHER: Boat rentals, fishing guides, tackle, licenses, bait, floating B & B floating restaurant/cafe, fully licensed, R/V resort.

CAUTION! This chartlet not intended for use in navigation.

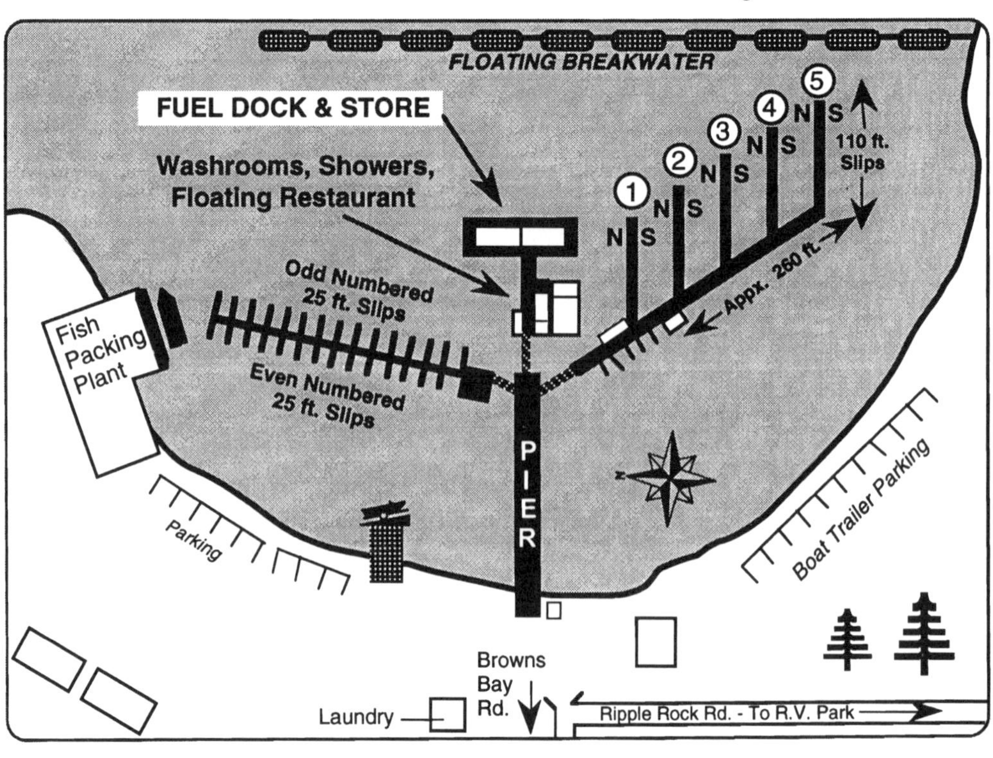

Heriot Bay

NAME OF MARINA: ***HERIOT BAY INN & MARINA*** RADIO: VHF 66A
TELEPHONE: 250-285-3322 MGR: Rolsin Doran
E-MAIL: info@heriotbayinn.com FAX: 250-285-2708 (Fax)
ADDRESS: P.O. Box 100 Heriot Bay, B.C. Canada V0P 1H0
SHORT DESCRIPTION & LOCATION: **www.heriotbayinn.com**

50°06.20' - 125°12.75' Located on S.E. side of Quadra Is. in Heriot Bay next to ferry terminal & Campground. Friendly older marina with historic Heriot Bay Inn & classic pub. Busy government dock just N. of marina & small shopping center in the area.

GUEST BOAT CAPACITY: ....Appx. 15-20 boats
DOCKSIDE DEPTH AT ZERO TIDE: ......15 ft. +
SEASON: ....All year
RESERVATION POLICY: ....Accepts
AMT W/ELECTRICITY: ....As available
FUEL DOCK: ....Gas, Dsl, & LP
MARINE REPAIRS: ....None
TOILETS: ....Yes
HOT SHOWERS: ....Yes
RESTAURANT: ....Yes
PICNIC AREA: ....Yes
BASIC STORE: ....Yes
BROADBAND/WI-FI: ....Yes
DAILY RATE: ....Moderate (70¢-$1.25/foot)
WINTER RATES AVAILABLE
Garbage: $4.00/Bag

GUEST DOCK: ....1800 ft. total
GUEST SLIPS: ....Docks only
WATER: ....Limited
AMPS: ....15-30 A
PUMP OUT STATION ....None
HAUL OUT: ....None
BOAT RAMP: ....Close by
LAUNDRY: ....Yes
BAR: ....Yes
POOL: ....None
GOLF: ....None
PET FRIENDLY: ....Good
OTHER: Rebecca Spit Prov. Park within walking distance.
Kayak & bike rentals.
Supermarket, liquor store, arts & crafts. RV Park.

CAUTION! This chartlet not intended for use in navigation.

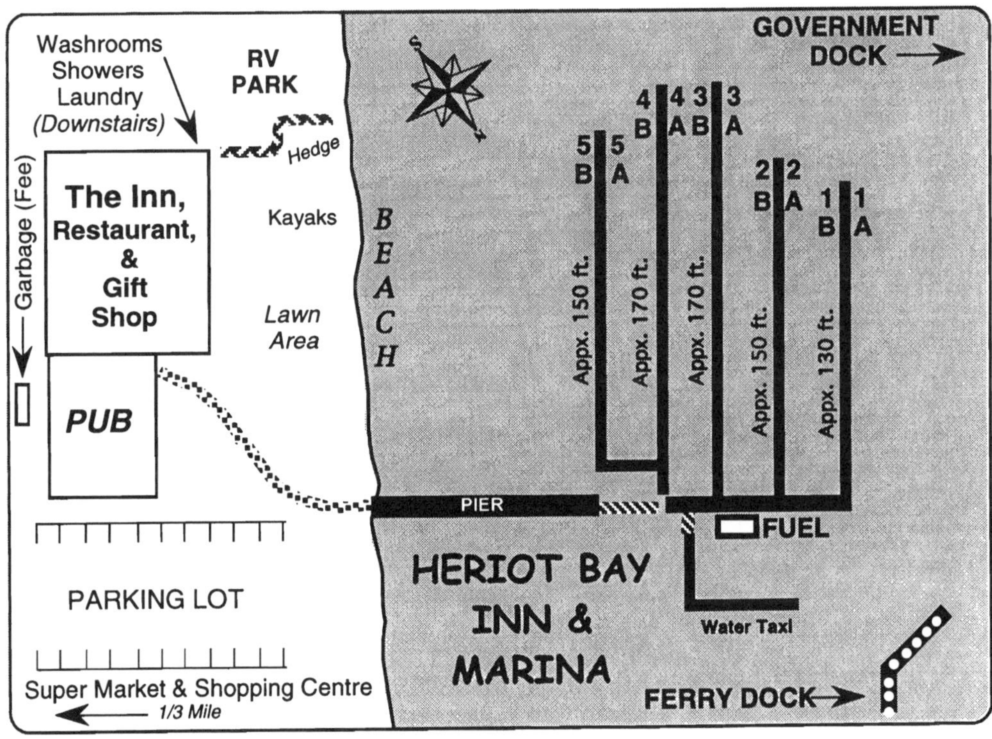

Heriot Bay

NAME OF MARINA: ***TAKU RESORT & MARINA*** RADIO: VHF 66A
TELEPHONE: 250-285-3031 MGR: Lynden McMartin
E-MAIL: info@takuresort.com FAX: 250-285-3712 (Fax)
ADDRESS: 616 Taku Road Heriot Bay, B.C. Canada V0P 1H0
SHORT DESCRIPTION & LOCATION: **www.takuresort.com**

50°06.13' - 125°12.06' Located on S.E. side of Quadra Is. in Heriot Bay & Drew Harbour, just east of ferry terminal. Excellent sturdy new docks & facilities in a park-like landscaped setting. Family & group oriented resort with hotel, cottages, and RV Park.

GUEST BOAT CAPACITY: ....Appx. 15-25 boats
DOCKSIDE DEPTH AT ZERO TIDE: 15 Ft. plus
SEASON: ....All year
RESERVATION POLICY: ....Accepts
AMT W/ELECTRICITY: ....All
FUEL DOCK: ....Close by
MARINE REPAIRS: ....Close by
TOILETS: ....Yes
HOT SHOWERS: ....Free hot showers
RESTAURANT: ....Close by
PICNIC AREA: ....Yes
BASIC STORE: ....Yes
BROADBAND/WI-FI: ....Limited
DAILY RATE: ....Moderate (75¢-$1.25/foot)

GUEST DOCK: ....1000 ft. Total
GUEST SLIPS: ....5 Small slips
WATER: ....Yes
AMPS: ....30-50 A
PUMP OUT STATION ....None
HAUL OUT: ....None
BOAT RAMP: ....Yes
LAUNDRY: ....Yes
BAR: ....Close by
POOL: ....None
GOLF: ....None
PET FRIENDLY: ....Excellent
OTHER: Rebecca Spit Prov. Park within walking distance. Internet access, hot tub, close to supermarket, liquor store, arts & crafts.

CAUTION! This chartlet not intended for use in navigation.

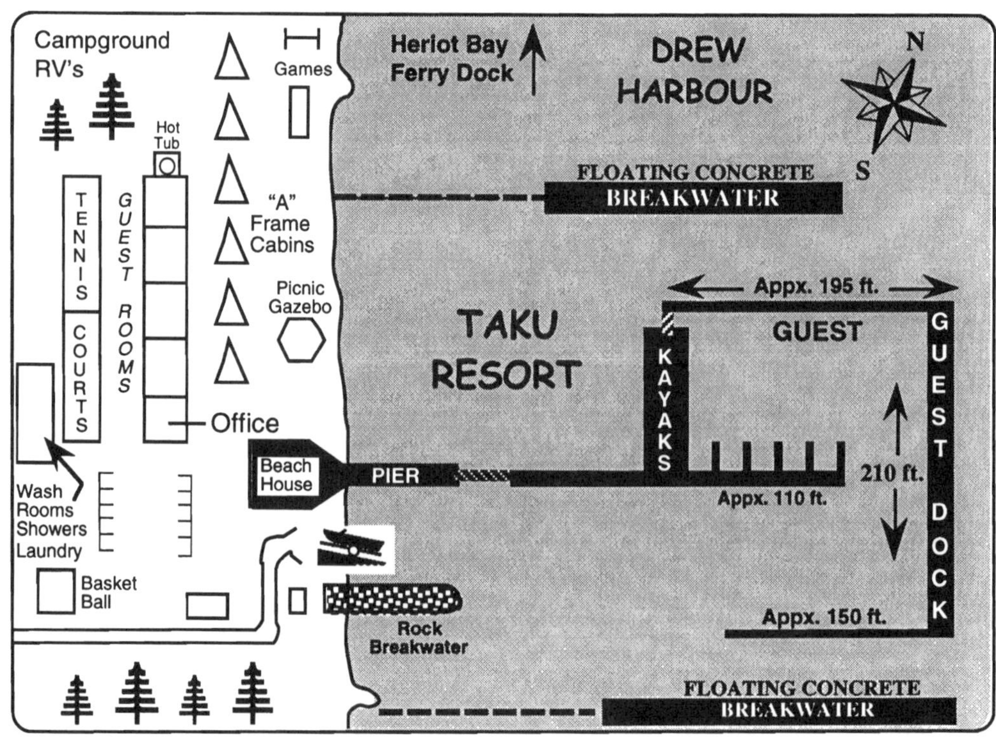

Cortes Island

NAME OF MARINA: ***GORGE HARBOUR MARINA RESORT*** **RADIO:** VHF 66A
TELEPHONE: **250-935-6433** **MGR:** Barb Hansen
E-MAIL: gorgehar@oberon.ark.com **FAX:** 250-935-6402 **(Fax)**
ADDRESS: P.O. Box 89 Whaletown, B.C. Canada V0P 1Z0
SHORT DESCRIPTION & LOCATION: **www.gorgeyharbour.com**

50°06.00' - 125°01.42' Full service marina resort located on SW Cortes Island at NW part of Gorge Harbour. Safe passage to harbor through "The Gorge". Very ample store w/fresh produce, meats, & groceries. Friendly Staff and outstanding restaurant.

GUEST BOAT CAPACITY: ....Appx. 35 boats
DOCKSIDE DEPTH AT ZERO TIDE: ......8-10 ft.
SEASON: ....All year
RESERVATION POLICY: ....Accepts
AMT W/ELECTRICITY: ....Appx. 60%
FUEL DOCK: ....Gas, Dsl, & LP
MARINE REPAIRS: ....Close by
TOILETS: ....Yes
HOT SHOWERS: ....Yes
RESTAURANT: ....Yes
PICNIC AREA: ....Yes
BASIC STORE: ....Yes
BROADBAND/WI-FI: ....Yes
DAILY RATE:.................Moderate (75¢-$1.25/foot)
FEE FOR GARBAGE DROP.
NO CHARGE FOR RECYCLING,

GUEST DOCK: ....1800 ft. total
GUEST SLIPS: ....Docks only
WATER: ....Limited
AMPS: ....15-30 A
PUMP OUT STATION ....None
HAUL OUT: ....None
BOAT RAMP: ....None
LAUNDRY: ....Yes
BAR: ....Licensed restaurant
POOL: ....None
GOLF: ....None
PET FRIENDLY: ....Good
OTHER: R.V. Park, campground, & lodge. Boat rentals. Many local activities, walks, & sightseeing.

CAUTION! This chartlet not intended for use in navigation.

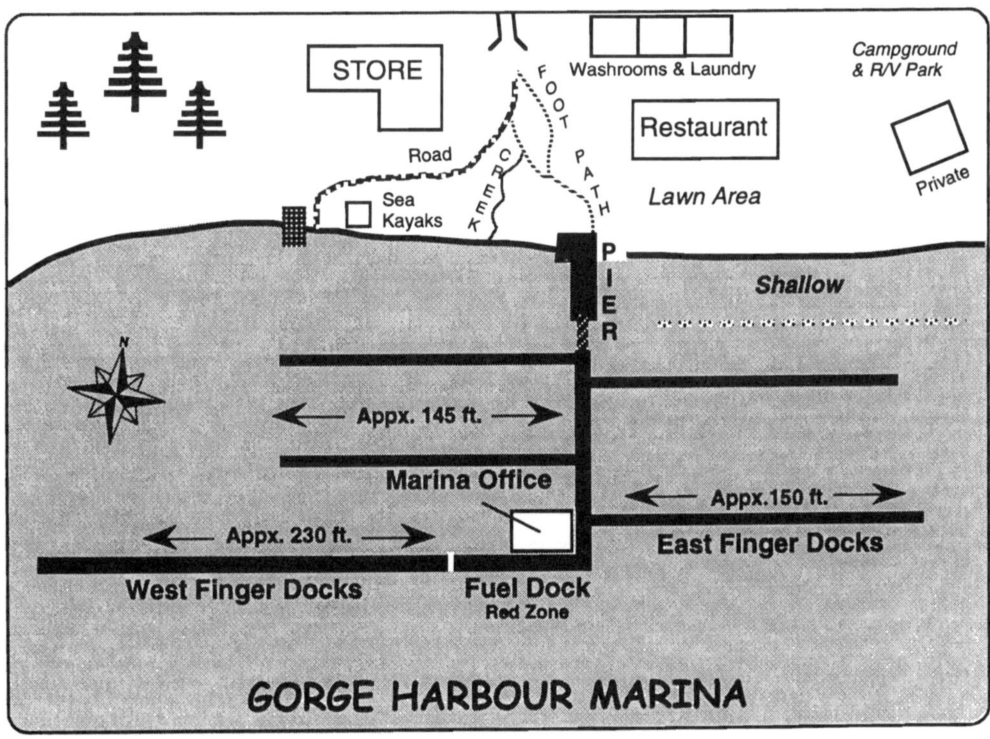

Cortes Island

NAME OF MARINA: ***MANSONS LANDING PUBLIC WHARF*** RADIO: None
TELEPHONE: 250-935-0180 MGR: Harbour Auth. of Cortes Is.
E-MAIL: corteshabour@hotmail.com FAX: 250-935-0182 (Fax)
ADDRESS: P.O. Box 243 Mansons Landing, B.C. Canada V0P 1K0
SHORT DESCRIPTION & LOCATION: http://www.dfo-mpo.gc.ca/sch/HB_BC_e.asp
50°06.00' - 124°59.00' Busy Govt. Wharf located on a spit fronting a drying lagoon along the E. side of Manson Bay. Dock is often filled with commercial boats and local small craft but plenty space for dinghies. The upland area is a wonderful BC Provincial Marine Park.

GUEST BOAT CAPACITY: ......Appx. 10-15 boats
DOCKSIDE DEPTH AT ZERO TIDE: ....17 ft.
SEASON: ....All year
RESERVATION POLICY: ....None
AMT W/ELECTRICITY: ....Most
FUEL DOCK: ....None
MARINE REPAIRS: ....None
TOILETS: ....Out-House
HOT SHOWERS: ....None
RESTAURANT: ....2 km
PICNIC AREA: ....Yes
BASIC STORE: ....2 km
BROADBAND/WI-FI: ....None
DAILY RATE: ....Economical
(Under 75¢/foot)
1 Hour Free

GUEST DOCK: ....Appx. 250 ft.
GUEST SLIPS: ....Dock only
WATER: ....None
AMPS: ....20 A
PUMP OUT STATION ....None
HAUL OUT: ....None
BOAT RAMP: ....None
LAUNDRY: ....None
BAR: ....None
POOL: ....None
GOLF: ....None
PET FRIENDLY: ....Excellent
OTHER: Park on saltwater lagoon has great swimming, shellfish harvesting, picnicking, trails. Short walk to lake.

CAUTION! This chartlet not intended for use in navigation.

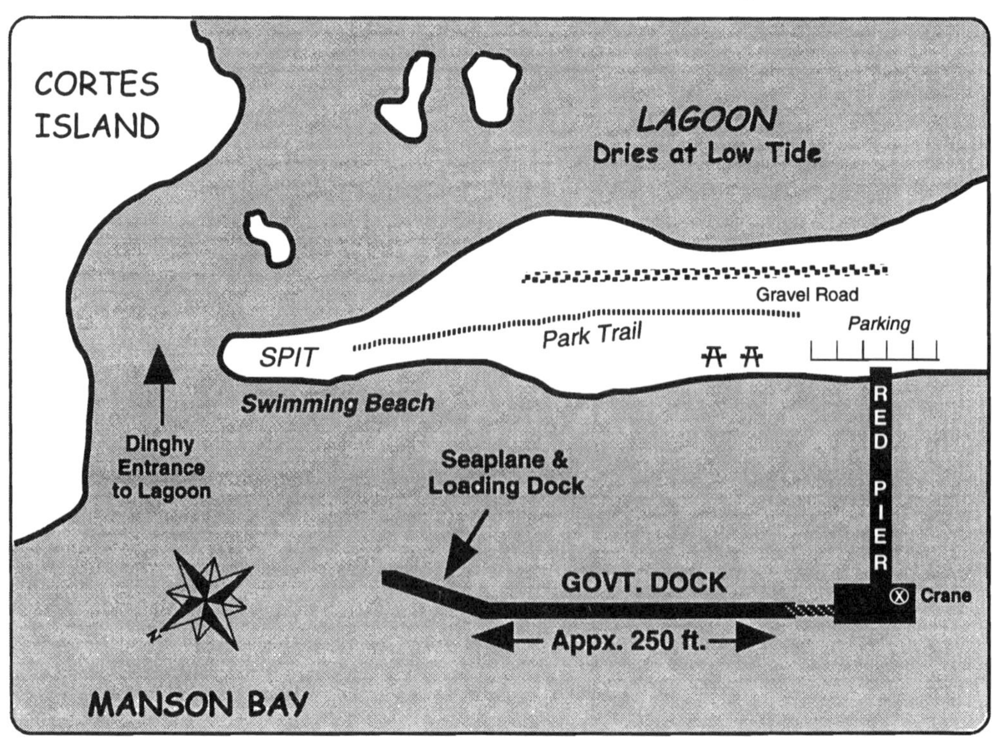

Cortes Island

NAME OF MARINA: ***SQUIRREL COVE PUBLIC WHARF*** RADIO: None
TELEPHONE: **250-935-0180** MGR: Harbour Authority of Cortes Island
E-MAIL: corteshabour@hotmail.com FAX: 250-935-0182 (Fax)
ADDRESS: P.O. Box 243 Mansons Landing, B.C. Canada V0P 1K0
SHORT DESCRIPTION & LOCATION: **http://www.dfo-mpo.gc.ca/sch/HB_BC_e.asp**

50°07.20' - 124°54.60' The busy sm. public dock is located in outer Squirrel Cove on E. side of Cortes Is. 15 Minute loading zones (marked in yellow) for good short provisioning stop - not much overnight moorage. Open to wakes & local boats occupy much of the dock.

GUEST BOAT CAPACITY: ....Appx. 0-6 boats
DOCKSIDE DEPTH AT ZERO TIDE: 6 Ft. +
SEASON: ....All year
RESERVATION POLICY: ....None
AMT W/ELECTRICITY: ....Limited
FUEL DOCK: ....None
MARINE REPAIRS: ....None
TOILETS: ....Yes
HOT SHOWERS: ....Yes
RESTAURANT: ....Yes
PICNIC AREA: ....None
BASIC STORE: ....Yes
BROADBAND/WI-FI: ....None
DAILY RATE: ....Economical (Under 75¢/foot)

Note: Can anchor out front in settled weather & use dinghy dock for provisioning.

GUEST DOCK: 200 ft. - Limited
GUEST SLIPS: ....Dock only
WATER: ....None
AMPS: ....15 A
PUMP OUT STATION ....None
HAUL OUT: ....None
BOAT RAMP: ....Yes
LAUNDRY: ....Yes
BAR: ....None
POOL: ....None
GOLF: ....None
PET FRIENDLY: ....Good
OTHER: Well stocked general store - hardware, ATM, garbage drop (fee), liquor store, Post Office, propane & auto gas. Dinghy dock.

CAUTION! This chartlet not intended for use in navigation.

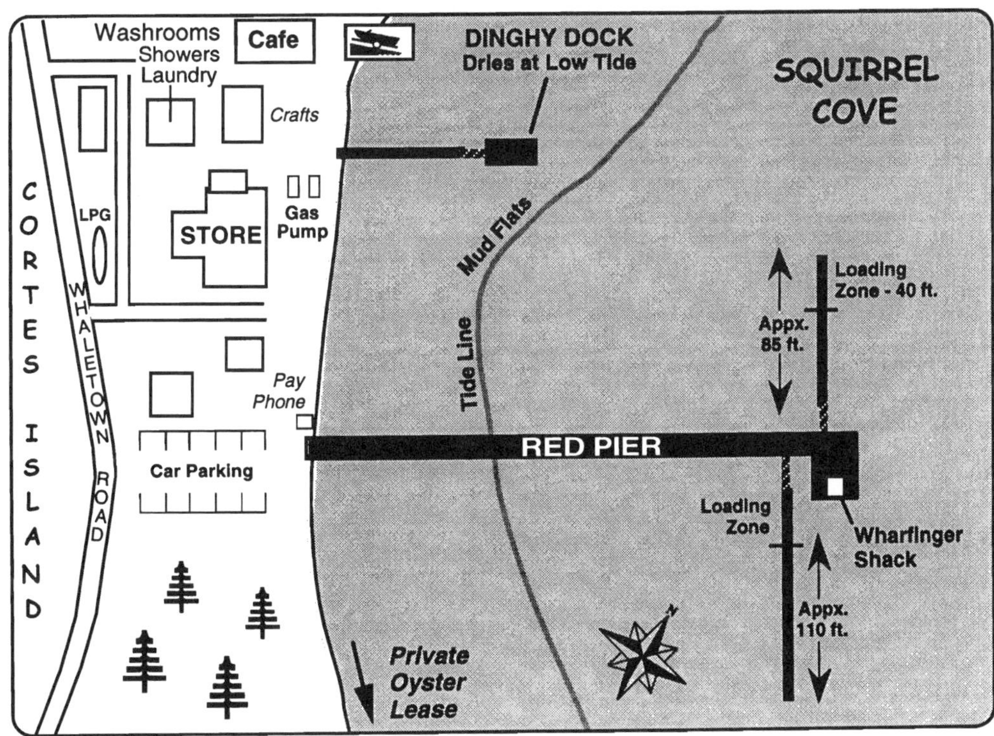

Cortes Island

NAME OF MARINA: ***CORTES BAY PUBLIC WHARF*** **RADIO:** None
TELEPHONE: **250-935-0180** **MGR:** Harbour Auth. of Cortes Is.
E-MAIL: corteshharbour@hotmail.com **FAX:** 250-935-0182 (Fax)
ADDRESS: P.O. Box 243 Mansons Landing, B.C. Canada V0P 1K0
SHORT DESCRIPTION & LOCATION: http://www.dfo-mpo.gc.ca/sch/HB_BC_e.asp

50°03.74' - 124°55.96' Located in protected Cortes Bay in between the private docks of Seattle Y.C. & Royal Vancouver Y.C. Public dock is directly in front of prominent red roofed house. Be sure to leave red marker to starboard upon entering Bay!

GUEST BOAT CAPACITY: ....Appx. 10-15 boats
DOCKSIDE DEPTH AT ZERO TIDE: 30 Ft. +
SEASON: ....All year
RESERVATION POLICY: ....None
AMT W/ELECTRICITY: ....All
FUEL DOCK: ....None
MARINE REPAIRS: ....None
TOILETS: ....None
HOT SHOWERS: ....None
RESTAURANT: ....4.5 miles
PICNIC AREA: ....None
BASIC STORE: ....4.5 miles
BROADBAND/WI-FI: ....BroadbandXpress
DAILY RATE:..............Economical (Under 75¢/foot)
1 Hour Free
GARBAGE: Fee

GUEST DOCK: ....440 ft. total
GUEST SLIPS: ....Dock only
WATER: ....None
AMPS: ....15 Amp
PUMP OUT STATION ....None
HAUL OUT: ....None
BOAT RAMP: ....None
LAUNDRY: ....None
BAR: ....None
POOL: ....None
GOLF: ....None
PET FRIENDLY: ....Excellent
OTHER: Pay phone, garbage (fee), general store 4.5 miles, prawns & oysters in area.

CAUTION! This chartlet not intended for use in navigation.

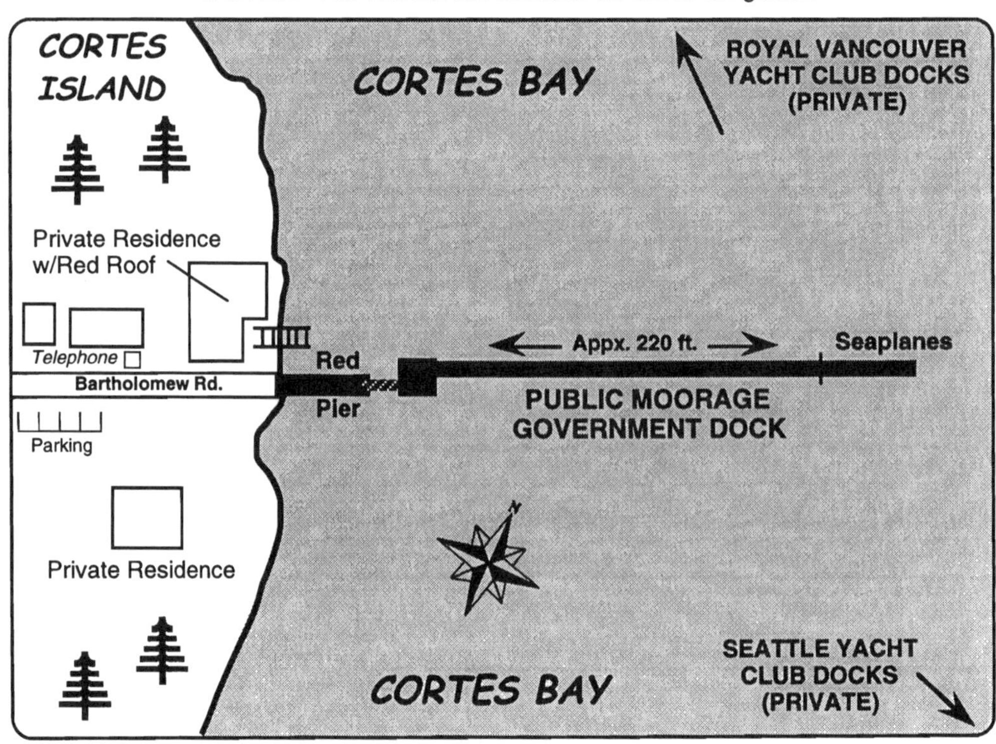

Refuge Cove

NAME OF MARINA: ***REFUGE COVE PUBLIC WHARF*** RADIO: None
TELEPHONE: **250-935-6659** MGR: Store
E-MAIL: refcov@twincomm.ca FAX: None
ADDRESS: General Delivery Refuge Cove, B.C. Canada V0P 1P0
SHORT DESCRIPTION & LOCATION: **http://www.upcoastsummers.com/refugecove/**

50°07.42' - 124°50.40 Located on SW corner W. Redonda Is. in the heart of Desolation Sound at the bottom of Lewis Channel. Popular resupply & rendezvous stop for cruising boaters. The store & shops are well stocked & the government docks offer visitor moorage.

GUEST BOAT CAPACITY: ....Appx. 25-35 boats
DOCKSIDE DEPTH AT ZERO TIDE: ....20 Ft. +
SEASON: ....Store - June 1 - Sept. 14
RESERVATION POLICY: ....None
AMT W/ELECTRICITY: ....Limited
FUEL DOCK: ....Gas, Dsl, & LP
MARINE REPAIRS: ....None
TOILETS: ....Yes
HOT SHOWERS: ....Yes
RESTAURANT: ....Burger stand & breakfast cafe
PICNIC AREA: Picnic deck
BASIC STORE: ....Yes
BROADBAND/WI-FI: ....BroadbandXpress
DAILY RATE: ....Economical (Under 75¢/foot)

NOTE: BEST TIME TO CHECK IN FOR OVERNIGHT MOORAGE IS AFTER 1500 HOURS.

GUEST DOCK: ....1400 ft. TTL
GUEST SLIPS: ....Dock only
WATER: ....Yes
AMPS: ....15 A
PUMP OUT STATION ....None
HAUL OUT: ....None
BOAT RAMP: ....None
LAUNDRY: ....Yes
BAR: ....None
POOL: ....None
GOLF: ....None
PET FRIENDLY: ....Fair
OTHER: Post Office, gift shop, baked goods, liquor agency, marine supplies, charts, float plane service. Garbage drop on Barge.

CAUTION! This chartlet not intended for use in navigation.

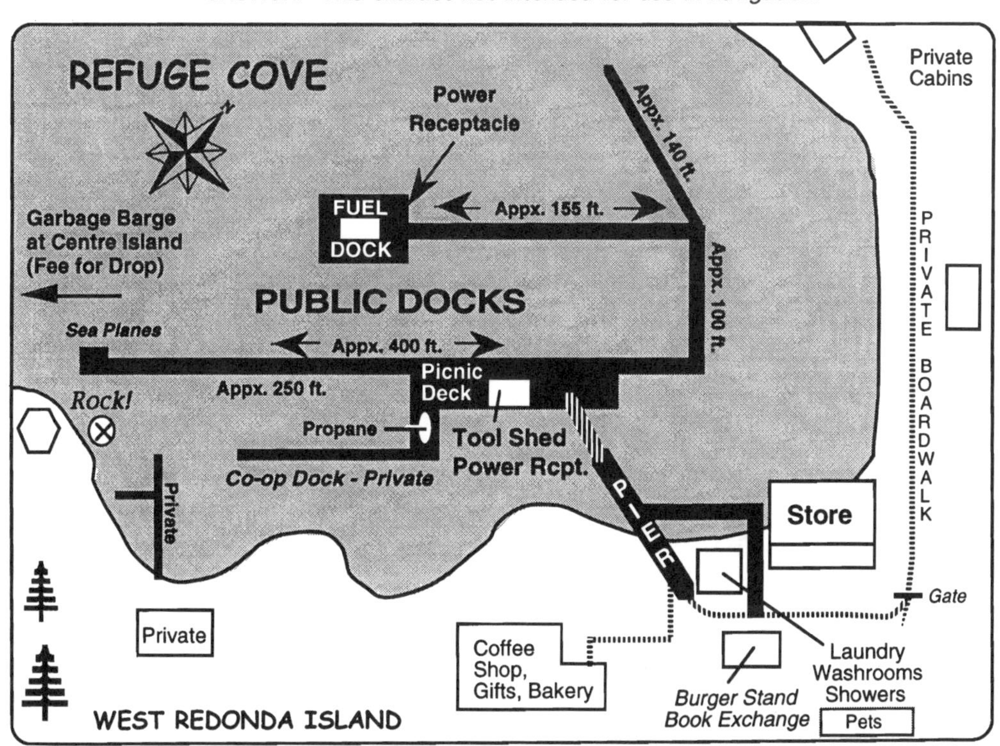

Toba Inlet

NAME OF MARINA: *TOBA WILDERNEST MARINA RESORT* RADIO: VHF 66 A
TELEPHONE: 250-830-2269 MGR: Kyle & Andrea Hunter
E-MAIL: tobawildernest@lincsat.com FAX: (403) 823-4156 (Fax)
ADDRESS: Mail: Toba Holdings, Box 9, Munson, Alberta Canada T0J 2C0
SHORT DESCRIPTION & LOCATION: **www.fishingwildernest.com**

50°19.50' - 124°47.70' Small wilderness marina resort located near the mouth of Toba Inlet, tucked inside of Double Island. Peace & quiet are plentiful along with spectacular Desolation Sound scenery & sportsman activities. Wonderful 15 minute hike to waterfall.

GUEST BOAT CAPACITY: ....Appx. 8-10 boats
DOCKSIDE DEPTH AT ZERO TIDE: 15 ft. plus
SEASON: ....All year
RESERVATION POLICY: ....Cabins only
AMT W/ELECTRICITY: ....None
FUEL DOCK: ....None
MARINE REPAIRS: ....None
TOILETS: ....Yes
HOT SHOWERS: ....Yes
RESTAURANT: ....None
PICNIC AREA: ....Yes
BASIC STORE: ....Yes
BROADBAND/WI-FI: ....None
DAILY RATE: ....Moderate (75¢-$1.25/foot)

GUEST DOCK: Appx 350 ft total
GUEST SLIPS: ....Dock only
WATER: ....Yes
AMPS: ....None
PUMP OUT STATION ....None
HAUL OUT: ....None
BOAT RAMP: ....None
LAUNDRY: ....None
BAR: ....None
POOL: ....Hot Tub for Rent
GOLF: ....None
PET FRIENDLY: ....Excellent
OTHER: 1 Mooring buoy, rustic log guest cabins, fishing & hiking trips, interesting self engineered hydro-electric plant.

CAUTION! This chartlet not intended for use in navigation.

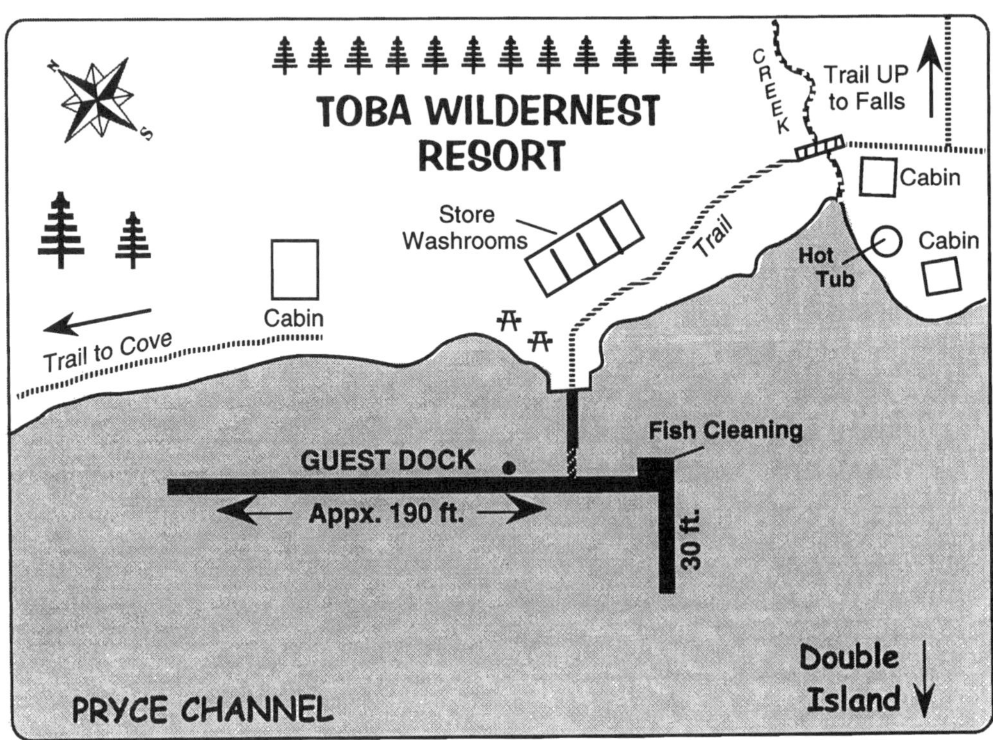

Big Bay-Stuart Island

NAME OF MARINA: ***STUART ISLAND COMMUNITY DOCK*** RADIO: VHF 66A
TELEPHONE: 250-202-DOCK(3625) MGR: Roger/Cathy Minor
E-MAIL: stuartislandca@aol.com FAX: None
ADDRESS: General Delivery, Stuart Island, B.C. Canada V0P 1V0

SHORT DESCRIPTION & LOCATION:

50°23.05' - 125°80.20' Located in Big Bay on west side of Stuart Is. near Yucultas rapids. The Docks were previously run by Transport Canada, but now divested to a progressive local organization making big improvements & trying hard to fill boaters needs.

GUEST BOAT CAPACITY: Appx 20-30 on docks
DOCKSIDE DEPTH AT ZERO TIDE: ....10 ft.+
SEASON: ....All year
RESERVATION POLICY: ....Recommended
AMT W/ELECTRICITY: ....None
FUEL DOCK: ....None
MARINE REPAIRS: ....None
TOILETS: ....Yes
HOT SHOWERS: ....Yes
RESTAURANT: ....Close by
PICNIC AREA: ....Covered outside picnic deck
BASIC STORE: ....Yes
BROADBAND/WI-FI: ....None
DAILY RATE: ....Moderate (75¢-$1.25/foot)

GUEST DOCK: ....Appx. 1100 ft.
GUEST SLIPS: ....Docks only
WATER: ....Yes
AMPS: ....None
PUMP OUT STATION ....None
HAUL OUT: ....None
BOAT RAMP: ....None
LAUNDRY: ....Yes
BAR: ....None
POOL: ....None
GOLF: ....None
PET FRIENDLY: ....Excellent
OTHER: New Docks, ice, Store, seafood, Post Office, fishing licenses & gear, walking trail on Island, probable Liquor Agency.

"WATCH US GROW!"

CAUTION! This chartlet not intended for use in navigation.

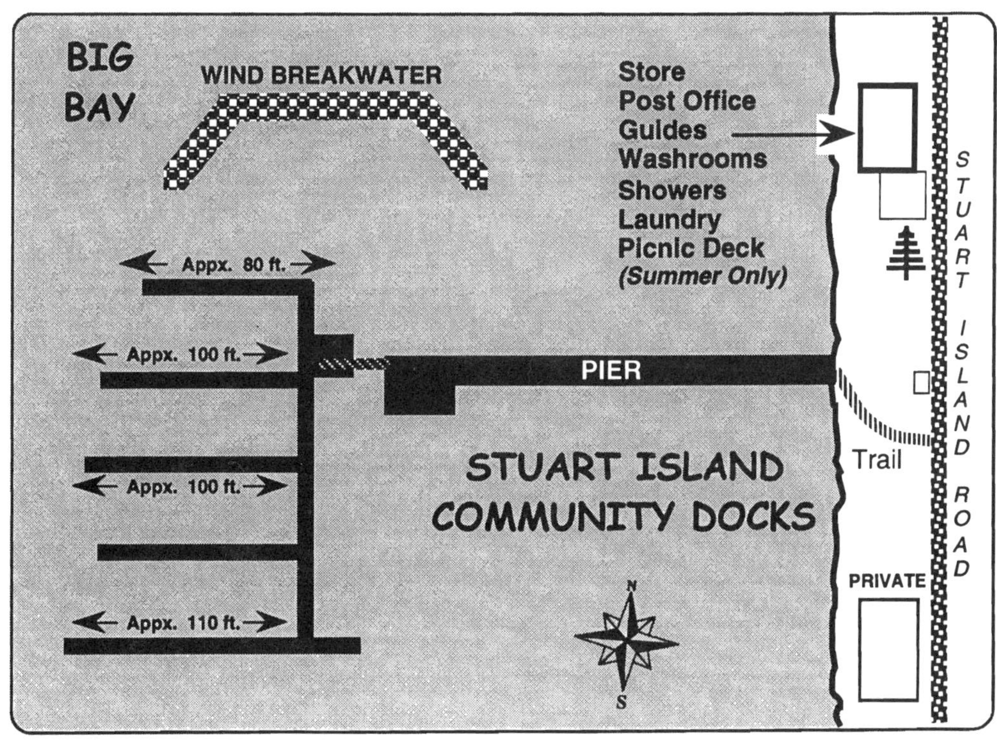

Dent Island

NAME OF MARINA: *DENT ISLAND LODGE* RADIO: VHF 66A
TELEPHONE: 250-203-2553 MGR: Henry Moll
E-MAIL: henry@dentisland.com FAX: 250-203-1041 (Fax)
ADDRESS: General Delivery Stuart Island, B.C. Canada V0P 1V0
SHORT DESCRIPTION & LOCATION: **www.dentisland.com**

50°24.40' - 125°10.90' Premier wilderness & fishing resort, yachting destination, & retreat on 2 sm. islets near N. entrance to Bute Inlet, close to Yuculta & Arran Rapids. Full boater amenities & moorage for boats up to 140 ft. Outstanding dining & accommodations.

GUEST BOAT CAPACITY: ....Appx. 12-14 boats
DOCKSIDE DEPTH AT ZERO TIDE: 10 ft. plus
SEASON: ....Open May thru October
RESERVATION POLICY: ....Accepts
AMT W/ELECTRICITY: ....All
FUEL DOCK: ....None
MARINE REPAIRS: ....None
TOILETS: ....Yes
HOT SHOWERS: ....Yes
RESTAURANT: ......Fine Dining - by reservation
PICNIC AREA: ....Yes
BASIC STORE: ....Close by
BROADBAND/WI-FI: ....Yes
DAILY RATE:.................Premium (Over $1.25/foot)
MOORAGE INCLUDES: Water, power, use of hot tub, sauna & exercise facility, comp. coffee & pastries.

GUEST DOCK: ....800 ft total
GUEST SLIPS: ....Docks only
WATER: ....Yes
AMPS: ....30-50 A
PUMP OUT STATION ....None
HAUL OUT: ....None
BOAT RAMP: ....None
LAUNDRY: ....Yes
BAR: ....Licensed Restaurant
POOL: ....None
GOLF: ....None
PET FRIENDLY: ....Good
OTHER: Lodge, cabins, & cottages. Fishing guides, wildlife viewing, fitness centre, sauna, Hot Tub overlooking the Rapids.

CAUTION! This chartlet not intended for use in navigation.

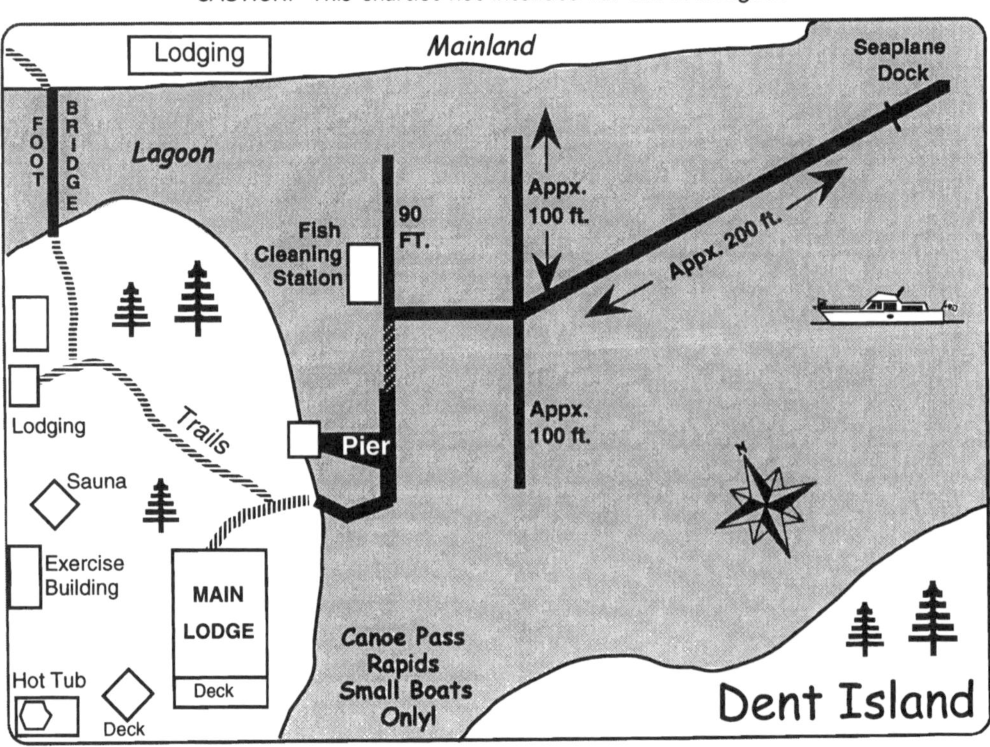

Frederick Arm

NAME OF MARINA: *OLEO'S RESTAURANT MOORAGE* RADIO: VHF 66A
TELEPHONE: 250-203-6670 MGR: Montoya Family
E-MAIL: None FAX: None
ADDRESS: P.O. Box 12, Stuart Island B.C. Canada V0P 1V0
SHORT DESCRIPTION & LOCATION:

50°28.02' - 125°15.58' Oleo's moorage & little restaurant is tucked behind a small islet adjacent to a fish farm, about 3/4 mile up Frederick Arm on the east side. Family operated restaurant with dock space for about 5 boats. Excellent food with European flair.

GUEST BOAT CAPACITY: ....Appx. 10 boats
DOCKSIDE DEPTH AT ZERO TIDE: ......40 ft. +
SEASON: ....April thru Sept.
RESERVATION POLICY: Required 1 day ahead
AMT W/ELECTRICITY: ....None
FUEL DOCK: ....None
MARINE REPAIRS: ....None
TOILETS: ....Yes
HOT SHOWERS: ....None
RESTAURANT: ....Yes
PICNIC AREA: ....None
BASIC STORE: ....None
BROADBAND/WI-FI: ....None
DAILY RATE: ....Moderate
(Moorage includes dinner)

GUEST DOCK: Appx. 350 ft. TTL
GUEST SLIPS: ....Docks only
WATER: ....None
AMPS: ....None
PUMP OUT STATION ....None
HAUL OUT: ....None
BOAT RAMP: ....None
LAUNDRY: ....None
BAR: ....Restaurant
POOL: ....None
GOLF: ....None
PET FRIENDLY: ....Challenging
OTHER: No shore access except dinghy. Restaurant serves up to 20 people each evening. Famous for Greek salad & cabbage rolls.

CAUTION! This chartlet not intended for use in navigation.

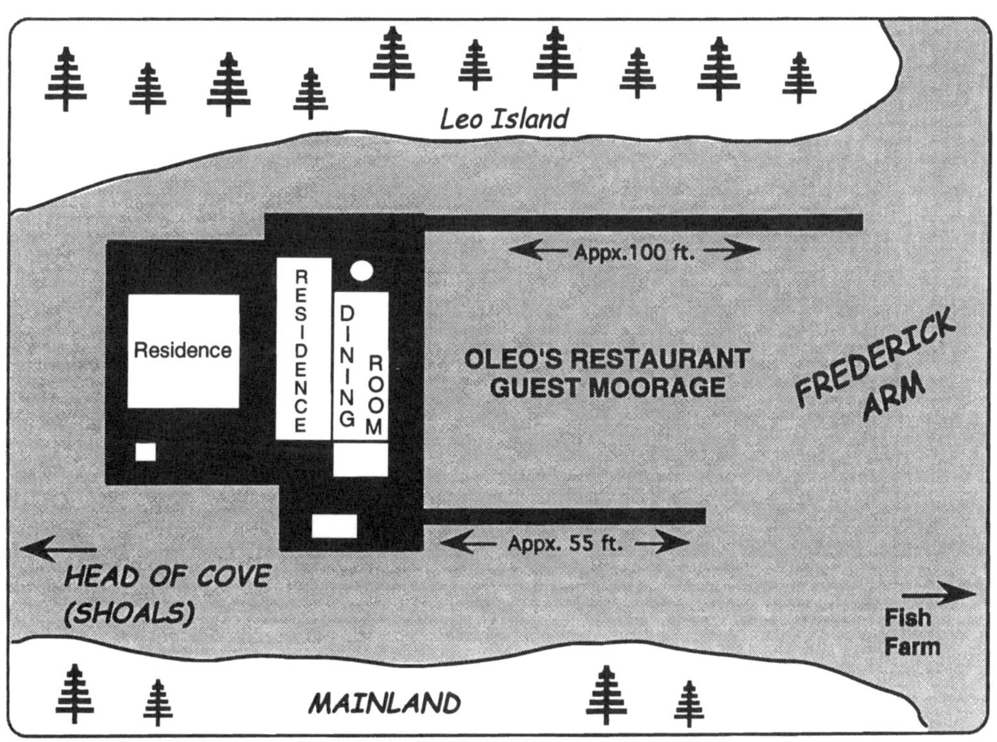

Shoal Bay

NAME OF MARINA: ***SHOAL BAY DOCK, LODGE & PUB*** RADIO: None
TELEPHONE: 250-287-6818 MGR: Mark MacDonald, Wharfinger
E-MAIL: None FAX: None
ADDRESS: General Delivery Blind Channel, B.C. Canada V0P 1B0
SHORT DESCRIPTION & LOCATION: **www.shoalbaylodge.com (Interesting!)**
50°27.50' - 125°21.95' Once the former settlement of Thurlow where gold was mined, there is a long DFO Government Dock, and a privately owned new lodge, cabins & pub. Located on NE side of E. Thurlow Island off Cordero Channel at base of Phillips Arm.

GUEST BOAT CAPACITY: 8-10 boats + rafting
DOCKSIDE DEPTH AT ZERO TIDE: ......10 ft. +
SEASON: ......Dock open all year
RESERVATION POLICY: ......None
AMT W/ELECTRICITY: ......None
FUEL DOCK: ......None
MARINE REPAIRS: ......None
TOILETS: ......At lodge
HOT SHOWERS: ......At lodge
RESTAURANT: ......Check at Pub
PICNIC AREA: ......Yes
BASIC STORE: ......None
BROADBAND/WI-FI: ......Yes
DAILY RATE: ......Economical (Under 75¢/foot)
MOORAGE PAID TO WHARFINGER AT PUB

GUEST DOCK: ......410 ft. total
GUEST SLIPS: ......Docks only
WATER: ......Seasonal
AMPS: ......None
PUMP OUT STATION ......None
HAUL OUT: ......None
BOAT RAMP: ......None
LAUNDRY: ......Yes
BAR: ...... Pub in summer
POOL: ......None
GOLF: ......None
PET FRIENDLY: ......Good
OTHER: Good anchorage off dock, good fishing in area. Lodge rebuilding in stages after fire of 2000.
RAFTING MANDATORY

CAUTION! This chartlet not intended for use in navigation.

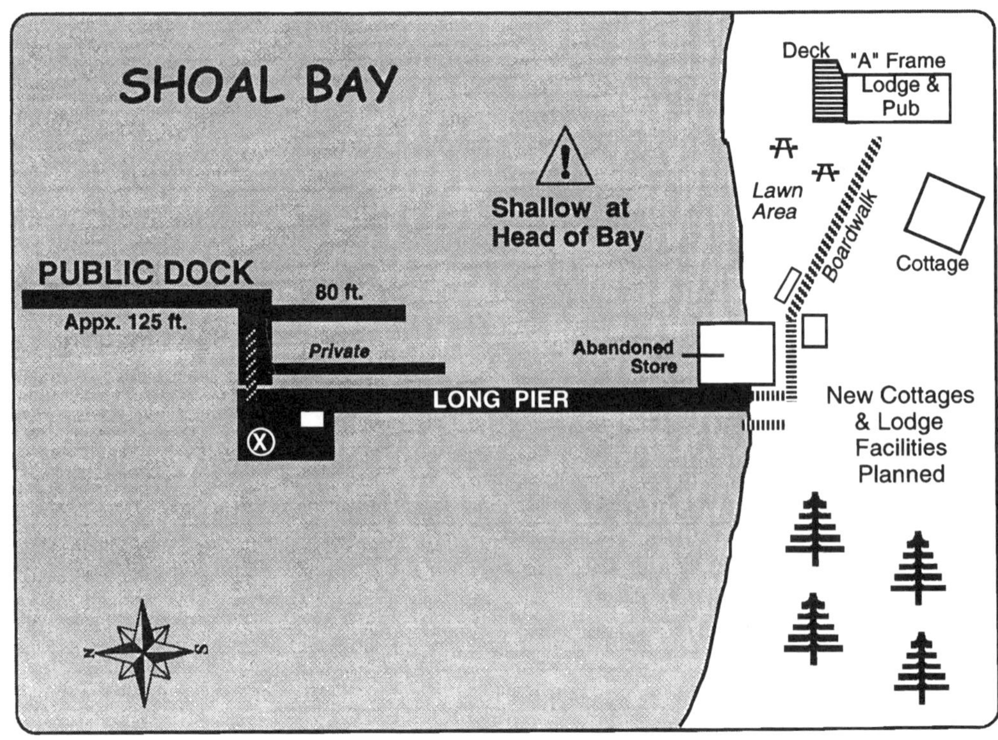

Cordero

NAME OF MARINA: ***CORDERO LODGE*** *(AKA CAMP CORDERO)* **RADIO:** VHF 66A
TELEPHONE: **250-287-0917** **MGR:** Reinhardt & Doris Kuppers
E-MAIL: info@corderolodge.com **FAX:** 250-287-8840
ADDRESS: General Delivery Blind Channel, B.C. Canada V0P 1B0
SHORT DESCRIPTION & LOCATION: **www.corderolodge.com**

50°26.70' - 125°27.10' Sm. picturesque floating marina, lodge, & restaurant situated on log floats behind Lorte Island on N side of Cordero Channel. **Note: Reduce speed when approaching docks & be aware of fast currents upon docking.**

GUEST BOAT CAPACITY: ....Appx. 15 boats
DOCKSIDE DEPTH AT ZERO TIDE: 30 ft. plus
SEASON: ....Open June to Mid Sept.
RESERVATION POLICY: ....Recommended
AMT W/ELECTRICITY: ....None
FUEL DOCK: ....Close by
MARINE REPAIRS: ....None
TOILETS: ....Yes
HOT SHOWERS: ....None
RESTAURANT: ....Yes
PICNIC AREA: ....Yes
BASIC STORE: ....None
BROADBAND/WI-FI: ....None
DAILY RATE: ....Moderate
(75¢-$1.25/foot)

GUEST DOCK: Appx. 700 ft TTL
GUEST SLIPS: ....Docks only
WATER: ....Limited
AMPS: ....None
PUMP OUT STATION ....None
HAUL OUT: ....None
BOAT RAMP: ....None
LAUNDRY: ....None
BAR: ....Yes
POOL: ....None
GOLF: ....None
PET FRIENDLY: ....Good
OTHER: Lodging-5 guest rooms, fishing charters, hiking trails, licensed restaurant, excellent German food.

CAUTION! This chartlet not intended for use in navigation.

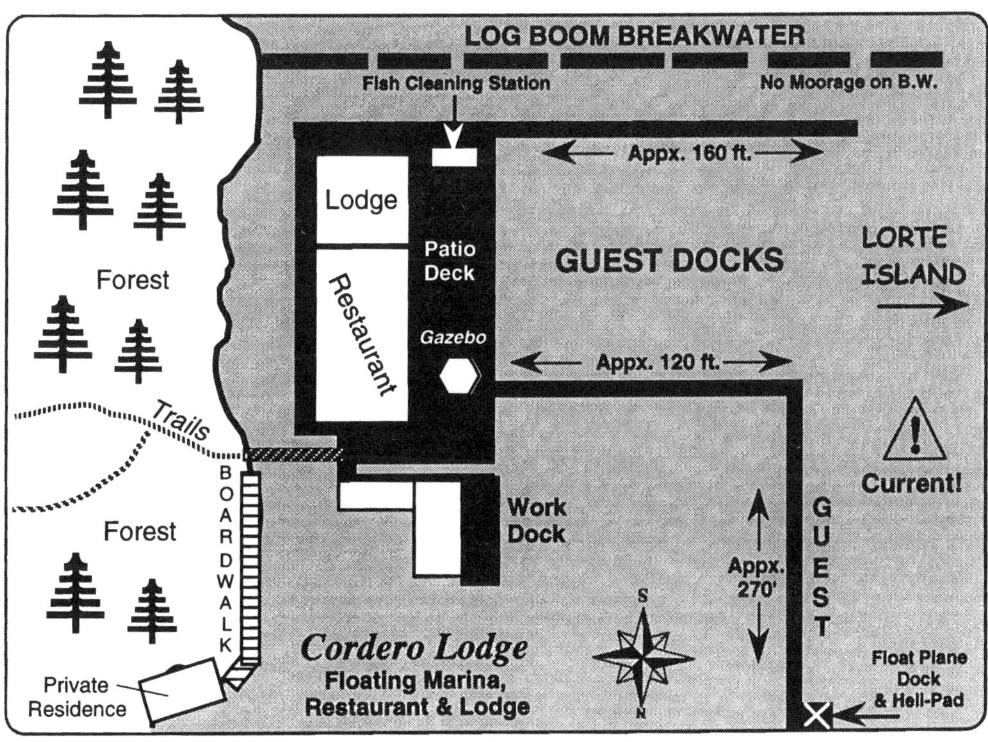

Blind Channel

NAME OF MARINA: ***BLIND CHANNEL RESORT*** **RADIO:** VHF 66A
TELEPHONE: **250-949-1420** **MGR:** Owners: The Richters
E-MAIL: info@blindchannel.com **FAX:**
ADDRESS: Blind Channel Trading Co. Blind Channel, B.C. Canada V0P 1B0
SHORT DESCRIPTION & LOCATION: **www.blindchannel.com**

50°24.85' - 125°30.00' Located on W side of Mayne Passage on E side of West Thurlow Island. Full service and well managed popular marina resort with ample sturdy dock space with many boater amenities. Excellent restaurant featuring German cuisine.

GUEST BOAT CAPACITY: ....Appx. 25-30 boats
DOCKSIDE DEPTH AT ZERO TIDE: ......10 ft. +
SEASON: ....All year
RESERVATION POLICY: ....Accepts
AMT W/ELECTRICITY: ....All
FUEL DOCK: ....Gas, Dsl, & LP
MARINE REPAIRS: ....None
TOILETS: ....Yes
HOT SHOWERS: ....Yes
RESTAURANT: ....Yes-Cedar Post Inn (Licensed)
PICNIC AREA: ....Yes
BASIC STORE: ....Yes
BROADBAND/WI-FI: ....Yes
DAILY RATE: ....Premium
(Over $1.25/foot)

INTERNET TERMINAL IN STORE

GUEST DOCK: ....2400 ft. total
GUEST SLIPS: ....7 - 80' slips
WATER: ....Yes-Spring Water
AMPS: ....15-20 A
PUMP OUT STATION ....None
HAUL OUT: ....None
BOAT RAMP: ....None
LAUNDRY: ....Yes
BAR: ....Yes
POOL: ....None
GOLF: ....None
PET FRIENDLY: ....Excellent
OTHER: Hiking trails, liquor store, Post Office, nice grocery store w/deli items,meats, & gifts, fishing tackle & licenses, rental cottage.

CAUTION! This chartlet not intended for use in navigation.

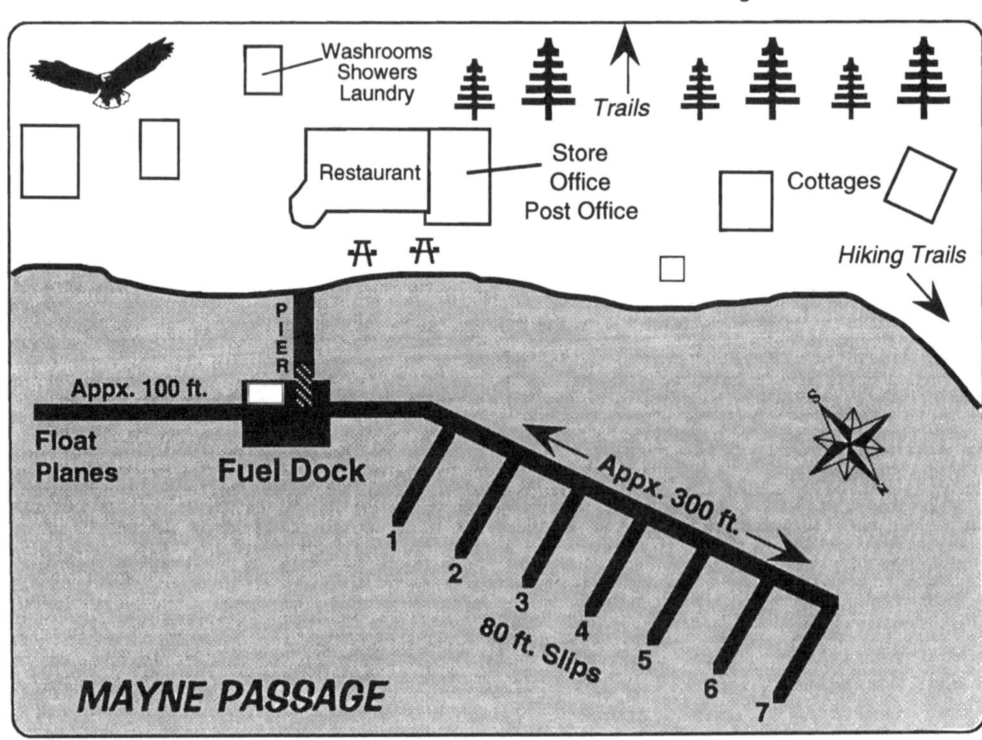

CHAPTER

11

THE BROUGHTON ARCHIPELAGO

Chapter Map - Page 2

Port Neville

NAME OF MARINA: ***PORT NEVILLE GOVERNMENT DOCK*** **RADIO:** VHF Ch. 06
TELEPHONE: 250-949-1535 **MGR:** Lorna Chesluk, Postmaster
E-MAIL: None **FAX:** None
ADDRESS: Port Neville P.O. Port Neville, B.C. Canada V0P 1M0

SHORT DESCRIPTION & LOCATION:

50°29.58' - 126°05.26' Port Neville is entered off Johnstone Strait between Ransom Pt. & Neville Pt. It provides good anchorage & small government dock plus an interesting & historic settlement on the E. side of entrance. Very friendly hosts welcome cruising boaters!

GUEST BOAT CAPACITY: ....Up to 9 boats
DOCKSIDE DEPTH AT ZERO TIDE: ....18 ft.
SEASON: ....All year
RESERVATION POLICY: ....Call ahead
AMT W/ELECTRICITY: ....None
FUEL DOCK: ....None
MARINE REPAIRS: ....None
TOILETS: ....None
HOT SHOWERS: ....None
RESTAURANT: ....None
PICNIC AREA: ....None
BASIC STORE: ....No provisions, Art & Gift shop
BROADBAND/WI-FI: ....None
DAILY RATE: Check with Wharfinger/Postmaster
OTHER: Art & Gift Shop, art gallery, book exchange, historic store/museum in the old building.

GUEST DOCK: Appx. 200 ft. TTL
GUEST SLIPS: ....Dock only
WATER: ....None
AMPS: ....None
PUMP OUT STATION ....None
HAUL OUT: ....None
BOAT RAMP: ....None
LAUNDRY: ....None
BAR: ....None
POOL: ....None
GOLF: ....None
PET FRIENDLY: Good - on leash
OTHER: Mail service 1 day ea. week (Wed). Summer potlucks & dessert nights often (No alcohol please). Beachwalks, wildlife, deer & bear sightings.

CAUTION! This chartlet not intended for use in navigation.

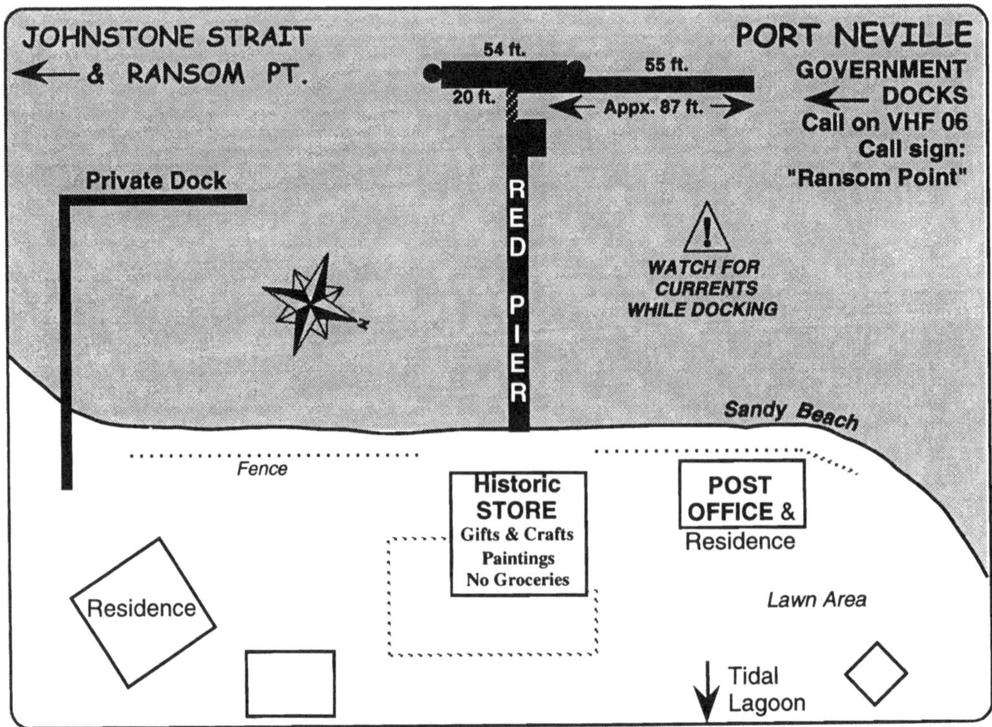

Lagoon Cove

NAME OF MARINA: *LAGOON COVE MARINA* RADIO: VHF 66A
TELEPHONE: None MGR: Bill & Jean Barber
E-MAIL: None FAX: None
ADDRESS: P.O. Box 42 Minstrel Is., B.C. Canada V0P 1L0

SHORT DESCRIPTION & LOCATION:

50°35.95' - 126°18.85' East Cracroft Island. Friendly & popular cruising marina resort with character, located just SW of Perley Island off Clio Channel & the Blow Hole. Most services for boaters available & amenities recently upgraded. Excellent fishing & prawning.

GUEST BOAT CAPACITY: ....Appx. 20-24 boats
DOCKSIDE DEPTH AT ZERO TIDE: ....Ample
SEASON: ....All year
RESERVATION POLICY: Same day - Call on VHF
AMT W/ELECTRICITY: Power available 9 hrs/day
FUEL DOCK: Gas, Dsl, LP, Stove Oil
MARINE REPAIRS: ....None
TOILETS: ....Yes
HOT SHOWERS: ....Yes
RESTAURANT: ....None
PICNIC AREA: ....Yes
BASIC STORE: ....Yes
BROADBAND/WI-FI: ....None
DAILY RATE: ....Moderate (65¢-$1.25/foot)
NOTE- 200 Ft. additional moorage away from main marina, no power or water, but within dinghy distance.

GUEST DOCK: Appx. 1000' Total
GUEST SLIPS: ....Docks only
WATER: ....Limited
AMPS: ....30 A
PUMP OUT STATION ....None
HAUL OUT: ....Rail
BOAT RAMP: ....None
LAUNDRY: ....None
BAR: ....None
POOL: ....None
GOLF: ....None
PET FRIENDLY: ....Good
OTHER: Gift Shop, fishing lic., tackle, charts, hiking trails, bears & eagles. Nightly happy hour potluck on Deck w/prawns provided.

CAUTION! This chartlet not intended for use in navigation.

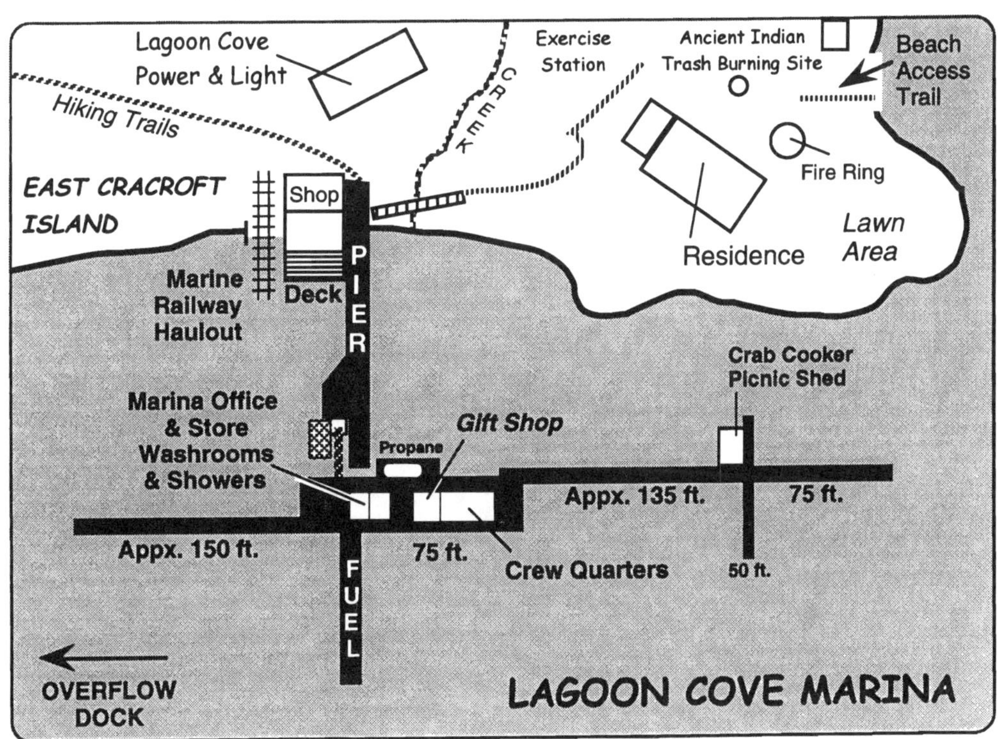

Echo Bay

NAME OF MARINA: ***ECHO BAY MARINA RESORT*** **RADIO:** VHF 66A
TELEPHONE: **250-974-7139** **MGR:** Bob & Nancy Richter
E-MAIL: echobay@island.net **FAX:** 866-861-9891 (Fax)
ADDRESS: Simoom Sound Post Office Echo Bay, B.C. Canada V0P 1S0
SHORT DESCRIPTION & LOCATION: **www.echobayresort.com**

50°45.50' - 126°29.85' Located inside Echo Bay on NW side of Gilford Island off Cramer Passage. This popular & busy marina resort is relaxed and comfortable in the heart of excellent fishing & cruising waters. Possible ownership change in the future.

GUEST BOAT CAPACITY: ....Appx. 25-30 boats
DOCKSIDE DEPTH AT ZERO TIDE: ......40 ft. +
SEASON: ....All year
RESERVATION POLICY: ....Accepts
AMT W/ELECTRICITY: ....All
FUEL DOCK: ....Gas, Dsl, & LP
MARINE REPAIRS: ....None
TOILETS: ....Yes
HOT SHOWERS: ....Yes
RESTAURANT: ....None
PICNIC AREA: ....Yes
BASIC STORE: ....Yes
BROADBAND/WI-FI: ....Yes
DAILY RATE: ....Moderate
(75¢-$1.25/foot)

GUEST DOCK: Appx 1900 Ft.TTL
GUEST SLIPS: ....Docks only
WATER: ....Yes
AMPS: ....15-30 A
PUMP OUT STATION ....None
HAUL OUT: ....None
BOAT RAMP: ....None
LAUNDRY: ....Yes
BAR: ....None
POOL: ....None
GOLF: ....None
PET FRIENDLY: ....Good
OTHER: Lg. picnic float, lodging, well stocked store with meat, produce. charts & licenses. Post Office, Internet. Park nearby - trails.

RESORT CHANGES POSSIBLE

CAUTION! This chartlet not intended for use in navigation.

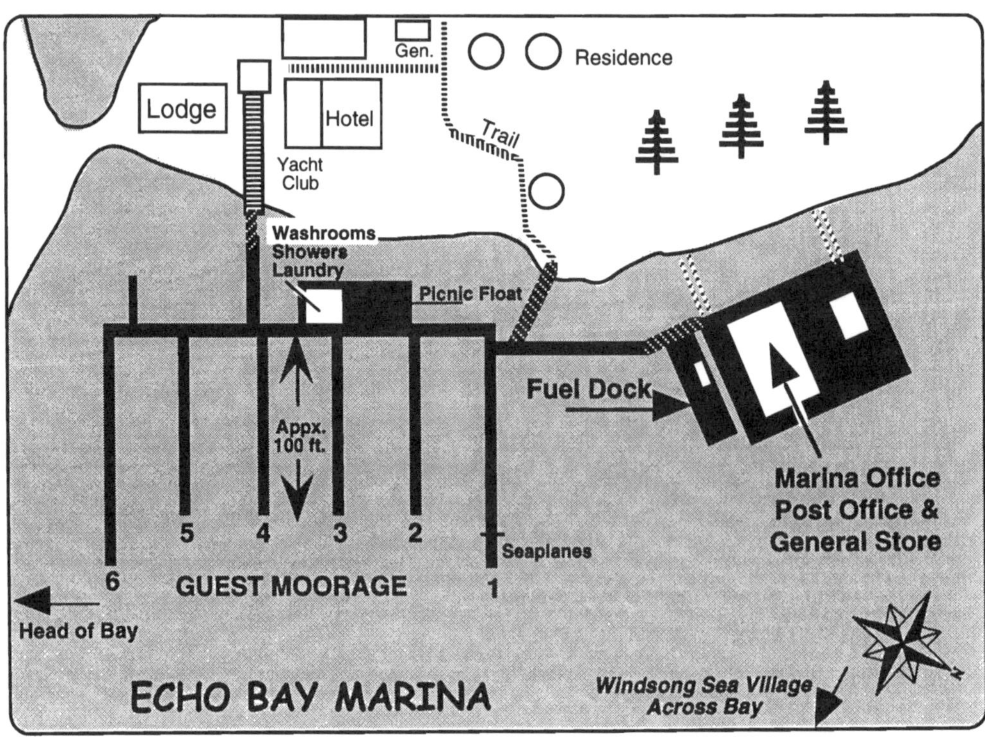

Echo Bay

NAME OF MARINA: ***WINDSONG SEA VILLAGE*** **RADIO:** VHF 66A
TELEPHONE: **250-956-3339** MGR: Jim O'Donnell
E-MAIL: windecho@island.net FAX: 250-956-3307 (Fax)
ADDRESS: P.O. Box 1487 Port McNeill, B.C. Canada V0N 2R0
SHORT DESCRIPTION & LOCATION: **http://www.alertbay.com/windsong/**

50°45.60' - 126°29.70' Located under the cliffs of Echo Bay on NW side of Gilford Island off Cramer Passage. Small quiet marina across the Bay from Echo Bay Marina, welcomes guest boaters. Friendly staff, mellow, and low key atmosphere. Possible ownership changes.

GUEST BOAT CAPACITY: ....Appx. 10-12 boats
DOCKSIDE DEPTH AT ZERO TIDE: 25 Ft. +
SEASON: ....Closed winter
RESERVATION POLICY: ....Accepts
AMT W/ELECTRICITY: ....None
FUEL DOCK: ....Close by
MARINE REPAIRS: ....Close by
TOILETS: ....Yes
HOT SHOWERS: ....Yes
RESTAURANT: ....None
PICNIC AREA: ....None
BASIC STORE: ....Close by
BROADBAND/WI-FI: ....None
DAILY RATE: ....Economical (Under 75¢/foot)

GUEST DOCK: Appx 700 ft. Total
GUEST SLIPS: ....Dock only
WATER: ....Yes
AMPS: ....None
PUMP OUT STATION ....None
HAUL OUT: ....None
BOAT RAMP: ....None
LAUNDRY: ....Close by
BAR: ....None
POOL: ....None
GOLF: ....None
PET FRIENDLY: ....Challenging
OTHER: Picnic dock & BBQ. Gallery featuring local artisans crafts & gifts, float houses for rent.

RESORT CHANGES POSSIBLE

CAUTION! This chartlet not intended for use in navigation.

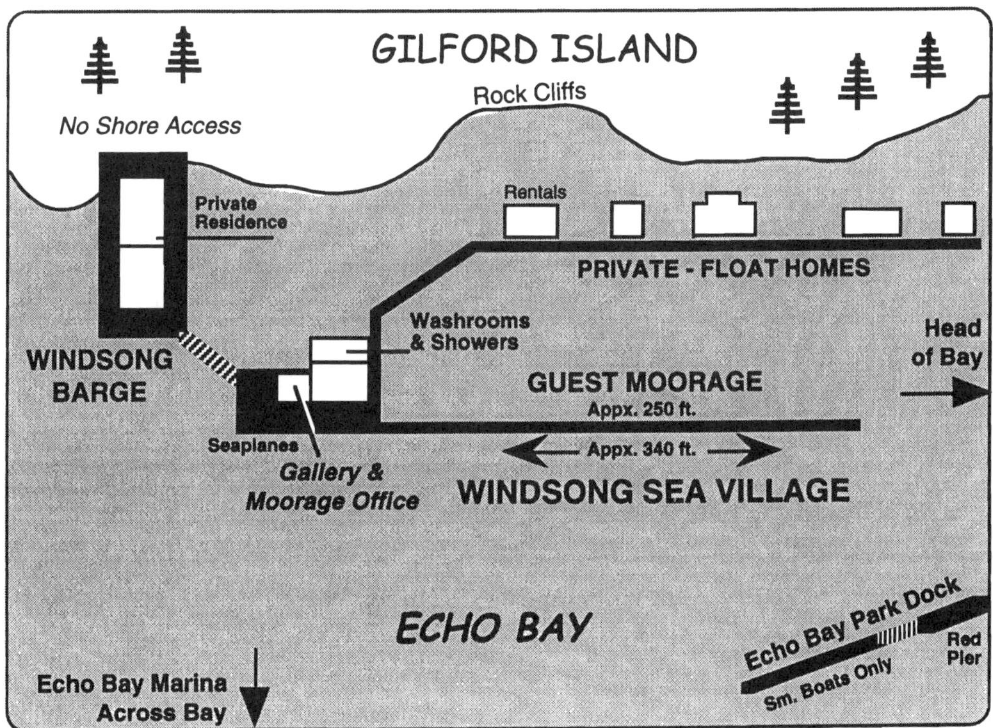

Pierre's Bay

NAME OF MARINA: ***PIERRE'S BAY LODGE & MARINA*** **RADIO:** VHF 66A
TELEPHONE: **250-949-2503** **MGR:** Pierre & Tove Landry
E-MAIL: info@pierresbay.com **FAX:** None
ADDRESS: Mail: Box 257 Gabriola Island, B.C. Canada V0R 1X0
SHORT DESCRIPTION & LOCATION: **www.pierresbay.com**

50°46.25' - 126°28.85 Located in a small bay on the W. shore of Scott Cove, E. of Powell Pt. One of the newest marinas in the Broughtons & highly acclaimed with excellent dining room. Good sturdy docks & very friendly hosts. The moorage is behind the Lodge.

GUEST BOAT CAPACITY: Appx. 30-35 boats
DOCKSIDE DEPTH AT ZERO TIDE: 20 ft. +
SEASON: ....All year
RESERVATION POLICY: ....Accepts
AMT W/ELECTRICITY: ....None
FUEL DOCK: ....None
MARINE REPAIRS: ....None
TOILETS: ....Yes
HOT SHOWERS: ....Yes
RESTAURANT: ....Yes
PICNIC AREA: ....Picnic float
BASIC STORE: ....None
BROADBAND/WI-FI: ....None
DAILY RATE: ....Economical (Under 75¢/foot)

GUEST DOCK: Over 1000 ft. TTL
GUEST SLIPS: ....Docks only
WATER: ....Limited
AMPS: ....None
PUMP OUT STATION ....None
HAUL OUT: ....None
BOAT RAMP: ....None
LAUNDRY: ....Yes
BAR: ....Yes
POOL: ....None
GOLF: ....None
PET FRIENDLY: ....Fair
OTHER: Scheduled PIG ROASTS (see website), Bakery, internet terminal, gift shop, suites, warm lounge, with satellite TV.

CAUTION! This chartlet not intended for use in navigation.

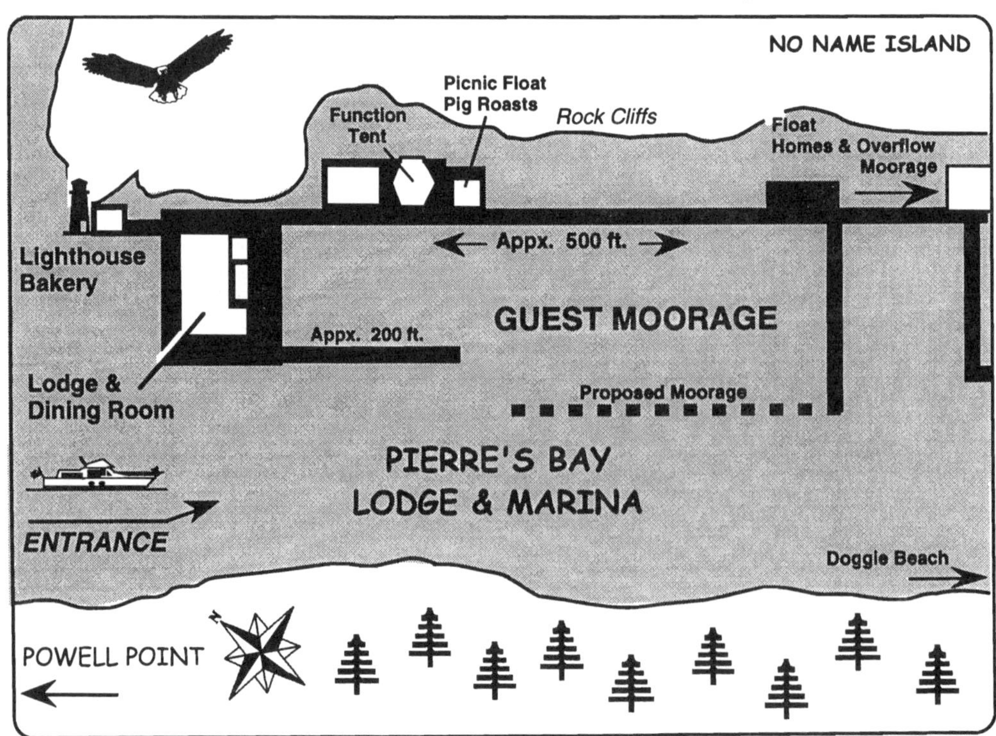

Shawl Bay

NAME OF MARINA: *SHAWL BAY MARINA* RADIO: VHF 66A
TELEPHONE: 250-483-4169 MGR: Lorne & Shawn Brown & Jo
E-MAIL: shawlbaymarina@hughes.net FAX: Fax Machine $2/page
ADDRESS: Simoom Sound Post Office, Simoom Sound, B.C. Canada V0P 1S0
SHORT DESCRIPTION & LOCATION: **http://www.shawlbay.com**

50°50.90' - 126°33.60' Enter off NE side of Penphrase Pass between Gregory Is. & Vigis Pt. on Wishart Pen. - Stay right & favor float homes. Floating marina is in the SE corner of back bay and is an old favorite cruising destination & popular meeting place for boaters.

GUEST BOAT CAPACITY: ....Appx. 25-30 boats
DOCKSIDE DEPTH AT ZERO TIDE: ......25 ft. +
SEASON: ....All year
RESERVATION POLICY: ....Accepts
AMT W/ELECTRICITY: ....Available 9 hrs/day
FUEL DOCK: ....None
MARINE REPAIRS: ....None
TOILETS: ....Yes
HOT SHOWERS: ....Yes
RESTAURANT: ....None
PICNIC AREA: ....Yes
BASIC STORE: Groceries, produce, home baking
BROADBAND/WI-FI: ....Yes (Donations)
DAILY RATE:........Moderate (75¢-$1.25/foot)
NOTE: MOORAGE FEE INCLUDES COMPLIMENTARY PANCAKE BREAKFAST.

GUEST DOCK: Appx 1000' Total
GUEST SLIPS: ....Docks only
WATER: ....Limited
AMPS: ....15-30 A
PUMP OUT STATION ....None
HAUL OUT: ....None
BOAT RAMP: ....None
LAUNDRY: ....Yes
BAR: ....None
POOL: ....None
GOLF: ....None
PET FRIENDLY: ....Challenging
OTHER: Internet terminal, rental cabins, covered boaters float, BBQ'S, gifts & drafts, book exchange, ice, health room, BBQ's, & water taxi.

CAUTION! This chartlet not intended for use in navigation.

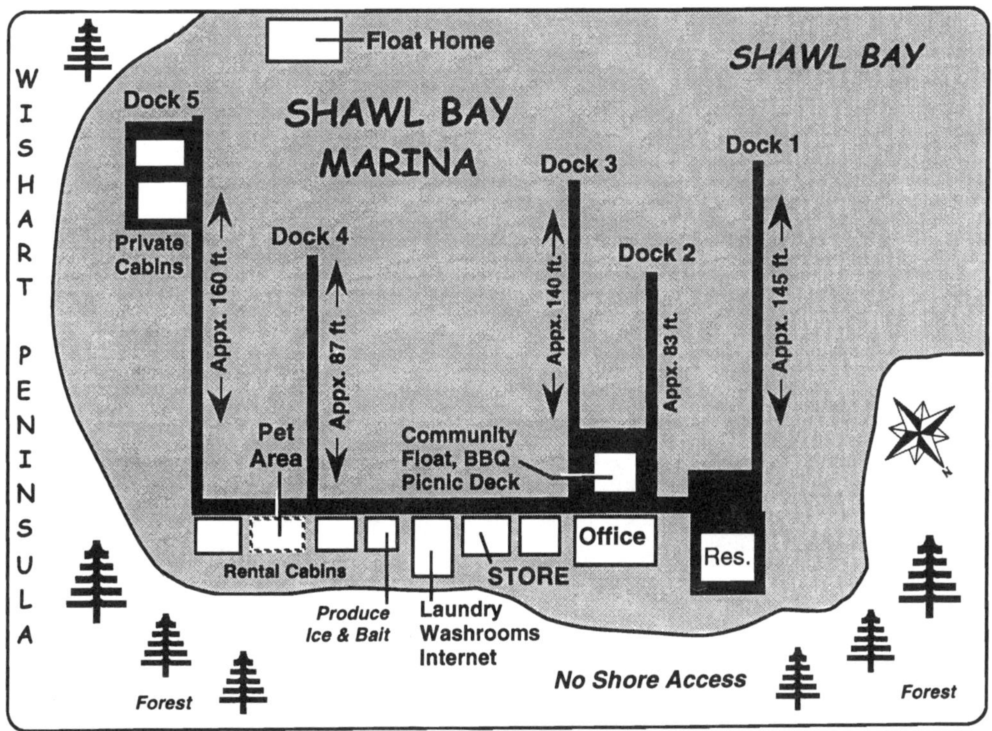

Kwatsi Bay

NAME OF MARINA: ***KWATSI BAY MARINA*** **RADIO:** VHF 66A
TELEPHONE: **250-949-1384** MGR: Anca Fraser & Max Knierim
E-MAIL: kwatsibay@hughes.net FAX: None
ADDRESS: Simoom Sound Post Office Simoom Sound, B.C. Canada V0P 1S0
SHORT DESCRIPTION & LOCATION: **www.kwatsibay.com**

50°52.07' - 126°15.05' Kwatsi Bay Marina is located at the head of Kwatsi Bay just off the north end of Tribune Channel. This is an interesting small wilderness marina with family atmosphere. Very boater friendly with quiet and beautiful moorage.

GUEST BOAT CAPACITY: ....Appx 10-12 boats
DOCKSIDE DEPTH AT ZERO TIDE: ......35 ft. +
SEASON: ....All year
RESERVATION POLICY: ....Accepts
AMT W/ELECTRICITY: ....None
FUEL DOCK: ....None
MARINE REPAIRS: ....None
TOILETS: ....None
HOT SHOWERS: ....Yes
RESTAURANT: ....None
PICNIC AREA: ....Picnic Dock
BASIC STORE: ....Gift shop only
BROADBAND/WI-FI: ....Planned
DAILY RATE:...Economical
(Under 75¢/foot)

GUEST DOCK: Appx 1000' Total
GUEST SLIPS: ....Dock only
WATER: ....Yes
AMPS: ....None
PUMP OUT STATION ....None
HAUL OUT: ....None
BOAT RAMP: ....None
LAUNDRY: ....None
BAR: ....None
POOL: ....None
GOLF: ....None
PET FRIENDLY: ....Good
OTHER: Gifts & crafts.
Guest cabin.
Spectacular walk to waterfall across the bay.

CAUTION! This chartlet not intended for use in navigation.

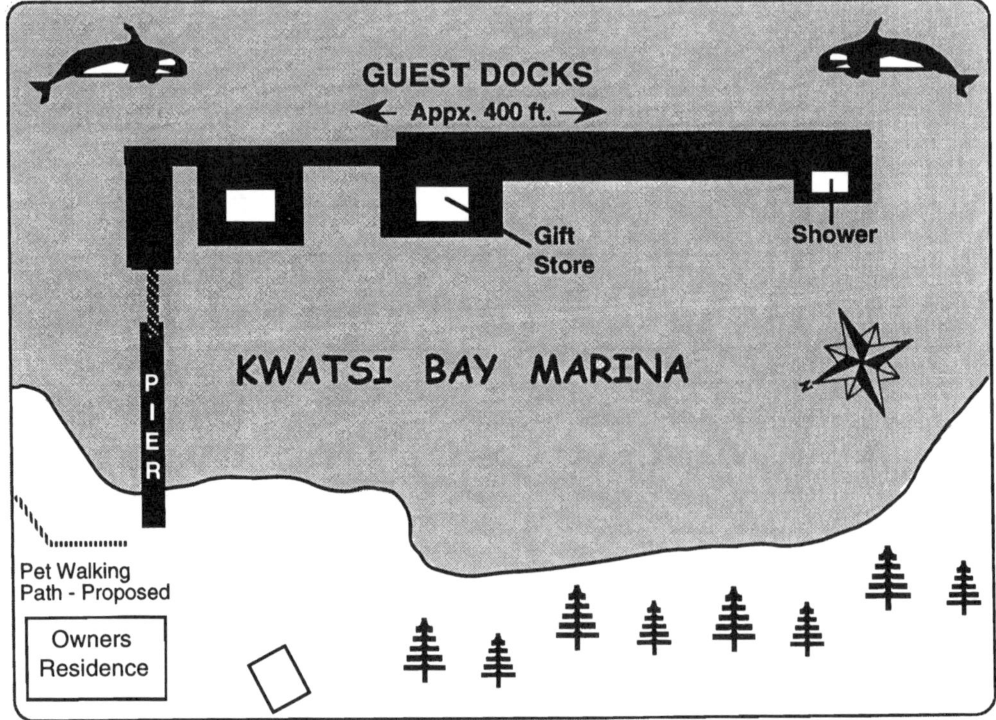

Greenway Sound

NAME OF MARINA: *GREENWAY SOUND MARINE RESORT* **RADIO:** VHF 66A
TELEPHONE: **250-974-7044**/360-466-4751 **MGR:** Tom & Ann Taylor, Owners
E-MAIL: tomann@ncia.com **FAX:** 250-974-7044
ADDRESS: P.O. Box 759 Port McNeilL, B.C. Canada V0N 2R0
SHORT DESCRIPTION & LOCATION: **www.greenwaysound.com** (good website)
50°50.30' - 126°46.30' Located inside Greenway Sound, Broughton Islands, just beyond 1st point to your port side. Extensive marina resort offering red carpeted docks & outstanding dining, plus all comforts of civilization. Very friendly & helpful atmosphere.

GUEST BOAT CAPACITY: ....Appx. 40-50 boats
DOCKSIDE DEPTH AT ZERO TIDE: 60 ft. plus
SEASON: ....Moorage late May to Mid Sept.
RESERVATION POLICY: ....Recommended
AMT W/ELECTRICITY: ....All
FUEL DOCK: ....Nearby at Sullivan Bay
MARINE REPAIRS: ....None
TOILETS: ....Yes
HOT SHOWERS: ....Yes
RESTAURANT: Call 1st - Reservations a must
PICNIC AREA: ....Nearby
BASIC STORE: ....Yes
BROADBAND/WI-FI: ....Yes
DAILY RATE:.................Moderate (75¢-$1.25/foot)
(Based on boat production length)

GUEST DOCK: ....2200 ft. TTL
GUEST SLIPS: ....Docks only
WATER: ....Yes
AMPS: ....15-30-50 A
PUMP OUT STATION ....None
HAUL OUT: ....None
BOAT RAMP: ....None
LAUNDRY: ....Yes
BAR: ....Dining Room
POOL: ....None
GOLF: ....None
PET FRIENDLY: ....Challenging
OTHER: Fully stocked store, fresh produce, gifts, floatplane service, boat sitting, mail holding, park & hiking trails nearby. Concierge service.

RESORT CHANGES POSSIBLE

CAUTION! This chartlet not intended for use in navigation.

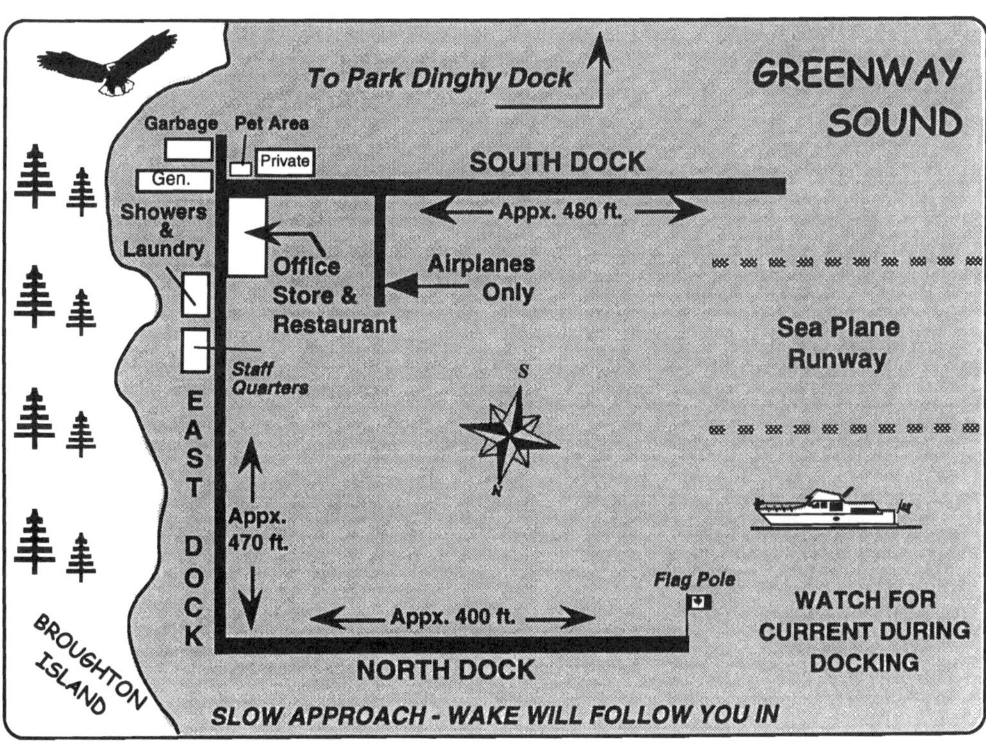

Sullivan Bay

NAME OF MARINA: ***SULLIVAN BAY MARINE RESORT*** RADIO: VHF 66A
TELEPHONE: **250-483-6881** MGR: Pat Finnerty & Lynn Whitehead
E-MAIL: palyn@telus.net FAX: None
ADDRESS: General Delivery Sullivan Bay, B.C. Canada V0N 3H0

SHORT DESCRIPTION & LOCATION:

50°53.20' - 126°49.55' Located just off Wells Passage on N. Broughton Is. in the midst of wilderness cruising. The uniquie floating village offers many service for boaters. Natural & popular turnaround point for boats cruising beyond Desolation Sound.

GUEST BOAT CAPACITY: ....Appx. 40-50 boats
DOCKSIDE DEPTH AT ZERO TIDE: 100 Ft. +
SEASON: ....Moorage seasonal, Fuel all year
RESERVATION POLICY: ....Accepts
AMT W/ELECTRICITY: ....All
FUEL DOCK: ....Gas, Dsl, & LP
MARINE REPAIRS: ....On call
TOILETS: ....Yes
HOT SHOWERS: ....Yes
RESTAURANT: ....Yes
PICNIC AREA: ....Yes
BASIC STORE: ....Yes
BROADBAND/WI-FI: ....Yes
DAILY RATE: ....Moderate
(75¢-$1.25/foot)

GUEST DOCK: Appx 4000' Total
GUEST SLIPS: ....Docks only
WATER: ....Yes
AMPS: ....15-50 A
PUMP OUT STATION ....None
HAUL OUT: ....None
BOAT RAMP: ....None
LAUNDRY: ....Yes
BAR: ....None
POOL: ....None
GOLF: ....None
PET FRIENDLY: ....Challenging
OTHER: Well stocked store, Post Office, liquor, historic float bldgs., boat sitting, float plane service to mainland. 32 miles to Port McNeil.

RESORT CHANGES POSSIBLE

CAUTION! This chartlet not intended for use in navigation.

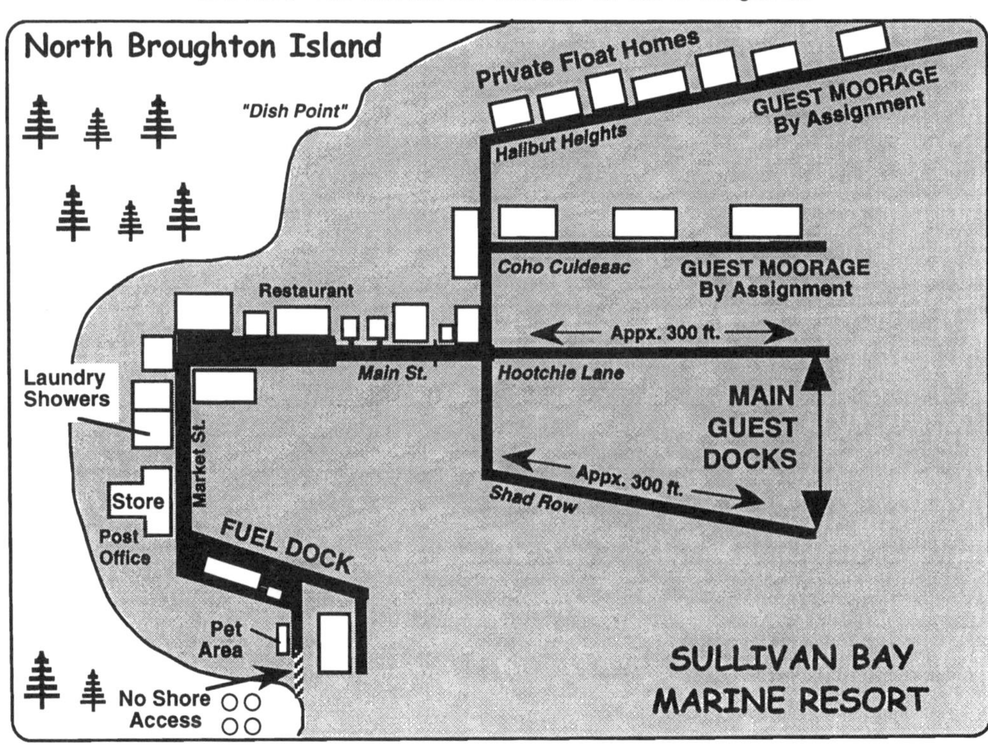

CHAPTER 12

VANCOUVER ISLAND
NORTH ISLAND

Chapter Map - Page 2

Telegraph Cove

NAME OF MARINA: ***TELEGRAPH COVE MARINA*** **RADIO:** VHF 66A
TELEPHONE: **250-928-3163 (Marina)** **MGR:** Elaine Sanford
E-MAIL: telcove@island.net **FAX:** 250-928-3162 (Fax)
ADDRESS: Box 2-8 Telegraph Cove, B.C. Canada V0N 3J0
SHORT DESCRIPTION & LOCATION: **www.telegraphcove.ca**

50°32.87' - 126°50.74' Ten nautical miles south of Port McNeil, the cozy Cove has 2 facilities with resort on historic boardwalk. Resort on NW side handles sm. boats under 25 ft. Marina on SE side has limited moorage for boats up to 60 ft. Limited maneuvering in Cove.

GUEST BOAT CAPACITY: ....Limited
DOCKSIDE DEPTH AT ZERO TIDE: ....4 ft. +
SEASON: ....All year
RESERVATION POLICY: ....Accepts
AMT W/ELECTRICITY: ....All
FUEL DOCK: ....Gas only
MARINE REPAIRS: ....None
TOILETS: ....Yes
HOT SHOWERS: ....Yes
RESTAURANT: ....Cafe & Resort Restaurant
PICNIC AREA: ....Yes
BASIC STORE: ....Yes
BROADBAND/WI-FI: ....Yes
DAILY RATE: ....Moderate (75¢-$1.25/foot)

GUEST DOCK: ....Limited
GUEST SLIPS: ....Limited
WATER: ....Limited
AMPS: ....15-30-50 A
PUMP OUT STATION ....None
HAUL OUT: ....None
BOAT RAMP: ....Yes
LAUNDRY: ....Yes
BAR: ....Pub
POOL: ....None
GOLF: ....None
PET FRIENDLY: ....Excellent
OTHER: Cafe, resort condos, lodging, kayak rentals & kayak launch, gift shop, R.V. Park, Whale Museum, General Store.

CAUTION! This chartlet not intended for use in navigation.

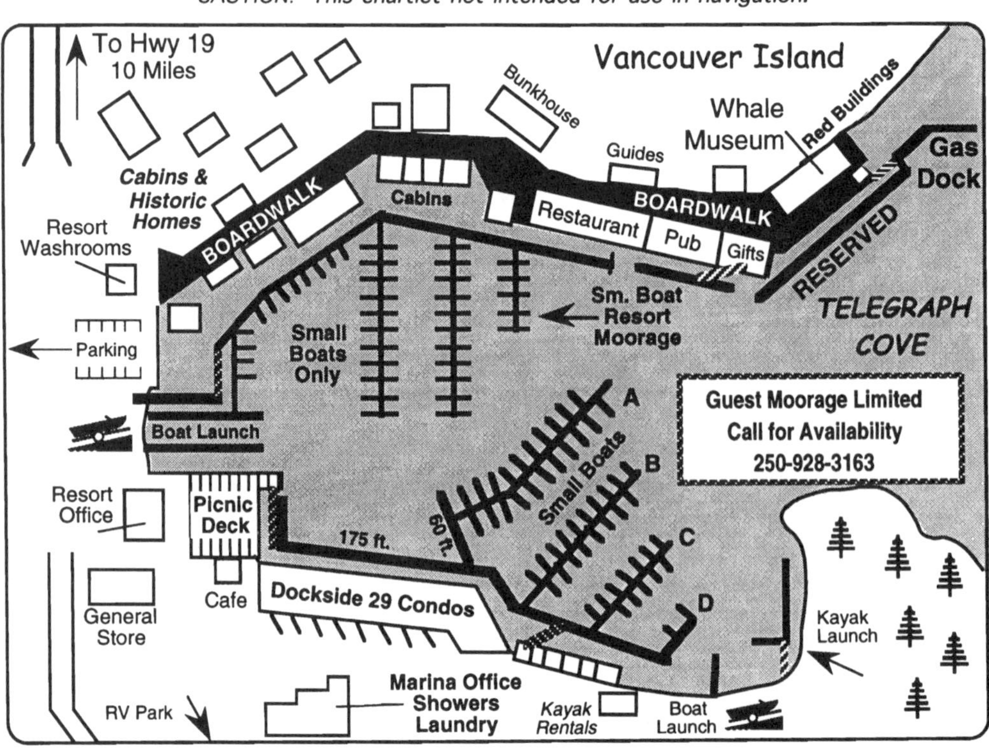

Alert Bay

NAME OF MARINA: ***ALERT BAY BOAT HARBOUR*** **RADIO:** VHF 66A
TELEPHONE: 250-974-5727/8255 **MGR:** Dan Kennedy, Harbour Manager
E-MAIL: boatharbour@alertbay.ca **FAX:** 250-974-5470 (Fax)
ADDRESS: Bay Service 2800 Alert Bay, B.C. Canada V0N 1A0
SHORT DESCRIPTION & LOCATION: **www.alertbay.ca**

50°34.70' - 126°54.84' Located on the S. side of Cormorant Island in Broughton Strait. The village is a fishing port rich with Indian cultural history. Full services for boaters. Best moorage is behind the breakwater protected sm. boat harbour just north of ferry terminal.

GUEST BOAT CAPACITY: ....Appx. 30-50 boats
DOCKSIDE DEPTH AT ZERO TIDE: ......15 ft. +
SEASON: ....All year
RESERVATION POLICY: ....None
AMT W/ELECTRICITY: ....All
FUEL DOCK: ....None
MARINE REPAIRS: ....None
TOILETS: ....None
HOT SHOWERS: ....None
RESTAURANT: ....Close by
PICNIC AREA: ....Yes
BASIC STORE: ....Yes
BROADBAND/WI-FI: ....None
DAILY RATE: ....Economical (Under 75¢/foot)
RAFTING IS MANDATORY
CALL AHEAD FOR MOORAGE AVAILABILITY

GUEST DOCK: ....2400 ft. total
GUEST SLIPS: ....Docks only
WATER: ....Yes
AMPS: ....20-30 A
PUMP OUT STATION ....None
HAUL OUT: ....None
BOAT RAMP: ....Close by
LAUNDRY: ....Close by
BAR: ....Pub
POOL: ....None
GOLF: ....None
PET FRIENDLY: ....Good
OTHER: Seafest 3rd weekend in July, U'mista Cultural Centre, Indian Dance Performance Thur.-Sat. Groceries, Hotel, Liquor.

CAUTION! This chartlet not intended for use in navigation.

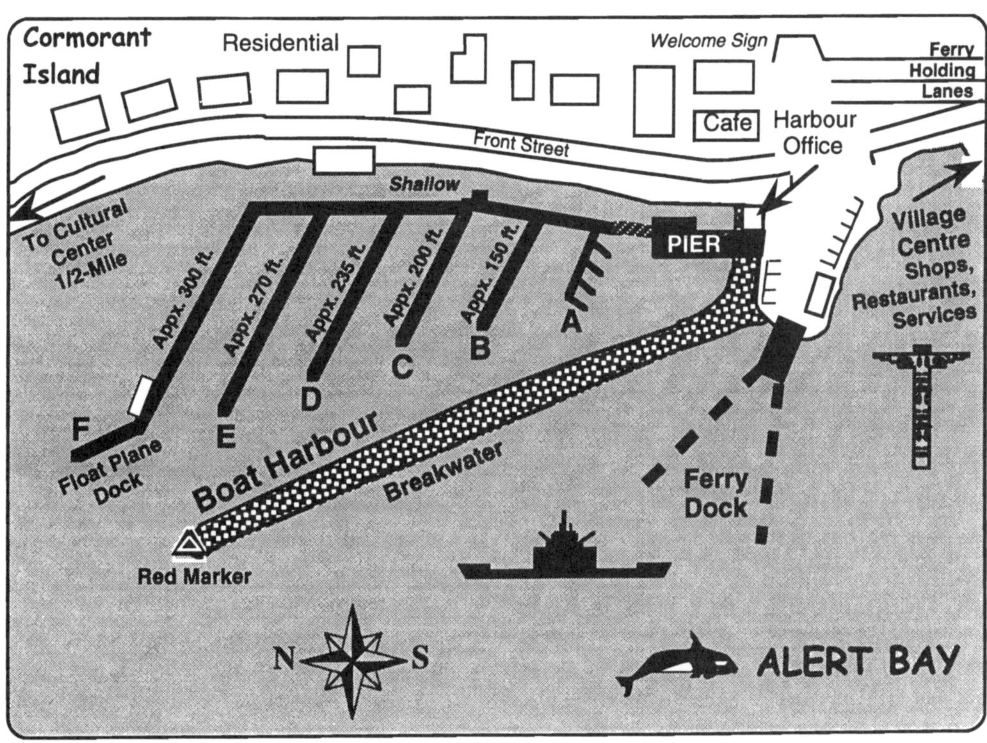

Port McNeill

NAME OF MARINA: *PORT McNEILL BOAT HARBOUR* RADIO: VHF 66A
TELEPHONE: 250-956-3881 MGR: Hilge Binner
E-MAIL: pmharbour@telus.net FAX: 250-956-2897
ADDRESS: P.O. Box 1389 Port McNeill, B.C. Canada V0N 2R0
SHORT DESCRIPTION & LOCATION: **http://www.town.portmcneill.bc.ca**

50°35.51' - 127°05.30' Located on Northern Vancouver Is. off Broughton Strait west of Deer Bluff. The modern town & marina offers nearly all services & amenities within walking distance. This is a popular provisioning stop for cruising boaters north. Helpful friendly staff.

GUEST BOAT CAPACITY: ....Appx 60-70 boats
DOCKSIDE DEPTH AT ZERO TIDE: ....9 ft. +
SEASON: ....All year
RESERVATION POLICY: ....None
AMT W/ELECTRICITY: ....All
FUEL DOCK: ....Gas, Dsl, & LP
MARINE REPAIRS: ....On premises
TOILETS: ....Yes
HOT SHOWERS: ....Yes
RESTAURANT: ....Yes
PICNIC AREA: ....Yes
BASIC STORE: ....Yes
BROADBAND/WI-FI: ....BroadbandXpress
DAILY RATE: ....Economical (Under 75¢/foot)

GUEST DOCK: Appx 1000' Total
GUEST SLIPS: ....Varies
WATER: ....Yes
AMPS: 20-100
PUMP OUT STATION ....Yes
HAUL OUT: ....None
BOAT RAMP: ....Yes
LAUNDRY: ....Yes
BAR: ....Yes
POOL: ....Close by
GOLF: ....Close by
PET FRIENDLY: ....Excellent
OTHER: Logging and History Museum, Supermarket, hardware & tackle, post office, liquor store, drug store, chandlery, hospital.

CAUTION! This chartlet not intended for use in navigation.

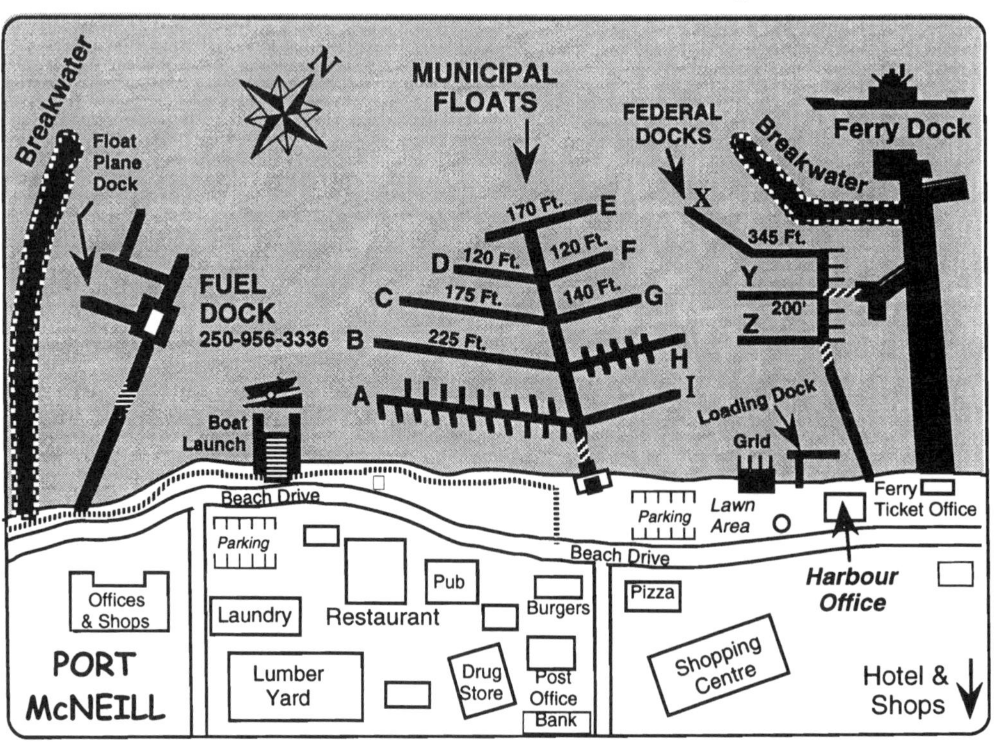

Sointula

NAME OF MARINA: ***MALCOLM ISLAND LIONS HARBOUR*** RADIO: VHF 66A
TELEPHONE: **250-973-6544** MGR: Lorraine Williams, Harbour Manager
E-MAIL: milha@island.net FAX: 250-973-6544
ADDRESS: P.O. Box 202 Sointula, B.C. Canada V0N 3E0
SHORT DESCRIPTION & LOCATION: **www.sointula.com**

50°38.39' - 127°02.05' Historic village located on Malcom Is. The boat harbour is in Rough Bay 1.5 miles north of the ferry dock & village. Excellent docks can accomodate pleasure boats among fishing vessels on either north or south side. Tie-up in any open spot.

GUEST BOAT CAPACITY: ....Varies
DOCKSIDE DEPTH AT ZERO TIDE: ......10 ft. +
SEASON: ....All year
RESERVATION POLICY: ....None
AMT W/ELECTRICITY: ....All
FUEL DOCK: ....None
MARINE REPAIRS: ....Close by
TOILETS: ....Yes
HOT SHOWERS: ....Yes
RESTAURANT: ....Take Out Cafe
PICNIC AREA: ....Yes
BASIC STORE: ....Close by
BROADBAND/WI-FI: ....None
DAILY RATE: ....Economical (Under 75¢/foot) Cash (Canadian Funds Appreciated), Cheque, or Visa/Mastercard

GUEST DOCK: ....Over 2400 ft.
GUEST SLIPS: ....Varies
WATER: ....Yes
AMPS: ....20-30 A
PUMP OUT STATION Planned
HAUL OUT: ....Close by
BOAT RAMP: ....None
LAUNDRY: ....Yes
BAR: ....Pub close by
POOL: ....None
GOLF: ....None
PET FRIENDLY: ....Excellent
OTHER: Good dinghy dock in town next to ferry dock. 1.5 Mi. walking to shopping liquor, bakery, museum. Friendly & interesting.

CAUTION! This chartlet not intended for use in navigation.

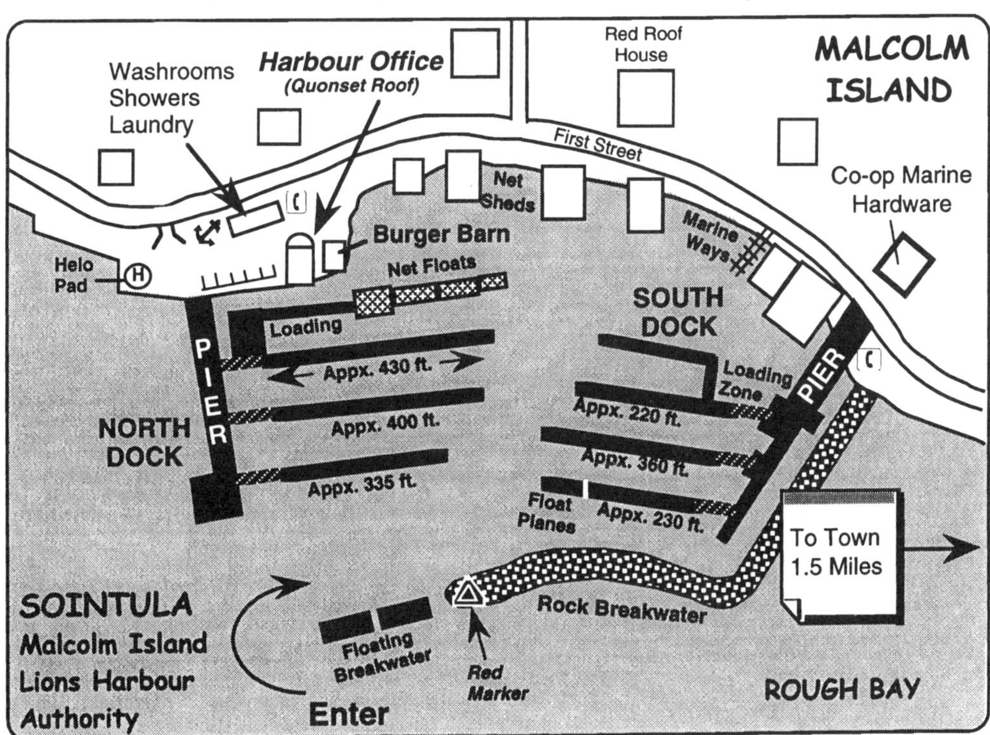

Port Hardy

NAME OF MARINA: ***QUARTERDECK INN & MARINA*** RADIO: VHF 66A
TELEPHONE: 250-949-6551 MGR: I.V. Villani, Owner
E-MAIL: info@quarterdeckresort.net FAX: 250-949-7777
ADDRESS: P.O. Box 910 Port Hardy, B.C. Canada V0N 2P0
SHORT DESCRIPTION & LOCATION: **www.quarterdeckresort.net**

50°42.83' - 127°29.32' Located in the Inner Basin of PortHardy, the northern most community on Vancouver Island. The busy & excellent marina resort is adjacent to the Harbour Authority Fisherman's Wharf. All city amenities available. Rents open slips.

GUEST BOAT CAPACITY: ....Varies
DOCKSIDE DEPTH AT ZERO TIDE: ....4-10 ft.
SEASON: ....All year
RESERVATION POLICY: ....None
AMT W/ELECTRICITY: ....All
FUEL DOCK: ....Gas, Dsl, & LP
MARINE REPAIRS: ....On premises
TOILETS: ....Yes
HOT SHOWERS: ....Yes
RESTAURANT: ....Yes
PICNIC AREA: ....Yes
BASIC STORE: ....Close by
BROADBAND/WI-FI: ....Yes
DAILY RATE: ....Moderate (75¢-$1.25/foot)

GUEST DOCK: ....Over 1000 ft.
GUEST SLIPS: ....Varies
WATER: ....Yes
AMPS: ....15-50 A
PUMP OUT STATION ....None
HAUL OUT: ....Yes
BOAT RAMP: ....Yes
LAUNDRY: ....Yes
BAR: ....Yes
POOL: ....None
GOLF: ....Close by
PET FRIENDLY: Good
OTHER: Courtesy vehicle, internet, charts, fishing tackle, liquor store,licenses, guides, boat rentals. Airport w/ service to YVR.

NOTES

Port Hardy

CAUTION! This chartlet not intended for use in navigation.

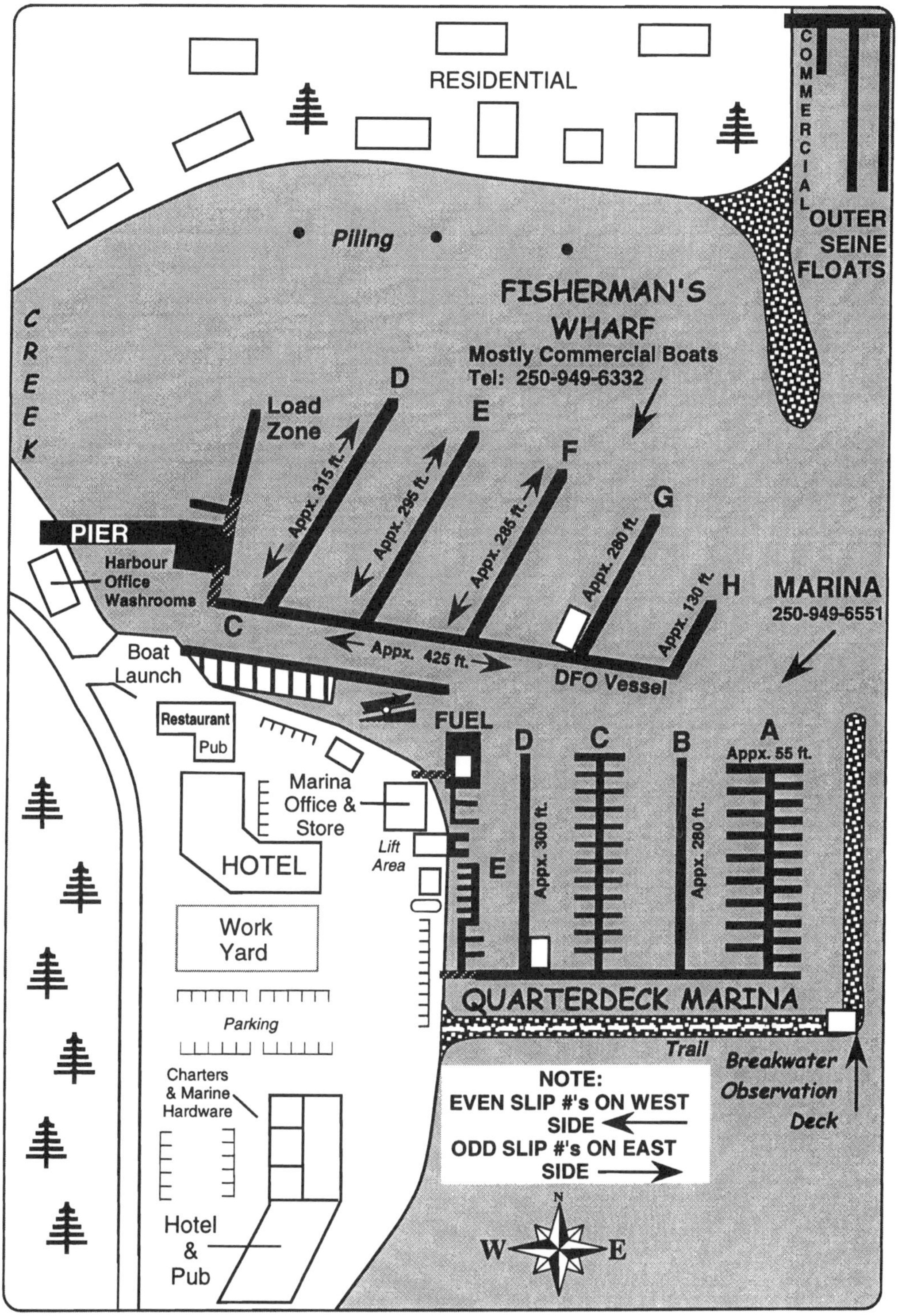

RESIDENTIAL

COMMERCIAL

OUTER SEINE FLOATS

Piling

CREEK

FISHERMAN'S WHARF

Mostly Commercial Boats

Tel: 250-949-6332

Load Zone

D

E

F

G

H

Appx. 315 ft.

Appx. 295 ft.

Appx. 285 ft.

Appx. 280 ft.

Appx. 130 ft.

PIER

Harbour Office Washrooms

C

Appx. 425 ft.

DFO Vessel

MARINA

250-949-6551

Boat Launch

Restaurant

Pub

FUEL

D

C

B

A

Appx. 55 ft.

Marina Office & Store

Lift Area

HOTEL

E

Appx. 300 ft.

Appx. 280 ft.

Work Yard

QUARTERDECK MARINA

Parking

Trail

Breakwater Observation Deck

Charters & Marine Hardware

NOTE:
EVEN SLIP #'s ON WEST SIDE ←
ODD SLIP #'s ON EAST SIDE →

Hotel & Pub

N

W

E

Port Hardy

NAME OF MARINA: ***PORT HARDY OUTER DOCK/Seagate*** **RADIO:** VHF 66A
TELEPHONE: **250-949-6332**/949-0336 cell **MGR:** Pat McPhee/Mary-Anne Smith
E-MAIL: phfloats@cablerocket.com **FAX:** 250-949-6332 Hbr Auth Fax
ADDRESS: 6600 Hardy Bay Road Port Hardy, B.C. Canada V0N 2P0
SHORT DESCRIPTION & LOCATION: **http://www.district.porthardy.bc.ca**

50°43.24' - 127°28.82' Located in Hardy Bay directly in front of town of Port Hardy, the most northern community on Vancouver Island. The public float is attached to a large pier with mooring buoys adjacent for pleasure boats. Convenient to shopping & services.

GUEST BOAT CAPACITY: ....Appx. 15 boats
DOCKSIDE DEPTH AT ZERO TIDE: ......10 ft. +
SEASON: ....Summer only
RESERVATION POLICY: ....None
AMT W/ELECTRICITY: ....All
FUEL DOCK: ....Close by
MARINE REPAIRS: ....Close by
TOILETS: ....Yes
HOT SHOWERS: ....Close by
RESTAURANT: ....Close by
PICNIC AREA: ....Yes
BASIC STORE: ....Yes
BROADBAND/WI-FI: ....None
DAILY RATE:..........Economical (Under 75¢/foot)

GUEST DOCK: Appx. 800 ft.Total
GUEST SLIPS: ....Docks only
WATER: ....Yes
AMPS: ....None
PUMP OUT STATION ....Yes
HAUL OUT: ....Close by
BOAT RAMP: ....Yes
LAUNDRY: ....Close by
BAR: ....Pubs close by
POOL: ....Close by
GOLF: ....Close by
PET FRIENDLY: ....Excellent
OTHER: Close to: Banks, business center, hospital, post office, supermarkert, malls, museum, whale watching, fishing guides & airport.

CAUTION! This chartlet not intended for use in navigation.

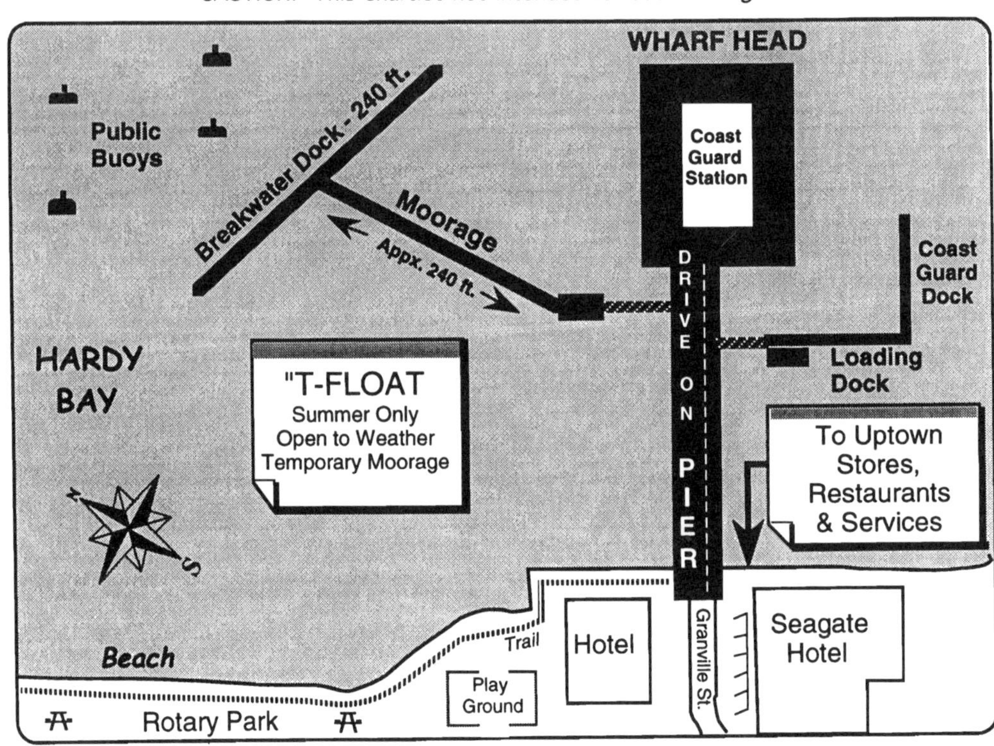

INDEX

INDEX

INDEX

INDEX

The Docking Flowchart

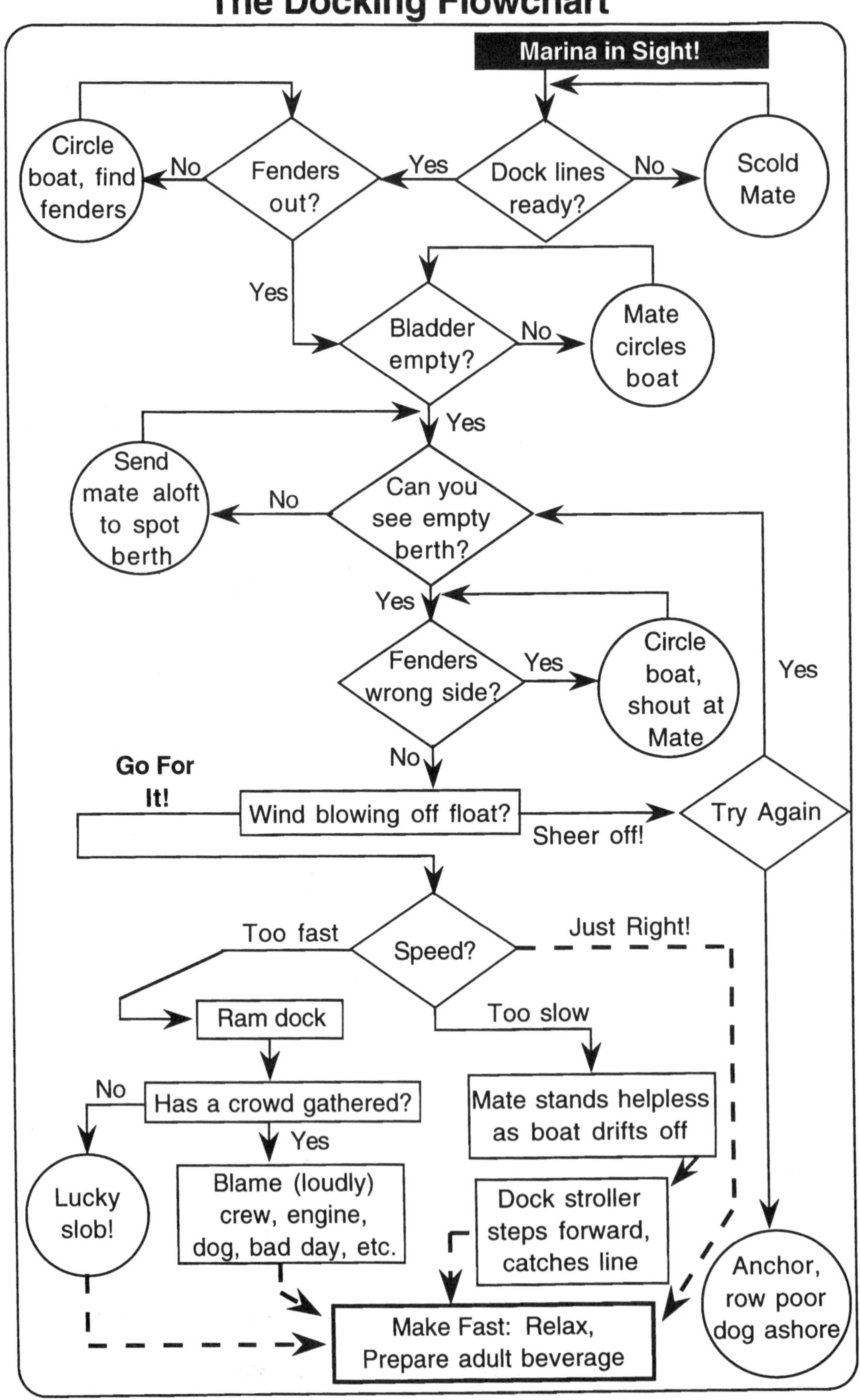